Petroleum Geochemistry and Basin Evaluation

AAPG Memoir 35

Petroleum Geochemistry and Basin Evaluation

Edited by
Gerard Demaison
and Roelef J. Murris

Published by
The American Association of Petroleum Geologists
Tulsa, Oklahoma 74101, U.S.A.

Association Editor: Richard Steinmetz
Science Director: Edward A. Beaumont
Project Editor: Douglas A. White
Design and Production: S. Wally Powell
Typographer: Eula Matheny

*"In the degree to which we understand
the habitat of oil, and thus make
shortcuts to its discovery and
reductions in its cost, we add to the
resources or reserves that can be
produced economically for
the benefit of mankind."*

L. G. Weeks

This passage is taken from *Habitat of Oil*, 1958, an AAPG Special Volume edited by L. G. Weeks.

Contents

Introduction

R.J. Murris
Shell Internationale Petroleum Mij B.V.
The Hague, Netherlands

During the past ten to fifteen years, organic geochemistry has become a widely accepted tool in oil and gas exploration. The concept that oil and most of the gas is generated in organic-rich source rocks, which are matured through elevated temperature and time, and from which the hydrocarbons are then expelled to migrate along carrier beds and other conduits to the traps, has proved to be a very fruitful one, with considerable impact on exploration thinking.

The earlier papers dealing with organic geochemistry and habitat of oil were essentially "after-the-fact" descriptions, pointing to the observable relation between source-rock distribution and the occurrence of hydrocarbon accumulations. The accumulated knowledge and increased insight in the geochemical processes and geological conditions that control oil and gas accumulations have more recently resulted in predictive models, which have as the main objective the reduction of exploration risk. At the same time, the emphasis has shifted from the chemical analysis of source beds and oil typing to integrated geological-geochemical habitat studies trying to come to grips with the intricate four-dimensional world of hydrocarbon generation, migration and accumulation.

The ability to reduce risk is the decisive criterion as to whether a given method or technique will be accepted by management as an effective exploration tool. Figure 1 shows an example of risk reduction through the application of geochemistry in prospect appraisal. This is an actual case, where 165 prospects were ranked in the pre-drilling stage based on a calibrated habitat modeling system (Sluijk and Nederlof, this volume) and then compared with the results after drilling.

As shown, ranking of the prospects on their trap size alone would have led, in this specific case, to a distinct improvement in forecasting efficiency (28 percent) over what would have happened if the prospects had been drilled in random order. However, by including geochemical charge and retention parameters, an actual ranking sequence of the prospect expectations was obtained, with an even better forecasting efficiency of 63 percent. It is clear that the area between the two stepped curves represents a considerable improvement in the deployment of exploration funds. This optimization can be ascribed to our current advanced understanding of the habitat of oil.

In selecting the papers for this volume, the editors have been strongly guided by the principle that the relation between the areas of mature, oil and gas generating source rock (hydrocarbon kitchens) and the derived accumulations should be clearly and unambiguously demonstrated, preferably with the aid of maps and sections. Detailed geochemical data and theory have been kept to the minimum, the emphasis being on the integration of regional geology and geochemistry. About half the papers are original contributions, the others are reprints often updated or expanded by the authors. Several "classics" are included, such as the Tissot et al paper on the Eastern Algerian Sahara.

The first six papers treat the general principles of organic geochemistry as applied to basin analysis and prospect evaluation. Demaison's introductory paper discusses the concept of the generative basin and gives an overview of the worldwide genetic relationship between generative areas and hydrocarbon accumulations. Furthermore, he stresses the main objective of such studies: reduction of exploration risk as, for instance, by analogical forecasting of high success ratios in areas associated with hydrocarbon-generative depressions.

Before the advent of geochemical methods, earlier attempts to evaluate the petroleum potential of basins or plays were essentially based on the descriptive geologic analog method: geological settings with known production were studied in detail, with the hope that comparable undrilled settings would also contain comparable oil and gas accumulations. The "Habitat of Oil" memoir, published in 1958 by AAPG, is one of the outstanding symposiums of this kind. However, with the advances in understanding of the basic processes which control generation, migration, accumulation, and retention, descriptive geologic analogs could be supplemented or even replaced by process-controlled analog models. These process-controlled methods can be approached in a strongly deterministic manner, whereby predictions are calculated according to an assumed mode of interaction of the controlling parameters, or in a more statistical manner, whereby known exploration outcomes are used to test the hypotheses underlying the applied habitat model. The five papers by Kontorovich, Ungerer et al, Bishop et al, Welte and Yukler, and Sluijk and Nederlof represent a spectrum from purely deterministic to statistically calibrated modeling.

The paper by Kontorovich presents in considerable detail the state-of-the-art of geochemical basin evaluation in the USSR, illustrated by examples from the rich Siberian hydrocarbon provinces. The papers by Ungerer et al, ·Bishop et al, and Welte and Yukler present systematic methods for the assessment of the petroleum potential of basins, parts of basins or individual prospects. These papers illustrate how far we have come in understanding the basic processes which lead to oil and gas accumulations, and how we can use this understanding to apply predictive integrated methods in petroleum exploration.

Sluijk and Nederlof use an approach whereby the hydrocarbon habitat model is tested against past exploration outcomes. In their system, the inherent uncertainties of the deterministic approach, i.e. the still incomplete knowledge of the complex system of interacting controlling parameters and the often fragmentary and

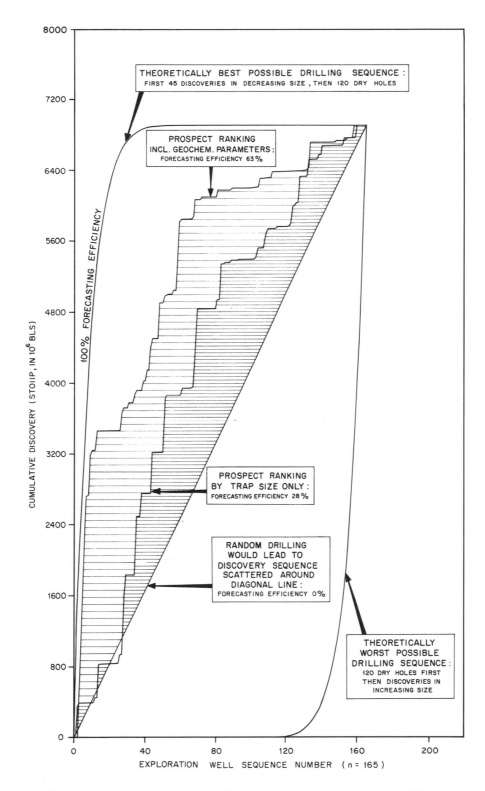

Figure 1. Improved forecasting efficiency through use of geochemical parameters in prospect evaluation. This graph represents an actual case of 165 prospects evaluated prior to drilling. Vertical axis = cumulative discovered volumes in place; horizontal axis = test well sequence number. The upper smooth curve represents the theoretical best possible ranking (100 percent forecasting efficiency) of the prospects, as plotted in hindsight from the actual outcomes; the lower right-hand smooth curve represents the worst possible prediction, i.e. drilling dry holes first (up to ±120) and then discoveries in increasing size order. The diagonal line represents the most likely discovery rate if the prospects had been drilled in random order. Ranking based on trap size alone, as assessed during prospect evaluation, would have led to the lower stepped curve (forecasting efficiency of 28 percent). Actual ranking established by complete prospect evaluation is represented by upper stepped curve (63 percent forecasting efficiency). Note that in the latter case, inclusion of geochemical parameters (hydrocarbon charge) and of retention parameters has about doubled the forecasting efficiency compared with ranking only by structural and reservoir data (= trap size), taken in isolation.

approximate knowledge of these parameters available in the exploration stage are, to a significant degree, overcome. The resulting prospect evaluation system produces statistically defined predictions, which can be meaningfully compared to the results after drilling, and then be used to update the system.

The second group of papers, seventeen in all, documents, with as wide a geographical spread as possible, a great diversity of geological settings and ages as well as depositional environments of source rocks. It is from observations of this scope that the fundamental aspects of source-migration-accumulation have been recognized as universal, permitting the evolution of predictive modeling.

Geographically the papers range from the North American continent over South America, the Atlantic, Europe, Africa, and the Middle East to Australia and New Zealand. The interior basins of North America are covered by Barrows and Cluff (Illinois Basin), Claypool et al (Green River Basin and Overthrust Belt), Meissner (Williston Basin), and Momper and Williams (Powder River Basin). The past and present margins of that continent are treated in the papers of Zieglar and Spotts (Great Valley, California) and Swift and Williams (Grand Banks). The latter paper is of great interest because it points to the potential of the Hibernia area, not yet discovered at that time. The papers on the margins form a link to the Atlantic basins and their Cretaceous source beds, described by Tissot et al.

The depositional environments of the source rocks span the entire range from lacustrine and alluvial plain to deep marine. Terrestrial source formations, often closely associated with coal measures, are described by Kantsler et al and Thomas from Australia, and by Pilaar from the Taranaki Basin, New Zealand. The Tertiary Niger Delta (Evamy et al) is a fine example of charge from deltaic and pro-delta source beds that interfinger closely with the multiple carrier beds so characteristic for that habitat. Zieglar and Spotts in their paper on the Great Valley of California also deal with source rocks deposited in an off-coastal to deltaic environment, but in this case along an active continental margin rather than the typical passive margin of the Niger delta.

Marine source rocks associated with carbonate and carbonate-evaporite sequences deposited in stable intracratonic or pericratonic basins are well represented: Illinois Basin (Cluff and Barrows), Phosphoria formation (Claypool et al), the Bakken of the Williston Basin (Meissner), La Luna of the Maracaibo Basin (Blaser and White) and the Upper Jurassic and Cretaceous source formations of the Middle East (Murris). Examples of marine source beds deposited during transgressive periods in clastic realms are found in the contributions of Momper and Williams on the Powder River Basin, Goff as well as Cooper and Barnard on the North Sea, Tissot et al on the Eastern Algerian Sahara, and Thomas on the Perth Basin (Kockatea Shale). The collection is completed with the deeper marine source beds described by Tissot et al (1980).

A special group in terms of tectonic settings are the rift basins, whereby the rifting may be followed by plate separation and drifting, such as is the case with the Espirito Santo Basin of Brazil (Estrella et al) and the Perth Basin, or whereby the separation may ultimately fail and the graben system is covered by a post-rift depression fill, such as the Central and Viking grabens of the North Sea. The rift grabens are restriction prone, and therefore provide a setting highly inducive to the deposition and preservation of organic-rich sediments. Prograding clastic wedges along the succeeding continental margins or filling in the post-rift depressions, such as the Tertiary of the North Sea, provide the necessary maturity.

As to geologic age, the oldest source rocks represented in this collection are the Cambrian formations of the Siberian platform (Kontorovich) and the Silurian shales of the Algerian Sahara (Tissot et al, 1973), ranging all the way up to the inferred late Tertiary source for the Niger Delta oil and gas. Worth noting is the apparent correlation between major global sea-level rises and the deposition of regional prolific source formations: the late Jurassic eustatic flooding produced the Hanifa of the Middle East, the Kimmeridge shale of the North Sea, and the Bazhenov shale of West Siberia, which between them account for a substantial part of the world's known reserves. Similarly the mid and late Cretaceous sea-level rises have their concomitant widespread organic-rich deposits, which are even found in the deep oceanic environment such as the Atlantic basins described by Tissot et al.

What generalizations can we extract from the great variety in hydrocarbon habitat so aptly documented by the papers included in this volume? First of all, that hydrocarbon generation and accumulation seems to be a universal process, which is bound to happen once the vital ingredients of source, maturity, reservoir, seal, and trap are present. Secondly, that considerable progress has been made in understanding the basic processes controlling generation, migration, and accumulation, and the geologic conditions favorable for these processes to have occurred in the right place and at the right time. Theories and models which are only valid for singular and often exceptional cases should therefore be regarded with due caution, an example being the hypothesis of oil migration through solution in hot formation water expelled from clays at great depths, from where it moves upward along faults and fractures. Though perhaps theoretically applicable to young Tertiary depocenters like the Niger or Mississippi deltas, such a mechanism is hard to envisage for the nevertheless oil-rich stable carbonate platforms, like those in the Middle East.

In assessing which of the various alternative theories on the generation, expulsion, migration, and accumulation of hydrocarbons are the more likely, the guiding words by Francis Bacon that, "... it is the peculiar and perpetual error of the human intellect to be more moved and excited by affirmatives than by negatives, whereas it ought to hold itself indifferently disposed towards both alike; indeed in the establishment of any true axiom, the negative instance is the more forcible of the two" (The New Organon, XLVI) remain today as valid as they were almost four centuries ago. The papers assembled in this volume, and the data and interpretations they contain about a wide spectrum of hydrocarbon habitats, may hopefully contribute to the reader's understanding of the processes controlling oil and gas occurrences. If, in doing so, this will ultimately lead to more effective and efficient exploration, this volume will have fulfilled its main purpose.

The Generative Basin Concept

Gerard Demaison
Chevron Overseas Petroleum Inc.
San Francisco, California

Recent progress in organic geochemical and other fields of earth sciences has made feasible the development of methods for evaluating the likeliness of hydrocarbon occurrence in an undrilled trap. A fundamental step in hydrocarbon charge prediction is to determine whether an undrilled trap has had access to hydrocarbon migration from mature source rocks.

Areas underlain by mature source rocks are called "petroleum generative depressions" or "hydrocarbon kitchens." A "generative basin" is defined as a sedimentary basin that contains one or more petroleum generative depressions. Mapping generative depressions is achieved by integrating geochemical data relevant to maturation and organic facies with structural and stratigraphic information derived from seismic surveys and deep wells.

Locales of high success ratios in finding petroleum are called "areas of high potential," "plays," or "petroleum zones." A rapid worldwide review of 12 sedimentary basins, described in order of geotectonic style, reveals the following regularities:

1. The zones of concentrated petroleum occurrence ("areas of high potential") and high success ratios are genetically related to oil generative depressions or basins. These depressions are mappable by integrated methods (geology, geophysics, and geochemistry).

2. The largest petroleum accumulations tend to be located close to the center of the generative basins or on structurally high trends neighboring deep generative depressions.

3. Migration distances commonly range in tens rather than hundreds of miles and are limited by the drainage areas of individual structures. Thus the outlines of generative depressions commonly include most of the producible hydrocarbon accumulations and the largest fields. Unusual cases of long distance migration are documented on certain foreland basin plates where stratigraphy and structure permitted uninterrupted updip movement of oil.

These three regularities provide powerful analogs for forecasting areas of high petroleum potential in undrilled or sparsely drilled basins.

It is important to keep in mind, however, that the predictive accuracy of geochemical mapping systems is dictated by the available data base. Hence, at early stages of exploration, geologists should guard from overtaxing geological and geochemical knowledge, and remain alert to the potentialities of the unknown.

INTRODUCTION

Since the turn of the century, one of the basic rules for finding petroleum has been to drill anticlinal structures in sedimentary basins. Straightforward application of the anticlinal theory has been totally successful: over 95 of the world's producible petroleum has been found stored in structural traps. Moreover, since the 1930s, most oil-bearing structural traps have been detected by geophysical methods. Seismic applications, have now reached a level of sophistication permitting the detection of traps in unusually difficult geologic conditions, thus opening new frontiers to exploration.

However, most explored anticlinal structures are found barren of hydrocarbons when drilled. Furthermore it has been observed that in some basins or portions of basins virtually all anticlinal traps are found barren of hydrocarbons although reservoirs and seals are present.

Lastly, the producible petroleum reserves of explored sedimentary basins follow an areal distribution which obeys temporal and paleolatitudinal considerations more than geotectonic style (Bois, Bouche, and Pelet, 1982). Bally (1980) even reached the earlier conclusion that "the classification of basins does little to improve our hydrocarbon volume forecasting ability." These observations point to regional inequalities in petroleum distribution caused by geologic factors other than just structure and reservoirs.

Geologic rationalizations heard in the 1950s and 1960s to explain the occurrence of barren traps and petroleum-poor basins were: 1) The oil and gas were once entrapped but have escaped through the seals, given geologic time; and 2) The oil has been "flushed out" by water movement.

Underlying these rationalizations were also the opinions that: 1) Entrapped oil can dissipate from the subsurface without leaving any residual traces; 2) Any dark shale is, a source rock; 3) Oil generation can occur at very shallow depth; and 4) There are neither limits nor restraints to migration distances in sedimentary basins.

These opinions supported the working concept that petroleum was ubiquitous in the subsurface, either because it was generated everywhere or because it would migrate anywhere. This optimistic outlook was necessary and useful in view of our lack of knowledge of petroleum formation mechanisms. Given such uncertainties, there was no alternative to systematic assumption of hydrocarbon presence in undrilled traps. This positive stand was an essential safeguard against the risk of prematurely condemning of viable exploration plays.

Since the mid-1960s, scientific breakthroughs in understanding petroleum formation and destruction processes have reduced many areas of uncertainty in petroleum geology. We owe the new perspectives to the post-war emergence of analytical techniques such as gas chromatography and mass spectrometry: they made possible the observation of geologic phenomena that, hitherto had been only a matter of speculation and controversy.

Because of these recent post-war technological advances, the following concepts are now recognized as useful in petroleum geology:

1. Evaporites are most efficient seals mainly because they offer very little or no pore space; however, the long term sealing properties of very fine grained, water-wet, porous rocks such as shales are also remarkably efficient *in the absence of open fractures.* This is due to the displacement pressure barrier effect created by the capillary pressure between oil and water in rock pores (Berg, 1975, Schowalter, 1976). Long-term sealing properties of very fine grained water-wet rocks are demonstrated by the excellent preservation of light oil and gas reserves in some very old sedimentary basins. For instance, shallow Paleozoic oil and gas in the Illinois, Michigan and Appalachian basins, major reserves in the Paleozoic Volga-Ural Basin (USSR) and giant Devonian and Ordovician fields in the southern Algerian Sahara demonstrate the sealing efficiency of very low permeability rocks, provided geologic history following entrapment has remained quiescent. All the above basins feature stable tectonic conditions and a lack of adverse thermal history since petroleum generation took place (during Late Paleozoic time, U.S.A. and Volga-Ural; and Cretaceous time, Algeria).

Rationalizations for barren traps and petroleum-poor basins by diffusion of *oil* through apparently tight seals, are difficult to reconcile with these geological observations, particularly where rocks have not suffered excessive thermal or tectonic stresses. These rationalizations also contradict theoretical considerations and laboratory experiments demonstrating that entry pressures are independent of time

(Schowalter, 1976). Because of high solubility in water, *natural gas* diffusion through undisturbed seals is possible, and has been observed in certain geologic settings (Leythauser et al, 1982).

Leakage from an oil accumulation through disturbed or poor quality seals, or in-situ degradation of entrapped oil, leaves residual traces in a reservoir because oil-wetting of the mineral grains does occur when oil saturations reach a level capable of sustaining water-free production (Schowalter and Hess, 1982). These traces can be observed by geochemical techniques or, in most cases, by modern hydrocarbon show detection methods as used in mud logging. *Hence, it is an inescapable conclusion that truly barren traps are found empty because no oil ever filled the reservoirs to begin with.*

2. Prolific oil producing basins, when geochemically evaluated, are shown to contain at least one adequately mature, deeply buried source rock system. It is often stratigraphically widespread and was deposited in an oxygen-depleted environment. Conversely, the bulk of dark, fine-grained sediments, in the sedimentary record, are *not* source rocks, and were deposited under oxygen-rich water, as in most of today's world oceans and lakes (Demaison et al, 1983).

3. Petroleum generation results from the transformation of kerogen (a solid organic substance) in the deep subsurface under the influence of both subsurface temperature and geologic time (Tissot and Welte, 1978).

4. Except in unusual cases of very long-range migration typically encountered on foreland basin plates, most entrapped oil in sedimentary basins originates from synclinal drainage areas that surround the trap itself. Thus migration distances commonly range in tens rather than hundreds of miles, particularly in strongly structured and/or faulted basins.

GEOLOGIC RISK REDUCTION BY GEOCHEMICAL METHODS

Exploration risk, being defined as the probability of spending exploration funds without economic success, has always been at the heart of the oil business. Geologic risk, which is a part of overall exploration risk, is fueled by uncertainties in subsurface geologic conditions, prior to drilling. It can also be expressed in terms of the probability of simultaneous occurrence of the key factors that determine the habitat of oil and gas in the subsurface.

Successful exploration for producible hydrocarbons in the subsurface depends on satisfying the following probabilities: 1) probability of existence of a trap (structure × reservoir × seal); 2) probability that the trap has received and physically retained a petroleum charge (source × maturation × migration paths × timing); and 3) probability that the entrapped petroleum has been preserved from the effects of thermal or bacterial degradation (temperature regime × meteoric water ingress).

Since these three main probabilities are independent of each other, the overall probability of discovering producible hydrocarbons at a given location is the product (not the sum) of the probabilities of these individual factors, that is, if any one of these three main factors is 0, the overall probability of success is 0, regardless of how favorable the other two remaining factors are.

Common sense agrees with probability concepts: no geologist will recommend drilling a syncline or a section void of reservoirs. However equally high risks, in a mathematical sense, are often taken by explorationists with respect to oil charge and oil preservation probabilities.

This is commonly the case if geochemical information is absent, poorly integrated, or ignored. In the face of uncertainty or skepticism regarding source and degradation regimes, explorationists tend to rely mainly on structural information to evaluate geologic risk prior to drilling. This exploration approach is justifiable when applied to proven petroliferous basins or plays. However, it can be improved by also taking into account measurable parameters relevant to oil and gas generation, migration, and destruction. This integrated approach is particularly effective in frontier basins, deep gas plays, and offshore environments, where exploration costs are inordinately high.

THE GENERATIVE BASIN CONCEPT

A fundamental step in the prediction of petroleum occurrence by geochemical methods is evaluation of whether an undrilled trap has had access to hydrocarbon migration from mature petroleum source beds. This statement holds the essence of the "generative basin concept."

Areas underlain by hydrocarbon generative source beds are called "petroleum generative depressions" or "hydrocarbon kitchens." The term "generative basin" describes a sedimentary basin containing one or more petroleum generating depression.

Recognition of generative depressions is achieved by overlaying organic facies maps and maturation maps of each key petroleum source horizon. Maturation maps are compiled from seismic depth maps, near the potential source horizons, and from maturation gradients derived from well data and calibrated time-temperature models (Waples, 1983). Organic facies maps reflect the stratigraphic distribution of organic matter types within a given source rock unit. They are compiled by integrating kerogen type data into the known paleogeographic and paleo-oceanographic context (Demaison et al, 1983).

The geochemical approach, in prospect appraisal, begins by investigating whether mature source beds are present in the drainage area of a trap. A further step consists of mapping areas of mature source beds and calculating both mature source rock volumes and petroleum yield. Lastly, migration pathways can be modeled between the mature source-rocks and the trap. This type of geologic exercise permits a ranking of prospects by the criterion of degree-of-access to mature source rocks.

The geochemical approach to basin evaluation consists of mapping oil generative depressions or basins and erecting a matrix of drilling success ratios, volumes of discovered hydrocarbons and "kitchen" potential. When these correlations have been established they may be used for comparative purposes and for future evaluation of geologic risk. Application of the "generative basin concept," leading to recognition and prediction of areas of high potential, is the object of this contribution.

RECOGNITION AND PREDICTION OF AREAS OF HIGH POTENTIAL

Historically recognized zones of high success ratios in finding petroleum in sedimentary basins are called "areas of high potential," "plays," "fair-ways," or "petroleum zones." Geochemical data, when mapped together with geology and structure derived from seismic information, permit early geographic and geologic delineation of these areas of high potential. Moreover, the generative depression mapping approach leads to a more realistic evaluation of geologic risk prior to drilling, than the blanket application of worldwide success ratios or arbitrary risk factors.

The following illustrations are meant to demonstrate that basin-wide geochemical mapping can help identify areas of high potential in sedimentary basins. The subject basins are rapidly reviewed in order of tectonic style.

RIFT BASINS

North Sea Basin

The heavily explored portion of the North Sea between the 55th and 62nd parallels is a major oil and gas province with estimated reserves of over 24 billion barrels of oil and over 55 Tcf of gas (Figure 1). Reservoirs range in age from Devonian to Eocene typically producing on tilted fault blocks and, to a lesser degree, on salt swells.

The principal oil sourcing system is of Late Jurassic Kimmeridgian-Volgian age. It is thermally mature and actively generating and expelling oil at formation temperatures higher than 200°F (93°C) corresponding to depths in excess of 10,000 ft (3,048 m). The generated oil moves to the nearest available reservoirs: older, fault-juxtaposed Middle Jurassic sands in the northern North Sea (Viking Graben); and Upper Cretaceous chalk in the southern North Sea (Ekofisk area and "Tail-end" graben).

A maturation map of the Kimmeridgian shale permits one to delineate both immature areas where temperatures are lower than about 200°F (93 °C) and no significant oil generation has taken place, and mature oil generative depressions (where present-day earth subsurface temperature exceeds 200°F [93°C]). This corresponds approximately to a vitrinite reflectance level of 0.6.

A statistical count of all dry holes and successful holes in the North Sea shows that: 1) Virtually all the oil and gas fields lie within or very near the oil generative depression containing mature Kimmeridgian source rocks; 2) The historical success ratio in the mature source fairway, or generative basin, is in the order of 1 in 3; 3) Outside the Kimmeridgian generative depression the historical success ratio is in the order of 1 in 30. The oil fields found thus far in this higher risk area (Beatrice, Briesling and Bream) are probably sourced from Middle or Lower Jurassic beds. The crude oils are different in composition from Kimmeridgian-sourced crudes; 4) Migration distances in the North Sea are commonly short and limited by the drainage areas of individual structures; and 5) The fields with the largest reserves tend to be close to the center of the generative depression (Statfjord, Piper, Forties, Ekofisk), in close proximity to the most thermally mature and thickest

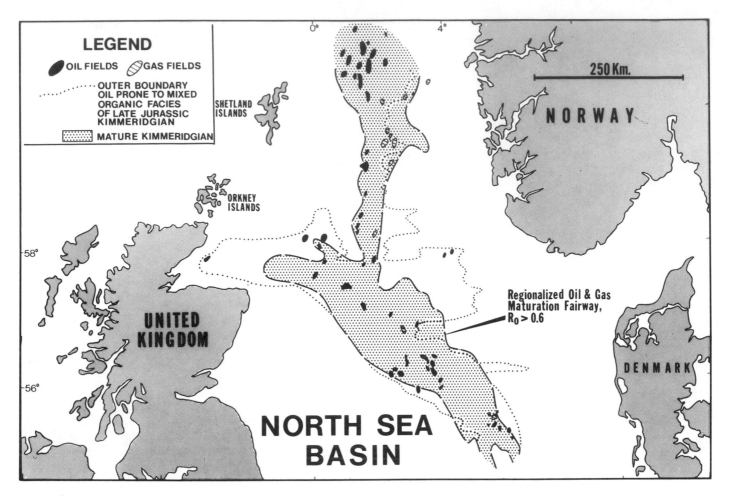

Figure 1. The North Sea Basin (Ziegler, 1980; Demaison et al, 1984).

Kimmeridgian shale depocenters.

Syrte Basin

The formline structure map at the base of the Upper Cretaceous in the Syrte Basin and Western Desert indicates that the oil fields are closely associated with regional structural depressions where Cretaceous and Paleogene source rocks are buried below about 2,500 m (8,202 ft) (Figure 2). Most of the largest fields (Sarir, Amal, Nafoora, Gialo) tend to occur close to the deepest parts of the generative depressions. Other large fields (Zelten, Waha, Defa, Dahra-Hofra) are on high horst blocks but in close proximity to deeply buried generative depressions. Oil occurrence in the western desert of Egypt relates to generative depressions involving Jurassic and lowermost Cretaceous source rocks to the north of the Quattara Ridge and Upper Cretaceous source rocks in the Abu Gharadig Basin (Parker, 1982).

Gippsland Basin

In the Gippsland Basin (Figure 3) only a relatively limited volume of source beds (Eocene, Paleocene and perhaps Upper Cretaceous paralic coal measures) is within the generative window, hence major oil and gas generation is confined to the central part of the basin. The oil fields (3 of them giants) are located close to the thickest and deepest hydrocarbon kitchens, within the oil window (Ro 0.6). The gas fields are located on the outer edge of the generative depression, in a fairway comprised between 0.45 and 0.6 vitrinite reflectance. However, carbon isotope data for the methanes of these gas fields are not available, therefore it is not known whether the gas is early or late thermal. Success ratios are spectacularly high in the oil and gas fairways which clearly coincide with source maturation patterns.

INTRACRATONIC DOWNWARPS

Western Siberian Basin

This mega-basin hosts the major share of Soviet oil reserves and about 6% of the world's reserves. It is a site of active exploration and production (Figure 4). The main source system is the Bhazenov Formation, a widespread laminated black shale of Late Jurassic age which is nearly synchronous with the Late Jurassic source beds of the North Sea. They are closely related to the same Late Jurassic "oceanic anoxic event."

It should be noted that the supergiant oil fields (Samotlor, Ust-Balisk, Surgut) are all located close to the center of this very large generative basin. The origin of the

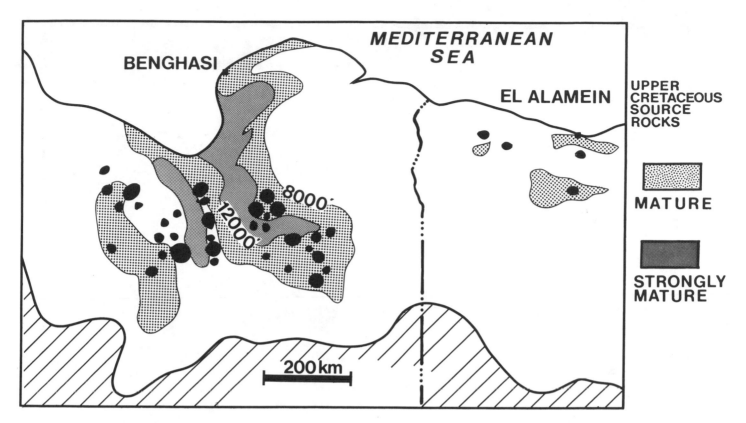

Figure 2. Syrte Basin, North Africa (Parsons et al, 1980; Parker, 1982).

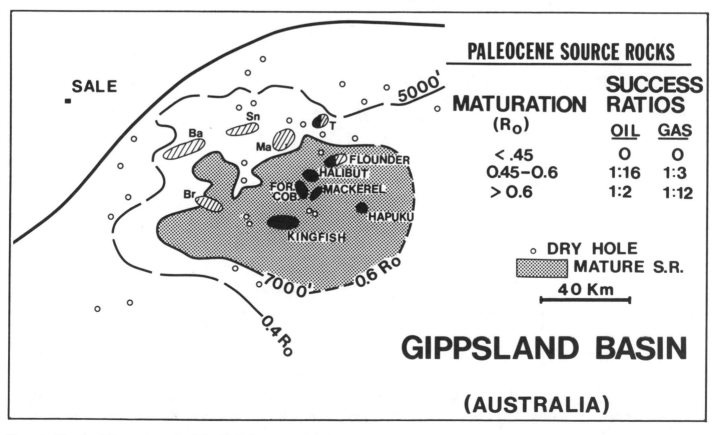

Figure 3. Gippsland Basin, Australia (Threllfall, Brown, and Griffith, 1976; Kantsler et al, 1978; Shibaoka, Saxby, and Taylor, 1978).

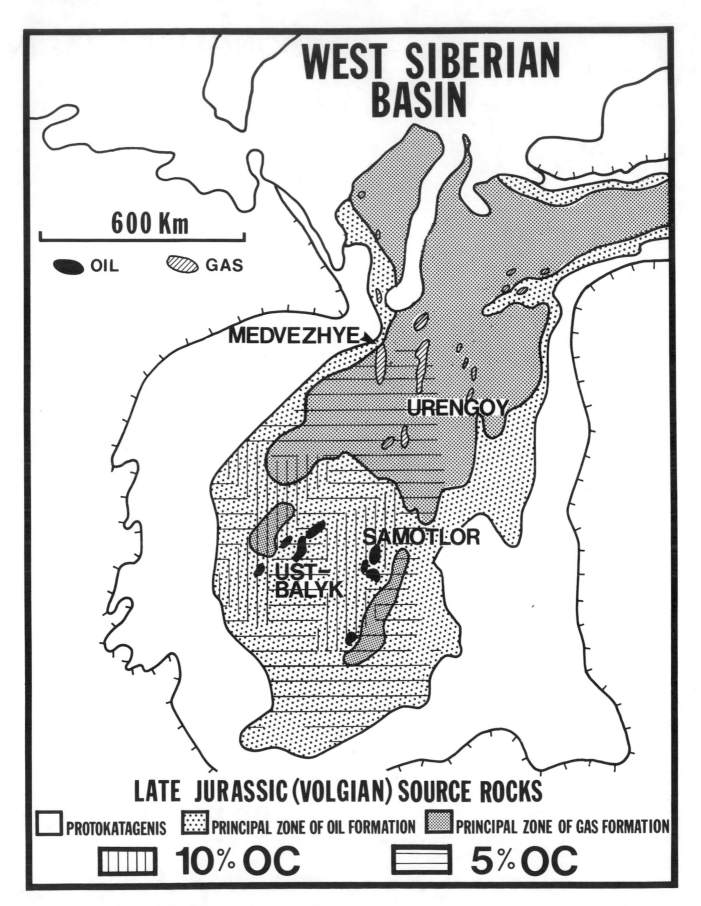

Figure 4. Western Siberia, USSR (Kontorovich, 1971, 1983).

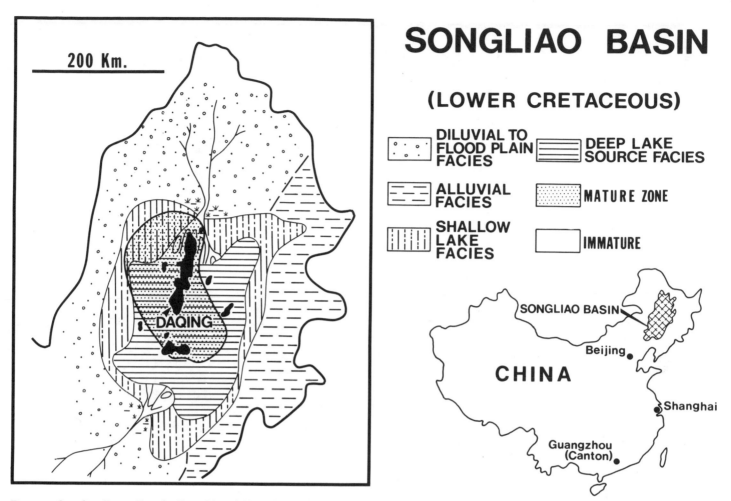

Figure 5. Songliao Basin, Peoples Republic of China (Wan Shangwen, Hu Wenhai, and Tan Shidian, 1982).

gas in the giant fields (for example, Urengoy) in the north is still controversial and could be partly "biogenic."

Songliao Basin

The Songliao Basin (Figure 5) is a large intracratonic rift-style basin in eastern China. It is filled with non-marine fluviatile to lacustrine sediments of Cretaceous to Tertiary age. The source beds are Early Cretaceous lacustrine, highly oil-prone, black shales. They are thermally mature and oil generative in the central part of the basin. The combination of deep lake facies, in Lower Cretaceous time, and favorable thermal maturation levels, geographically outlines that area containing the bulk of the oil reserves in the Songliao Basin. Geologists from the PRC have recently expressed the view that the oil generating depressions contain over 80 percent of the oil reserves of Songliao, Domying and other depressions of East China (Hu, 1983).

The largest oil-bearing structural complex is the giant Daqing oil field. It is located immediately updip from the center of the generative basin, the Pijia-Gulong generative depression. The giant Daqing field complex is believed to be the largest oil accumulation in existence in a nonmarine basin, anywhere in the world. It accounts for 80 percent of the total discovered oil in place in the Songliao Basin. The

maximum distance of hydrocarbon migration is less than 25 mi (40 km).

Illinois Basin

A generalized isoreflectance (maturation) contour map on top of the Lower Carboniferous (New Albany) oil source shale sequence in the Illinois Basin (U.S.A.) shows more than 3/4 of the basin's oil occurring in the Lower Carboniferous (Mississippian) above the New Albany shale (Figure 6). Also, the bulk of Devonian and Silurian oil yet found in the Illinois Basin occurs within a few hundred feet of the base of the New Albany source shale. Note that over 90 percent of the oil produced so far from the Illinois Basin lies in a fairway comprised between 0.7 Ro and 30 mi (48 km) updip from 0.6 Ro (approximately 0.55 Ro) as measured in the New Albany shale. Migration distance does not exceed 30 mi (48 km) and most of the large accumulations, such as the La Salle anticline fields, are immediately updip from the center of the generative depression.

Cooper Basin

Figure 7 is a simplified and schematized isoreflectance

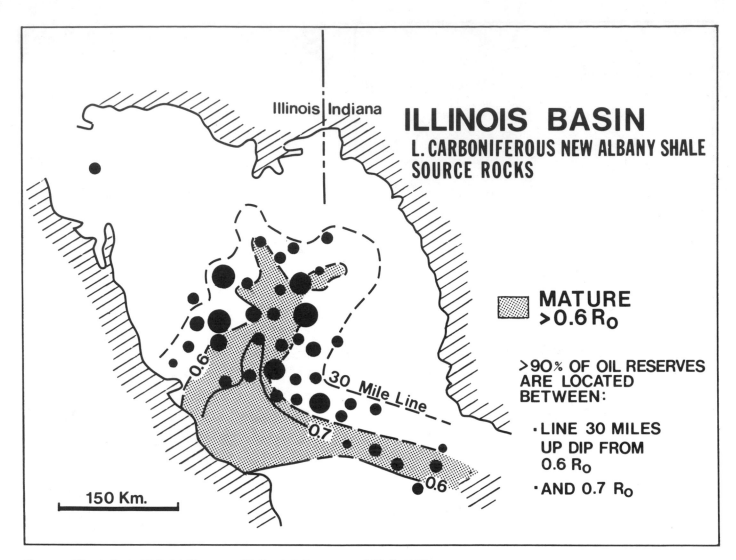

Figure 6. Illinois Basin (U.S.A.) (Swann and Bell, 1958; Barrows and Cluff, 1983).

maturation contour map on top of the Permian coal measures in the Cooper Basin of Central Australia. Giant gas fields sourced by the Permian coals and coaly shales occur in or near the zone of optimum maturation (reflectance 0.9 to 2.0). Permian sands on anticlinal structures which are both in the immature zone, and not within drainage reach from the gas generative zone, have been found to be water wet.

Anticlines with Permian sands in the post-mature zone have yielded natural gas high in carbon dioxide: sandstone porosities in these postmature zones are very low due to extensive silicification. The largest gas field (Moomba) lies immediately updip of the main gas generative depression.

Paris Basin

The Paris Basin (Figure 8) has minor oil production, mainly from Middle Jurassic limestones. Reportedly the co-sources for most of these oil accumulations are black shales at the base of the Middle Jurassic, and the Liassic (Toarcian) shales. However, the latter are isolated by a thick nonsource shale from the mid-Jurassic reservoirs. As earlier

recognized by Tissot et al (*in* Tissot and Welte, 1978, p. 510), producing fields are in, or very close to, the zone where the source shales are most deeply buried and oil generative in the central part of the basin. The Paris Basin contains many anticlines outside of the generative depression, some large, with adequate well-sealed reservoirs. So far, they have been found barren. Short migration distances can be explained by a combination of mediocre carrier continuity and rather limited oil volumes available for migration.

PASSIVE ATLANTIC-TYPE MARGINS

Barrow-Dampier Basin

Apparently the major source unit in the stratigraphic section of the northwest shelf of Australia (Figure 9) is in marine shales of Late Jurassic to Lower Cretaceous age which tend to be overpressured. Although marine, these rocks are mainly gas-prone, with, however, subordinate, sporadic potential for oil. The outline of the mature Jurassic generative depression delineates virtually all the commercial

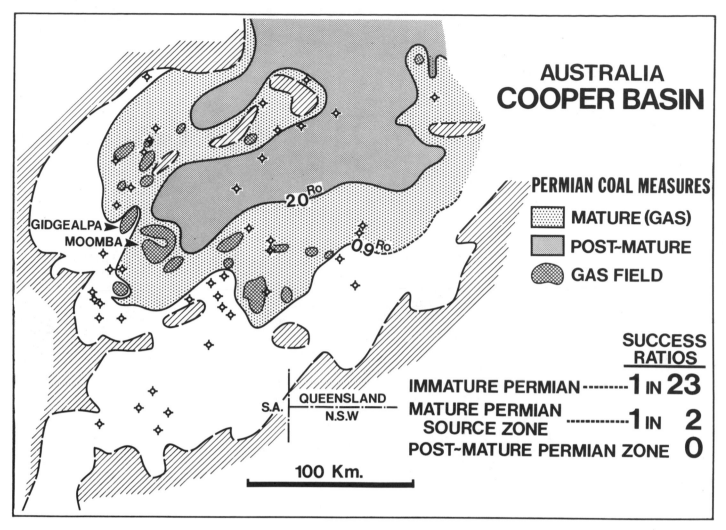

Figure 7. Cooper Basin, Australia (Kantsler et al, 1983).

gas and oil accumulations. Migration distances are short. Lower Jurassic and Triassic coals are rich in inertinite and give mediocre hydrocarbon yields. The suggestion that they are only a secondary co-source is supported by the observation that there are no commercial gas accumulations in Triassic reservoirs, except where these are within drainage reach from the Jurassic generative depression.

COMPLEX MULTICYCLE BASINS

Illizi and Ahnet Basins

The regional patterns of oil and gas occurrence in the Illizi and Ahnet Basins (Figure 10) were recognized as far back as the late 1960s. Lack of oil occurrence in the Devonian and Ordovician of the Ahnet Basin relates to post-maturity of the Silurian source shales, reflectances being in excess of 2%. In the Illizi Basin the high success ratio oil fairway, which contains the four largest oil fields (Zarzatine, Tin Fouye, Edjeleh and Tiguentourine) is located in the zone of more moderate maturation (0.6 to 1.0 Ro) of the Paleozoic source shales.

FORELAND BASINS

Middle East Petroleum Zones

Figure 11 shows the clear geographic association of Cretaceous fields in the Middle East with zones of deep burial of Cretaceous rocks, which include several source horizons. Several large oil fields, in the shallow Cretaceous burial area on the south side of the Gulf, seem to have been sourced from the Upper Jurassic (Murris, personal communication). Gas fields coincide with those areas where Cretaceous rocks are buried below 5,000 m (16,404 ft) in the post mature, metagenetic stage. It is interesting to note that Cretaceous organic-rich rocks are widespread in the Levant (Syria, Lebanon, Jordan, Israel) where they often reach oil-shale levels of richness. With the exception of parts of Syria, and perhaps under or near the Dead Sea, the general lack of maturity due to insufficient burial of the source beds precludes the occurrence of major petroleum producing areas.

Williston Basin

About half of the discovered oil in the Williston Basin

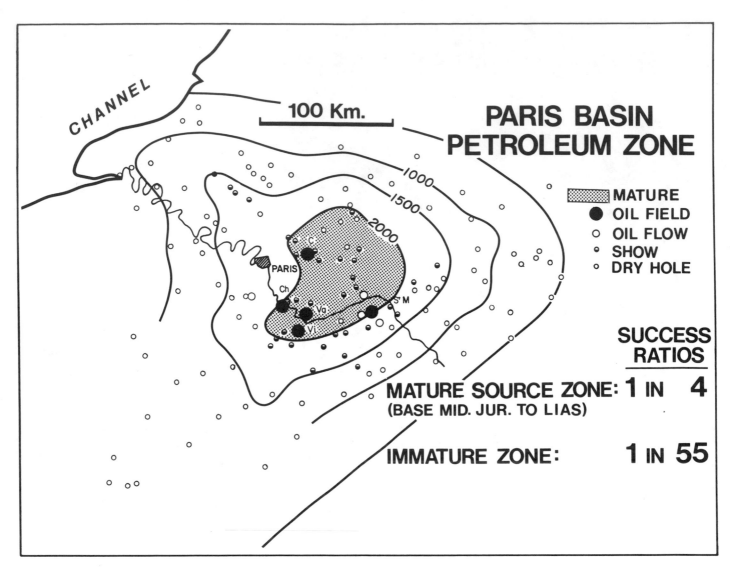

Figure 8. Paris Basin (France) (Tissot and Welte, 1978).

(Figure 12) has been entrapped within the generative depression, and the other half has migrated a rather long distance updip (in excess of 60 mi [97 km]) on the northeastern flank. This has been made possible by adequate lateral reservoir continuity, enhanced by "shunting" from one permeable carrier to another. Moreover the structural style which is homoclinal, with little interruptions, has allowed extensive lateral drainage to take place. The largest accumulation in the Williston Basin is on the Nesson Anticline. It contains about 500 million barrels of oil. It is noteworthy that this major anticlinal trend is plunging into the heart of the Bakken generative depression, and drains its deepest and most mature portion.

CONCLUSIONS

A rapid review of the petroleum generation aspects of 12 basins around the world, for which stratigraphic source bed occurrence and maturation gradients are known, reveals the following regularities:

1. In all reviewed cases, regardless of sediment age, basin size, and tectonic style, the zones of preferred petroleum occurrence ("areas of high potential") are genetically correlated to geochemically identified oil or gas generative depressions. The latter are geographically outlined by the favorable thermal maturation fairways relevant to key petroleum source horizons. The success ratios within, or close to, the generative depressions are in the order of 1 in 3, wherever such ratios have been statistically evaluated.

2. The largest petroleum accumulations tend to be located close to the center of the generative depressions. One common denominator to the giant accumulations, besides structural size, is the large volumes of hydrocarbon-generating sediments, drained for long spans of geologic time.

3. Migration distances for most reviewed cases commonly range in tens rather than hundreds of miles and are limited to the drainage areas of individual structures. Thus the outlines of the mature hydrocarbon generative depressions commonly include most of the basin's hydrocarbon reserves and the largest fields. These conclusions apply mainly to rift basins, intra-cratonic

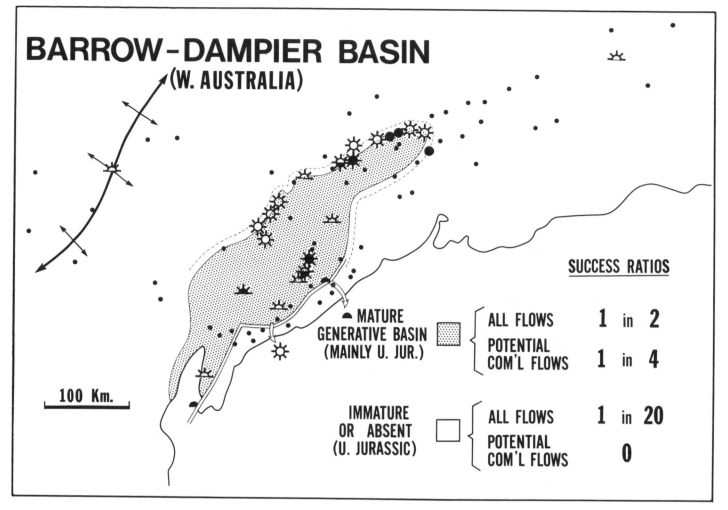

Figure 9. Northwest Shelf of Australia.

platform-type basins, and passive oceanic margins, including deltas, and wherever migration distances are limited by structural style. Cases of long-distance migration are documented in certain foreland basins of North America (for example, the Denver Basin and Williston Basin), and in the peri-Andean basins of Ecuador and Peru, or for the heavy oils of Western Canada and Eastern Venezuela (Demaison, 1977). In all cases, both stratigraphy and structure favored uninterrupted updip movements of oil.

In conclusion, these observations demonstrate the fundamental reason for geographic patterns of regional petroleum occurrence: that being their association with widespread, petroleum generative, source beds in the subsurface. The generative basin concept may also be applied in a predictive mode, since the three regularities, listed earlier, provide powerful analogs in forecasting areas of high potential in undrilled or sparsely drilled basins.

It is important to keep in mind, that the implementation of generative basin mapping methods, requires a sufficient data bank: regional seismic compilations are needed, with geochemical profiles from existing deep wells: together they permit realistic basinwide reconstruction of generation, migration, and accumulation processes. Furthermore, oil-to-oil and oil-to-source correlation studies, by biological

markers and other methods, are often indispensable.

In summary, the predictive accuracy of geochemical mapping systems is dictated by the available data base, while final interpretation is as dependent as ever on geologic skills and experience. Thus at early stages of exploration, when data are sparse, geologists should guard from overtaxing geochemical knowledge. The words of Wallace Pratt: "The oil-finder...need always be always alert to the potentialities of what he does not know," still apply today, as they did some 30 years ago.

ACKNOWLEDGMENTS

The writer thanks the management of Chevron Overseas Petroleum Inc., and Standard Oil Company of California, for permission to publish this paper. The writer is also indebted to R.G. Alexander Jr., A.J.J. Holck, R.J. Murris, and M.H. Nederlof for critical reviews of the manuscript.

REFERENCES

Bally, A.W., and S. Snelson, 1980, Realms of subsidence, *in* Facts and principles of world petroleum occurrence:

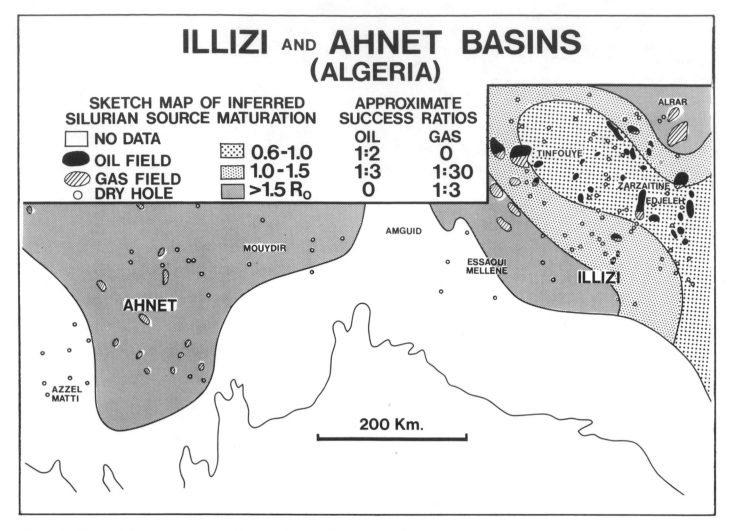

Figure 10. Algerian Sahara (Correia, 1967; Tissot et al, 1973; Perrodon, 1980).

Canadian Society of Petroleum Geologists Memoir 6, p. 9-94.

Barrows, M.H., and R.M. Cluff, 1983, New Albany shale group (Devonian-Mississippian) source rocks and hydrocarbon occurrence in the Illinois Basin, *in* G. Demaison and R.J. Murris, eds., Geochemistry and basin evaluation: AAPG Memoir 35, this volume.

Berg, R.R., 1975, Capillary pressure in stratigraphic traps: AAPG Bulletin, v. 59, p. 939-956.

Bois, C., P. Bouche, and R. Pelet, 1982, Global geologic history and distribution of hydrocarbon reserves: AAPG Bulletin, v. 66, p. 1248-1270.

Correia, M., 1967, Relations possibles entre l'etat de conservation des elements figures de la matiere organique et l'existence de gisements d'hydrocarbures: Revue Institute Francais de Petrole 22, n. 9, p. 1285-1306.

Crostella, A., and M.A. Chaney, 1978, The petroleum geology of the outer Dampier Sub-basin: Australian Petroleum Association Journal, v. 18, pt. 1, p. 13-33.

Demaison, G.J., 1977, Tar sands and supergiant oil fields: AAPG Bulletin, v. 61, p. 1950-1961.

——— , et al, 1984, Predictive source bed stratigraphy; a guide to regional petroleum occurrence: London, Proceedings 11th World Petroleum Congress, John Wiley and Sons.

Dow, W.G., 1974, Application of oil-correlation and source rock data to exploration in Williston Basin: AAPG Bulletin, v. 58, p. 1252-1262.

Hu, 1983, Panel discussion 1: London, Proceedings 11th World Petroleum Congress, John Wiley and Sons.

Kantsler, A.J., et al, 1983, Hydrocarbon habitat of the Cooper-Eromanga Basin, Australia, *in* G.J. Demaison and R.J. Murris, eds., Geochemistry and basin evaluation: AAPG Memoir 35, this volume.

Kontorovich, A.E., ed., 1971, Regional geochemistry of petroleum bearing Mesozoic formations in Siberian basins: Ministry of Geology of the USSR, Memoir 118 (in Russian).

——— , 1983, Geochemical methods for quantitative evaluation of petroleum potential in sedimentary basins; examples from Siberian basins (USSR), *in* G.J. Demaison and R.J. Murris, eds., Geochemistry and basin evaluation: AAPG Memoir 35, this volume.

Leythaeuser, D., R.G. Schaefer, and A. Yukler, 1983, Role of diffusion in primary migration of hydrocarbons: AAPG Bulletin, v. 66, p. 408-429.

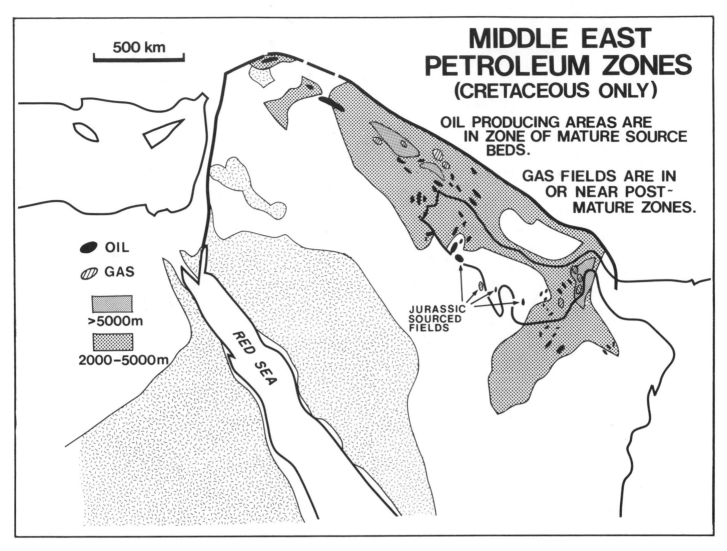

Figure 11. Middle East (Riche and Prestat, 1978; Murris, 1983).

Meissner, F.F., 1983, Petroleum geology of the Bakken Formation, Williston Basin, North Dakota and Montana, *in* G.J. Demaison and R.J. Murris, eds., Geochemistry and basin evaluation: AAPG Memoir 35, this volume.

Murris, R.J., 1983, Middle-East stratigraphic evolution and oil habitat, *in* G.J. Demaison and R.J. Murris, eds., Geochemistry and basin evaluation: AAPG Memoir 35, this volume.

Parker, J.R., 1982, Hydrocarbon habitat of the western desert of Egypt: Abstract, EGPG Sixth Exploration Seminar, Cairo.

Parsons, M.G., A.M. Zagaar, and J.J. Curry, 1980, Hydrocarbon occurrence in the Syrte Basin, Libya, *in* Facts and principles of world petroleum occurrence: Canadian Society of Petroleum Geologists Memoir 6, p. 723-732.

Perrodon, A., 1980, Geodynamique petroliere Genese et repartition des gisements d'hydrocarbures: Paris, New York, Masson.

Riche, P.H., and B. Prestat, 1979, Paleogeographie du Cretace de proche et moyen orient et sa signification petroliere: Proceedings 10th World Petroleum Congress,

Heyden and Son Ltd., London.

Schowalter, T.T., 1976, The mechanics of secondary migration and entrapment: Wyoming Geological Association Earth Science Bulletin, v. 9, no. 4, p. 1-43.

—— and P.D. Hess, 1982, Interpretation of hydrocarbon subsurface shows: AAPG Bulletin, v. 66, p. 1302-1327.

Swann, D.H., and A.H. Bell, 1958, Habitat of oil in the Illinois Basin, *in* L.G. Weeks, ed., Habitat of Oil: AAPG Special Publication.

Shibaoka, M., J.D. Saxby, and G.H. Taylor, 1978, Hydrocarbon generation in Gippsland Basin, Australia; comparison with Cooper Basin, Australia: AAPG Bulletin, v. 62, p. 1159-1170.

Threlfall, W.F., B.R. Brown, and B.R. Griffith, 1976, Gippsland Basin offshore, *in* Economic geology of Papua-New Guinea: Australasian Institute of Mining and Metallurgy Monograph 7, p. 41-67.

Tissot, B., et al, 1973, Origine et migration des hydrocarbures dans le Sahara oriental (Algerie), *in* Advances in organic geochemistry: Paris, Technip.

——, and D.H. Welte, 1978, Petroleum formation and occurrence: New York, Springer-Verlag.

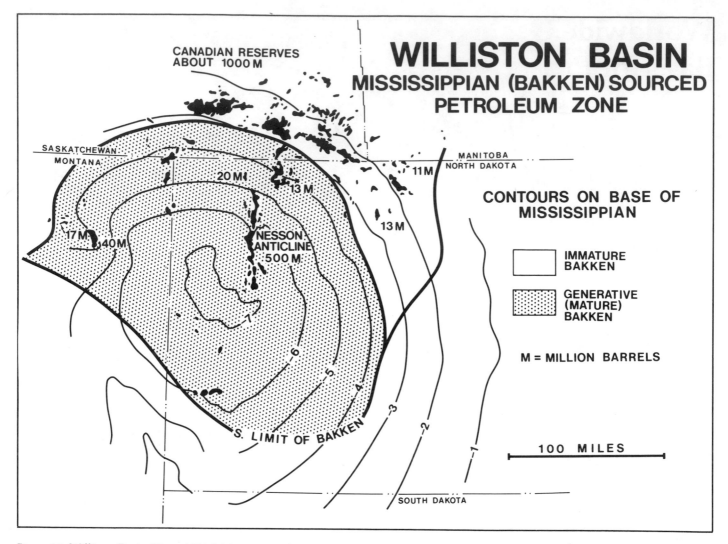

Figure 12. Williston Basin (Dow, 1974; Meissner, 1983).

Wang, Shangwen, Hu Wenhai and Tan Shidian, 1982, Habitat of oil and gas fields in China: Oil and Gas Journal, v. 80, n. 24, p. 119-128.

Waples, D.W., 1983, Physical-chemical models for oil generation: Colorado School of Mines Quarterly, v. 78, no. 4, p. 15-30

Ziegler, P.A., 1980, Northwest European Basin geology and hydrocarbon provinces, *in* Facts and principles of world petroleum occurrence: Canadian Society of Petroleum Geologists Memoir 6, p. 653-706.

Worldwide Geological Experience as a Systematic Basis for Prospect Appraisal

D. Sluijk and M.H. Nederlof
Shell Internationale Petroleum Mij
The Hague, Netherlands

To evaluate the uncertain hydrocarbon potential of single exploration prospects, use is made of a computer model to simulate the processes of hydrocarbon generation, expulsion and migration (leading to "hydrocarbon charge"), trapping and retention, and finally recovery.

Simulation of the charge and retention processes is based on the results of various "calibration studies". Such studies imply the statistical analysis of extensive data sets, which represent worldwide exploration experience. The statistically highly significant results of these studies are built into the simulation model.

For the estimation of the probability of charge, worldwide prior probabilities are updated in two steps: firstly, on the basis of prior observations of charge in the area of the prospect; and secondly, on the basis of the generation and migration parameters specific to the prospect in question. The latter update is made in line with the results of the statistical analysis of the charge calibration file, which is a representative collection of numerous case histories. This learning set also forms the basis for the estimation of the charge volumes.

In the simulation of the complex process of trapping and retention, the sealing capacity plays a role. Based on the analytical results of another calibration file, the seal properties as described by the appraiser can be used to estimate the retention potential of the seal (i.e., the maximum differential pressure, or hydrocarbon column, the seal can withstand).

The conceptual models followed in the calibration studies mentioned above are deliberately simple, so that the parameters needed for the description of the model can be readily estimated during actual prospect appraisal. This simplicity, and the implicit incompleteness of the models, contributes to uncertainties in the resulting formulas. The residual errors, however, are taken into consideration when the formulas are applied in the simulation model, the less significant parameters receiving less weight in the end result.

INTRODUCTION

The aim of prospect appraisal is to obtain an estimate of the recoverable hydrocarbon volumes in an exploration prospect before drilling. Much more than with reserve estimations after drilling, prospect appraisal must take the uncertainties into account. It is possible to do this if all parameters involved in the estimation are expressed as probability distributions rather than as single numbers. A convenient way to work in such a probabilistic manner is to apply repetitive calculations, each time using equally likely choices for all the input parameters (random sampling). This so-called Monte Carlo procedure does not result in a single number, but in a series of equally likely outcomes that reflect the uncertainties both in the input and in the applied formulas.

Our appraisal system[1], designed for the evaluation of single exploration prospects, is based on a simple geochemical/geological model. This model follows two complex natural processes and one technical process:

1. hydrocarbon generation, expulsion and migration, leading to "hydrocarbon charge" (i.e., hydrocarbon volumes available for entrapment).
2. trapping and retention of oil and gas.
3. recovery.

The system employs a computer simulation of these processes, whereby the formulas for charge and retention result from calibration studies. These studies are based on worldwide collections of well-described case histories covering a wide range of geological conditions including validly tested hydrocarbon-bearing as well as dry traps, the so-called calibration file, or learning set.

---

[1]Since both basic data and results are, to a high degree, proprietary, only the principles of the system can be divulged.

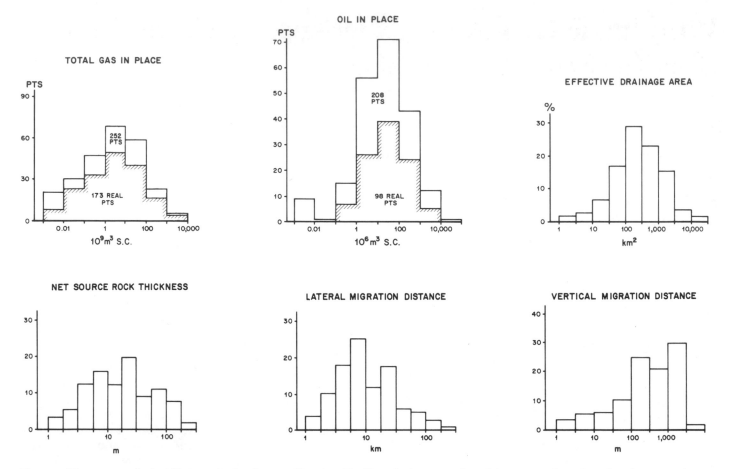

Figure 1. Histograms of a few file items in the charge calibration file. Note the log-normality of these examples: oil in place (208 points); total gas in place (252 points); a few geological parameters (346 points).

Such a data base should reflect exploration experience and should also be representative of the widest variety of occurrences of oil and gas. Thus, it comprises a prior distribution, a starting point for the estimate of hydrocarbon volumes. A successful statistical analysis of the data base (calibration study) leads to the recognition of certain geochemical/geological variables which have a bearing on the outcome. By taking such variables into account in the simulation formulas, the prior uncertainty can be reduced in a process of updating, which leads to a posterior estimate, the so-called Bayesian approach (Winkler, 1975; Nederlof, 1979 and 1981). The latter reduction of prior uncertainty is the main aim of the procedure.

It is good practice for the calibration study to follow a causal, explicit model, which incorporates both geochemical and geological parameters. These parameters should be known for all points of the calibration file. For the design of an appraisal system, it is important to restrict the parameters to only those which can, with some degree of confidence, be estimated before drilling. This pragmatic requirement most likely implies that the model is simple and incomplete. Such a model can never lead to a complete understanding of the processes involved and can only partly reduce the prior uncertainty. The residual uncertainty is taken into account when the statistical formulas are applied in the Monte Carlo simulation model.

Our calibration, with regard to hydrocarbon charge and retention, proved to be quite successful; procedures and results form the essence of this paper.

CHARGE CALIBRATION

The Calibration File

The learning set for charge calibration contains several hundred case histories ("calibration points") for which the charge model is thought to be reasonably well-known. These data were assembled from 37 geological basins spread over the whole world.

For each point the learning set contains qualitative and quantitative descriptions of a variety of geological parameters which are thought to be relevant to hydrocarbon charge. Moreover, it is known for each point whether hydrocarbons are trapped and retained, and if so, how much oil, solution gas, and free gas are (initially) in place. Finally, the calibration file contains data needed to judge, in each case, whether or not the trap is full to its structural spill-point, and whether or not the seal is charged to its full holding capacity. The latter judgement is based upon the results of a separate seal calibration.

The proprietary nature of the data set prevents extensive publication. Nevertheless, it is possible to include a few histograms which show the ranges of some of the important parameters contained in the charge calibration file (Figure 1).

The Charge Volume

The charge volume, the dependent or Y-variable in the present study, can be deduced from the observed hydrocarbon volumes. If a trap is present with a seal capable of holding at least some hydrocarbons and the structure is dry, the charge is equal to zero (Y = 0). If a trap is not full to its structural spill-point and the seal is able to hold more than it actually does, there appears to have been a charge constraint and the observed hydrocarbon volume in place is equal to the charge (Y = observed volume, or in other words, the observed volume is a real value for charge). If, however, the trap is full to spill-point, or is filled to the full retention potential of the seal, we are dealing respectively with a trap or seal constraint. In such cases the observed hydrocarbon volume is a minimum value for charge (Y > observed volume). In statistics such an observation is termed "censored." For the purpose of analysis it should be labelled with a "censoring code": in this case REAL (= uncensored), or MINIMUM, respectively.

Objectives of the Charge Calibration

The objective of the charge calibration exercise is two-fold. The first aim is to find a combination of geological parameters which permits the classification of hydrocarbon-bearing calibration points (Y > 0) versus dry ones (Y = 0); i.e., the combination of X-variables which is most significant in answering the question "are hydrocarbons present?". The second aim is to find a combination of geological parameters which shows a quantitative relationship to the hydrocarbon volumes observed; i.e., the combination of X's which contributes most to answering the question "if hydrocarbons are present, how much?".

The first objective is achieved through discriminant analysis; the second through regression analysis. Viewing the presence of both real and minimum values for Y, a special regression technique is used which allows the introduction of censored Y-values.

Geochemical Material Balance

A geochemical material balance formula for charge was postulated to serve as a guide for the selection of the possibly significant qualitative and quantitative geological parameters, and for the translation of these into quantities which could serve as X-variables.

In this equation the hydrocarbon charge is considered to be the product of four factors. The first two factors cover generation and expulsion, the last two migration:

1. The volume of the source rock contributing to the charge of a given trap, which in turn is the product of two factors:
   DRAR = that part of the effective drainage area (sq km) that is below the top of mature source rocks (in our study the onset of oil generation is taken at a Lopatin value of 21, which in our maturity calibration corresponds to a vitrinite reflectance (R_o) of 0.62; see also Waples, 1980);
   SRTH = the net thickness (m) of the source rock in this part of the drainage area.
2. The actual hydrocarbon yield per unit volume of source

rock, which is the product of:
   PUY = the potential ultimate yield (litres oil, or cu m gas at standard conditions) per cu m source rock (that is, the total amount of hydrocarbons the source rock can expel when brought from an immature to a post-mature state);
   YFM = the fraction of this potential yield generated and expelled at the given maturity of the drainage area.
3. The fraction of the yield that was generated, expelled, and migrated after trap formation:
   YFT = yield fraction for timing
4. The fraction of the yield that survives migration:
   MFR = migration fraction.

With CHARGE expressed in millions of cu m oil or billions of cu m gas at standard conditions, the following material balance equation can be written out:

$$\text{CHARGE} = 10^{-3} \times \text{DRAR} \times \text{SRTH} \times \text{PUY} \times \text{YFM} \times \text{YFT} \times \text{MFR} \qquad (1)$$

The last four factors should be defined specifically for oil or for gas, as appropriate.

The Independent Variables

The independent or X-variables used in the statistical analysis of the calibration file are geochemical/geological parameters defined in such a way that they fit the material balance formula. Some factors in the generation/expulsion part of the formula are directly measurable:

1. The *effective drainage area* (DRAR), or fetch area that feeds the trap in question. Firstly, this requires structural maps at source rock and cap rock levels to determine the drainage pattern; secondly it requires an estimation of the shallowest depth at which maturity occurs (see above).
2. The *net thickness* (SRTH) of the source rock in the drainage area, applying a cut-off of a certain minimum richness, say 1 percent organic carbon. This figure (together with type, see below) is usually derived by extrapolation from observations outside the drainage area in question.

Other figures have to be derived indirectly:

1. The *potential ultimate yield* (PUY) can be considered to be a function of original (= immature) source-rock type and richness. The type can be translated into the C_R/C_T ratio (the ratio of residual carbon after pyrolysis, C_R, to the total insoluble organic carbon before pyrolysis, C_T). The change in this ratio, when the source rock goes from immature to post-mature, is a measure for the potential ultimate oil yield per cu m per 1 percent organic carbon (see Table 1). No such relationship is known for gas.
2. The *yield fraction for maturity* (YFM) is basically dependent on the geochemical knowledge of the hydrocarbon yield of the various source-rock types as a function of maturity. The yield curves employed in this study (Figure 2) are based upon unpublished laboratory results (Royal Dutch/Shell Exploration and Production

OIL :

FRACTION OF POTENTIAL ULTIMATE YIELD ⟶

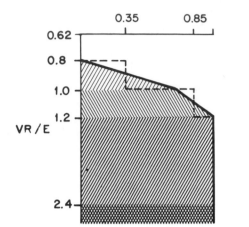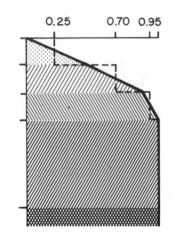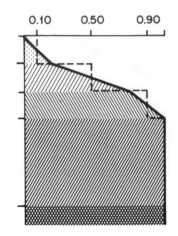

THREE TYPES OF SOURCE ROCK:

| MAINLY- HUMIC TO MIXED | MAINLY- KEROG. TO KEROG. BACT. | KEROGENOUS ALGAL |

GAS :

FRACTION OF POTENTIAL ULTIMATE YIELD ⟶

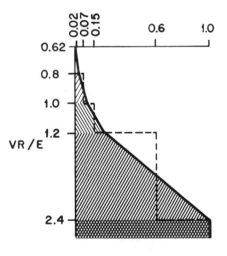

ALL SOURCE ROCK TYPES

Figure 2. Hydrocarbon yield fraction as a function of maturity for oil (dependent on the source-rock type) and for gas. The maturity intervals given are those used to describe the maturity in the charge calibration file and consequently also in the simulation model. The maturity scale is expressed in estimated vitrinite reflectance (VR/E, equivalent to R_o in percent).

Laboratories, Rijswijk, The Netherlands). The determination of the yield fractions also requires an estimate of the percentage distribution of the various maturity intervals in the effective drainage area, which by definition has an estimated vitrinite reflectance (VR/E or R_o) of at least 0.62; the boundaries of the five VR/E intervals employed in the present study are at 0.62, 0.80,

1.00, 1.20, and 2.40. This VR/E distribution is calculated from the burial and temperature history of the source rock at various points in the drainage area (Figure 3) by applying Lopatin's time-temperature integral method. An in-house calibration against measured vitrinite reflectance values is used. In principle, our maturity calculations are very similar to those published by Waples (1980).

Table 1. Source rock type, C_R/C_T and potential ultimate oil yield (PUYO).

Source rock type Shell (KSEPL, Rijswijk)	Tissot and Welte, 1978	C_R/C_T	PUYO $1/m^3$ SR/1% C
Humic		0.88	0.0
Mainly humic	III	0.75	3.7
Mixed		0.65	6.6
Mainly kerogenous		0.50	10.9
Kerogenous, predominantly bacterial	II	0.40	13.8
Kerogenous, predominantly algal	I	0.20	19.6

3. The *yield fraction for timing* (YFT) requires an estimate of the time of trap formation, together with estimates of the time of the start, the peak and the end of oil and free gas generation/expulsion. These milestones in the generation history can be deduced from the analysis of burial graphs at various points of the kitchen, taking into account the source-rock type and the yield curves for maturity mentioned earlier.

No simple model exists for the migration fraction. It is assumed that migration efficiency is inversely proportional to the length and the intricacy of the migration path, the "resistivity" per unit length. In the migration study both parameters have been split in two, for migration parallel and perpendicular with the stratigraphy:

1. LMD = *lateral migration distance*, the distance (km) on the map between the center of the drainage area and the

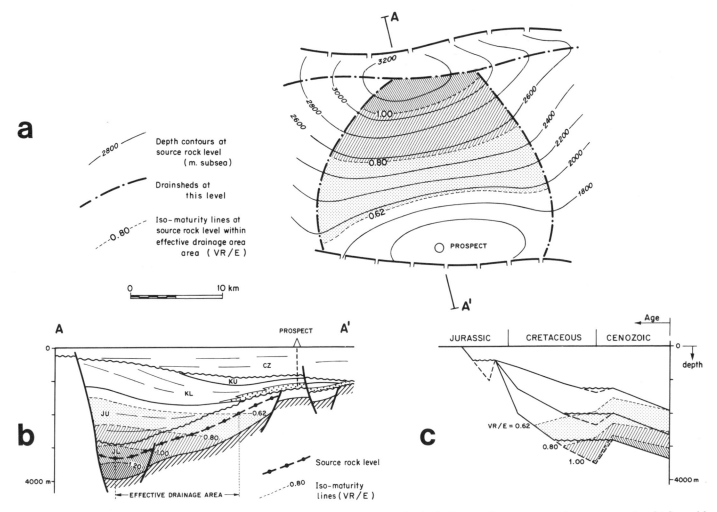

Figure 3. An example of a charge model, applied to a structural closure at the level of a Lower Cretaceous sandstone reservoir which could have been charged from an area to the north of the trap containing a Lower Jurassic source rock.

a. Drainage map, shown by the depth contours at the source rock level. The effective drainage area is bounded by 'synclinal divides' and by the contour at the depth at which the source rock has just reached maturity (VR/E = 0.62). The relevant deeper iso-maturity lines (0.80 and 1.00) are also indicated, and describe the VR/E distribution of the effective drainage area.

b. Cross section through drainage area and prospective structure.

c. Burial graphs for the source rock at three important points: one at the deepest point, one in the center of the effective drainage area, and one at a point which has just reached maturity. The VR/E iso-lines, based upon maturity calculation, indicate the timing of the generation/migration process.

| PROBABILITIES (Discriminant Analysis) | | | | VOLUMES (Censored Regression) | |
P (HC)	P (O/HC)			OIL	GAS (Free Gas + solution Gas)
272+/58-	208+/45-	Sample Size	Sample Size	98 REAL/110 MIN	173 REAL/79 MIN
34%	19%	Misclassification	Residual Standard Deviation (log)	0.79	0.85
1.2	1.7	Separation of means in terms of pooled Standard Deviations	Degrees of Freedom	133.8	201.9
			R-Square = Explained Variation	63%	67%
		X-VARIABLES:			
+	+	DRAR - Effective Drainage Area		+	+
+	+	SRTH - Net Source Rock Thickness		+	+
	+	PUY - Potential Ultimate Yield		+	+
+	+	YFM - Yield Fraction for Maturity		+	+
+	+	YFT - Yield Fraction for Timing		+	+
		LMD - Lateral Migration Distance			−
		VMD - Vertical Migration Distance			−
−	−	LMF - Lateral Migration Factor			
−	−	VMF - Vertical Migration Factor			

Figure 4. Summary of results of the charge calibration; "+" means positive correlation; "−" means negative correlation; blank means no significant correlation.

center of the trap.
2. VMD = *vertical migration distance*, the average stratigraphic separation between source rock and reservoir (in m).
3. LMF = *lateral migration factor*, a quantification of the resistivity along the lateral path, based upon the presence and intensity of features supposedly influencing the migration. Fracturing and faulting parallel with the path are thought to enhance migration, and they make the factor smaller. Cross-faulting, or the presence of lithological or structural barriers are thought to impede migration; they make the factor greater.
4. VMF = *vertical migration factor*, a quantification of the migration resistivity perpendicular with the stratification. It is based on the average lithology of the stratigraphic section between source rock and reservoir, quantified according to a scale which attributes higher values to better "sealing" lithologies. For all lithologies there is some reduction if fracturing and faulting occur. This quantification of vertical migration resistivity follows the same rules as the quantification of lithology, fracturing, and faulting for the determination of the retention potential of a seal, as subsequently described under seal calibration.

The relationship between the migration fraction and the above parameters is as follows:

$$MFR = C \times \frac{1}{LMD} \quad \frac{1}{VMD} \quad \frac{1}{LMF} \quad \frac{1}{VMF} \qquad (2)$$

whereby C is a constant, different for oil and gas.

Charge Volume Regression

Multivariate censored regression has been carried out to find the relationship between the dependent variable Y (charge volume) and a number of independent variables X_1, X_2, ..., X_n, the generation/migration parameters. Since all variables involved tend to have a log-normal distribution they are all handled after log-transformation.

Two dependent variables have been analyzed: oil charge and total gas charge (i.e., free gas plus solution gas).

All X-variables described in the foregoing paragraphs have been tested; not all of these gave significant correlations in all exercises. The "successful" X-variables for each individual exercise are indicated in a summary of the results (Figures 4, 5, and 6).

For oil charge, the learning set contains 98 points with a REAL value for charge and 110 points with a MINIMUM value. The regression equation for oil charge reads as follows:

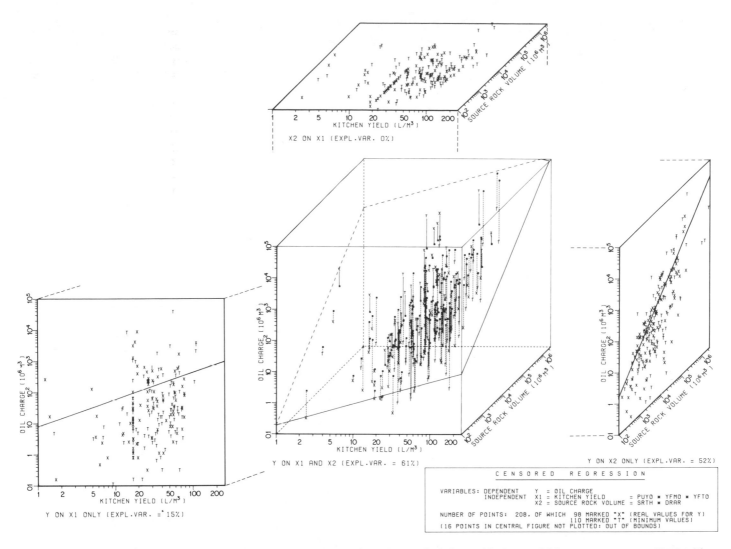

Figure 5. The central plot is a three-dimensional display of a censored regression of oil charge (the Y-variable) on two composite X-variables ("kitchen" yield in litres oil per cu m source rock—X_1; and net source rock volume in millions of cu m—X_2). Data are taken from the calibration file. Positive residuals are shown as solid lines, the negative residuals as broken lines. This figure illustrates the strong correlation of oil charge with "kitchen" volume (that is, effective source rock volume—the product of effective drainage area and net source rock thickness). Partial regressions are shown on the diagram surrounding the central plot.

$$\log(\text{OIL CHARGE}) = \beta_o + \beta_1\log(\text{DRAR}) + \beta_2\log(\text{SRTH})$$
$$+ \beta_3\log(\text{PUYO}) + \beta_4\log(\text{YFMO}) + \beta_5\log(\text{YFTO}) \quad (3)$$

The β_o, β_1, etc., are the regression coefficients. The equation can easily be brought in a form which allows comparison with the geochemical material balance formula (equation 1):

$$\text{OIL CHARGE} = 10^{\beta_o} \times \text{DRAR}^{\beta_1} \times \text{SRTH}^{\beta_2} \times \text{PUYO}^{\beta_3}$$
$$\times \text{YFMO}^{\beta_4} \times \text{YFTO}^{\beta_5} \quad (4)$$

Substituting -3 for β_o and 1 for each of the other regression coefficients would make equation 4 equivalent to equation 1, apart from the absence of the factor related to migration. The latter absence is discussed below.

For total gas charge the file contains 173 points with REAL Y-values and 79 points with MINIMUM values. All these points have a source rock which is not purely humic. The regression equation for total gas charge in a form which allows comparison to equation 1 reads as follows:

$$\text{GAS CHARGE} = 10^{\beta_o} \times \text{DRAR}^{\beta_1} \times \text{SRTH}^{\beta_2} \times \text{PUYG}^{\beta_3}$$
$$\times \text{YFMG}^{\beta_4} \times \text{YFTG}^{\beta_5} \times \text{LMD}^{\beta_6} \times \text{VMD}^{\beta_7} \quad (5)$$

The above regression equations are very similar to the geochemical material balance formula. All regression coefficients related to generation parameters (DRAR, SRTH, PUY, YFM, YFT) differ less than one standard deviation from the ideal value 1, with only one exception. The β_4 for oil is considerably lower than 1; this is thought to reflect either the poor description of YFMO on our file or simply our inability to properly estimate this variable.

It is remarkable that out of the four variables related to migration only the distances gave significant (negative) correlation, and then only in the cases of gas charge. The migration factors, which express the resistivity of the migration path, do not play a role in the volume regressions, contrary to their response in the discriminant analysis (see Figure 4).

The observation that none of the migration parameters

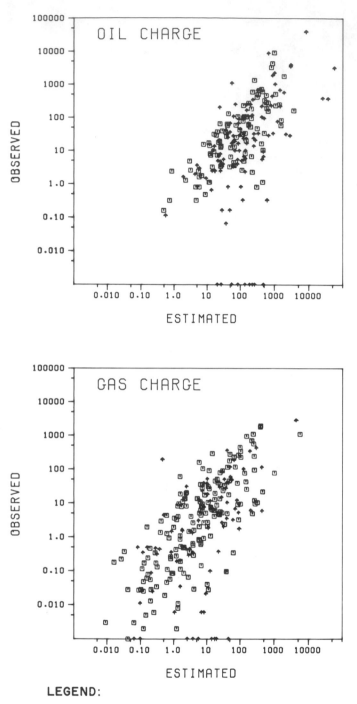

LEGEND:

◻ **REAL VALUE**   ✦ **MINIMUM VALUE**

Figure 6. Scatter plots of estimated charge volumes versus observed charge volumes for oil and total gas for all points of the charge calibration file (in millions and billions of cu m at standard conditions, respectively). The estimated volumes result from the application of the regression formulas for charge to the calibration points.

play a role in the volume regression for oil charge suggests that, if a migration path exists, no appreciable oil loss takes place during migration. This is contrary to the findings for gas, of which on average three-quarters appear to get lost during migration. These observations apparently reflect

differences in the migration process between oil and gas. Possibly oil migrates in the continuous liquid phase along discrete paths. Most of the migrating oil apparently takes the same path so that the residual oil along this path only forms an insignificant fraction of the total oil charge. Gas, on the other hand, is more soluble in formation water and can migrate only after having saturated a large body of water between source and trap. This implies a considerable loss.

Calibration points with coal as the only source rock are not included in the above regression for total gas charge. These points have been analyzed separately in a study which led to slightly different but comparable results.

Charge Probability Analysis

Multivariate discriminant analysis has been applied to seek a combination of quantitative geological parameters which permits classification of the points from the learning set in two subsets, one with hydrocarbon charge and one without.

When considering the presence/absence of hydrocarbon charge, two consecutive questions have to be answered: (1) "Is there any hydrocarbon charge?", and (2) "If so, is there oil?"

Both questions have been the subject of a separate discriminant analysis, each one referring to a charge probability: P(HC), the probability of having hydrocarbon charge, and P(O/HC), the probability of having oil charge, given the presence of hydrocarbons.

For the analysis of P(HC), 272 points with hydrocarbons were available, and 58 without: in this exercise dry points with a poor seal (unable to hold any appreciable gas) had to be excluded. For the analysis of P(O/HC), 208 oil-bearing and 45 non-oil-bearing points were available; in this case points with a structure full to spill-point with free gas had to be excluded, as oil might once have been present. All previously described X-variables have been tested; those which gave significant correlations are summarized in Figure 4.

It is remarkable that for both probabilities the *migration factors* gave a significant correlation whereas the *migration distances* did not. It appears that (within the dimensions represented in the calibration file) the migration distance cannot prevent charge, whereas the migration resistivity can become so great that it does.

Discriminant analysis leads to a pseudo-regression equation (Van de Geer, 1971; Nederlof, 1981) which allows calculating discriminant scores. It would be ideal if all success cases in the calibration file (class 1, hydrocarbon charge) had a positive score; and all failure cases (class 2, no charge) a negative score. In the reality of the calibration study the mean of class 1 is indeed positive and that of class 2 negative, but the two distributions are partly overlapping. This implies a certain percentage of misclassification; 34 percent in the case of P(HC) and 19 percent in the case of P(O/HC). The separation of the means is statistically significant at the 0.5 percent level.

The partly overlapping probability density functions for class 1 and class 2 can be used to update a prior probability (= previously established probability; see Winkler, 1975; also Nederlof, 1981). The calibration file for charge is not considered an adequate source for such a prior probability, since it contains too few failure cases as compared to the reality of exploration.

Another data source is required which is more representative in terms of probabilities, and for this purpose, worldwide exploration drilling data have been analyzed, based on a large number of unit areas. For each unit area a figure has been established for the prior probabilities, P(HC) and P(O/HC). This has led to the establishment of a probability distribution for each of the two charge probabilities, which reflects the variation in success rate in the world as a whole. These probability density functions (beta distributions) are used in the simulation in the appraisal stage as "prior probability distributions."

SEAL CALIBRATION

The Calibration File

A study of the trapping efficiency of cap rocks was made on 160 calibration points with a wide geographic distribution (Nederlof and Mohler, 1981). This learning set includes both structural and stratigraphic traps with oil, gas, or mixed accumulations, sealed by unfaulted cap rocks of varying thickness and lithology. For all points an oversupply of hydrocarbon charge was assumed on the basis of circumstantial geological evidence. The majority of the cases are underfilled, which is attributed to a "seal constraint."

The Retention Potential

A hydrocarbon accumulation causes a differential pressure across the seal which is proportional to the height of the hydrocarbon column and the density difference between hydrocarbons and formation water. The retention potential of a seal can be expressed as the maximum differential pressure the seal can hold. If that maximum is exceeded, the seal starts leaking, and continues leaking until the pressure has dropped to that maximum. The retention potential is the dependent, or Y-variable, in this calibration exercise.

The nature of the data requires the application of a regression program which permits the use of censored values for Y. In the case of an inferred seal constraint (that is, an underfilled trap in a geological setting with an oversupply of hydrocarbons) the actual differential pressure is taken to be equal to the maximum the seal can hold, and the observed Y is a REAL value for the retention potential.

If the trap is filled to spill-point, however, the actual pressure is most likely less than the maximum possible and the observed Y gives a lower limit for the retention potential; the censoring code is MINIMUM.

If, on the other hand, we are dealing with a stacked reservoir section with a common oil-water contact, the intermediate seals are suspected of not holding the underlying column, the effective sealing being provided by the topseal. In such a situation the calculated pressure for an intermediate seal is probably too high and the Y-value gives an upper limit; the censoring code is MAXIMUM.

The Independent Variables

The following geological parameters have been used as independent or X-variables. and have contributed to the explanation of the variation of Y, the retention potential:

1. The effective *seal thickness* (that is, the minimum thickness of the sealing formation taking into account reductions due to minor faulting). True fault seals have not been included in the learning set.
2. The *seal lithology*. For the purpose of this study a quantification scale had to be designed for lithology. It is the same empirical scale which has been used for the definition of the vertical migration factor (VMF, see above). It contains five main lithological classes, which in increasing order of sealing capacity are: sand, silt, marl clay, and salt. These classes are placed in this order on a logarithmic scale. The system permits determination of a lithofactor for every lithology or mixture of lithologies.
3. The *depth*.
4. The *degree of fracturing and faulting*.

Regression Results

The multivariate statistical analysis has led to a regression equation of retention potential on the above mentioned seal parameters which gives a 60 percent explained variation.

Results of this study have been applied in the determination of the censoring codes for Y in the charge calibration, as discussed in the preceding section. The regression equation is also incorporated in the simulation model for prospect appraisal, whereby the retention potential of the seal may lead to a constraint in the trapping and retention phase.

THE SIMULATION MODEL

The simulation process followed in our prospect appraisal system is based on the calibration studies discussed in the foregoing sections. It employs the various regression equations whereby the appropriate standard deviations are taken into account.

The Input

For each prospect to be appraised, the user should define a play concept in line with the conceptual model on which the charge calibration is based. The input relevant to the prospect in question should contain the following elements:

1. Those parameters relevant to generation/expulsion and migration which are needed to define the X-variables to be used in the application of the regression formulas for charge volumes and discriminant scores.
2. Local exploration results, in the form of statements regarding earlier tests in the area that concerned the play concept in question: the number of relevant, validly-tested structures which were found to be devoid of charge, the number of structures which had a hydrocarbon charge, and of these the number of structures which had an oil charge.
3. The seal parameters needed to define the X-variables to be used to estimate the retention potential of the seal, according to the regression formula in question.
4. Data describing the size and the shape of the prospective trap, the thickness of the gross reservoir, the net to gross

ratio, the porosity of the net reservoir, the hydrocarbon saturation, and the reservoir pressure and temperature. This category of data is needed to calculate the effective pore volume within the closure. For some of these input data, like hydrocarbon saturation, the system contains back-up data, representing a worldwide range which becomes effective if the appraiser cannot give his own estimate.

5. Data describing the recovery efficiency for oil and for free gas, to enable calculation of recoverable volumes.

In prospect appraisal, no certainty can be expected for the large number of input data to be provided. Therefore, input can be made in the form of probability distributions.

Monte Carlo Procedure

Both the application of regression equations with their standard errors and the handling of input data in the form of probability distributions require a Monte Carlo approach. For a combination of random draws for each of the input parameter distributions, a complete simulation is carried out. Repeated calculation (in a number of "cycles") produces a thousand equally likely outcomes representing an expectation curve.

Charge Simulation

The starting point for determinating the charge probability for each prospect to be appraised is formed by the distribution of the worldwide prior probability. The input for local exploration results is used to update the worldwide prior into a local prior, which again is in the form of a probability distribution. Then, in every single cycle, the answer to the question of presence or absence of charge is given through: (1) random sampling from the distribution of local prior probabilities; (2) updating of this prior, based upon the discriminant score for this cycle; and, (3) binominal simulation. The outcomes of the 1,000 cycles form a yes/no vector which gives the probability of having charge in the prospect in question.

The charge simulation as carried out for every single cycle is sketched in Figure 7. If hydrocarbons are present in a cycle the volume of total gas is calculated. If oil is present, its volume is also calculated and the amount of solution gas must be established. For that purpose, the maximum possible gas/oil ratio is calculated; this requires knowledge of the pressure and temperature of the possible accumulation (which is derived from the input for trap description) and the density of the oil. The latter is estimated based upon the appraiser's input (if any) and/or on the worldwide distribution of oil densities as a function of depth.

The gas is "stored" in the oil as solution gas either until the available gas is used up, or until the oil is saturated, in which case free gas is left over. The resulting gas/oil ratio (GOR) is calculated. The percentage of the cycles which end with gas and no oil, or with saturated oil and free gas, gives the probability to have free gas.

Up to this point in the course of the simulation process, the oil and free gas volumes are at standard conditions. In preparation for the trapping and retention phase, the volumes are converted to subsurface conditions, whereby

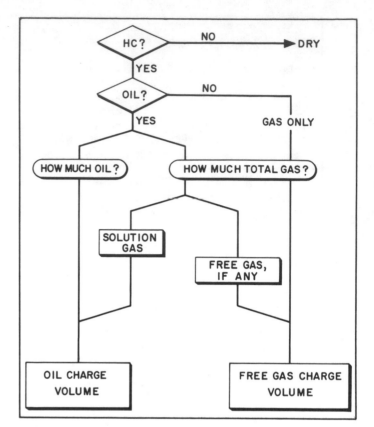

Figure 7. Diagram of the charge simulation.

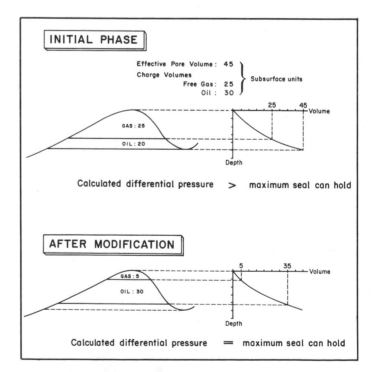

Figure 8. An example of some of the steps taken in the trapping and retention simulation. In the initial phase the differential pressure across the seal, resulting from the hydrocarbon column(s), is compared with the estimated maximum pressure the seal can hold according to the seal calibration results. If the calculated pressure exceeds the retention potential, modification of the fill must take place until equilibrium is attained.

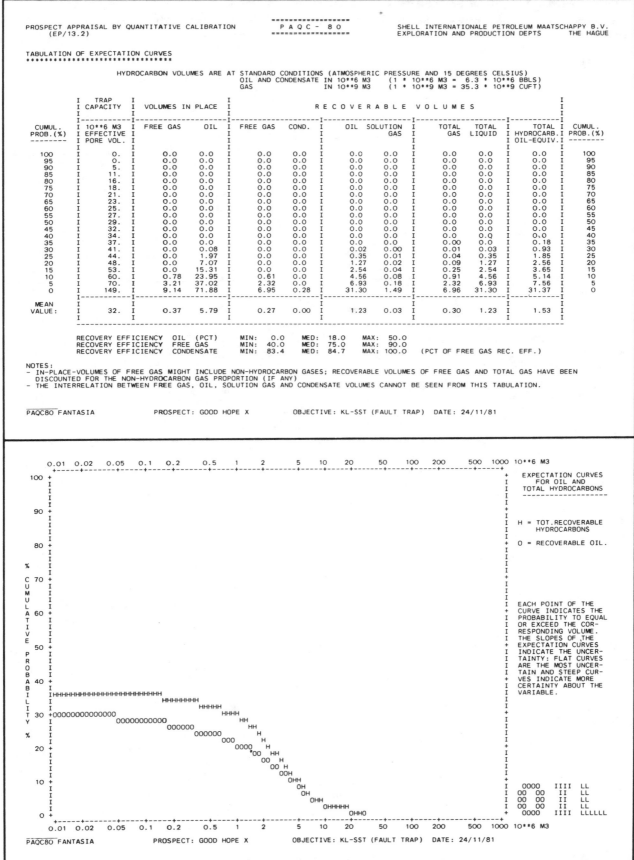

Figure 9. Specimen of two of the computer-generated output pages of the prospect appraisal system.

use is made of the GOR (see above) and of data on subsurface pressure and temperature derived from the input.

Trapping and Retention Simulation

This phase of the simulation is carried out assuming that the trap was charged in its present-day configuration and depth and that the oil and free gas were simultaneously available. Such effects as subsidence and uplift of the trap after its charge, flushing under hydrodynamic conditions, and thermal effects are neglected.

The volume of the trap is calculated in each cycle from the relevant input parameters (i.e., not supported by any calibration). On the other hand, the retention potential is estimated in a calibrated way. In each cycle, the trap volume is compared with the volumes of free gas and oil resulting from the charge simulation in the same cycle.

In cases of no charge, these volumes are equal to zero. In case of an available charge, the free gas is first put in the trap and thereafter the oil to the limit of the trap capacity. The next step is to calculate the differential pressure caused by this initial fill; for this purpose the heights of the gas and oil columns have to be calculated taking into account the shape of the structure. The resulting differential pressure is compared with the retention potential, the maximum pressure allowed by the seal; if it is exceeded, a further modification to lower the pressure must take place (Figure 8). The first step of this modification is to "bleed off" gas through the seal and, as long as the oil charge permits, fill the space that becomes available with oil. If all the free gas has been released and the differential pressure is still too high, the last step is to bleed off oil.

This entire procedure is followed a thousand times. Finally, the volumes of oil and free gas trapped and retained are converted to surface conditions (volumes of oil and free gas in place: STOIIP and GIIP).

Recovery Simulation

In the last phase of the simulation, recovery efficiencies are applied according to the input made by the appraiser. The final result is in the form of expectation curves for recoverable oil, solution gas, free gas, and total hydrocarbons in oil equivalent (Figure 9).

The output also contains data on probabilities to have condensate and on condensate volumes. These data are based on calibration studies not discussed in this article. Also given are probabilities of success and mean success volumes for a range of cut-off values for both oil and free gas. These data are important for subsequent economic analysis and decision-making.

CONCLUSIONS

The charge and retention calibrations have been successful in explaining about two-thirds of the variations in

the Y-variables of the calibration files, and in general suggest that the models are reasonable, although not perfect. The study has also shown the weak points, which are useful guides for future improvements.

We stress that the expectation curves resulting from the simulation represent a translation of the geologists' input into probabilities of occurrence and hydrocarbon volumes, which is in accordance with the accumulated exploration experience contained in the calibration file. The geologists' judgement on the values of the input parameters remains the determining factor for the outcome of the analysis. The simulation program ensures that the input factors receive the appropriate weights and are combined in a sensible way to form the expectation curve.

Although subjective assessment of input parameters remains unavoidable, subjective judgement is reduced in this system to such a level that it is unlikely to go completely astray.

Practical experience in using the system has shown satisfactory correspondence between predictions and after-drilling results (Nederlof, 1979).

ACKNOWLEDGMENTS

The authors wish to express their gratitude to many Shell colleagues in research and operations who cooperated in the studies described, especially H.P. Mohler (seal calibration) and A.B. Baak (programming). They thank Shell Internationale Petroleum Mij, The Hague, for granting permission to publish this article.

REFERENCES

Nederlof, M.H., 1979, The use of habitat of oil models in exploration prospect appraisal: Proceedings, 10th World Petroleum Congress, Panel Discussion 1, paper 2, p. 13-21.

———, 1981, Calibrated computer simulation as a tool for exploration prospect assessment: U.N. ESCAP CCOP Technical Publication 10, p. 122-138.

———, and H.P. Mohler, 1981, Quantitative investigation of trapping effect of unfaulted caprock, *in* San Francisco, Annual AAPG-SEPM-EMD-DPA Convention: AAPG Bulletin, abstract, v. 65, p 964.

Tissot, B.P., and D.H. Welte, 1978, Petroleum formation and occurrence: Berlin, Springer-Verlag, 538 p.

Van de Geer, J.P., 1971, Introduction to multivariate analysis for the social sciences: San Francisco, W.H. Freeman and Co., 293 p.

Waples, D.W., 1980, Time and temperature in petroleum formation; application of Lopatin's method to petroleum exploration: AAPG Bulletin, v. 64, p 916-926.

Winkler, R.L., 1975, An introduction to Bayesian inference and decision: New York, Holt Rinehart and Winston Inc., 563 p.

Petroleum Origin and Accumulation in Basin Evolution—A Quantitative Model

D.H. Welte
M.A. Yukler
Institute for Petroleum and Organic Geochemistry
D-5170 Julich
Federal Republic of Germany

Basin data—geologic, geophysical, geochemical, hydrodynamic, and thermodynamic—can be combined for quantified hydrocarbon prediction. A three-dimensional, deterministic dynamic basin model can be constructed to calculate all the measurable values with the help of mass- and energy-transport equations and equations describing the physical and/or physicochemical changes in organic matter as a function of temperature. Input data consist of heat flux, initial physical and thermal properties of sediments, paleobathymetric estimates, sedimentation rate, and amount and type of organic matter. Subsequently, the model computes pressure, temperature, physical and thermal properties of sediments, maturity of organic matter, and the hydrocarbon potential of any source rock as a function of space and time.

Thus the complex dynamic processes of petroleum formation and occurrence in a given sedimentary basin can be quantified. For example, hydrocarbon potential maps for any given source rock and any geologic time slice of the basin evolution can be provided as computer printouts. The computer model can be applied to any stage of an exploration campaign and updated as more information becomes available.

INTRODUCTION

Increasing demand and decreasing supply of hydrocarbons require increased activities in petroleum exploration and an improvement in the exploration success ratio. New oil and gas fields must be found and explored areas should be reassessed for additional oil and/or gas pools.

The systematic search for petroleum accumulations started toward the end of the 19th century with the acceptance of the "anticlinal theory." The vertical movement of petroleum in a static medium was considered. Studies on multiphase flow systems of gas, oil, and water resulted in the "hydrodynamic theory." The search focused on the detection of suitable subsurface structures which could host petroleum accumulations. Geophysical methods have been developed and improved to help locate these structures. However, the timing and the amount of petroleum generation were seldom considered. Later, particularly over the last 10 years, organic geochemical studies supplied the most needed chemical data on the generation, migration, and accumulation of petroleum. From these data, new concepts were developed on the temperature-time dependence of petroleum generation and on the other complex processes of migration and accumulation.

Because of the enormous amount of data required to describe this complex basin system, quantification of the processes was not possible. Therefore, only qualitative or semiquantitative studies were made and presented as case studies. With the invention of large and fast computers, quantitative studies can be initiated.

Until now statistics have played a major role in exploration for hydrocarbons. The success ratio was directly related to the complexity of the system studied and how well it fit a predetermined frequency distribution. Our studies show that most processes in a sedimentary basin are not only time-dependent but are also strongly interrelated to a degree never considered in the past. A small error in the determination of one process can result in a totally erroneous answer. On the basis of experience gained in basin studies, we, therefore, chose a three-dimensional, dynamic deterministic model to quantify the previously mentioned processes. This approach has been successfully tested with existing sedimentary basins. For reasons of confidentiality no details on the areas studied can be given.

SYSTEM CONCEPT AND SIMULATION

To understand the complex natural phenomena of petroleum occurrences in sedimentary basins, surface and subsurface samples are systematically collected and analyzed. The various processes are described and the interrelations are determined. Thereafter one tries to comprehend the "system" in which these processes occur. Once an understanding of the system is reached, practical goals are sought which go beyond a purely scientific or theoretical description of the qualitative nature of the problem. To reach these practical goals a quantification is needed to answer questions as to when and how much petroleum was formed and where it has accumulated.

About 1890 the "anticlinal theory" was widely accepted as the controlling principle of petroleum accumulation. The main idea was that petroleum is driven in a water-saturated environment by buoyancy forces, which are vertical, and accumulates in crestal positions of anticlines or, in general, in the highest local positions to which it can migrate in any structure. At the beginning of the 20th century the "hydraulic theory" of oil and gas accumulations was developed (Munn, 1909a, 1909b; Shaw, 1917; Mills, 1920; Rich, 1921, 1923, 1931, 1934; Illing, 1938a, 1938b, 1939). Hubbert (1940, 1953) further developed the hydraulic theory with a mathematical basis and this theory then received wide acceptance. These theories, however, addressed only the question of where accumulations were located.

With the acceptance of the anticlinal theory geologists started to search for anticlines or for "geologic highs" in the subsurface. These structures commonly are not visible from the surface and, therefore, geophysics became an important tool in petroleum exploration.

Gravimetric, magnetic, and geoelectric methods were developed, applied, and improved to satisfy the needs of oil exploration. The invention of refraction and reflection seismics in the 1920s was a big step forward in the search for petroleum. Every improvement in geophysical methods resulted in better understanding of the subsurface formations and structures and enabled detection of series of new finds. In the past 20 years remote sensing techniques were developed and their use has become popular. However, all these methods are concerned only with the possible location of a petroleum accumulation.

Therefore, it is clear that to answer not only the questions "where," but also, "why, when, and how much," we have to establish the generation, migration, and accumulation processes quantitatively. In the past few decades, extensive studies on sediments and petroleum have shown that petroleum originates from finely disseminated organic matter buried within sediments that have been subjected to elevated temperatures (about 50°C and higher). A sediment may be considered a good source rock if it meets certain criteria. These include the amount of organic matter, both soluble (bitumen) and insoluble (kerogen), the type of kerogen, and the maturity of the organic matter (Tissot and Welte, 1978). Identification of specific source rocks is dependent also on the correlation of the composition of extractable hydrocarbons and nonhydrocarbons.

The release of petroleum compounds from kerogen and their movement within and through the pores of a source rock are defined as primary migration. The movement of petroleum expelled from a source rock through the wider pores of more permeable and porous carrier and reservoir rocks before final emplacement, being mainly controlled by

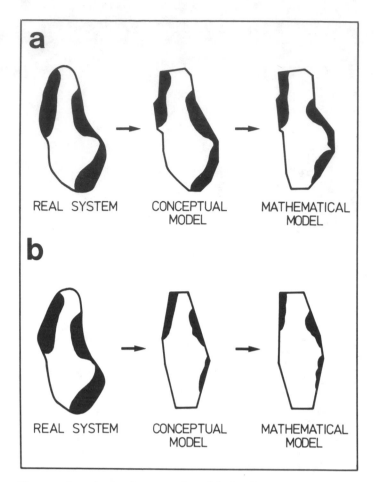

Figure 1. Steps in development of model of real-world system: a, conceptual and mathematical models successfully represent real system; b, owing to poor understanding of real system, models do not represent it.

buoyancy, is called secondary migration. Petroleum may then be collected in reservoir rocks in various types of structural or stratigraphic traps and form accumulations.

The science of organic geochemistry or petroleum geochemistry is now adequately developed to be applied quantitatively to problems of the generation, migration, and accumulation of petroleum. During the past decade organic geochemistry has become a useful tool in petroleum exploration. New and rapid methods have been developed to fulfill the main requirements of petroleum exploration—assessment of source rock potential, source rock maturity, and source rock/oil correlation (Welte et al, 1981).

The system in which the generation, migration, and accumulation of petroleum occurs is the three-dimensional dynamic geologic framework. The geometry and location of the system are determined from paleogeography and seismic information about the basin. The sediment inputs (type, source, rate, etc.), depositional environments, paleobathymetric estimates, mineralogic changes, and tectonic movements are determined from geology. The direction and rate of fluid movement and hydraulic properties of fluids and sediments are determined from

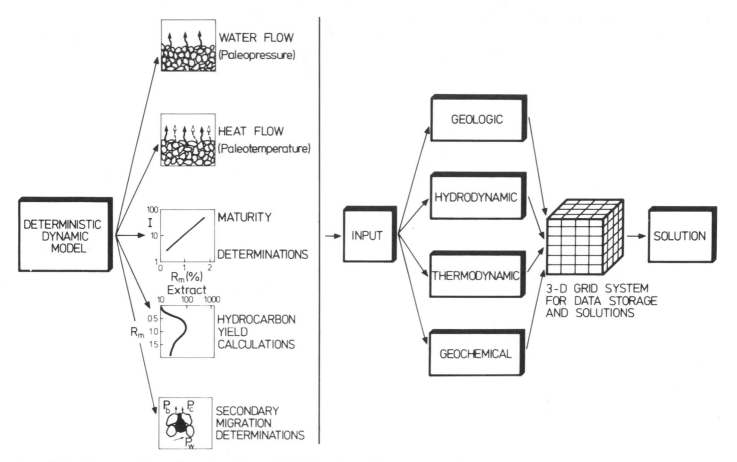

Figure 2. Development of three-dimensional deterministic dynamic model.

hydrodynamics. The direction and rate of heat flow and thermal properties of fluids and sediments are determined from thermodynamics. Finally, type, amount, maturity, and generation potential of organic matter are determined from organic geochemistry. A "qualitative" determination of these processes and the relations among them have been obtained through numerous temporal and spatial observations in the field and in the laboratories. Now, we unite all these data and their interrelations to comprehend the real system which is called the conceptual model (Figure 1a). When our determination of the processes and the relations among them is incorrect or incomplete the conceptual model does not represent the real system (Figure 1b).

Once an understanding of the system is reached, we try to make quantitative petroleum exploration predictions, that is, answer the questions, why, when, how much, and where. This requires the simulation of the various processes in the system by either physical or mathematical models. Because the complex nature of the problem is unsuitable for a physical model, a mathematical model should be used.

Simulation is a class of techniques that involves setting up a model of a real system and then performing experiments on the model. In this sense it is a simplification of the real world and, thus, leads to errors—conceptual and mathematical. A mathematical model is based on the conceptual model and therefore inherits the errors from the conceptual model. These errors are due to poor knowledge of the real system and its behavior (Figure 1a and 1b). Incorrect original input will result in errors in the conceptual model.

Certain assumptions are made in the mathematical formulation of a real system. The assumptions and simplifications made in the mathematical formulation and/or in the solution techniques lead to mathematical errors. These errors should be quantitatively determined so that one can obtain reliable answers from simulation studies. Sensitivity analysis aids in the computation of such errors (Yukler, 1976, 1979).

Petroleum generation, migration, and accumulation occur in a three-dimensional framework as a function of time, and are controlled by very complex and interrelated mechanisms (Tissot and Welte, 1978; Yukler et al, 1978; Welte and Yukler, 1980). Furthermore, the distribution and the characteristics of organic matter do not show any fixed pattern: Under these conditions we have chosen a three-dimensional deterministic dynamic model. This model can be used at any stage during an exploration program. It is applicable to new areas, as well as to old prospects with abundant data where it provides a more quantitative and detailed fresh appraisal.

Figures 2, 3, and 4 illustrate the flow chart of the three-dimensional dynamic deterministic model. The details are explained in the following.

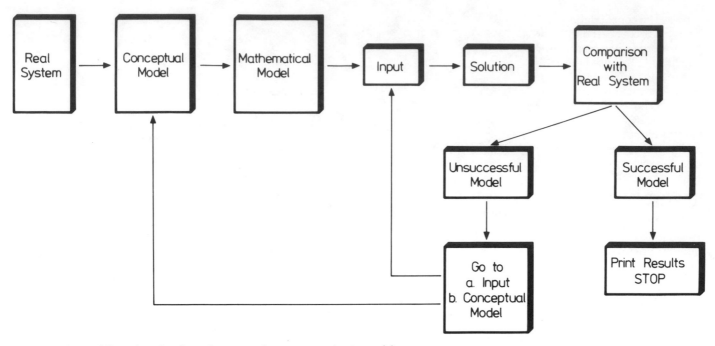

Figure 3. General flow chart for three-dimensional quantitative basin model.

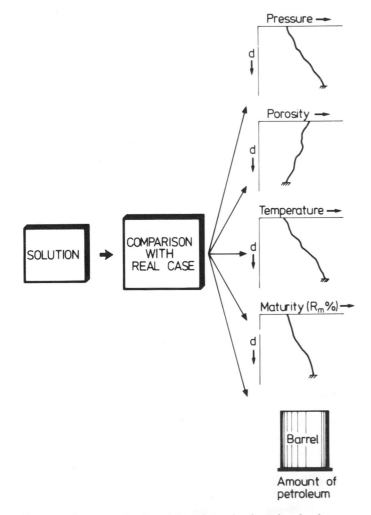

Figure 4. Comparison of model results with selected real values.

QUANTITATIVE BASIN ANALYSIS—THREE DIMENSIONAL MODEL TO SIMULATE GEOLOGIC, HYDRODYNAMIC, AND THERMODYNAMIC DEVELOPMENT

A "qualitative" understanding of petroleum generation, migration, and accumulation has been obtained from numerous temporal and spatial reconstructions of basin histories and laboratory studies. As a result, various theories and hypotheses have been developed to explain these complex phenomena. In a normal qualitative approach the validity of these theories and hypotheses are demonstrated by case histories. Unfortunately, case histories which do not prove the validity of the theories and hypotheses are either not considered or lead to speculations. To overcome these difficulties or to improve these misleading approaches, a "quantitative" evaluation of the total system and mechanisms prevailing therein is required. Only then can the validity of the theories and the hypotheses applied be effectively checked.

Knowledge of paleopressures and paleotemperatures during the evolution of a sedimentary basin is important for the solution of many problems such as (1) stratigraphic and structural development; (2) changes in physical properties of fluids (density and viscosity) and sediments (compaction, permeability, porosity, etc); (3) changes in thermal properties of fluids and sediments (heat capacity, thermal conductivity, etc.); (4) mineralogic changes affected by temperature and pressure; (5) generation, migration, and accumulation of petroleum; (6) fluid flow mechanism, such as fluid flow directions and rates and determination of abnormal fluid pressure zones.

Gibson (1958) studied the excess pressures assumed to be generated by a moving boundary condition, such as continuous sedimentation, and developed an equation to compute the consolidation of a clay layer. Using Gibson's equation, Bredehoeft and Hanshaw (1968) examined pressure-producing mechanisms in a basin. Sharp and Domenico (1976) used the same equation to compute energy transport in compacting sedimentary sequences. Bishop (1979) determined compaction of thick, abnormally pressured shales using the same equation, but added the effect of external loading (Taylor, 1948). All these quantitative studies are carried out in one dimension which requires the highest degree of symmetry in a three-dimensional framework. However, Gibson's equation cannot accurately determine compaction of sediments, for it does not handle the compressibility of sediments rigorously. A new equation for fluid flow in sediments with moving boundary conditions (sedimentation, compaction, and erosion) was derived by Yukler et al (1978). The hydraulic head (or pore pressure) in sediments can be computed in three-dimensions (x, y, and z) and as a function of time with the following equation (Welte et al, 198), where the inflow-outflow is equal to the net accumulation due to grain and fluid compressibility plus the net accumulation due to the change in sediment density, change in rate of sedimentation, and change in water depth:

$$\frac{1}{\rho}\left[\frac{\partial}{\partial x}\rho K\frac{\partial h}{\partial x}+\frac{\partial}{\partial y}\rho K\frac{\partial h}{\partial y}+\frac{\partial}{\partial z}\rho K\frac{\partial h}{\partial z}\right]$$

$$=S_s\frac{\partial h}{\partial t}+\alpha\left[-(L-z)\frac{\partial \gamma_s}{\partial_t}-(\gamma_s-\gamma_w)\frac{\partial_L}{\partial_t}-\gamma_w\frac{\partial H}{\partial t}\right] \quad (1)$$

Here,

L	=	length
M	=	mass
T	=	time
h	=	hydraulic head, L
H	=	water depth, L
L	=	sediment thickness, L
K	=	hydraulic conductivity, L/T
S_s	=	storativity, 1/L
t	=	time, T
x, y, z	=	three orthogonal vectors
α	=	compressibility of solid skeleton, LT^2/M
γ_s	=	specific weight of bulk sediment, M/L^2T^2
γ_w	=	specific weight of fluid, M/L^2T^2
ρ	=	density of water, M/L^3

The term with the compressibility of fluid is neglected, since the error is found to be negligible.

The heat flow equation for the simultaneous transfer of heat both by conduction and convection (due to water flow) was introduced by Stallman (1963):

$$\frac{\partial}{\partial x}K\frac{\partial T_m}{\partial x}+\frac{\partial}{\partial y}K\frac{\partial T_m}{\partial y}+\frac{\partial}{\partial z}K\frac{\partial T_m}{\partial y}$$
conduction

$$-\rho_wC_{pw}\left[\frac{\partial}{\partial x}V_xT_m+\frac{\partial}{\partial y}V_yT_m+\frac{\partial}{\partial z}V_zT_m\right]+Q$$
convection   source/sink

$$=\rho_{ws}C_{ws}\frac{\partial T}{\partial t} \quad (2)$$
net accumulation

where

E	=	energy
°C	=	temperature in degrees Celsius
C_{pw}	=	specific heat of fluid, E/M°C
C_{ws}	=	specific heat of bulk sediment, E/M°C
K	=	thermal conductivity, E/LT°C
Q	=	sink ($-$) or source ($+$) term, E/L^3T
T_m	=	temperature, °C
V_x, V_y, V_z	=	fluid flow in x, y, and z directions, respectively, L/T
ρ_w	=	density of fluid, M/L^3
ρ_{ws}	=	density of bulk sediment, M/L^3

At this step, the dynamic model simulates the formation and evolution of a sedimentary basin by the quantification of the processes taking place within the basin. The processes are strongly interrelated and observed by the variation of the variables in the real basin, pressure and temperature, and in the parameters, compressibility, porosity, permeability, density, viscosity, thermal conductivity and heat capacity. As explained, in the section under "System Concept and Simulation," the model is a fully integrated dynamic deterministic model based on the physical and chemical principles that govern the processes in any sedimentary basin. Thus, the model requires constant checks both in the input parameters and in the quantification of the processes.

All physical and thermal parameters of the system are pressure- and temperature-dependent and are, therefore, always recalculated with each computation of pressure and temperature by a special iterative technique (Yukler et al, 1978). The preceding equations are also integrated from the bottom of a sedimentary unit to the top and the resulting equations are solved by a suitable numerical analysis technique (Yukler, 1976).

QUANTITATIVE APPRAISAL OF HYDROCARBON POTENTIAL

The generation and emplacement of petroleum is a time-dependent, dynamic process linked to the evolution of a sedimentary basin. Hence, the quantitative appraisal of the hydrocarbon potential is integrated into the three-dimensional deterministic dynamic basin model (Figures 2 and 3) which is constructed according to the regional geologic framework.

Basic data for the assessment of the hydrocarbon potential of a basin consist of the regional geology, hydrodynamics, and geothermics and knowledge on generation, migration, and accumulation of petroleum. The identification of source rocks throughout the basin and the amount, type, and maturity of their organic matter are important parameters used for this part of the model (Welte and Yukler, 1980). A decisive role is played by the regional geothermal gradients and their variation with time.

Effect of Temperature on Maturation of Organic Matter

Thermal evolution of source rocks changes many physical and chemical properties of the organic matter. The changes in these properties are used as indicators for maturation. The parameters most commonly used in petroleum exploration are optical examination of kerogen, physicochemical analysis of kerogen, and chemical analysis of extractable bitumen (Tissot and Welte, 1978). All these

measurements and studies are done on the present end products of thermal evolution. Our main objective is to determine changes in maturity as a function of time to the present and then to compare the computed end results with the observed values.

Thermal evolution of source rocks is directly related to the geologic and hydrodynamic processes. Studies by Yukler et al (1978) showed that temperature and pressure distributions are interrelated and depend on the geologic development of a basin. Hence, temperature not only affects the maturity of organic matter, but also affects the physical and thermal properties of fluids and sediments. This means that such parameters as density, viscosity, porosity, permeability, and thermal conductivity, as well as compaction of sediments and the specific yield of hydrocarbons are influenced.

Heat flow in a sedimentary system is composed of two components. There is heat transport by conduction and by convection (equation 2). Heat distribution by convection is directly related to fluid movement. The direction and rate of fluid flow are determined from hydraulic head and hydraulic conductivity in the system. Water, however, has

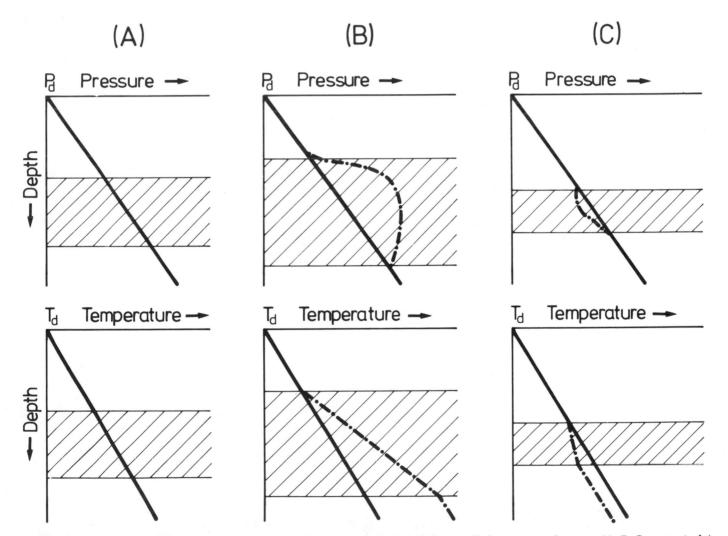

Figure 5. Pressure and temperature distribution in normal, abnormally high, and abnormally low pressured systems (A, B, C, respectively). P_d and T_d are pressure and temperature at arbitrary depth d. Solid lines show changes in abnormally high and low pressure systems.

a low thermal conductivity, but a high specific heat capacity with respect to sediments. Therefore, water is a poor heat conductor, but has a large heat-holding capacity. These properties of water are pressure- and temperature-dependent.

Pressure distribution in the deposited sediments is influenced by the hydrodynamic properties of sediments, especially by porosity and permeability, and by the changes in the overburden pressure due to additional sedimentation, erosion, and changes in water levels. Increase in the overburden pressure on a sedimentary unit, that is, the increase in load, is supported by water pressure and by grain-to-grain stress distribution. With increase in overburden pressure the sedimentary unit compacts because of the bleed-off of compaction water and as a result of solid-matrix and water compressibility. As the unit compacts, the water leaves the system and the overburden pressure is supported by the grain-to-grain stress distribution and by the pressure in the remaining water in the unit. If the permeability in the sedimentary unit is not sufficient to release water with respect to increase in overburden pressure, for example, in fine-grained sediments such as shales, most of the water stays in the unit and the overburden pressure is then mainly supported by the water

pressure. This is termed an abnormally high pressure medium. Other factors, such as aquathermal pressuring and generation of hydrocarbons, may also contribute to pressure increases. When the amount of water escaping from the system is abnormally high owing to high permeability such as in coarse sands, the overburden is supported mainly by the solid skeleton, and the pore space in the system decreases sharply. This is called an abnormally low pressure medium. As a result there is a sharp increase in the temperature gradient on top of an abnormally high pressure medium and a sharp decrease in the temperature gradient on top of an abnormally low pressure medium (Figure 5). The abnormally high-pressured sedimentary unit represents an insulator because of low thermal conductivity of the water-filled pore spaces. As the water cannot easily move out of the system, the heat coming into the unit is largely stored in the water of the unit and results in a sharp increase in temperature gradient. In an abnormally low pressure medium, pore space is reduced so thermal conductivity is kept higher than in a normal pressure medium (hydrostatic) and specific heat capacity is kept low. Hence, the abnormally low pressure medium represents a conductor.

From 0 to 100°C, changes in density of water can be neglected, whereas decrease in viscosity with increasing temperature is important (Paaswell, 1967). Hydraulic conductivity of a medium is the ability of a medium to transmit water and is defined as a product of permeability of the medium and the specific weight of the fluid divided by the viscosity of the fluid. Therefore, decrease in viscosity with increase in temperature will yield higher hydraulic conductivities which will result in higher water flow rates. Higher water flow rates will yield higher compaction and changes in temperature, pressure, and physical and thermal properties versus depth relations. Figure 6 illustrates the changes in porosity versus depth relations with changes in heat flux (Yukler et al, 1978).

The system concept which is formulated by the conceptual model and the resulting mathematical model in this study allows us to determine the temperature distribution in a sedimentary basin with greater accuracy than any other indirect method available. With the knowledge of the spatial and temporal distribution of temperature, the thermal evolution of the organic matter can be determined.

Our mathematical model allows us to develop, at least, a time-dependent three-dimensional relative-temperature-distribution pattern even in unexplored basins without any well data. This relative pattern can easily be gauged as soon as samples are available to measure physical properties of the sediments, temperature, and vitrinite reflectance.

Determination of Maturity of Organic Matter from Lopatin's Method

Lopatin (1971) analyzed the Ruhr coals and especially the coal seams in the Munsterland-1 borehole. In coalification reactions the reaction rate, in general, doubles with every increase of 10°C in temperature. Lopatin mathematically analyzed the geologic and petrologic findings of the Munsterland borehole and introduced the temperature-time

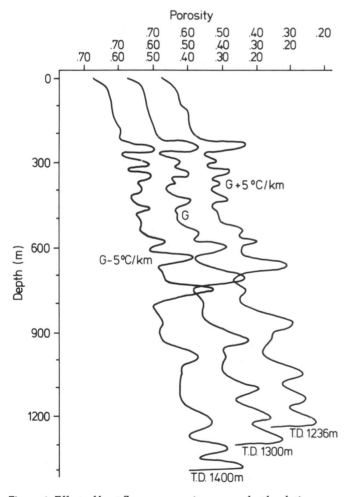

Figure 6. Effect of heat flow on porosity versus depth relation (from Yukler et al, 1978).

index (I) which is the sum of the products of the effective geologic heating time (G), and the temperature correction factor (T):

$$I = T_1G_1 + T_2G_2 + \ldots + T_nG_n \ldots \qquad (3)$$

Subsequently, a correlation equation between vitrinite reflectance, $R_m\%$, and the temperature-time index was found,

$$R_m\% = 1.301 \lg I - 0.5282 \ldots \qquad (4)$$

From simultaneous solutions of equations 1 and 2, the temperature is computed as a function of time and space. Then T and G values are determined. From equation 3 the temperature-time index is computed and replaced in equation 4. The vitrinite reflectance is then calculated as a function of space and time.

Lopatin's method is used as a first approach. It needs further improvement to be applicable to the different types of organic matter and at different levels of maturation. Despite its "oversimplified" approach, however, Lopatin's method in our experience has worked well.

Quantification of Hydrocarbon Generation Potential

At the next step the calculated vitrinite reflectance values will be used to determine the amount of hydrocarbons to be expected from possible source rocks within the three-dimensional sedimentary basin as a function of time.

For this purpose a plot of a hydrocarbon generation curve for type II and type III kerogen is given in Figure 7. In the example treated in this paper the source rock in question contains organic matter of type II kerogen. The peak hydrocarbon generation for this curve is shown at a vitrinite reflectance value of 0.70 percent. In this connection it must be realized that the hydrocarbon generation curves represent observed maximum values of hydrocarbons which are not always reached. Because hydrocarbon generation curves also have other deficiencies due to migrational phenomena, parameters such as "kerogen to hydrocarbon conversion" should be employed to appraise a source rock. Nevertheless, for the time being, hydrocarbon generation curves as given in Figure 7 are used for calculations of the hydrocarbon potential.

Let's assume that three different vitrinite reflectance values ($R_m\%$) of 0.35, 0.70, and 1.04 percent are computed for three different source rocks. These vitrinite reflectance values will result in ratios of hydrocarbons/organic carbon of about 22, 147, and 16 mg/g, respectively (Figure 7). In this way, prior to any exploratory drilling and before any actual source rock analyses are available a relative distribution of the regional hydrocarbon potential for a given source rock can be established. At a later stage, when rock samples are available the hydrocarbon potential can be quantified more accurately. Along this line, with the assumption or estimation of expulsion efficiency, amount of petroleum probably expelled from source rocks can be given.

PRIMARY AND SECONDARY MIGRATION OF HYDROCARBONS

As a first approach we assumed that primary migration is mainly a pressure-driven hydrocarbon-phase movement. Primary migration, therefore, is a normal process that occurs in any mature source rock accompanying the generation of hydrocarbons. It takes place through available pores or by microfracturing of dense source rocks. Hydrocarbon-phase migration is in agreement with the empirical geologic and geochemical data (Tissot and Welte, 1978).

There are three major parameters that control secondary migration and the subsequent formation of oil and gas pools. These are buoyancy, capillary pressure, and hydrodynamics. The buoyancy is computed by subtracting the petroleum density from the formation water density and multiplying by the height of the petroleum column. At present, in the absence of reliable data, the estimation of the height of a petroleum column is very problematic. The pore sizes in the capillary pressure equation for sandy layers are computed from Berg's (1975) equation. The interfacial tension is corrected for temperature as given by Schowalter

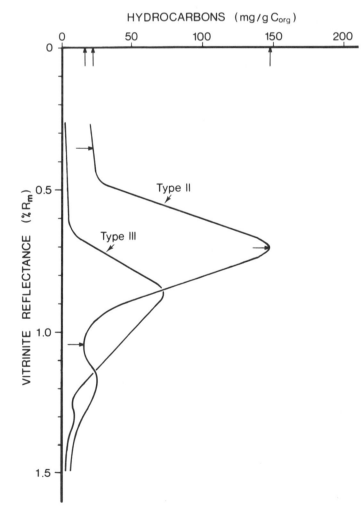

Figure 7. Hydrocarbon generation curve for type II and type III kerogen.

(1979). The pore pressures or hydrodynamic conditions are computed from equation 1. With the combination of all these parameters, as a first approach, possible secondary migration directions of petroleum and traps most likely to contain petroleum can be indicated.

THREE-DIMENSIONAL MODEL APPLICATION

The model has been applied successfully to real basins as shown by the example in Figure 8. The objectives of this study are to determine the geologic history, paleopressure, paleotemperature and generation, migration, and accumulation of the hydrocarbons in a given sedimentary basin. On the basis of all the available data we have an understanding of the basin, that is, the conceptual model. The mathematical model is based on the conceptual model.

Figure 8 shows continuous sedimentation of a sequence of different sediments (shale, sandy shale and sand, shale and sandy shale, limestone, sandy shale, shale and sandy shale layers) as an input. The sedimentation occurred on top of a practically impermeable basement during a total time span of 195 million years. The lowest shale unit is the source rock and the overlying sand layer is the reservoir rock. The sedimentation took place in shallow water (20 to 100 m) except for carbonates where water depth was approximately 500 m. The structural pattern is also very simple with some normal faults that have very small throws. The faults are discontinuous and are mainly at the edges of the basin owing to local stress anomalies.

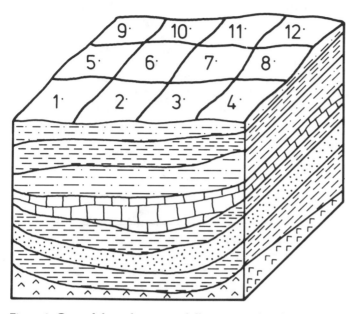

Figure 8. General three-dimensional illustration of study area.

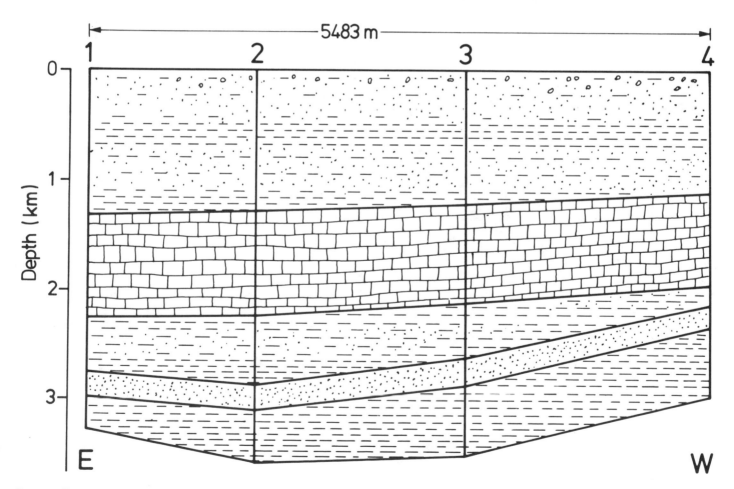

Figure 9. East-west cross section through study area.

Initial porosities (pore volume/bulk volume) of 0.62, 0.40, 0.28, and 0.54 are assumed for shale, sand, limestone, and sandy shale, respectively. Permeability versus porosity relations are determined from available literature data and from core analysis. The heat flux is 1.1×10^{-6} cal/cm^{-2}/sec^{-1} and the average initial temperature at the sediment-water interface is 15°C. Initial physical and thermal parameters are chosen depending on the lithologic descriptions (Yukler et al, 1978). The errors in the initial estimates of sediment input and heat flow are minimized as discussed by Yukler et al (1978).

A time step of 100,000 years is chosen with a vertical grid interval of 10 m and lateral grid interval of 1,000 m. The initial time step can be very small and can be increased by a certain factor at successive time steps based on the numerical analysis technique used (Halepaska and Hartman, 1972; Yukler, 1976).

The basin studied is 3,384 sq km with a maximum sediment thickness of 4,500 m, generating 1.5228×10^6 node points. Equations 1, 2, 3, and 4 are solved for this grid network and the pore pressure, temperature, physical, and thermal properties, maturation of organic matter, and amount of hydrocarbons generated are determined at each grid point as a function of time. These results are then used to determine possible secondary migration directions. This latter information is related to the existence of traps. The solution of this problem can be obtained by a fast computer with large memory. We used an IBM 370/168 for this problem.

Figure 9 illustrates an east-west cross section computed by the model through the center of the sedimentary basin. For simplicity and discretion, we will present the model results only along this cross section. The porosity versus depth relation is shown in Figure 10 at locations 1, 2, 3, and 4. The low porosity and permeability of the limestone layers decrease the subsurface fluid movement into or out of the formations below it, thus forming a closed system. Therefore, petroleum generated in this system cannot migrate out of it. This is important for mass balance studies on the generation of petroleum. The temperature-time relation for the source rock at locations 1, 2, 3, and 4 is shown in Figure 11. From Lopatin's method as described previously, the maturation of the organic matter, in terms of vitrinite reflectance (R_m), is calculated as a function of time for the source rock (Figure 12). The time of peak oil generation, where R_m is about 0.7, is 20, 25, and 23 million years at locations 1, 2, and 3, respectively. Vitrinite reflectance of 0.7 is not reached at location 4. Figure 13 is a simplified example showing the computed secondary migration directions and possible locations of oil accumulations.

CONCLUSIONS

A computer model to quantify generation, migration, and accumulation of oil has been developed and applied to existing sedimentary basins. The model results were computed with all available data on sedimentary

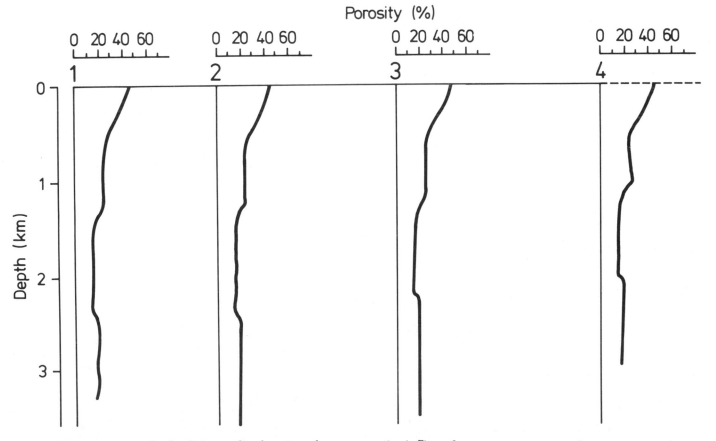

Figure 10. Porosity versus depth relations at four locations along cross section in Figure 9.

thicknesses, pressure, temperature, porosity, density, thermal conductivity, specific heat, maturity of organic matter, and amount of petroleum in place. Necessary corrections in sedimentation rates, initial physical and thermal parameters, heat flux, and coefficients of the maturity equation were made using sensitivity equations.

Necessary corrections in sedimentation rates, initial physical and thermal parameters, heat flux, and coefficients of the maturity equation were made using sensitivity equations. The sensitivity analysis technique and the constant checks in the system according to the guidelines dictated by the physical and chemical principles enabled this fully

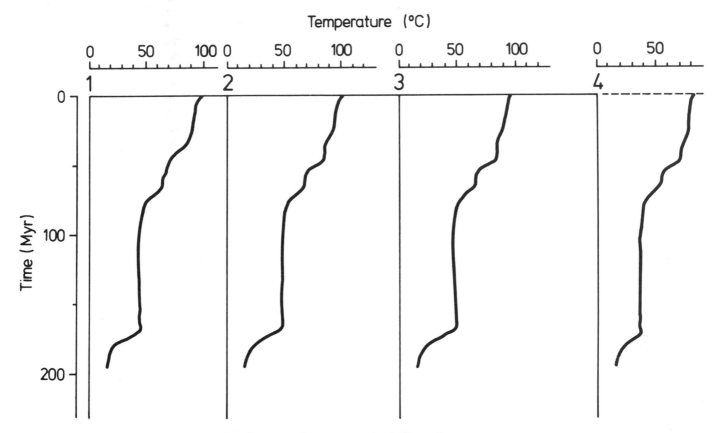

Figure 11. Temperature-time relations at four locations along cross section in Figure 9.

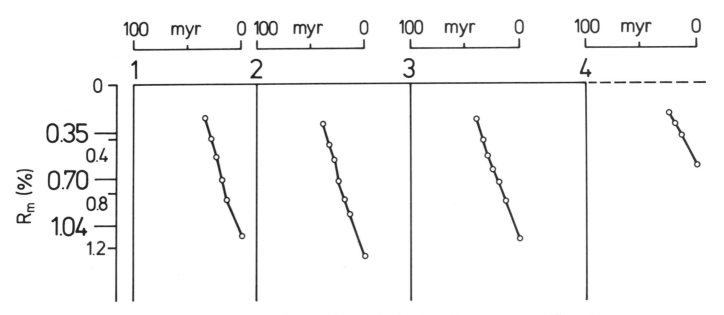

Figure 12. Evolution of maturity of organic matter as function of time at four locations along cross section in Figure 9.

integrated dynamic model to converge to the correct answer. Thus the input parameters into this model are fully calibrated for each basin being investigated. The allowable errors were ±8 percent in physical and thermal parameters, ±2°C in temperature, ±10 percent in maturity, and ±15 percent of petroleum in place.

ACKNOWLEDGMENTS

Financial support for the study came from German Federal Ministry for Research and Technology (BMFT) Grant No. ET 3070. B.M. Radke gave valuable advice with respect to hydrocarbon generation curve.

REFERENCES CITED

Berg, R.R., 1975, Capillary pressure in stratigraphic traps: AAPG Bulletin, v. 59, p. 939-956.
Bishop, R.S., 1979, Calculated compaction states of thick abnormally pressured shales: AAPG Bulletin, v. 63, p. 918-933.
Bredehoeft, J.D., and B.B. Hanshaw, 1968, On the maintenance of anomalous fluid pressure. I. Thick sedimentary sequences: Geological Society of America Bulletin, v. 79, p. 1097-1106.
Gibson, R.E., 1958, The progress of consolidation in a clay layer increasing in thickness with time: Geotechnique, p. 172-182.
Halepaska, J.C. and F.W. Hartman, 1972, Numerical solution of the 3-dimensional heat flow, *in* Short papers on research in 1971: State Geological Survey of Kansas Bulletin 204, pt. 1, p. 11-13.
Hubbert, M.K., 1940, The theory of ground-water motion: Journal of Geology, v. 48, p. 785-944.
——— , 1953, Entrapment of petroleum under hydrodynamic conditions: AAPG Bulletin, v. 37, p. 1954-2026.
Illing, V.C., 1938a, The migration of oil, *in* A.E. Dunstan et al, eds., The science of petroleum, v. 1: London, Oxford University Press, p. 209-215.
——— , 1938b, An introduction to the principles of the accumulation of petroleum, *in* A.E. Dunstan et al, eds., The science of petroleum, v. 1: London, Oxford University Press, p. 218-220.
——— , 1939, Some factors in oil accumulation: The Institute of Petroleum Journal, v. 25, p. 201-225.
Lopatin, N.V., 1971, Temperature and geologic time as factors in coalification: Akademiya Nauk Uzbekskoy

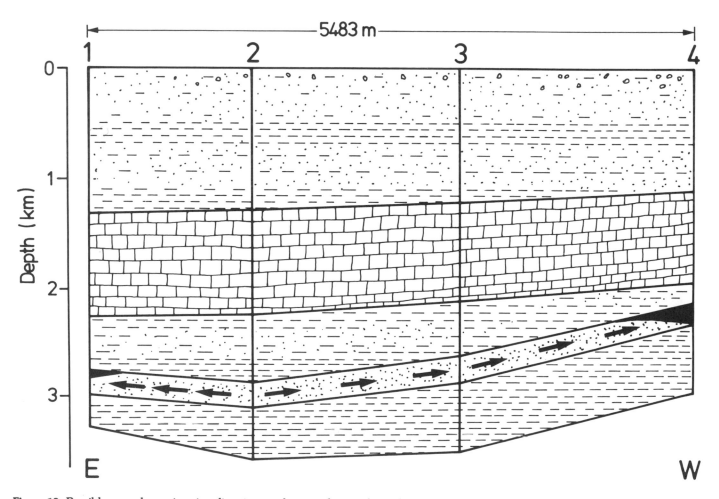

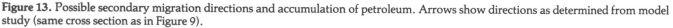

Figure 13. Possible secondary migration directions and accumulation of petroleum. Arrows show directions as determined from model study (same cross section as in Figure 9).

SSR, Izvestiya, Seriya Geologichrskaya, v. 3, p. 95-106.

Mills, R. van E., 1920, Experimental studies of subsurface relationship in oil and gas fields: Economic Geology and the Bulletin of the Society of Economic Geologists, v. 15, p. 398-421.

Munn, M.J., 1909a, Studies in the application of the anticlinal theory of oil and gas accumulation—Sewickley quadrangle, Pa.: Economic Geology, v. 4, p. 141-157.

——, 1909b, The anticlinal and hydraulic theories of oil and gas accumulation: Economic Geology, v. 4, p. 509-529.

Paaswell, R.E., 1967, Thermal influences on flow from a compressible porous medium: Water Resources Research, v. 3, p. 271-278.

Rich, J.L., 1921, Moving underground water as a primary cause of the migration and accumulation of oil and gas: Economic Geology, v. 16, p. 247-371.

——, 1923, Further notes on the hydraulic theory of oil migration and accumulation: AAPG Bulletin, v. 7, p. 213-225.

——, 1931, Function of carrier beds in long-distance migration of oil: AAPG Bulletin, v. 15, p. 911-924.

——, 1934, Problems or the origin, migration, and accumulation of oil, *in* Problems of petroleum geology: AAPG, p. 337-345.

Schowalter, T.T., 1979, Mechanism of secondary hydrocarbon migration and entrapment: AAPG Bulletin, v. 63, p. 723-760.

Sharp, J.M., Jr., and P.A. Domenico, 1976, Energy transport in thick sequences of compacting sediments: Geological Society of America Bulletin, v. 87, p. 390-400.

Shaw, E.W., 1917, The absence of water in certain sandstones of the Appalachian oil fields: Economic Geology, v. 12, p. 610-628.

Stallman, R.W., 1963, Computation of ground-water velocity from temperature data, *in* Methods of collecting and interpreting ground-water data: U.S. Geological Survey Water—Supply Paper 1544-H, p. 36-46.

Taylor, D.W., 1948, Fundamentals of soil mechanism: New York, John Wiley and Sons, 510 p.

Tissot, B., and D.H. Welte, 1978, Petroleum formation and occurrence: Berlin, Springer-Verlag, 538 p.

Welte, D.H., and M.A. Yukler, 1980, Evolution of sedimentary basins from the standpoint of petroleum origin and accumulation—an approach for a quantitative basin study: Organic Geochemistry, v. 2, p. 1-8.

—— et al, 1981, Application of organic geochemistry and quantitative analysis to petroleum origin and accumulation—An approach for a quantitative basin study, *in* G. Atkinson and J.J. Zuckerman, eds., Origin and chemistry of petroleum: Elmsford, N.Y., Pergamon Press, p. 67-88.

Yukler, M.A., 1976, Analysis of groundwater flow systems and an application to a real case, *in* D. Gill and D.F. Merriam, eds., Geomathematical and petrophysical studies in sedimentology, computery and geology, v. 3: Elmsford, N.Y., Pergamon Press, p. 33-49.

——, C. Cornford, and D.H. Welte, 1978, One-dimensional model to simulate geologic, hydrodynamic and thermodynamic development of a sedimentary basin: Geologisches Rundschau, v. 67, p. 960-979.

Concepts for Estimating Hydrocarbon Accumulation and Dispersion

Richard S. Bishop[1]
Harry M. Gehman, Jr.
Allen Young
Exxon Production Research Co.
Houston, Texas

The present volume and composition of trapped hydrocarbons are temporary events controlled by the interacting volumes of trap, oil, and gas, all of which change through time. Thus, both the volume and type of trapped hydrocarbon result from the interacting effects of kerogen amount, type, and maturity; gas and oil migration efficiencies; trap growth; and the possibility that free gas may displace trapped liquids. By estimating the historical evolving effects of each factor, we can estimate trapped hydrocarbon volume as well as trap content (i.e., only oil, oil leg and gas cap, or only gas).

The procedures that lead to such estimates are: (1) estimate the quantities of oil and gas provided by the source within the drainage area; (2) estimate the quantities of gas lost by dissolution and diffusion; (3) estimate the trap volume; and (4) compare hydrocarbon volumes with trap volume to find the limiting volume.

INTRODUCTION

The accumulation of hydrocarbons depends upon numerous physical and chemical processes that, for the most part, operate independently. Although such processes as hydrocarbon generation, trap development, and reservoir porosity reduction operate independently, they commonly control other variables which change through time and during burial. For example, as some rocks become buried more deeply, we typically (and correctly) would interpret this to mean that nearby traps would be more likely to contain gas. Although this is generally true, it is not axiomatic because at the same time that gas generation is increasing, the removal of gas is also increasing owing to the increase of gas solubilities in both water and oil.

Additionally, we see that present day accumulations of hydrocarbons depend not only on intrinsic rock properties (e.g., kerogen type, maturity, and rock porosity), but also on absolute volumes (e.g., drainage area and source thickness, trap closure area and reservoir thickness). Recently, several authors have shown how burial and thermal histories may be used to predict petroleum generation and rock properties, particularly source maturities and yields (Hood et al, 1975; McDowell, 1975; Tissot and Espitalie, 1975; Nalivkin et al, 1976; Spevak et al, 1976; Saxby, 1977; Tissot and Welte, 1978; DuRochet, 1980; Waples, 1980; Angevine and Turcotte, 1981; Nakayama and Van Siclen, 1981; Welte and Yukler, 1981; Welte et al, 1981).

Volumes of hydrocarbons generated by source rocks, however, cannot be used alone to predict uniquely trapped hydrocarbon volumes and (except for the overmature trap) whether the trap contains oil, gas, or both. Rather, such predictions also require other estimates, particularly trap volumes and volume of hydrocarbons lost during secondary migration. Thus, we suggest that such seismically estimatable dimensions of drainage area and trap configuration interact with calculatable intrinsic physical properties including kerogen maturity, hydrocarbon yields, gas solubility, gas diffusional losses, and reservoir porosity to determine both probable trapped hydrocarbon volumes and whether the trap contains oil, free gas, or both.

To explain the accumulation of hydrocarbons in this way, we need to estimate the fundamental volumes of trap, available oil, and available free gas. These appear to be the primary variables controlling the volumes of trapped hydrocarbons and whether the trap contains only oil, oil and gas cap, or only gas. By "available" we mean the volume of oil and gas that possibly arrives in the immediate vicinity of the trap. These volumes obviously are less than the total gross volume of liquid and gaseous hydrocarbons generated within the source beds in the drainage area. They are less because not all hydrocarbons generated migrate out of the source rocks, and because not all hydrocarbons that enter the carrier aquifer beds reach the trap. We estimate the

---

[1]Now with Exxon Co., U.S.A., Houston, Texas.

Table 1. Method of Calculation

I. FUNDAMENTAL VOLUMES
   1. Calculate fundamental volumes (see Table 2 for variables and Table 3 for method of calculation)
      Total oil migrated within the drainage area
      Total gas migrated within the drainage area
      Trap volume
      Formation water volume exposed to gas within the drainage area
   2. Add any spilled hydrocarbons received from downdip traps to the indigenous hydrocarbons generated within the drainage area. This is the total volume of oil and gas in aquifer within drainage area. Step II determines the quantity of hydrocarbons actually available to a trap.
II. SECONDARY MIGRATION EFFICIENCY: This is the calculation of hydrocarbon volumes lost by migration in an aquifer.
   1. Dissolve gas in both exposed formation water and total oil
   2. Determine gas volumes lost by diffusion and adjust dissolved volumes accordingly
   3. Determine volume of free (undissolved) gas available to prospect
   4. Disperse oil along secondary migration pathway
III. *VOLUME COMPARISON: This is a search for the smallest (i.e., limiting) volume in order to determine volume of *in place* hydrocarbons and whether gas has displaced oil (i.e., to determine whether the trap contains oil or gas or both). First, adjust oil and gas volumes to subsurface temperature and pressure conditions. Second, compare trap volume to oil and gas volume using following sequence of logic.
   1. If Free Gas is Available to Trap:
      A. If Free Gas Volume > Trap Volume
         Trap Content = Gas Only
         Gas Volume In Place = Trap Volume
         Spilled Gas = Trap Volume − Total Gas Volume
         Spilled Oil = Total Oil Available
      B. If Free Gas Volume < Trap Volume
         Trap Content = Gas Cap and Oil Leg (if oil is available)
         Gas Volume = Total Free Gas Available
         Oil Volume in Place (Oil Leg) is determined by:
         (1) If Total Oil > (Trap Volume − Gas Cap Volume)
            Oil Leg = Trap Volume − Gas Cap Volume
            Spilled Oil = Total Oil − Oil Leg
         (2) If Total Oil < (Trap Volume − Gas Cap Volume)
            Oil Leg = Total Oil Available
      C. If Free Gas Volume < Trap Volume *and* No Oil Is Available
         Trap Content = Gas Only
         Gas In Place = Total Gas Available
   2. If Free Gas Is Not Available To Trap (Trap Content = Oil Only)
      A. If Oil Volume > Trap Volume
         Oil In Place = Trap Volume
         Spilled Oil = Trap Volume − Total Oil
      B. If Oil Volume < Trap Volume
         Oil In Place = Total Oil Available

*Trap volume is the absolute amount of pore space actually capable of containing hydrocarbons. Trap content refers to whether trap contains oil only, oil leg and gas cap, or gas only.

net volumes that migrate out of source rocks by subtracting from the total gross generated volumes the amounts of hydrocarbons retained within source rocks. The latter amounts are estimated from total organic carbon contents, organic matter types, and maturities of the source intervals (Hunt, 1979). The net volumes of oil and gas that may arrive in the vicinity of the trap are estimated by subtracting from the net "migrated" volumes that enter the carrier aquifer interval the amounts that may possibly get "dispersed" or lost during transit to the trap. The latter amounts are difficult to estimate, but involve consideration of factors such as secondary migration path length, bed continuity, and interruptions by nonsealing and nontrapping faults and fractures. Amounts of gas lost by dissolution in the aquifer pore waters are calculated.

The reader will appreciate that the estimation of these corrections to get the net volumes of oil and gas that may be "available" to traps is subject to judgment and is not precise.

Table 1 describes our method for calculating trapped hydrocarbon volumes. The procedure used to predict present-day trap content requires estimating historical volumes: (1) estimate the quantities of indigenous* oil and gas released from the source and migrated in the carrier aquifer intervals within the drainage area, (2) estimate the quantities of gas lost during secondary migration (e.g., by dissolution and diffusion), (3) estimate the trap volume, and (4) compare the remaining net "available" hydrocarbon volumes (at trap conditions) with trap volume to find the limiting volume.

Although these calculations are discussed as though they occur only at an instant in time, this is for convenience of expression. We recognize they do not develop

*By "indigenous" we mean that the source is stratigraphically adjacent to and usually very similar in age to the carrier or reservoir-type bed that carried the hydrocarbons to the trap.

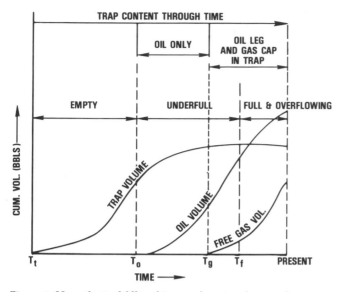

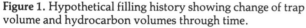

Figure 1. Hypothetical filling history showing change of trap volume and hydrocarbon volumes through time.

instantaneously in nature, but over long periods of time. In idealized form, Figure 1 shows the changes of volumes as we envision they could occur through time. The figure shows the trap started growing (T_t) and had largely completed its growth when hydrocarbon migration began and oil began accumulating (T_o). Filling of the trap with oil proceeded between T_o and T_g, when free gas first appeared. Oil and gas filled the trap to capacity at T_f, and since then the trap has been spilling liquids. Note that since T_f, both oil and gas have migrated to the trap, but only gas has been trapped.

Colloquial use of the term "timing" commonly describes events as occurring "before" or "after" (e.g., hydrocarbons were generated before or after trap growth). Such use suggests instantaneous events and may lead to incorrect interpretations of processes that actually occur during geologically significant lengths of time. Thus, attempts to calculate fundamental volumes clearly should be modeled through geologic time. The term "integrated" aptly describes this procedure because the physical dimensions (e.g., source drainage area and thickness, and trap closure area and reservoir thickness) are combined with less definable intrinsic properties (e.g., hydrocarbon yields and porosity) to define the interactions of source, migration, and trap. These dimensions and properties are set out in Table 2.

This procedure provides insights to several aspects of geologic histories of prospects that previously may have been unrecognized or neglected. For example, maturities of source and of trap are guides to whether the trap may contain oil or gas. However, consideration of the fundamental volumes also allows one to predict whether a thermally immature trap might contain only gas, being free (undissolved) gas which migrated upward from mature and overmature sources or reservoirs in sufficient quantity to displace any oil previously contained in the immature trap as Gussow (1954) proposed. Such calculations also show whether trap volume or source rock yield limits the volume of trapped hydrocarbons. As suggested, such comparisons lead to estimates of volumes of hydrocarbons which may have been spilled from a mature trap and, therefore, would

Table 2. List of variables.

Indigenous Oil and Gas Volume
(Used in Table 1, Step I, Fundamental Volumes)

Drainage area
Effective source thickness
Original total organic carbon
Average liquid yield
Average gas yield
Gas formation volume factor
API of trapped oil*
Trap temperature*
Gravity of trapped gas*

Nonindigenous Oil and Gas Volumes

Gas spilled from downdip traps
Oil spilled from downdip traps

Trap Volume

Trap area
Gross reservoir thickness
Net/Gross ratio
Factor to adjust for trap shape
Reservoir porosity
Hydrocarbon saturation

Dispersed Oil and Gas Volumes
(Used in Table 1, Step II, Migration Efficiency)

Gas Dissolved in Oil and Water
(and appropriate variables listed above)

Source-rock porosity
Nonindigenous water volume
Gas solubility in indigenous water
Gas solubility in oil
Gas solubility in nonindigenous water

Gas Diffused from Drainage Area

Seal thickness
Seal diffusion coefficient
Drainage area
Time since significant gas generation

---

*Or substitute Oil FVF for those variables. The formation volume factor (FVF) is the volume occupied in the subsurface by one surface stock tank barrel of hydrocarbon.

be available to accumulate in updip (commonly immature) traps.

The possibility that no free gas has accumulated in a trap originates not only from insufficient yield from the source (possibly the result of immaturity), but also from the opportunity for both oil and formation water to dissolve gas, thus possibly preventing generated and migrated gas from being sufficient in quantity to form a free-gas phase in a trap. In addition, gas from the drainage area also may escape by diffusion as Leythaeuser et al (1982) suggested, thereby further reducing the quantity of gas that might be trapped.

CONCEPTUAL BASIS

This section describes the methodology of calculating the volumes of dispersed and accumulated hydrocarbons. Table 1 summarizes the procedure and Table 2 lists the variables needed to make the calculations. Three volumes—oil, free gas, and trap—interact to determine both the volume and the composition of trapped hydrocarbons.

The steps involved are: (1) estimate the fundamental volumes, (2) estimate the quantities of gas lost during the secondary migration of the trap, and (3) compare "available" hydrocarbon volumes with trap volume to find the smallest or limiting volume. The fundamental volumes required are the trap volumes (i.e., the absolute pore volume which can contain hydrocarbons), the oil and gas volumes expelled from the presumed source within the drainage area, the volume of formation water which was exposed to hydrocarbons along the indigenous migration pathway, and the volume of probable losses of gas via diffusion. The migration loss is an estimate of the quantities of gas lost by dissolution in both formation water and in migrated oil and that lost by diffusion through the sealing lithology. Any remaining gas is free gas and is available to fill (and perhaps spill from) the trap.

The volume comparison (Table 1, step III) shows whether the trap volume or source rock yield limits the volume of trapped hydrocarbons. Of course, the quantities of available hydrocarbons are adjusted to trap conditions of pressure and temperature before making the comparison.

If the trap volume is less than the volume of free gas available, no oil remains trapped and both the originally trapped oil and the excess free gas are spilled. The spilled oil and gas are presumed to be available to any updip traps. Thus, any oil generated at earlier times within the drainage area that once migrated into now gas-filled traps will also have been spilled updip.

The last step in describing the prospect includes converting in-place hydrocarbons to recoverable volumes, expressing recoverable volumes as oil-equivalent barrels, estimating total gas-oil ratio under surface conditions, and calculating percent of trap filled.

The following text describes some of these procedures in more detail.

PROCEDURES

Fundamental Volumes

Several procedures can be used to determine both the volumes of the hydrocarbons available to the trap and the volume of the trap (e.g., Tissot and Welte, 1978; DuRochet, 1980; Mitra and Beard, 1980; Nakayama and Van Siclen, 1981; Welte and Yukler, 1981; Welte et al, 1981). Variables that need to be considered in estimating the volume of the trap are illustrated in Figure 2. Note particularly that how and whether the fault is sealing may greatly affect the trap volume.

The types of data listed in Table 2 are of two kinds: data that lead rather directly to quantities and sizes, and data about intrinsic rock or fluid properties. Generally, the quantity-size data depend upon the interpretation of the geology of the drainage area and of the trap. Sample analyses provide important checks on total organic carbon contents, organic matter types, and maturities. Parrish (1982) showed how paleo-upwelling patterns may be used to predict petroleum source beds. Estimates of intrinsic rock and fluid properties may be calculated by methods mentioned in the preceding paragraph. These calculations

Table 3. Equations used to determine fundamental volumes.

Indigenous Oil and Gas Volumes Migrated from Source

(1) Kerogen Quantity = Drainage Area × Effective Source Rock Thickness × Original TOC

(2) Oil Volume (in res. bbl) = Kerogen Quantity × Oil Yield × Units Constant × Oil FVF

(3) Gas Volume (in res. bbl) = Kerogen Quantity × Gas Yield × Units Constant × Gas FVF

Trap Volume

(4) Barrels = Trap Area × Reservoir Thickness × Porosity × $(1 - S_w)$ × Trap Geometry Correction × Net/Gross × Units Constant

Gas Volume Dispersed

Formation Water Volume

(5) Barrels = Drainage Area × [(Reservoir Thickness × Porosity) + (Effective Source Thickness × Porosity)] × (Units Constant)

Dissolved Gas

(6) Volume of Dissolved Gas if water and oil are both saturated = (Gas Solubility in water × Water Volume) + (Gas Solubility in Oil × Total Oil Volume)

(7) If oil and water are not saturated with gas, the calculation of dissolved gas volume in (6) cannot be used. See text.

Diffused Gas

(8) Volume of Diffused Gas = [Residence Time × Gas Concentration in Water × Seal Diffusion Coefficient × Drainage Area × Units Constant] ÷ (Seal Thickness)

also depend upon interpretations of local stratigraphy and thermal history (e.g., DuRochet, 1980; Angevine and Turcotte, 1981; Nakayama and Van Siclen, 1981). It is quite important to check that calculated temperature profiles agree with the best information about actual present-day temperature profiles and to check that calculated kerogen maturities agree with measured kerogen maturities (e.g., Hood et al, 1975; Steckler and Watts, 1978; Waples, 1980).

It is important also to recognize that most of the calculated volumes are the result of multiplying (Table 3) the pertinent variables listed in Table 2. This means, for example, that calculated hydrocarbon volumes depend equally on all these variables. None is more significant than another. Thus, doubling the intrinsic liquid yield has the same effect as doubling the drainage area or doubling the total organic carbon.

Indigenous Oil and Gas

We define the indigenous volumes of hydrocarbons as the quantities of oil and gas that have migrated from a

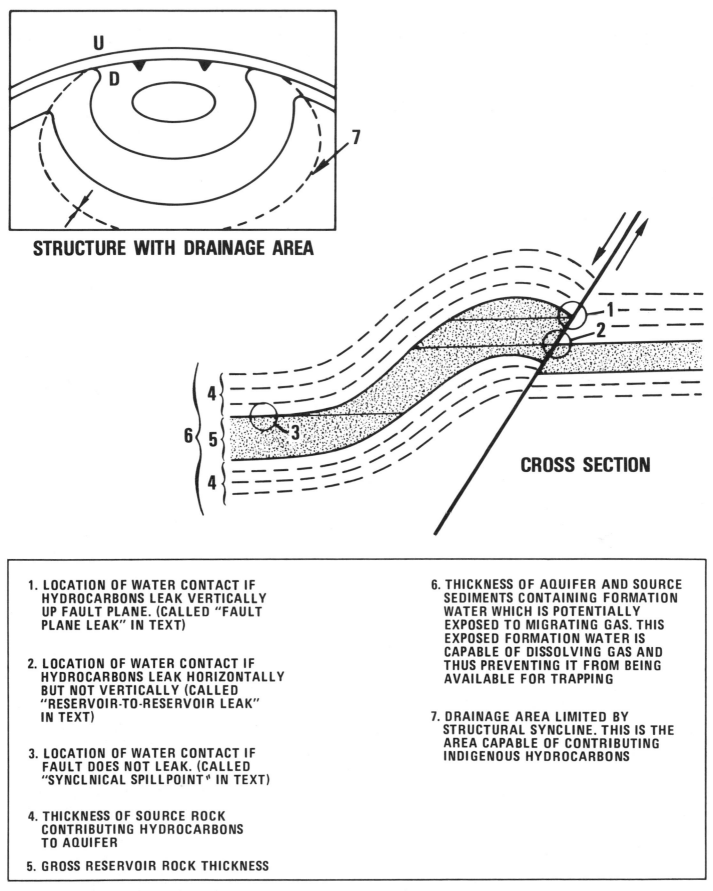

STRUCTURE WITH DRAINAGE AREA

CROSS SECTION

1. **LOCATION OF WATER CONTACT IF HYDROCARBONS LEAK VERTICALLY UP FAULT PLANE. (CALLED "FAULT PLANE LEAK" IN TEXT)**

2. **LOCATION OF WATER CONTACT IF HYDROCARBONS LEAK HORIZONTALLY BUT NOT VERTICALLY (CALLED "RESERVOIR-TO-RESERVOIR LEAK" IN TEXT)**

3. **LOCATION OF WATER CONTACT IF FAULT DOES NOT LEAK. (CALLED "SYNCLNICAL SPILLPOINT" IN TEXT)**

4. **THICKNESS OF SOURCE ROCK CONTRIBUTING HYDROCARBONS TO AQUIFER**

5. **GROSS RESERVOIR ROCK THICKNESS**

6. **THICKNESS OF AQUIFER AND SOURCE SEDIMENTS CONTAINING FORMATION WATER WHICH IS POTENTIALLY EXPOSED TO MIGRATING GAS. THIS EXPOSED FORMATION WATER IS CAPABLE OF DISSOLVING GAS AND THUS PREVENTING IT FROM BEING AVAILABLE FOR TRAPPING**

7. **DRAINAGE AREA LIMITED BY STRUCTURAL SYNCLINE. THIS IS THE AREA CAPABLE OF CONTRIBUTING INDIGENOUS HYDROCARBONS**

Figure 2. Geometry of source, trap, and drainage area.

specific source interval within a defined drainage area. The quantities of migrated hydrocarbons depend both upon the total amount of kerogen within the specified source interval and upon the portions of it converted to oil and gas, called here the "intrinsic yields." These yields vary with depth and maturity from place to place within the specified source interval. Average yields are estimated by making the necessary calculations at numerous locations within the drainage area. La Plante (1974), Tissot and Espitalie (1975), Dow (1977), Leythaeuser et al (1979), and others have indicated that intrinsic yields depend also upon kerogen type. Thus, variations of the organic matter type, as well as the variations in total organic carbon and maturity within the drainage area, must be considered in arriving at estimated average intrinsic yields of a drainage area. Each of several interpretations may result in quite different averages, just as differing interpretations of spillpoint may significantly affect estimated trap size.

Where the source interval is thick and uninterrupted by more permeable beds or streaks, and fractures or faults, it may be inappropriate to assume that hydrocarbons effectively could migrate out of the entire source interval. McAuliffe (1979), Leythaeuser et al (1982), and other have discussed possible mechanisms of primary migration. Smith et al (1971), Vandenbroucke (1971), Chetverikova (1979), and Nikolenko et al (1979) all have suggested that hydrocarbons migrate more efficiently from portions of source rocks closet to permeable "collector" beds.

Nonindigenous Hydrocarbons

Nonindigenous hydrocarbons are hydrocarbons made available to the trap from other than the specified source interval or local drainage area. Thus, nonindigenous hydrocarbons include hydrocarbons received from other traps and hydrocarbons migrating into the drainage area through nonindigenous pathways (e.g., faults). Schowalter (1979) provided an excellent discussion of secondary hydrocarbon migration and entrapment.

Even though the migration of hydrocarbons through complex nonindigenous pathways may be poorly understood, one should attempt to estimate both the volumes of water exposed to gaseous hydrocarbons and the average solubility of gas in that water. Culbertson and McKetta (1951), Sultanov et al (1972), and Bonham (1978) provide useful solubility information. (Although most of this solubility information is for methane in fresh water, corrections for the reduction of solubilities in saline waters are made; see Barkan and Yakutseni, 1982.) This dissolution decreases the possibility of free gas in the trap and lowers the total gas-oil ratio.

Trap Volume

The trap volume estimate is the cornerstone of prospect and trap evaluation, not only because it is the last stop for the hydrocarbons, but also because, in contrast to most other variables discussed here, the trap usually is definable from seismic data. Additionally, trap size may control trap content because gas can displace oil; thus, if both oil and free gas are available, smaller traps are more apt to contain gas than oil.

Although most variables determining trap volume are easily estimatable, the assumptions made about the trap seal can significantly influence the estimated trap volume (Figure 2) (e.g., Smith, 1966; Weber et al, 1978; Schowalter, 1979). Whether leaks above obviously synclinal spillpoints are present is not commonly known before drilling. Nonetheless, evaluating them can be quite revealing because it can show the considerable control of differently defined trap sizes not only on volume of trapped hydrocarbons, but also on composition of that hydrocarbon.

Migration Efficiency

Dissolved Gas

As mentioned above, formation waters dissolve hydrocarbon gases and thus reduce the amount of gas available to form a separate gas phase in the trap. Dissolution of gas in oil also reduces the chance of free gas if oil is present, or if it once was present in the trap, but has been spilled. We use the formation water volume to mean the volume of water exposed to migrating gas. This includes water both within the sediments interval and in the permeable aquifer sediments leading to the trap. Thus, it is not all of the water in the stratigraphic section, but just the water within the drainage area that was exposed to indigenous gas. In addition to this static water, dynamic water may enter the drainage area and remove additional gas. The introduction of, and the timing of, dynamic waters may significantly influence whether traps contain oil, gas, or nothing. For example, early introduction may prevent free gas from forming and displacing already trapped oil. Late introduction to traps containing only free gas might dissolve the gas and leave nothing.

In addition to the quantities of gas that dissolve in formation waters, gas also dissolves in oil. Thus, free (or undissolved) gas can form only when both the water and the trapped oil have been saturated. Where there is no free gas, dissolution occurs in proportion to the ratios of the solubilities of gas in water and in oil. In other words, gas dissolves simultaneously both in water and in oil in proportion to the ability of each to dissolve the gas. In consequence, the total gas-oil ratio even of the undersaturated "only oil" case depends not only upon the trap's and source's unique volumetrics, but also on the maturities of the source, the trap, and on the kerogen type.

Obviously, in addition to the water solubilities mentioned above, solubilities in oil are needed. These data are available in standing (1947), McCain (1978), an Glasö (1980). In addition to temperature and pressure, their correlations and charts require an estimate of the API gravity of oil. Waples (1980) suggested that API gravity may be estimated from time-temperature integrals that also are used to calculate maturity. The input for both calculations includes stratigraphic information and thermal histories.

It is worth noting that the solubilities of gas in water and in oil are much more affected by pressure than by temperature. Therefore, the effects of dissolution on the availability of free gas will be greater for deep traps, traps in deep water, and traps having abnormally high pore-fluid pressures.

Diffused Gas

Diffusion calculations estimate the quantities of gas lost

by diffusing from the entire drainage area through the overlying sealing lithology and into the next aquifer or to the surface. Generally, only small quantities of gas are lost by diffusion, but the quantities can become significant in situations where thin seals are combined with large drainage areas, long residence times, weak sources, and coarser grained seals (such as siltstones). Furthermore, some faulted and fractured seals that retain oil quite well may transmit (leak) gas rather freely by processes other than diffusion (Smith, 1966). Antonov (1963), Smith et al (1971), Sokolov et al (1971), Pandey et al (1974), and Leythaeuser et al (1982) give examples of diffusion calculations and guidelines on diffusion coefficients.

Volume Comparison

Thus far, we have discussed the trap volume and the available hydrocarbon volumes as they apply to integrated trap-prospect assessment. The next step is the search for the limiting volume. The following paragraphs detail the volume comparison procedure (see also Table 1).

Three quantities—the available volume of oil, the available volume of free gas, and the volume of the trap—interact to determine the volume of the trapped hydrocarbons and their composition (i.e., oil versus gas). If the trap volume is less than the volume of total free gas available, there is excess free gas to be spilled. This spilled gas presumbly is available to enter updip traps. In addition, any oil that occupied now gas-filled traps at earlier times within the drainage area also will have spilled updip.

Table 1 describes the comparison procedure. After determining the volume of free gas, if any (step I), and adjusting both oil and gas volumes to subsurface conditions (step II), we compare the free gas volume to the trap volume (step III). If the free gas volume exceeds the trap volume (step III 1A), then the trap should contain only gas (i.e., it is the "only gas" case). The difference between trapped volume and total free gas is the volume of spilled free gas.

If the free gas is less than the trap volume (step III 1B), the remaining trap space may contain oil (i.e., this is the "oil leg and gas cap" case). Any excess liquids are spilled updip (step III 1B). If no liquids are present, however, as in overmature traps (step III 1C), and the gas volume is less than the trap volume, the trap should contain only gas and nothing will have been spilled.

If no free gas is present (step III 2), the trap may contain only oil. The amount trapped again is estimated by comparing the amount of oil available to the trap with the trap volume. If the oil volume exceeds the trap volume (step III 2A), the amount trapped equals the trap volume; the difference is spilled. If the oil volume is less than the trap volume (step III 2B), the trap should contain all of the oil available and is not full.

This comparison of volumes of available oil, free gas, and trap is necessary because source rock volumetric yields often exceed trap capacity. In these cases, oil may be spilled as free gas displaces reservoired liquids. Therefore, the purpose of comparing volumes is to determine the limiting volume and to determine whether free gas could have displaced oil.

Lastly, gas displacement also can affect the total gas-oil ratio of the trap. The total gas-oil ratio is the ratio of trapped, in-place, gas volume (both that dissolved in the oil and that in any gas cap, on a surface-condition, standard cubic foot basis) to the surface, stock-tank barrels of all of the oil in place in the trap. Hence, the gas-oil ratio is determined not only by kerogen type and source maturity, but also by the fundamental volumes.

Recoverable Hydrocarbons

After the in-place volumes of hydrocarbons are estimated, the next step is to determine the volume of recoverable hydrocarbons. Multiplying the in-place volumes by estimated recoveries for oil and gas shows the quantities of recoverable hydrocarbons. These also are expressed on the basis of surface conditions.

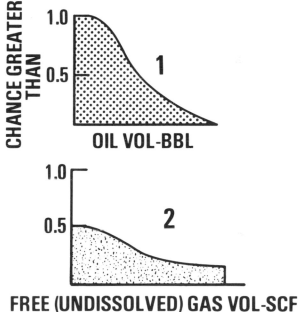

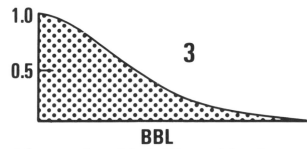

Figure 3. Summary of essential calculations made from the methodology described in the text. These curves describe the source-rock productivity and trap volume.

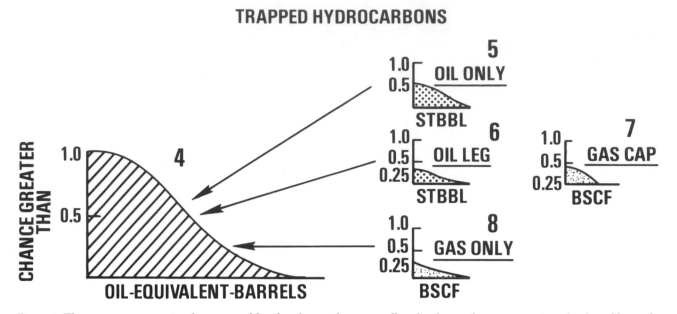

Figure 4. These curves summarize the recoverable oil and gas volume as well as the chance that trap contains oil only, oil leg and gas cap, and gas only.

Table 4. Mean values used in example calculation.

I.   Fundamental Volumes

Total indigenous oil migrated	=	245. × 10⁶ STB
Total indigenous gas migrated	=	1,089. × 10⁹ SCF
Trap volume	=	174. × 10⁶ bbl
Static (indigenous) formation water volume	=	19,239. × 10⁹ bbl

II. Nonindigenous Hydrocarbons

Nonindigenous oil (saturated)	=	0.0 × 10⁶ STB
Nonindigenous free gas	=	0.0 × 10⁹ SCF

III. Migration Efficiency

Gas dissolved in indigenous oil, indigenous water, and nonindigenous water	=	814. × 10⁹ SCF
Gas diffused from drainage area	=	81. × 10⁹ SCF
Total lost gas	=	895. × 10⁹ SCF
Net free gas	=	194. × 10⁹ SCF

IV. Formation Volume Factors

Oil	=	1.6 Res. bbl/STB
Gas	=	0.00075 Res. bbl/SCF

V.   Recovery Efficiencies

Oil	=	0.32 (fraction)
Gas	=	0.80 (fraction)

VI. Gas-Oil Ratio

Total GOR	=	2,203. SCF/bbl
Solution GOR	=	1,200. SCF/bbl

EXAMPLE CALCULATION

The reader will recognize that what we have presented so far is a description of a series of single calculations. They are "single" because each variable used has been described as being a best estimate or best average value. Few of these variables are known with great accuracy, and there are considerable uncertainties about most. Therefore it is more realistic to specify each variable by a most probable value, along with a range of possible values indicating the range of uncertainty about the most probable value. Thus, in our actual calculations, each variable is entered in this way, and the entire calculation procedure described in the preceding sections is repeated many times. Each time, a single value of each variable is selected by a "Monte Carlo" sampling process from among the range of probable values for that variable. Thus, each time the calculation is repeated, a different value of each variable is entered. The result is that the calculated volumes from each calculation differ. In this manner, we get in the calculated results a reflection of the effects of our uncertainties about the variables involved. The resulting distributions of calculated probable outcomes express our uncertainties about these geologic variables. That is, we obtain a range of calculated results rather than just single values. Thus, whether the trap is calculated to contain only oil, only gas, or oil leg plus gas cap differs among the many calculated trials. The distribution of these outcomes is expressed as probability of oil versus gas. Other calculated results, including percent of trap fill, also show ranges of values, indicating uncertainties that arise from our uncertainties about the many input variables.

Figures 3 and 4 summarize the most significant aspects of predictions that are made with the methods described here. Curves 1 and 2 (Figure 3) describe the volumes of hydrocarbons available to a prospect, and curve 3 (Figure 3) describes the volume of the trap itself. In the example shown, both oil and free gas are available. Note, however, that the free gas curve does not start at a probability of 1.0. Rather, it starts at approximately 0.5, meaning that the trap has only a 50-50 chance of having free gas available to it. Formation water and oil may have dissolved all of the migrated gas.

In Figure 4, curve 4 describes the recoverable oil-equivalent barrels (OEB) the trap may contain. This OEB curve is a summary of the information shown in curves 5 through 8, which show both the probable volumes of the three possible combinations of oil and gas (i.e., oil only, oil leg and gas cap, and gas only). Note that none of the four curves has an initial probability of 1.0. This is because each curve describes the possibility of just one possible combination of oil and gas. In other words, curve 5 shows a 50 percent chance that the trap contains only oil; it also describes the probable range of the trapped volume of oil. Curves 6 and 7 show that the trap has a 30 percent chance of having both an oil leg and a gas cap. Curve 8 shows that the

trap has a 20 percent chance of containing gas only. The sum of the three possible combinations of trapped hydrocarbons is 100 percent (50 + 30 + 20), thus the OEB curve has an initial probability value of 1.0. The trapped volumes for each of the three trap content combinations may be described as either in-place or recoverable volumes.

Table 4 lists the mean values used in this example calculation and Table 5 (which is an expansion of Table 1) uses those values to show the basic procedure of the calculation.

Note that, in our example, the trap contains oil leg and gas cap, yet Figure 4 shows the trap content is slightly less likely than the oil only case. Such a result originates from some mean values not always being the most probable values. In addition, the dependencies of some variables makes it even more difficult to estimate the most probable trap outcome without going through the entire process of calculation.

Table 5. Example Calculation.

I. HYDROCARBONS AVAILABLE TO TRAP

    1. Available oil = indigenous oil + nonindigenous oil - dispersed oil
        = 245. + 0.0 − 0.0
        = 245. $\times 10^6$ STB

    2. Available free gas = (indigenous gas − dissolved − diffused) + nonindigenous gas
        = (1,089. − 814. − 81.) $\times 10^9$ + 0.0
        = 194. $\times 10^9$ SCF

II. ADJUST HYDROCARBON VOLUMES TO SUBSURFACE CONDITIONS

    1. Available oil = STB oil $\times$ FVF oil
        = 245. $\times$ 1.6
        = 392. $\times 10^6$ res. bbl oil

    2. Available gas = SCF gas $\times$ FVF gas
        = 194. $\times 10^9 \times$ 0.00075
        = 145.5 $\times 10^6$ res. bbl gas

III. TRAP VOLUME

    1. Trap Volume = 174. $\times 10^6$ bbl

IV. VOLUME COMPARISON

    1. If free gas is present and:

        A. Free gas vol. > trap vol.

Trap content	= gas only
Gas in place (bbl)	= trap vol.
Gas in place (SCF)	= trap vol. ÷ gas FVF
Spilled gas (bbl)	= trap vol. − tot. free gas
Spilled gas (SCF)	= spilled gas (bbl) ÷ gas FVF
Spilled oil (bbl)	= total oil available

        B. Free gas vol. < trap vol.
        Trap content = gas cap and oil leg (if oil is available)

        B-1. Oil is available

Trap content	= oil leg and gas cap
Gas cap	= Total free gas
	= 145.5 $\times 10^6$ res. bbl of gas
Trap volume not filled	= trap volume − free gas volume
	= (174. − 145.5) $\times 10^6$
	= 28.5 $\times 10^6$ res. bbl

(CONTINUED PAGE 50)

(TABLE 5 CONTINUED)

Compare oil vol. to unfilled trap vol.:
If tot. oil > unfilled trap volume

Oil leg in place (res. bbl)   = unfilled trap vol.
  = 28.5×10^6 res. bbl

Oil leg in place (STB)   = unfilled trap vol. ÷ FVF oil
  = $(28.5/1.6) \times 10^6$
  = 17.8×10^6 STB

Spilled oil (STB)   = tot. oil − oil leg
  = $(245. - 17.8) \times 10^6$
  = 227.2×10^6 STB

If tot. oil < unfilled trap vol.

Oil leg in place (STB)   = tot. oil available
Spilled oil (STB)   = 0.0

B-2. Oil is not available

Trap content   = gas only
Gas in place   = tot. gas available (SCF or res. bbl)
  = 0.0

2. Free gas is not present:
Trap content = oil only

A. Oil vol. > trap vol.

Oil in place (res. bbl)   = trap vol.
Oil in place (STB)   = trap vol. ÷ FVF oil
  = 0.0

Spilled oil (STB)   = (tot. oil − trapped oil) ÷ FVF oil

B. Oil vol. < trap vol.

Oil in place   = tot. oil available
Spilled oil   = 0.0

V. RECOVERABLE HYDROCARBONS

1. Oil only case

Recoverable Oil (STB)   = [Oil in place (res. bbl) × oil rec. eff.] ÷ FVF oil
  = 0.0

Recoverable gas (SCF)   = rec. oil × total GOR
  = 0.0

2. Gas only gas

Recoverable gas (SCF)   = [gas in place (res. bbl) × gas rec. eff.] ÷ FVF gas
  = 0.0

3. Oil leg and gas cap case

Recoverable free gas (SCF)   = gas in place (SCF) × gas rec. eff.
  = $(145.5 \times 0.8) \times 10^9$
  = 116.4×10^9 SCF

Recoverable oil (STB)   = oil in place (STB) × oil rec. eff.
  = $(17.8 \times 0.32) \times 10^6$
  = 5.7×10^6 STB

Recoverable dissolved gas (SCF)   = recoverable oil × solution GOR
  = $5.7 \times 10^6 \times 1,200.$
  = 6.84×10^9 SCF

VI. OIL-EQUIVALENT BARRELS

Oil only case   = rec. oil (in STB) + [rec. gas (in SCF)] ÷ 5,600
  = 0.0 oil-equivalent bbl

Gas only case   = [rec. gas (in SCF)] ÷ 5,600
  = 0.0 oil-equivalent bbl

Oil leg and gas cap case   = rec. oil (in STB) + (rec. free gas + rec. diss. gas) ÷ 5,600
  = $5.7 \times 10^6 + [(116.4 + 6.84) \times 10^9] ÷ 5,600$
  = 27.7×10^6 oil-equivalent bbl

CONCLUSIONS

The accumulation of hydrocarbons is a transient and complicated event, influenced by not only source rock productivity and trap volume but also by migration efficiency and trap efficiency. The factors controlling the oil versus gas content include not only kerogen type (oil prone or gas prone) and maturity of both source rock and trap, but also gas displacement of oil and selective gas leakage from traps. Traps which selectively leak gas but not oil might be recognized by having oil saturated with gas but no gas cap.

The dissolution and diffusion of gas may reduce the gas-oil ratio of any trapped oil and thereby also influence the shrinkage factor of the oil. Oils with low gas-oil ratios therefore may result not only from sources of low maturity but also from having their dissolved gas content reduced by either diffusion and/or exposure to large volumes of water.

ACKNOWLEDGMENTS

The ideas presented have evolved through several years of applying predictive methods to exploration problems of Exxon affiliate companies. We are very pleased to acknowledge the important contributions of our many colleagues who have experimented with the ideas.

In addition, we express appreciation for editorial assistance provided by R.O. Grieve, J.P. Shannon, Jr., and D.A. White. We also thank Exxon Production Research Company for permission to publish.

REFERENCES CITED

Angevine, C.L., and D.L. Turcotte, 1981, Thermal subsidence and compaction in sedimentary basins: AAPG Bulletin, v. 65, p. 219-225.

Antonov, P.L., 1963, Range and extent of gas diffusion from deposits into water outside the oil-water contact: Gasovaia Promphlennost, v. 8, p. 1-6, translated by R.T. Schweisberger.

Barkan, Ye.S., and V.P. Yakutseni, 1982, Gas prospects at great depths: International Geology Review, v. 24, p. 253-259.

Bonham, L.A., 1978, Solubility of methane in water at elevated temperatures and pressures: AAPG Bulletin, v. 62, p. 2478-2481.

Chetverikova, O.P., 1979, Calculating primary migration coefficients for liquid and gaseous hydrocarbons, *in* Petroleum source rocks and principles of their diagnostics: Papers presented at National Seminar, Moscow, 1977, N.B. Vassoyevich and P.P. Timofeyev, eds., 1979, translated by R.T. Schweisberger.

Culbertson, O.L., and J.J. McKetta, 1951, The solubility of methane in water at pressures to 10,000 psia: AIME Petroleum Transactions, v. 192, p. 223-226.

Demaison, G.J., and G.T. Moore, 1980, Anoxic environments and oil source bed genesis: AAPG Bulletin, v. 64, p. 1179-1209.

Dow, W.G., 1977, Kerogen studies and geological interpretations: Journal of Geochemical Exploration, v. 7, p. 76-99.

——— , 1978, Petroleum source beds on continental slopes and rises: AAPG Bulletin, v. 62, p. 1584-1606.

DuRochet, J., 1980, The DIAGEN program, two procedures to calculate the diagenetic evolution of organic matter: Bulletin, Center of Research Exploration and Production, Elf-Aquitaine, v. 4, p. 813-831.

Glasö, O., 1980, Generalized pressure-volume-temperature correlations: Journal of Petroleum Technology, v. 32., no. 5, p. 785-864.

Gussow, W.G., 1954, Differential entrapment of oil and gas—a fundamental principle: AAPG Bulletin, v. 38, p. 816-853.

Hood, A., C.C.M. Gutjahr, and R.L. Heacock, 1975, Organic metamorphism and the generation of petroleum: AAPG Bulletin, v. 59, p. 986-996.

Hunt, J.M., 1979, Petroleum geochemistry and geology: San Francisco, W.H. Freeman, 617 p.

La Plante, R.E., 1974, Hydrocarbon generation in Gulf Coast Tertiary sediments: AAPG Bulletin, v. 58, p. 1281-1289.

Leythaeuser, D., R.G. Schaefer, and A. Yukler, 1982, Role of diffusion in primary migration of hydrocarbons: AAPG Bulletin, v. 66, p. 408-429.

——— , et al, 1979, Hydrocarbon generation in source beds as a function of type and maturation of their organic matter; a mass balance approach: Proceedings, 10th World Petroleum Congress, v. 2, p. 31-41.

McAuliffe, C.D., 1979, Oil and gas migration—chemical and physical constraints: AAPG Bulletin, v. 63, p. 761-781.

McCain, W.D., Jr., 1978, The properties of petroleum fluids: Tulsa, Petroleum Publishing Company, 325 p.

McDowell, A.N., 1975, What are the problems in estimating the oil potential of a basin?: Oil and Gas Journal, v. 73, no. 23, p. 85-90.

Mitra, S., and W.C. Beard, 1980, Theoretical models of porosity reduction by pressure solution for well sorted sandstones: Journal of Sedimentary Petrology, v. 50, p. 1347-1360.

Nakayama, K., and D.C. Van Siclen, 1981, Simulation model for petroleum exploration: AAPG Bulletin, v. 65, p. 1230-1255.

Nalivkin, V.D. et al, 1976, Criteria and methods for the quantitative assessment of oil and gas occurrence in large, poorly-studied territory: Sovetskaya Geologiya, no. 1, p. 28-39.

Nederlof, M.H., 1980, The use of habitat of oil models in exploration prospect appraisal: Proceedings, 10th World Petroleum Congress, v. 2, p. 13-22.

Nikolenko, V.A., L.A. Polster, and P.I. Sadykova, 1979, The relationship between oil source rocks and reservoir rocks in syngenetic oil- and gas-bearing sections, *in* Petroleum source rocks and principles of their diagnostics: Papers presented at national seminar, Moscow, 1977, N.B. Vassoyevich and P.P. Timofeyev, eds., 1979, translated by R.T. Schweisberger.

Pandy, G.N., M.R. Tek, and D.L. Katz, 1974, Diffusion of fluids through porous media with implications in petroleum geology: AAPG Bulletin, v. 58, p. 291-303.

Parish, J.T., 1982, Upwelling and petroleum source beds, with reference to Paleozoic: AAPG Bulletin, v. 66, p. 750-774.

Saxby, J.D., 1977, Oil-generating potential of organic matter in sediments under natural conditions: Journal of Geochemical Exploration, v. 7, p. 373-382.

Schowalter, T.T., 1979, Mechanics of secondary hydrocarbon migration and entrapment: AAPG Bulletin, v. 63, p. 723-760.

Smith, D.A., 1966, Theoretical considerations of sealing and nonsealing faults: AAPG Bulletin, v. 50, p. 363-374.

Smith, J.E., J.G. Erdman, and D.A. Morris, 1971, Migration, accumulation and retention of petroleum in the earth: Proceedings, 8th World Petroleum Congress, v. 2, p. 13-26.

Sokolov, V.A., et al, 1971, Migration processes of gas and oil, their intensity and directionality; Proceedings, 8th World Petroleum Congress, v. 2, p. 493-505.

Spevak, Yu.A., M.S. Burshtar, and S.P. Malkin, 1976, Method of appraisal of potential resources and predicted reserves of oil and gas based on gas-geochemical studies: Sovetskaya Geologiya, no. 6, p. 112-119.

Standing, M.B., 1947, Pressure-volume-temperature correlation for mixtures for California oils and gases: API Drilling and Production Practices, p. 275-287.

Steckler, M.S., and A.B. Watts, 1978, Subsidence of the Atlantic-type continental margin off New York: Earth and Planetary Science Letters, v. 41, p. 1-13.

Sultanov, R.G., V.G. Skripka, and Yu.A. Namiot, 1972, Solubility of methane in water at high temperatures and pressures: Gasovaia Promyshlennost, v. 17, no. 5, p. 6-7.

Summerhayes, C.P., 1981, Organic facies of middle Cretaceous black shales in deep North Atlantic: AAPG Bulletin, v. 65, p. 2364-2380.

Tissot, B., and J. Espitalie, 1975, L'evolution thermique de la matiere organique des sediments: applications d'une simulation mathematique: Revue de l'Institut Francais du Pétrole, v. 30, p. 743-777.

—— and D.H. Welte, 1978, Petroleum formulation and occurrence: a new approach to oil and gas exploration: Berlin, Springer-Verlag, 521 p.

Vandenbroucke, M., 1971, Study of primary migration: variation in rock extracts from a transition zone from source rock to reservoir rock, *in* H.F. von Gaertner and H. Wehner, eds., Advances in organic geochemistry: Pergamon Press, Oxford-Braunschweig (1972), p. 547-565.

Waples, D.W., 1980, Time and temperature in petroleum formation: application of Lopatin's method to petroleum exploration: AAPG Bulletin, v. 64, p. 916-926.

Weber, K.J., et al, 1978, The role of faults in hydrocarbon migration and trapping in Nigerian growth fault structures: Proceedings, 10th Annual Offshore Technology Congress, v. 4, p. 2643-2653.

Welte, D.H., and M.A. Yukler, 1981, Petroleum origin and accumulation in basin evolution: AAPG Bulletin, v. 65, p. 1387-1396.

—— and —— and D. Leythaeuser, 1981, Applications of organic geochemistry and quantitative analysis to petroleum origin and accumulation—an approach for a quantitative basin study, *in* G. Atkinson and J.J. Zuckerman, eds., Origin and chemistry of petroleum: Elmsford, N.Y., Pergamon Press, p. 67-88.

Geological and Geochemical Models in Oil Exploration; Principles and Practical Examples

P. Ungerer
F. Bessis
P.Y. Chenet
B. Durand
E. Nogaret
Institut Francais du Pétrole
Rueil-Malmaison, France

A. Chiarelli
SNEAP
Pau, France

J.L. Oudin
Compagnie Francaise des Pétrole
Paris, France

J.F. Perrin
Université de Bordeaux, France

It is now possible to develop mathematical models that make quantitative predictions of the geological phenomena leading to oil accumulations. The models presented in this study are deterministic, that is, based on physical or chemical laws and not on statistical analysis.

A first set of models may be used during the initial stage of exploration of a sedimentary basin, when little geological data are available. Their purpose is to determine the general geological evolution and the overall petroleum potential. A backstripping model makes an automatic reconstruction in time of sedimentary basins, taking into account the progressive compaction of the sediments. It also computes the sediment load on the basement together with the subsidence variations in time, enabling a geodynamic model to be used for some basins to simulate the history of heat flow. It is then possible to reconstruct the temperature history of each sedimentary unit. A kinetic model of organic matter maturation is used to compute the possible area and timing of petroleum formation.

A second set of models may be applied when exploration is more advanced, to estimate the importance of various traps in a given petroleum province. These models, which need more extensive data than the first set, take into account the influence of migration processes on oil accumulation. The migration model gives a quantitative description of the formation of hydrocarbons and computes the pressure regime of the fluids. Thus it determines the amount of petroleum expelled from the source rocks by compaction and its possible accumulation in traps. A thermodynamic model computes hydrocarbon migration in the gas phase and its consequences for the composition of oils.

The Gulf of Lion passive margin in France, the Viking Graben in the North Sea, and the Mahakam Delta in Indonesia are used as examples.

INTRODUCTION

The increasing cost of petroleum exploration in frontier areas requires optimization of the opportunities to find commercial discoveries. On the other hand, finding more hydrocarbons in already producing zones requires the discovery of subtle traps.

The purpose of the mathematical models presented is to help in exploration for this double objective. In a given basin, these models can provide an interpretation of the processes leading to oil accumulation. Thus they enable the selection of the most favorable zones on the basis of the available geological data.

Up to now, the petroleum potential of a given area was estimated mostly in a qualitative way, looking at the possible structures, reservoirs, cap rocks, and source rocks. The processes of hydrocarbon formation, migration and accumulation were only roughly understood and the geologists would assess their chances of success from experience. This kind of approach led to the development of statistical models, based on correlations between geological parameters and oil accumulation. However, this approach needs a very large amount of data that can only be supplied by larger oil companies.

It is now widely accepted that the accumulation of oil results through a long chain of geological processes: the sedimentary organic matter (OM) is progressively buried during basin subsidence; this burial causes a temperature increase, under which influence the OM progressively generates hydrocarbons; once sufficient maturation is reached, the hydrocarbons are expelled from the source rocks, and they may migrate in carrier beds and accumulate in traps, provided a caprock is present.

During the past decade, much progress has been made in the understanding of these processes. They have been described by physical or chemical laws that enable us to set up mathematical models. These models may be applied to a given area and provide quantitative results.

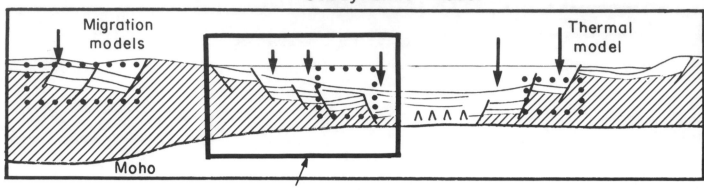

Figure 1a. General scheme showing the scale on which the various models are applied. The arrows represent the possible applications of the thermal model, while the dotted frames are related to the migration models.

Each of the models presented here (Figure 1) deals with a specific geological process:

- A backstripping model makes an automatic reconstruction of the sedimentary basin in time through successive cross sections.
- A geodynamic model is used for the basins formed by extension. It reconstructs the heat flow history.
- A thermal model computes the temperatures in the basin. It is coupled to a kinetic model of OM maturation.
- A migration model computes the amounts of petroleum generated in the source rocks and the amounts accumulated in traps through time.
- A thermodynamic model determines the amount and composition of the hydrocarbons that may migrate in gaseous solution.

VALIDITY CONDITIONS OF THE MODELS

Before discussing the various models, it is useful to review the conditions that have to be respected for their validity in practical cases of basin exploration.

1) The processes that are described by the model must be the actual processes occurring in the basin evolution and they have to be represented by correct equations. For instance, it is important for migration purposes to assume that the migration of oil occurs mainly in a separate phase from water, and not in an aqueous solution. Generally speaking, the processes are selected on the basis of theoretical considerations, typical basin examples and/or laboratory experiments.

2) The computer program has to give a good approximation of the exact mathematical solution. This numerical difficulty is easily solved in one dimensional problems. However, many uncertainties may occur in the case of two- or three-dimensional problems, and they generally lead to some error in the quantitative results.

3) The natural complexity of the basin has to be considered without excessive simplification. In each case, a conceptual system of the basin is elaborated using the available geological data (Welte and Yukler, 1981). This system should give a consistent reproduction of the

stratigraphical, lithological, and geochemical data at the scale considered, and must account for small size details if necessary. For instance, thin permeable beds may have a considerable importance for the migration of hydrocarbons and therefore should not be neglected. Generally, the validity of the conceptual system is limited by the lack of data at great depths (poor resolution of seismic profiles, scarcity of deep wells).

4) The amount and quality of the input data must be sufficient to allow reasonably accurate results. Some of the models require data that are frequently obtained with low accuracy. For instance, the temperature and source rock data that are needed by the migration model are often subject to great uncertainties, resulting in important possible errors in the calculated amounts of accumulated oil.

It may be argued that these conditions are never fully satisfied in practice. They constitute, however, a goal that we should strive to reach.

THE USE OF DETERMINISTIC MODELS IN OIL EXPLORATION

In a poorly known sedimentary basin, such as an undrilled continental margin, a first set of models may be used to describe the general pattern of basin evolution. This allows in some cases an estimate of the possible maturation of the source rocks. The models are used in the following sequence: first, the burial and compaction history is reconstructed and basin subsidence through time is determined by the backstripping method; second, the temperature history of each sedimentary unit is computed from the heat flow variation with time (in some basins, for example passive continental margins, the heat flow may be derived from the results of subsidence studies based on a geodynamic model); finally, the temperature results are used as input data for the maturation model enabling the reconstruction of the oil window through time.

This first set is particularly applicable to sedimentary basins that have undergone a simple tectonic history, limited to extensional tectonics followed by general subsidence.

INPUT DATA
(Poor accuracy data may be
reassessed if necessary for the
consistency of the results).

**OUTPUT
RESULTS**

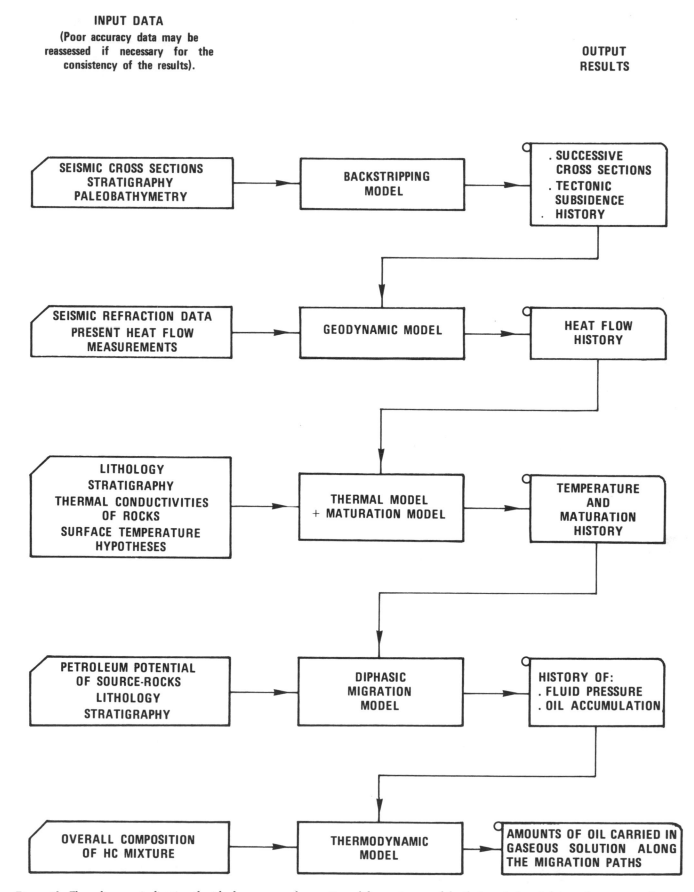

Figure 1b. Flow diagram indicating the ideal sequence of operation of the various models, their main input data and output results.

At more advanced stages of exploration the well data enable us to have a fairly reliable idea of source rock distribution and richness and thus to estimate quantitatively the amounts of hydrocarbons generated in the basin. It is then possible to use the second set of models, that study the migration, accumulation and dysmigration of hydrocarbons. The migration model that represents these processes for a cross section may use the paleotemperature reconstruction produced by the previous models. Its results show pressure and hydrocarbon saturation within the sediments through time. Thus it allows us to quantitatively determine the volume of hydrocarbon accumulations. In addition, the pressure reconstruction allows the prediction of probable high pressure zones. The pressure results are also used as input for the thermodynamic model, together with temperature data. This model determines the amount and composition of the hydrocarbons that migrate dissolved in the gas phase and that are left behind by exsolution along the migration path of the gas. The thermodynamic model may also be used to predict the maximum condensate content of a wet gas as function of temperature and pressure.

BURIAL OF SEDIMENTS AND SUBSIDENCE: THE BACKSTRIPPING METHOD

The reconstruction of the progressive burial of sediments and the corresponding basin subsidence is usually carried out in a qualitative way through the construction of palinspastic cross sections. These are used to predict the most favorable environments for the occurrence of source rocks and permeable units. In order to determine more precisely the thickness of the sedimentary column through time, compaction may be taken into account, as achieved by Perrier and Quiblier (1974) or Magara (1978). The backstripping method, as proposed first by Watts and Ryan (1976) is still more elaborate. It consists of the computation and automatic drawing of the successive shapes of the basin since the beginning of sedimentation. The model elaborated by the present authors accounts for the thickness variation of the sedimentary units through burial, and for paleobathymetry. The weight of the sediments on basement is computed by the model. It is thus possible to estimate what part of the subsidence that is due to the loading effect of sediments and water. The residual subsidence, not due to the sedimentary load, is computed by the model and is called tectonic subsidence (Watts and Steckler, 1981). Thus the vertical movements caused by deep seated tectonic processes may be estimated (Figure 2).

Compaction
The thickness evolution of the sedimentary column during compaction depends on the type of sediment, rate of sedimentation and loading, possibilities of drainage of expelled fluids, and mineral transformation, cementation and dissolution processes during diagenesis.

For sands, clay, and marls, compaction is mainly due to the expulsion of interstitial fluids and the volume of solid matrix remains relatively constant. For carbonates, cementation may be important but is generally compensated by dissolution. Diagenetic studies show that the dissolution

fluids enriched with calcium precipitate in the vicinity of the dissolution zone (De Charpal and Devaux, 1981). Thus, the hypothesis of solid matrix volume conservation during compaction may be considered valid for all lithologies.

Compaction is computed from average relationships between porosity and depth of burial. The porosity data may be obtained either from well logs or from core analysis. Following Magara (1978), it is assumed that the final state of compaction (or porosity) is obtained at maximum depth of burial, that is, no appreciable rebound occurs during uplift and erosion.

Paleobathymetry
Once the changes in thickness of the sedimentary column are determined, the reconstruction of the total subsidence requires a reference level for the top of the sediments. For this purpose, the paleobathymetric evolution and eustatic sea level changes have to be considered.

The paleobathymetry is estimated from biostratigraphic studies, based on the paleontological record from well data. The precision is good in shallow-water sedimentary facies but much less accurate for pelagic sediments. Through use of seismic stratigraphy techniques (Vail et al, 1977a, b), the paleocontinental shelves, slopes, and deep basins can be reconstructed, giving a first estimate of the paleobathymetry at a given time. Coupled to adequate hypotheses about eustatic sea level changes, as will be discussed later, this allows computation of total subsidence.

Additional hypotheses on the mechanical behavior of the basement have to be made for the subsequent determination of the tectonic subsidence.

Mechanical Behavior of Basement
The main effect of the sedimentary loading is to amplify the subsidence caused by deep seated thermomechanical processes. In plate tectonic theory, the lithosphere consists of rocks in a solid state overlying a molten asthenosphere. In this paper, the term lithosphere corresponds to that part of this solid layer which has an elastic behavior and is generally thinner than in the plate tectonic model.

In order to remove the separate loading effect of sediments and water on the lithosphere (backstripping), it is necessary to consider a model describing the behavior of the lithosphere under loads of large lateral extension. The first of these "isostatic models" was proposed by Airy in 1855, who assumed that the lithosphere had no strength at all. Consequently the crust behaves as a set of columns moving independently of each other (Figure 4a).

Although this very simple model was found accurate enough to account for the observed gravity anomalies in some basins, it is now believed that the lithosphere behaves rather like a rigid plate overlying a fluid asthenosphere. The main arguments supporting this "regional isostatic" behavior come from studies on oceanic features such as seamounts and ridges (see for instance Watts et al, 1975). These studies indicate that the oceanic lithosphere deforms as an elastic plate, of which the flexural rigidity (or the equivalent elastic thickness), which controls the amount and lateral distribution of the deformations, is mainly dependent on the thermal structure of the lithosphere at the time of loading. The elastic thickness of oceanic lithospheres has

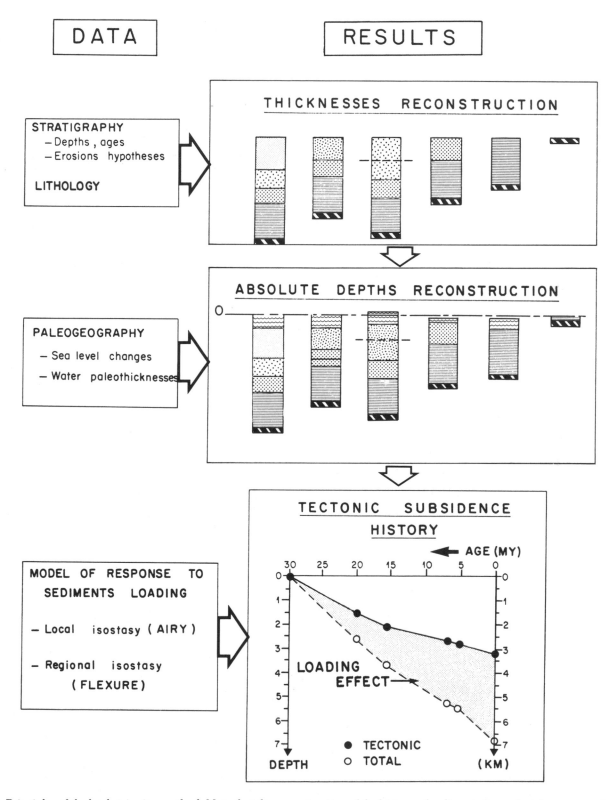

Figure 2. Principles of the backstripping method. Note that the reconstruction of the history of sedimentation only takes compaction into account and does not assume any geodynamic processes.

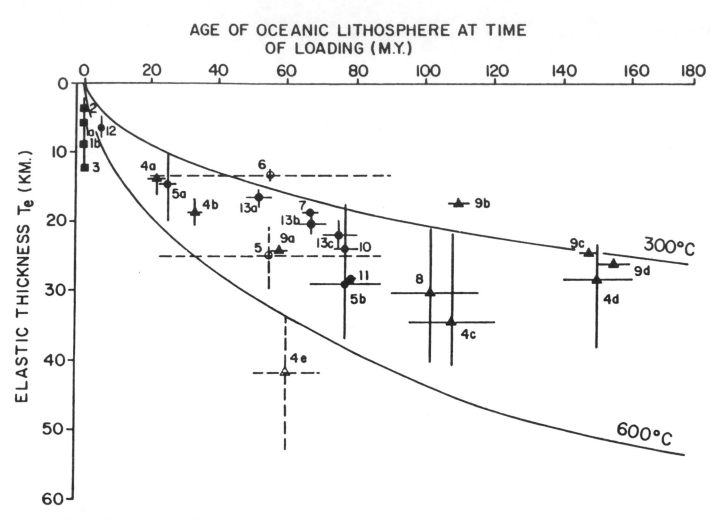

Figure 3. Plot of effective elastic thickness T_e of oceanic lithosphere against age of the lithosphere at the time of loading (from Watts et al, 1980). The estimates of T_e are based on oceanic crustal topography (solid squares), seamounts and oceanic islands (solid circles) and deep sea trench-outer rise systems (solid triangles). The 300 and 600°C isotherms are based on the cooling plate model.

been found to increase with age of loading according to the time-depth relation of an isotherm given by a simple cooling plate model (450°C ± 150°C, see Watts and Steckler, 1981; and Figure 3). Recently, Karner et al (1983) have shown that the continental lithosphere behaves similarly.

For our purpose, we use such estimates of elastic thickness evolution through time (see Beaumont et al, 1982; Watts et al, 1982) to compute the tectonic subsidence, and so estimate the crustal heat-flow with a higher accuracy. Since the lithosphere deflection at a given location in the basin depends on the whole sedimentary load, it is necessary to consider both vertical and lateral distribution of the sediments through time for the subsidence computations (Figure 4).

Eustatic Sea Level Changes

The estimate of sea level changes during basin evolution permits us to compute all crustal vertical movements with respect to a reference level common to its whole history. Many authors have proposed "eustatic curves" on the basis of worldwide correlations of stratigraphic features for various margins (Vail et al, 1977a, b). Watts et al (1982)

have shown that Vail's curves include tectonic effects and so are not exclusively related to eustatic processes. Computing subsidence for several margins in the northern Atlantic, and assuming exponential decay related to thermal contraction of the lithosphere, Watts and Steckler (1981) found that eustatic sea level changes are not so irregular as suggested by Vail. The general trend is a high sea level during the Cretaceous (100 to 300 m; 328 to 984 ft) and a slight decrease up to the present. These variations appear to be mainly controlled by the variation in accretion rate of oceanic ridges.

TECTONIC SUBSIDENCE AND THERMAL STATE OF BASINS RESULTING FROM EXTENSIONAL TECTONICS

Numerous sedimentary basins, such as rift basins and Atlantic-type continental margins, have undergone a rather simple tectonic history. Their structural evolution is characterized by an extensional tectonic phase (rifting), followed by a general subsidence period without active faulting.

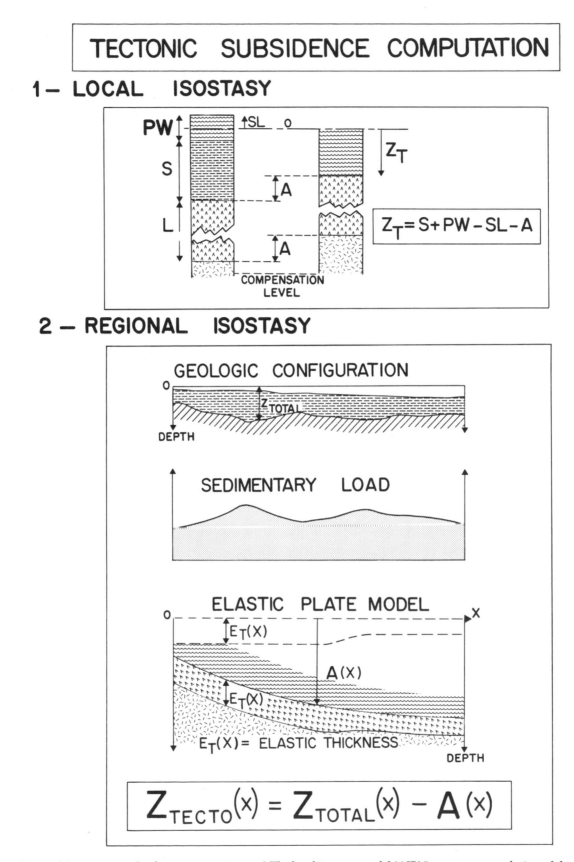

Figure 4. Models used for tectonic subsidence computations: 1) The local isostasy model (AIRY) assumes no evolution of the mechanical properties of the lithosphere; 2) The regional isostasy model assumes a possible increase of flexural rigidity through time. Thus, at a given step, the actual deflection of the lithosphere (A[x]) is inherited from the past history of sedimentary loading. Note that the flexure rigidity is allowed to vary through space.

Passive margins studied by Keen and Barett (1981, Eastern Canada), and Montadert et al (1979, Ireland) exhibit a tectonic subsidence decaying exponentially with time. Similarly, Sclater et al (1980) have shown that the heat flow decreases with time after the rifting phase. Quantitative predictions of subsidence and heat flow patterns have been made based on thermal models of the lithosphere (Sleep, 1971; McKenzie, 1978; Falvey and Middleton, 1981; and Royden and Keen, 1980).

A practical example is the margin offshore Ireland, where for the deepest part of the basin the subsidence history has been studied by Chenet et al (1983). They have shown that subsidence is explained satisfactorily by a 50 percent stretching of the continental crust and a strong thinning of the lithosphere, including the lower part of the crust. These model predictions seem to be in good agreement with the seismic reflection and refraction data.

In order to get a better estimate of the heat flow than obtained by the model of Royden and Keen (1980), our model considers also the heat production by radioactive decay in the continental crust. Basic parameters are the stretching ratio of the continental crust, the thinning ratio of the lithosphere, and the rate of radioactive heat production within the crust.

These parameters are estimated by comparing the tectonic subsidence predicted by the model with the results obtained by the backstripping method. Available data on present heat flow and on the thinning of the continental crust from seismic refraction experiments are also taken into account. These heat flow measurements must be corrected for the effect of sedimentation. Attention must be paid to the fact that the geodynamic model described here deals only with vertical heat transfer within the lithosphere. For this reason, its predictions should only be used where tectonic subsidence does not display sudden lateral changes.

THERMAL HISTORY OF A SEDIMENTARY BASIN

Heat flow evolution, as predicted by geodynamic models or by empirical studies, may be used to determine the temperature history of a basin.

Heat Transfer

The temperature evolution of a sedimentary slice is controlled by its burial, by the heat flow across the sedimentary pile, and by the mode of heat transfer. In most sedimentary basins, conductive heat transfer predominates. Thus the temperature distribution depends on heat flow, burial rate, and thermal conductivity of the sediments.

However, when fluid circulation is important, convective heat transfer may influence the thermal gradient within the sediments. Since the geological data are generally not abundant when the model is used at an early stage of basin exploration, these processes are difficult to quantify. Therefore the application of the thermal model is limited to basins that have not been subjected to tectonic phases leading to important fracturation. Additionally, it is assumed that there is no important convective heat transfer by lateral drainage.

Thermal Model

For the basins meeting these requirements, we built a thermal model that takes vertical heat transfer by conduction and by expulsion of fluids during compaction into account. The variations of the thermal properties of the sediments with burial are also taken into account. The model provides the temperature history of any level of the sedimentary column and is coupled to a kinetic model of OM maturation (Tissot and Espitalie, 1975). The application range is restricted to sedimentary basins of simple structure and large dimensions, and beds that are not strongly dipping. These restrictions are caused by the monodimensional character of the model. A bidimensional model using the same basic principles will be developed which will extend the application possibilities.

Possible Effects of Convection and Rapid Sedimentation Rates

The input of cold sediments at the top of the sedimentary column causes a decrease of the surface heat flow with respect to the bottom heat flow, a transient process which is important in case of rapid sedimentation (1,000 m/million years). Even at lower sedimentation rates the compaction fluids could locally modify the temperature distribution if they are expelled through preferential drainage zones.

Preliminary studies have also been carried out by Perrin (1983a, b) on the possible impact of natural convection. In this process, fluid circulation is caused by temperature related density variations of water. The two-dimensional model that has been developed shows that convection cells are likely to occur in porous and permeable media. The fluid flow greatly disturbs the geothermal gradient and the temperature could remain almost constant, vertically, over hundreds of meters.

APPLICATION TO THE CENOZOIC WESTERN MEDITERRANEAN BASIN

The first set of models—backstripping, geodynamic and thermal—has been applied to the Cenozoic Western Mediterranean basin. This basin includes the onshore Camargue basin, the Gulf of Lion continental margin and the opposite margins of Sardinia and Corsica (Figure 5a). Exploration is more advanced onshore than offshore, where few wells have been drilled. The basin opening results from an initial rifting phase during the Oligocene-lower Miocene (Biju-Duval, Montadert, 1977).

Subsidence History of the Camargue

The tectonic subsidence in the Camargue basin (up to 5,000 m, or 16,404 ft, of Cenozoic sediments, see Figure 5b) is reconstructed by the backstripping model from data on 20 exploratory wells. The subsidence pattern appears to be very similar for the different wells (Figure 5c).

The main tectonic phases affecting the basin may be assessed from the interpretation of the tectonic subsidence. Thus, the Oligocene-lower Miocene phase is characterized by a rather high subsidence rate and active normal faulting. The subsidence rate is much lower during the middle-late Miocene when the whole basin displays a general subsidence without active faulting. The complex pattern from the Pliocene to the present corresponds to the reactivation of faulting in this area, where tilting occurred due to overthrusting in the Alpine chain.

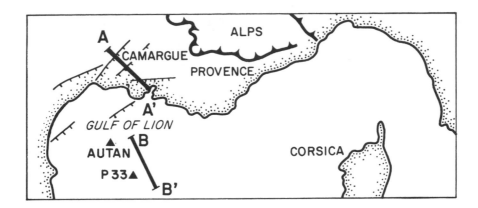

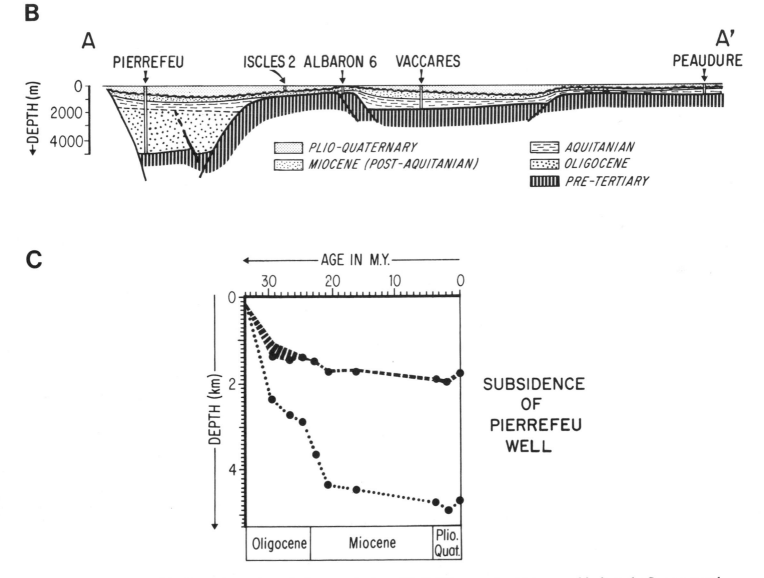

Figure 5a. Location map of the Cenozoic intracratonic Camargue basin and Gulf of Lion margins. Main normal faults in the Camargue and main overthrusts in the Provence are shown.
Figure 5b. Schematic cross section A-A' of the Camargue basin. Note that many normal faults die out in the Aquitanian.
Figure 5c. Subsidence curves of the Pierrefeu well and reconstitution of the progressive burial of the various sedimentary units. Heavy dashed lines—tectonic subsidence; dotted lines—total subsidence of pre-Cenozoic substratum.

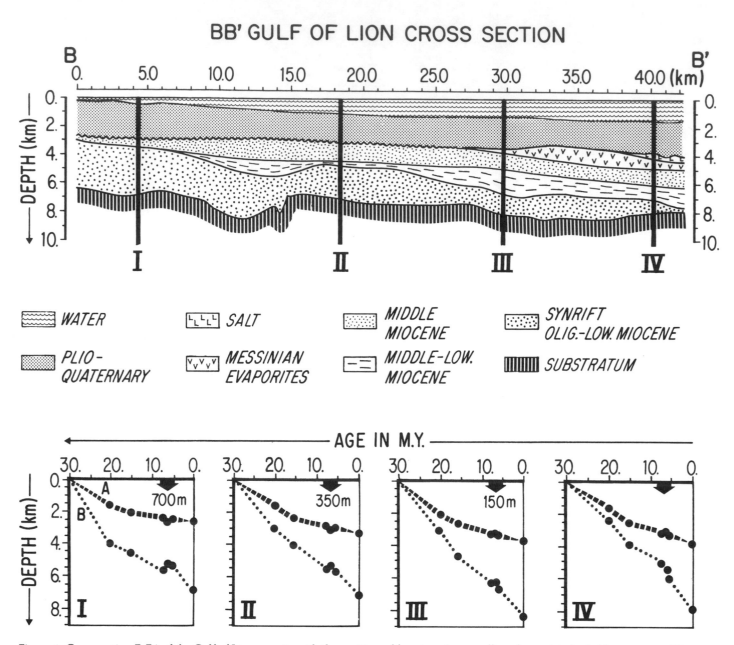

Figure 6. Cross section B-B′ of the Gulf of Lion margin with the position of four imaginary wells and associated subsidence curves. The amount of erosion during the regression at the end of the Miocene is indicated below the arrow.

Subsidence History of the Gulf of Lion Margin

The evolution of the Gulf of Lion passive continental margin is reconstructed at the hand of cross sections evolving through time. The geological cross section of Figure 6 is based on a seismic profile from the continental shelf to the deep basin. The pre-Tertiary basement shows a typical rift structure with horsts, grabens, and tilted blocks. Normal faults die out in the overlying layers. The rifting is completed at the end of the Aquitanian and followed by a sedimentary prism of prograding Miocene deposits. This unit is truncated at the top by an unconformity which corresponds to a major regression at the end of the Miocene, caused by the closure of the Mediterranean basin (Cita, 1982). At this time, evaporitic sediments are deposited in the deepest part of the basin. The thick uppermost unit of

Pliocene to Quaternary age corresponds to the rapid infilling of the basin after the Pliocene transgression. Apart from local salt or clay movements, neither faults nor folds of tectonic origin affecting post-Aquitanian deposits have been detected on the seismic profiles.

The successive cross sections are based on seismostratigraphic interpretation, assuming that basement geometry is conserved from one step to the other during the postrift subsidence period. We also assumed that the tectonic subsidence did not undergo any sudden change in rate between Aquitanian and early Pliocene times, since no active tectonics have been recorded during this period. This hypothesis is consistent with the hypothesis of thermal control of postrift subsidence (McKenzie, 1978). We thus computed the amount of erosion and regression at upper

SECTION B B′

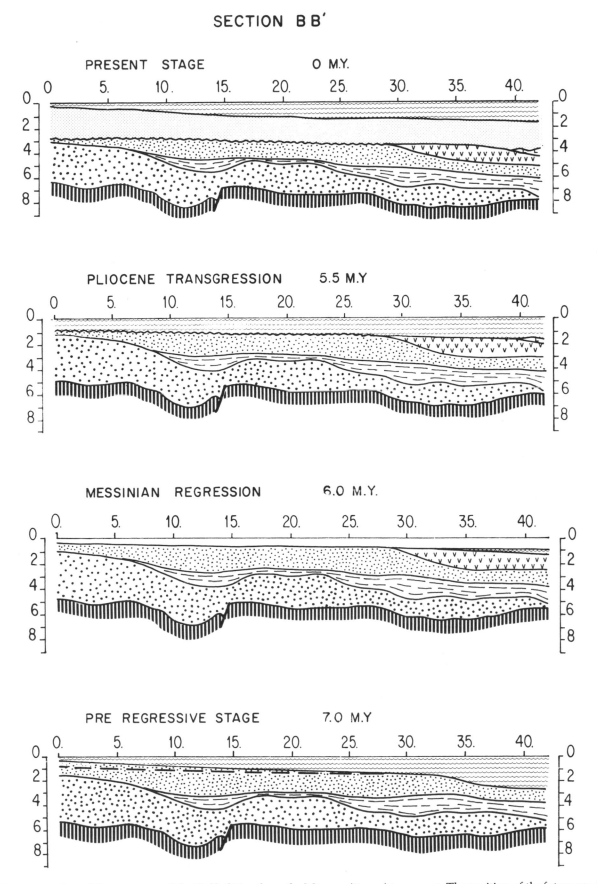

Figure 7. Reconstruction of the geometry of the Gulf of Lion from the Miocene (7 m.y.) to present. The position of the future erosion surface at 6 m.y. is indicated with a dotted line.

Miocene times (Figure 6, below). The successive shapes of the basin—before, during and after the regression event—are the final result (Figure 7).

Heat Flow and Temperature History

The geodynamic model was applied to the Gulf of Lion margin, calibrating its parameters from the tectonic subsidence reconstruction and from surface heat flow measurements. The resulting heat flow variations with time have been used as input data in the thermal model. We computed the current geothermal gradient and the temperature history of the sediments for two locations: the Antan well in the upper part of the margin, and the imaginary well P33 in the lower part (see Figures 8 and 6 for location).

As shown by Figure 8, the computed present-day geothermal gradient matches satisfactorily the measurements in the Autan well. The thermal model also indicates that the effects of the upper Miocene erosion on the present geothermal gradient have completely disappeared. However, the underlying series was cooled during the erosional events (Figure 8c) and the maturation of the organic matter was consequently slowed.

In the case of the P33 "imaginary" well, the computed thermal gradient at the top of the sediments is similar to the measured one. The theoretical gradient at depth is lower than near the surface, due to the high sedimentation rate during the Plio-Pleistocene period and the change with depth of the thermal properties (Figure 8a). High temperatures, around 250°C, are expected at the base of the sedimentary column. This suggests that if any hydrocarbons were formed here, they are presently cracked into gas.

These applications of the backstripping, geodynamic, and thermal models indicate their potential use in oil exploration. First, they enable us to compute the subsidence history of the basin, to estimate the magnitude of the main regressive events, and to determine the corresponding thickness of eroded sediments. Second, they allow us to assess the temperature history with greater confidence, which is important for maturation studies.

This kind of study has already been applied to various sedimentary basins formed by extension. Royden (1982), using similar models, has shown that the study of the subsidence of the Pannonian basin was useful for the reconstruction of the tectonic history and that the maturation predictions satisfactorily matched the geochemical data. Encouraging results have also been obtained by Royden and Keen (1980) on the eastern U.S. Atlantic margin. Of course the temperature results provided by such models may be used in further studies about the migration and accumulation processes, once more geochemical data have become available.

OIL GENERATION AND MIGRATION

As shown by Durand et al (1980) the ultimate recoverable reserves of oil amount to only 1 percent of the oil formed in all the sedimentary rocks presently found on earth. Besides, it is well known that a substantial number of apparently valid traps are not filled with hydrocarbons, even in rich zones. This suggests that the migration process

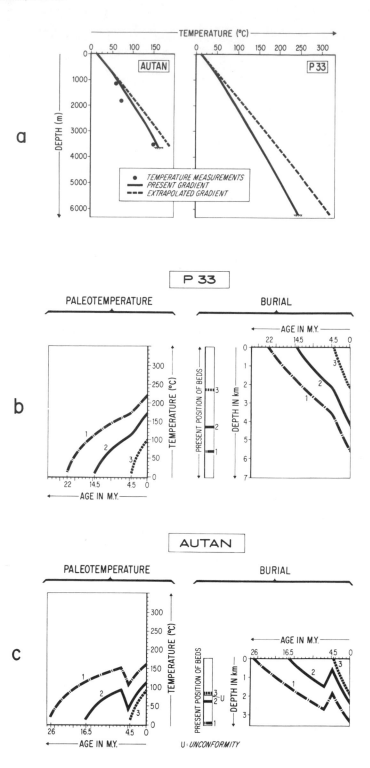

Figure 8a. Computed present-day geothermal gradient for the Autan well, and the P33 imaginary well (location Figure 4). The dotted line represents the temperatures extrapolated from surface measurements.

Figure 8b, c. Reconstruction of the burial history and paleotemperature for three horizons in each well. The main unconformity in Autan corresponds to the strong erosion at the end of the Miocene. The amount of erosion has been computed with the backstripping model. Notice the decrease of temperatures for beds 2 and 3 at the time of erosion, concomitant with a decrease in the overburden.

plays a very selective and dominant role in the occurrence of oil pools.

Physical and Chemical Processes Involved

It is generally accepted that the generation of petroleum results from maturation of the sedimentary organic matter, as controlled by temperature and time. There is no evidence that pressure plays an important role (Tissot and Welte, 1978, p. 183). The modelling of maturation has been described by several authors: Huck and Karweil (1955), Lopatin and Bostick (1973), and Tissot and Espitalie (1975).

On the other hand, migration of oil and especially expulsion from the source rocks is still not well known, although a wide range of mechanisms have been proposed: 1) migration in aqueous solution; 2) migration by diffusion; and 3) migration of hydrocarbons and water in separate phases.

Though the first two monophasic processes have been proposed by various authors in the past, current evidence and data strongly favor the third, biphasic process, which has consequently been adopted for our migration model.

The main driving force for expulsion is thought to be the compaction of the source rock under the sediment load. In the first stage of maturation, the oil saturation within the pore system of the source rock is low, so that the oil or gas phase is discontinuous, and migration is prevented by capillary forces. Later on, provided the petroleum potential is high enough, the hydrocarbon phase becomes continuous, and expulsion can start. Once the hydrocarbons have reached a permeable bed, buoyancy becomes the most important driving force.

This scheme is valid if the hydrocarbon phase is unique, which is probably the case for most source rocks within the oil generating zone (Ungerer, Behar and Discamps, 1981). However, it may not be the case at shallower depth, where the decrease in temperature and pressure can cause separation into a gas phase and an oil phase. Once individualized the gas phase moves much more easily than the oil (Nogaret, 1983). The experimentally measured solubilities of oil in gas are such that important amounts of oil may be carried in solution (Zhuze, 1974). During the upward migration, the oil moved in solution undergoes retrograde condensation when temperature and pressure decrease. At first, the least soluble components are left behind, whereas the light hydrocarbons (more soluble) condense at shallower depths.

Generation and Migration Modelling

Based on the considerations above, the modelling of migration should include the computation of: oil and gas generation, compaction, triphasic fluid flow, and phase behavior of the hydrocarbon mixtures. These four processes are interrelated, and as a consequence the migration model should intergrate all four into a unique system. It should also be at least two-dimensional, since lateral migration of petroleum is generally needed for its accumulation in traps.

This problem is even more complex than the triphasic and three-dimensional computer models that have been developed in reservoir engineering. Effectively, the porous medium is not static, but changing as a result of burial and compaction, and the process takes place on a million-year scale. This high complexity coupled to the desire for separating the different processes involved, made us develop two models instead of one:

- A biphasic migration model (oil/water) accounts for processes 1, 2, and 3, under the assumption that the hydrocarbon mixture is monophasic.
- A thermodynamic model computes the composition of the hydrocarbon phase(s), assuming that they are in equilibrium (process 4).

We have developed basic equations for the oil generation and migration. The generation of oil is computed from the kinetic model published by Tissot and Espitalie (1975), with the difference that our model makes no provision for secondary degradation of hydrocarbons (dismutation into gas and carbonaceous residue). In one of the examples we simulated gas formation, modifying the initial oil potential, multiplying them by a coefficient which is assumed to account for the secondary cracking yield.

Hydrocarbon expulsion is influenced by pressure in the compaction process. Thus it is not possible to simply use porosity versus depth relationships as done by backstripping, as this method does not allow for possible undercompaction. Following Smith (1971), compaction is represented in our model by the law of effective stress, whereby effective stress is considered as a function of porosity. The function differs from one lithology to another, reflecting the fact that different rocks compact in different ways.

Darcy's law, extended to polyphasic flows through the relative permeability concept, has been adopted for our model. According to Matheron (1967), application of Darcy's law to very slow monophasic flows is perfectly justified. Its extension to polyphasic flows by the relative permeability concept has, on the other hand, no theoretical justification and we have to consider it as a pragmatic solution only.

Pressures in the hydrocarbon and water phases are assumed to be linked by the capillarity equation, the capillary pressure sign depending on the direction of flow. In our case it can be taken as positive, for we generally find a growing oil saturation, whether in source rocks where hydrocarbons are generated, or carrier beds. The intrinsic permeability of the porous medium is computed from the porosity and the average grain size of the rock. Source rocks or permeable beds through which fluids pass in the course of secondary migration are treated similarly, taking anisotropy into account. Furthermore, in the source rock kerogen is converted into oil, ultimately leading to an increased hydrocarbon pressure within its pores (Du Rouchet, 1981).

When pressure (P) increases in a porous rock, rupture occurs when one of the effective stress components becomes negative. This is evidenced either by creep (diapirism), or by the opening of fissures. These fissures, through which fluids may flow, close as soon as pressure drops. To account for this process, we assumed that if pressure exceeded the minimum stress S_3, rock permeability increased in proportion to the square of the difference $(P-S_3)$.

The data required by the model are mainly the geometry of the cross section, the age and lithology of the stratigraphic levels, the subsidence and temperature history, and finally the initial petroleum potential of the various

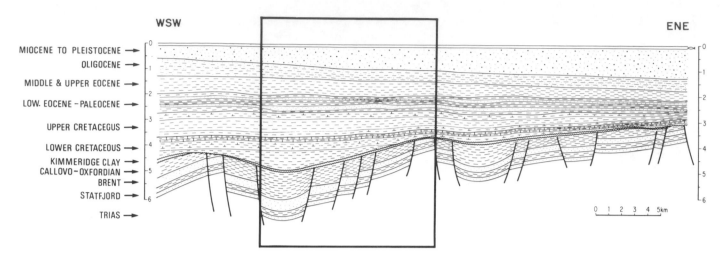

Figure 9. Schematic section of part of the Viking graben in the North Sea. The framed zone was selected for application of the migration model. It is bounded by the two largest faults which define a tilted block below the major unconformity.

sediments. The results consist of a reconstruction of compaction (porosity, pressure) of oil generation (maturation level) and of migration (oil saturations in source rocks and reservoirs).

APPLICATION EXAMPLES

Fault Block in Viking Graben, North Sea

Figure 9 represents a typical situation in the southern part of the Viking Graben (Frigg). Source rock units correspond to the Kimmeridge Clay, Heather and Dunlin formations, as well as the Brent coals, while three permeable units correspond to the Brent, Statfjord, and Paleocene sands. The section is classified by a finite element grid, using six types of lithology: sand, sandy silt, silty clay, clay, marl, and coal, with lithological and geochemical characteristics shown in Table 1.

A relatively high oil potential is assigned to the Brent coals, for analysis indicates that they are richer in hydrogen than is type III of Tissot et al (1974). A geothermal gradient of 30°C/km and a surface temperature of 5°C have been taken. It is also assumed that the faults bounding the block form impermeable barriers, but that the faults inside the block do not interrupt the continuity of the Jurassic sands.

The model plots through time the depth, pressure, and oil saturation for each element of the grid, thus presenting a scenario for oil accumulation over geological time. Figure 10 compares the state of oil saturation (Figure 10a) and pressure (Figure 10b) in the Maastrichtian, when according to the model, hydrocarbons were starting to accumulate in the upper part of the Brent sands, with the present-day situation.

Four classes of oil saturation are represented: 0-5 percent, 5-10 percent, 10-20 percent, 20 percent and over. The model indicates for the present time the existence of hydrocarbon accumulations in the upper part of the Middle Jurassic sandstone reservoirs, and also high oil saturations in source rocks in synclinal positions. The source rocks in which saturation is highest are the Brent coals, because these have the highest organic carbon content. These are also the first to expel oil, which occurs mainly where oil saturation

Table 1. Lithological and geochemical characteristics for Viking Graben, North Sea, formations.

Formation	Lithology	Mean Organic Carbon Content (% wt)	Mean Initial Oil Potential (mg/g OC)
Eocene-Paleocene	Sand	0	0
	Silty sand	0.5	210
	Silty clay	0.7	210
	Clay	1	
Senonian	Marl	0.7	70
	Clay	0.7	70
Turonian	Marl	1.5	70
Cenomanian	Marl	1.5	70
Lower Cretaceous	Clay	1.5	70
Upper Jurassic Kimmeridge	Clay	7	310
Heather	Clay	6.5	210
Middle Jur. Brent	Sand	0	0
	Coal	30	220
Lower Jurassic Dunlin	Clay	3.5	210
Statfjord	Sand	0	0

exceeds 20 percent (Figure 15). Subsequently, the Kimmeridge Clay, Heather and Dunlin source rocks in turn supply the reservoirs. In sands with high intrinsic permeability, a low relative permeability will suffice to cause oil displacement. Secondary migration, under the

Figure 10a, b. Representation of the main results of the model applied to the Viking graben: a) oil saturation; b) excess pressure relative to hydrostatic. The first stage (at the top) corresponds to the beginning of oil expulsion (Maastrichtian). The top of the oil generation zone corresponds to a 10 percent conversion ratio of the potential. At the present-day stage, a pool has formed in the permeable levels of the Jurassic.

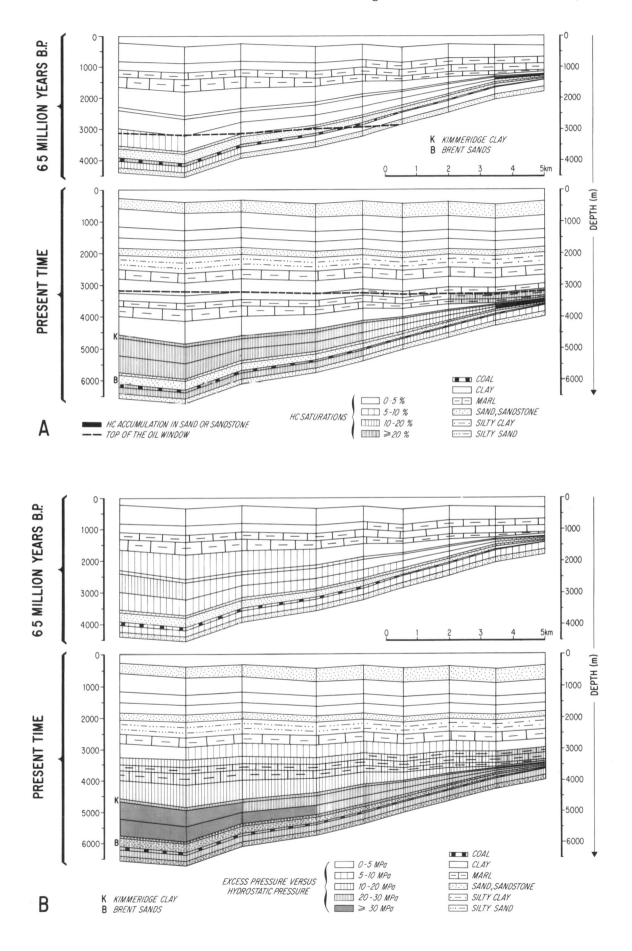

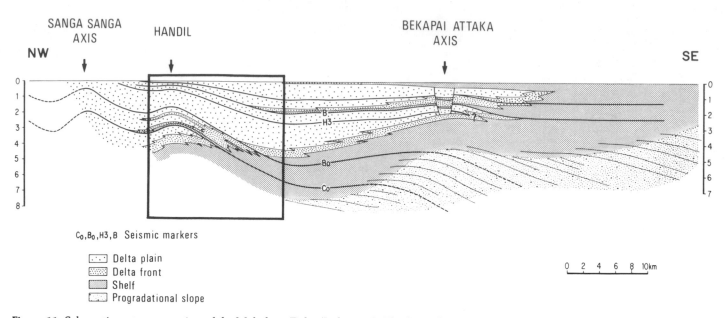

C₀,B₀,H3,B Seismic markers

Delta plain
Delta front
Shelf
Progradational slope

Figure 11. Schematic east-west section of the Mahakam Delta (Indonesia). The framed zone was selected for application of the migration model.

effect of buoyancy, therefore occurs with low residual saturation in the beds passed through (1 to 2 percent). The model also forecasts a dysmigration through the clay overburden of the upper part of the Brent reservoirs, which is only several tens of meters thick.

Five classes of excess pressure relative to hydrostatic are calculated: 0-5 MPa, 5-10 MPa, 10-20 MPa, 30 MPa and over. Imposed boundary conditions are: zero flow at base and through the faults bounding the block, lateral flow possible by gradual reduction of hydraulic resistance above the Lower Cretaceous, and a hydrostatic system at the upper boundary of the grid. The model calculates a hydrostatic system in the Paleocene sands and slight excess pressure in their overburden, as well as considerable excess pressure from 3,000 m (9,843 ft) down. These forecasts are in agreement with observational data. Due to the transmission, toward the top of the structure, of excess pressure treated by source-rock compaction and hydrocarbon generation, dysmigration is focused there.

Absolute excess pressure is highest in the synclinal part of the Callovo-Oxfordian Heather formation. The model thus indicates that, with the geological assumptions made at the outset, expulsion from this formation should be downward into the Brent sands. However, if there would be a lightly more permeable bed within the Kimmeridge Clay-Heather unit, this would be enough to cause lateral migration toward the structural top of the source-rock unit. This sensitivity stresses the impact of slight lithological or geometric variations on the direction of flow. Such minor variations are unfortunately impossible to predict if no wells have been drilled in the synclinal zones.

In this example, we also varied the hydrocarbon density and viscosity values to simulate a gas phase. Our objective was to understand what consequences these variations could have on the amount of possible dysmigration through the Brent reservoir seals. At the same time, we reduced the oil generating potential of the various source rocks to produce approximately the same amounts of hydrocarbons

as would have been generated if gas had been produced. Without modifying the other parameters, the model calculates a migration into the Paleocene sands of quantities of "gas" of the same order of magnitude as found in Frigg. This method, though open to criticism from the theoretical viewpoint, does reveal that the possibilities of gas displacement are much higher than for oil, all other things being equal.

The Mahakam Delta, Indonesia, a Tertiary Delta

The geological section of Figure 11 represents a situation described by Durand and Oudin (1979). The middle Miocene to Pliocene sequence is composed of interbedded sands, silty clays, and coals. The facies pattern shows a continuous progradation of the delta from west to east, with the exception of a small transgressive episode. Sand continuity is high and conditions for secondary migration are therefore favorable. On the Handil axis and in the surrounding synclinal areas, sufficient depths are reached to enable large quantities of hydrocarbons to be generated. Since the number of elements in the grid is restricted (25 vertically and 8 horizontally) we could not represent the small (meter) scale lithological features, which are common in such deltaic formations. In generalizing, we assumed that the delta plain sands were interconnected up to a present

Figure 12a, b. Representation of the main results of the model in the case of the Handil structure. a) Oil saturation. The coals, too thin to be properly represented, have oil saturations over 20 percent in the oil generation zone. The top of this zone corresponds to a 10 percent conversion ratio of the oil potential. At the present-day stage, pools have been formed in the sands and limestones at the top of the Handil structure between 1,500 and 3,000 m (4,921 to 9,843 ft), as well as in a stratigraphic trap. b) Excess pressure relative to hydrostatic.

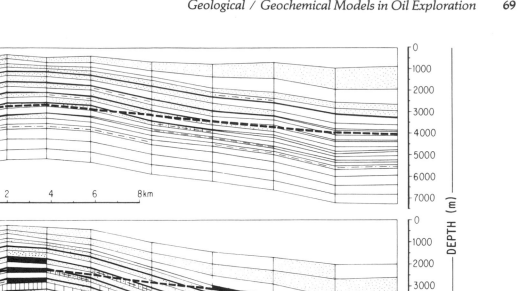

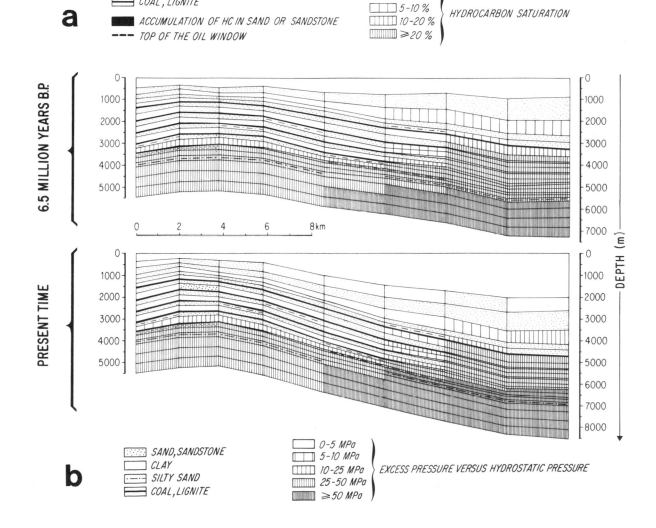

SAND, SANDSTONE
CLAY
SILTY SAND
COAL, LIGNITE
ACCUMULATION OF HC IN SAND OR SANDSTONE
TOP OF THE OIL WINDOW

0-5 %
5-10 %
10-20 %
≥20 %
} HYDROCARBON SATURATION

a

SAND, SANDSTONE
CLAY
SILTY SAND
COAL, LIGNITE

0-5 MPa
5-10 MPa
10-25 MPa
25-50 MPa
≥50 MPa
} EXCESS PRESSURE VERSUS HYDROSTATIC PRESSURE

b

Table 2. Lithological and geochemical characteristics of Handil axis and surrounding synclinal areas, Mahakam Delta, Indonesia.

Facies	Lithology	Mean Organic Carbon Content (% wt)	Mean Initial Oil Potential (mg/g OC)
Prodelta	Clays	0.0	0
Delta front	Clays	1.5	310
Delta plain	Silty sands	0.0	
	clays	2.5	310
	Clays (+ coals)	5.5	310
	Coals	35.0	310

depth of 2,800 m (9,186 ft) across the Handil structure and deeper on its flanks. The proportion and distribution of coals has been plotted to the limits of the model. Four lithologies have been defined: clay, silty sand, sand, and coal, with the lithological and geochemical characteristics shown in Table 2.

Based on available data, the geothermal gradients decline from the top of the Handil structure (31°C/km) shown to the synclinal axis (23°C/km). A surface temperature of 27°C is assumed. Applied boundary conditions are as follows: zero flow at the base of the grid and at its right hand vertical limit (syncline between the Handil and Bekapai structures), hydraulic resistance low at the left hand vertical limit down to about 2,500 m (8,202 ft). It proved much more difficult to put this section together than the previous one, because the vertical dimensions of the lithological elements are small compared to the grid dimensions.

Figure 8a shows oil saturations at −6.5 million years, the time at which the model forecasts the beginning of hydrocarbon expulsion from the source rocks, as well as present time. As in the Viking Graben case, expulsion starts first in the coals. The model suggests that the accumulated oil comes mainly from the coal beds (which again are richer in hydrogen than type III of Tissot et al, 1974), a logical consequence not only of their quantitative importance, but also because of a basic assumption on which our model is based—that expulsion is more rapid and efficient from an organic-rich sediment than from a lean one.

Repeated runs of the model calculated the present-day existence of several superimposed accumulations at the top of the structure or in flank positions, at depths corresponding to those of the real accumulations. The calculated quantities of accumulated hydrocarbons, extrapolated from the two-dimensional model to the three-dimensional real world, are of the same order of magnitude as those observed.

The calculated excess pressure (Figure 12b) shows two distinct zones: 1) the delta plain where sand continuity ensures a fully hydrostatic system; and 2) the delta front and prodelta where excess pressure is forecast.

The pressure contrast is very large at the top of the Handil structure at the transition from delta plain to delta front, where pressures reach values causing hydraulic fracturation. Figure 13 shows that the computed pressures are in agreement with the observations at Handil, with geostatic pressures below 3,000 m (9,842 ft). This matching

has been obtained by assuming that the sand beds were interconnected in the sand-rich facies, and allows us to predict the depth at which the geostatic pressures can be found in undrilled areas, such as in the syncline. The fluids move within the undercompacted zone from the syncline toward the top of the structure, because permeability is higher parallel to stratification. Vertical expulsion of these fluids from the high pressure zone occurs mainly across the anticline, where hydraulic fracturation is focused.

A large number of model runs have been made on this example in order to assess the role of the main parameters. We were thus able to quantify the impact of modification, such as permeability restrictions in migration channels, or changes in source rock distribution, or the values assigned to hydrocarbon viscosity. We were also able to determine the sensitivity of the thermal system: a 10 percent increase in the geothermal gradient results in a doubling of oil accumulation.

Possible Applications of the Migration Model

In these two modelled cases it has proved possible to account for observed data such as: location of accumulations, quantities of hydrocarbons in place, and current pressure systems, without the need for substantial modification of the parameters when switching from one case to the other. This seems to indicate that the physical and numerical approximations used are relatively satisfactory. However, in a model like ours, which involves a large number of parameters, it is always possible to use it to explain a variety of situations, as only a slight modification in lithology, geothermal gradient, or viscosity results in markedly different outcomes. We must therefore take care not to use it in its present state to describe situations where little geological knowledge and no data to check results are available, especially if quantitative results are desired. In particular, geochemical data have to be sufficient for a reasonably accurate assessment of the petroleum potential of the source rocks. For the time being, this is also necessary for any model claiming to describe migration. Furthermore, its resolving power is limited and it is illogical to try to process examples in which lithological features are much smaller in size than the possible grid units considered. In this respect, the Mahakam Delta is a borderline case.

The value of a model like ours is that in the case of well understood examples, it enables the researcher to synthesize the various types of data and to deduce their consequences with respect to generation and migration of oil. It is in this way that the examples processed have revealed the considerable importance of the thermal system. The interest in describing the thermal history of sedimentary basins better than is usually done is clearly apparent.

THERMODYNAMIC MODEL

The purpose of the thermodynamic model is to determine the phase behavior of the HC mixtures as a function of pressure and temperature along the migration paths: it determines the composition of the gas and liquid phase, provided the pressure, temperature, and bulk composition of the mixture are known. The pressure in the subsurface is

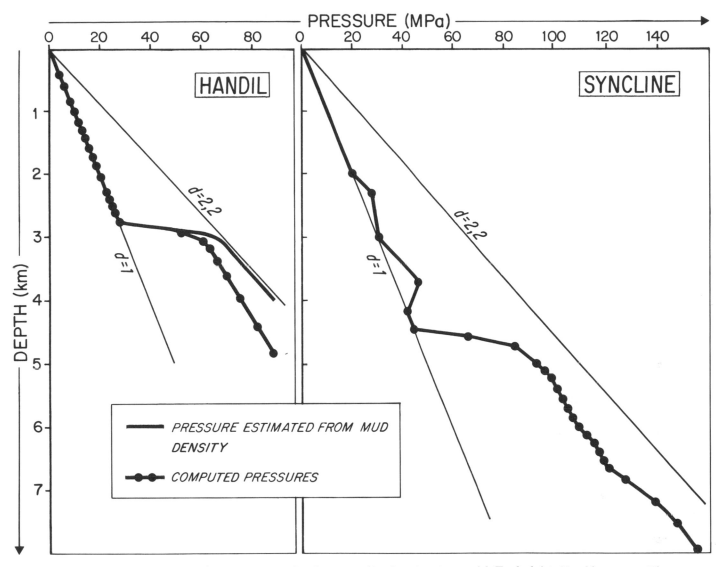

Figure 13. Comparison of pressure gradients as measured and computed by the migration model. To the left is Handil structure. The abnormal pressure zone occurs at 2,800 m (9,186 ft). To the right is synclinal location. The high pressure zone is predicted at 4,600 m (15,092 ft).

generally so high that the gas phase has liquid-like properties. For instance, at P = 30 MPa, and T = 70°C, the density of methane is about 0.18 g/cu cm, which is higher than the density of liquid methane below boiling point (0.15 g/cu cm). This justifies the use of the Regular Solution Theory, which was developed by Hildebrand (1923) to describe the solubility of liquids. The resulting model is mostly applicable to nonpolar hydrocarbons for pressures in excess of 1,000 to 2,000 m (3,281 to 6,562 ft).

Application of the Thermodynamic Model to Handil Structure, Mahakam Delta

The thermodynamic model is used to determine at what depth a typical oil/gas mixture is monophasic. At Handil, it appears that for an oil/gas ratio of 1:3 in weight, the mixture is monophasic for depths in excess of 2,800 m (9,186 ft). This is due to the pressure increase, which favors miscibility. For other locations, the boundary of the monophasic and biphasic state matches roughly with the

upper limit of the high pressure zone, which occurs between 2,800 m (9,186 ft) at the top and 4,500 m (14,764 ft) in the syncline (Figure 13).

Another use of the model is to compute the composition of the oil and gas phases during vertical migration. As mentioned earlier, the condensed oil is thought to remain mostly trapped while the gas undergoes dysmigration. The results (Figure 14) show that the initial mixture which is monophasic at 3,000 m (9,842 ft) in the high pressure zone, separates out into a heavy oil phase and a gas phase when the temperature decreases; upward, successively lighter oils are condensing. Their amount is far from negligible, since 12 percent of the initial oil is condensed between 1,000 and 2,000 m (3,281 to 6,562 ft). A consequence is that the migration of oil in gaseous solution is quantitatively important, at least in some cases. These results are also considered to explain: 1) the origin of some very light oils at low depths; 2) the increase with depth of condensate content in natural gases (Durand and Oudin, 1979); and, 3) the

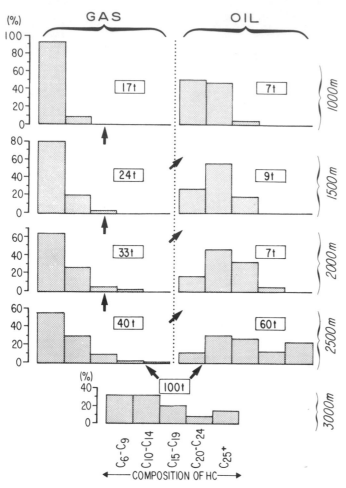

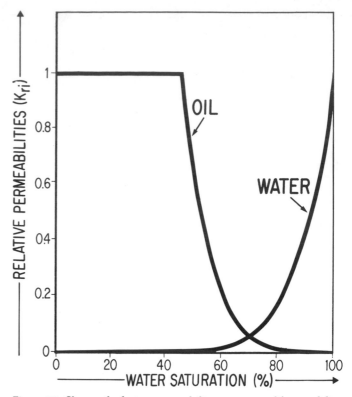

Figure 15. Shape of relative permeability curves used for modeling. For low saturation values, relative permeability is very low but not nil.

Figure 14. Evolution of the composition of the oil and gas phases during vertical migration, as computed by the thermodynamic model. The initial composition of the mixture is represented by the deep condensates that are found in the Mahakam Delta in the high pressure zone. Important amounts of successively lighter oils are condensed during the upward migration of the gas phase.

amount of the distillate fraction in reservoired oils, which shows a slight but unusual decrease with increasing depths (Oudin and Picard, 1982).

Finally it should not be forgotten that this kind of migration is subordinate to the generation and migration of important amounts of gas from the deepest parts of the basin. The reliability of the results depends presently on the estimation of these amounts rather than on the accuracy of the thermodynamic model itself.

CONCLUSIONS

The reduction of exploration risks implies better prediction of the zones, structures, or stratigraphic horizons favorable for oil and gas accumulation, so that the location of expensive geophysical surveys or exploratory wells can be optimized.

During early stages of exploration, the available geological information is scanty and incomplete. There are almost no well data and few seismic cross sections where some reflectors may be traced from not too distant explored zones. Seismostratigraphic interpretation may give an idea of the lithologies, and knowledge of worldwide anoxic

events may help in locating possible source rocks. At this stage, the models can operate in two ways:
- In the first place, the backstripping model helps to restore the evolution of the geological cross sections with time. The shape of the basin is determined at each step and the position as well as the time of formation of the traps is better understood.
- In the second place, our models allow us to depict the geodynamic and thermal evolution of the basin. The maturation model then gives an estimation of the oil generating zone on the geological cross sections, at each step of basin evolution.

These early models allow us to answer some important questions: 1) Which are the zones of the basin where the generation of oil is probable?; and, 2) If determined, does the generation occur later than trap formation?

At this stage, a selection can be made of the zones which have been charged by the possible source rocks. However, more information is needed, mainly through exploratory wells, in order to predict the most favorable structures. Once temperature data are available and the reservoirs, source rocks and the type of organic matter are better known, the migration model may be applied and helps in answering the following questions: 1) What is the time span between oil generation and its possible accumulation in the traps?; 2) What are the key parameters that control the volume of accumulated HC?; 3) Considering the time of expulsion of HC from the source rocks, are the accumulations oil or gas?; and, 4) Where and at what depth is there a risk for high pressures? In an ultimate stage, the thermodynamic model helps in determining the condensate

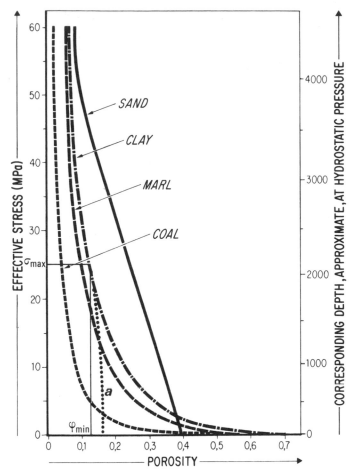

Figure 16. Effective stress versus porosity curves used in studying a North Sea block. If effective stress declines, the curve followed (a) is different from the initial one and joins it at the point of maximum compaction reached ($\emptyset_{min}$, σ_{max}).

content of the gas and the zones where retrograde condensation of oil is most probable.

However, referring to the four validity conditions mentioned in the introduction, it should be emphasized that the predictions given by the various models strongly depend on the input data and on the assumptions made when some geological parameters are not available. One of the main advantages of the mathematical models is their ability to test the influence of uncertainties in the input. They allow us to predict the structures which are in any cases favorable, whatever the uncertainties of the geological data may be. On the other hand, data acquisition can be directed toward those key parameters which control the prediction of the more favorable zones.

REFERENCES

Beaumont, C., C. Keen, and R. Boutilier, 1982, On the evolution of rifted continental margins; comparison of models and observations for the Nova Scotia Margin: Geophysical Journal of the Royal Astronomical Society, v. 70, p. 667-715.

Biju-Duval, B., and L. Montadert, eds., 1977, Structural history of the Mediterranean Basin, *in* 25th Congress of the International Commission for the Scientific

Exploration of the Mediterranean Sea (Split, Yugoslavia): Paris, Ed. Technip, 448 p.

Chenet, P.Y., et al, 1983, Extension ratio measurements on the Galicia Portugal and N. Biscay continental margins; implications for some evolution models of passive continental margins, *in* J.S. Watkins and C.L. Drake, eds., Studies in continental margin geology: AAPG Memoir 34, p. 703-716.

Cita, M.B., 1982, The Messinian salinity crisis in the Mediterranean, a review, *in* H. Buckhemer, ed., Alpine Mediterranean geodynamics: Washington, D.C., Geodynamic Series, American Geophysical Union, v. 7, p. 113-140.

De Charpal, D., and Devaux, 1981, Diagenèse des calcaires granulaires de Bassin Parisien et du Nord de l'Aquitaine; conséquences sur les propriétés du réservoir: l'Institut Francais du Pétrole Internal Report.

Durand, B., and J.L. Oudin, 1979, Exemple de migration des hydrocarbures dans use série deltaique; le delta de la Mahakam, Indonésie: Bucarest, Communication au 10eme Congress Mondial du Pétrole.

—— et al, 1980, Kerogen, insoluble matter from sedimentary rocks: Paris, Technip, 519 p.

Du Rouchet, J., 1981, Stress fields, a key to oil migration: AAPG Bulletin, v. 65, p. 74-85.

Falvey, D., and M.F. Middleton, 1981, Passive continental margins; evidence for a prebreakup deep crustal metamorphic subsidence mechanism: Oceanologica Acta, n° sp., Colloque C3, 26ème C.G.I., p. 103-104.

Hildebrand, J.M., 1929, Regular solution theory: Journal of the American Chemical Society, p. 51-66.

Huck, G., and J. Karweil, 1955, Physikalische problem der inkohlung: Brennst.-Chem., v. 36, p. 1-11.

Karner, G.D., M.S. Steckler, and J.A. Thorne, 1983, Long-term thermomechanical properties of the continental lithosphere: Nature, v. 304, p. 250-253.

—— and A.B. Watts, 1982, On isostasy at Atlantic-type continental margins: Journal of Geophysical Research, v 87, p. 2923-2948.

Keen, C.E., and D.L. Barett, 1981, Thinned and subsided continental crust on the rifted margin of eastern Canada; crustal structure, thermal evolution and subsidence history: Geophysical Journal of the Royal Astronomical Society, v. 65, p. 443-465.

Lopatin, N.V., and N.H. Bostick, 1973, Geology in coal catagenesis, *in* Nature of organic matter in recent and ancient sediments: Moscow, Symp. Nauka, p. 79-90 (in Russian).

Magara, K., 1978, Compaction and fluid migration, *in* Practical petroleum geology developments in petroleum science 9: Amsterdam, Elsevier.

Matheron, G., 1967, Eléments pour une théorie des milieux poreux: Masson, 164 p.

McKenzie, D., 1978, Some remarks on the development of sedimentary basins: Earth and Planetary Science Letters, v. 40, p. 25-37.

Montadert, L., et al, 1979, North east Atlantic passive margins; rifting and subsidence processes, *in* M. Talwani, ed., Deep drilling results in the Atlantic ocean; continental margins and paleo environment: Washington, D.C., American Geophysical Union, Series 3.

Nogaret, E., 1983, Solubilité des hydrocarbures dans les gaz comprimés; application à la migration de pétrole dans les bassins sédimentaires: Paris, Thesis, May, Ecole des Mines.

Oudin, J.L., and P.F. Picard, 1982, Genesis of hydrocarbons in the Mahakam Delta and the relationship between their distribution and the over pressured zones: Jakarta, Indonesia, Presented at the XIth Annual Indonesian Petroleum Association Convention.

Perrier R., and S. Quiblier, 1974, Thickness changes in sedimentary layers during compaction history: AAPG Bulletin, v. 58, p. 507-520.

Perrin, J.F., 1983a, Trasferts thermiques dans les bassins sédimentaires: Revue de l'Institut Francais du Petrole, v. 38.

——— , 1983b, Modélisation du champ thermique dans les bassins sédimentaires; application au bassin de la Mahakam, Indonésie: University of Bordeaux, Doctoral Thesis.

Royden, C., 1982, Subsidence and heat-flow in the Pannonian Basin: Boston, MA, Massachusetts Institute of Technology, Ph.D. Thesis.

——— and C.E. Keen, 1980, Rifting processes and thermal evolution of the continental margin of eastern Canada determined from subsidence curves: Earth and Planetary Science Letters, v. 51, p. 343-361.

——— , J. Sclater, and R.P. Von Herzen, 1980, Continental margin subsidence and heat-flow; important parameters in formation of petroleum hydrocarbons: AAPG Bulletin, v. 64, p. 173-187.

Sclater, J.G., C. Jaupart, and D. Galson, 1980, The heat-flow through oceanic and continental crust and the heat loss of the earth: Review of Geophysics and Space Physics, v. 18, p. 269-311.

Sleep, N.H., 1971, Thermal effect of the formation of Atlantic continental margins by continental break-up: Geophysical Journal of the Royal Astronomical Society, v. 24, p. 325-350.

Smith, J.E., 1971, The dynamics of shale compaction and evolution of pore fluid pressure: Mathematical Geology, v. 3, p. 239-269.

Tissot, B., et al, 1974, Influence of nature and diagenesis of organic matter in formation of petroleum: AAPG Bulletin, v. 58, p. 499-506.

——— and J. Espitalie, 1975, L'évolution thermique de la matière organique des sédiments; applications d'une simulation mathématique: Revue de l'Institut Francais du Petrole, v. 30, p. 743-777.

——— and D.H. Welte, 1978, Petroleum formation and occurrence: Springer-Verlag.

Ungerer, P., F. Behar, and D. Discamps, 1981, Tentative calculation of the overall volume of organic matter; implications for primary migration: Bergen, Proceedings of the 10th International Meeting of Organic Geochemistry.

Vail, P.R., R.M. Mitchum, Jr., and S. Thompson, III, 1977a, Seismic stratigraphy and global changes of sea level, *in* C. Payton, ed., Seismic stratigraphy—applications to hydrocarbon exploration: AAPG Memoir 26, p. 83-97.

——— , ——— , and ——— , 1977b, Relative changes of sea level from coastal onlap, *in* C. Payton, ed., Seismic stratigraphy—applications to hydrocarbon exploration: AAPG Memoir 26, p. 36-71.

Waples, D.W., 1980, Time and temperature in petroleum formation; application of Lopatin's method to petroleum exploration: AAPG Bulletin, v. 64, p. 916-926.

Watts, A., and M.S. Steckler, 1981, Subsidence and tectonics of Atlantic type continental margins: Oceanologica Acta, N° SP., Colloque C3, Géologie des Marges Continentales, 26éme C.G.I., p. 143-154.

——— , J.R. Cochran, and G. Selzer, 1975, Gravity anomalies and flexure of the lithosphere; a three-dimensional study of the great metear seamount, northeast Atlantic: Journal of Geophysical Research, v. 80, p. 1391-1398.

——— , G.D. Karner, and M.S. Steckler, 1982, Lithospheric flexure and the evolution of sedimentary basins: Philosophical Transactions of the Royal Society of London, A305, p. 249-281.

——— and W.B.F. Ryan, 1976, Flexure of the lithosphere and continental margins basins: Technophysics, v. 36, p. 25-44.

——— and M.S. Steckler, 1979, Subsidence and eustasy at the continental margin of eastern North America: Washington, D.C., American Geophysical Union Maurice Ewing Symposium Series, v. 3, p. 218-234.

Welte, D.H., and M.A. Yükler, 1981, Petroleum origin and accumulation in basin evolution; a quantitative model: AAPG Bulletin, v. 65, p. 1387-1396.

Woodside, M., 1971, The thermal conductivity of porous media: Journal of Applied Physics, v. 32.

Zhuze, T.P., and Bourova, 1977, Influence des différents processus de la migration primaire des hydrocarbues sur la composition des pétroles dans les gisements, *in* J. Goni and E. Campos, eds., Advances in organic geochemistry: Madred, ENADIMSA, p. 493-499.

APPENDIX 1: BACKSTRIPPING METHOD[1]

Reconstruction of Burial

The volume of solid matrix Vs in a porous sediment of volume V with porosity Ø may be written as: $Vs = (1 - Ø)V$. When dealing with the evolution of the thickness Δh of slice of sediments during its burial, Perrier and Quiblier (1974) have shown that the thickness Δh is related to the porosity/depth relationship $Ø(Z)$ and the thickness of the solid matrix: Δh_s in the following form:

$$\Delta h_s = \int_Z^{Z + \Delta h} \{'1 - Ø(Z)\} \, dz \qquad (1)$$

Assuming Δh_s remains constant during the burial history, the Δh variations with depth may be computed after formula (1).

---

[1] The equations used in the backstripping method have been taken from Perrier and Quibler (1974) concerning the reconstitution of burial of a slice of sediments and were inspired from Watts and Steckler (1981) and Timoschenko et al (1959) about the tectonic subsidence computation.

Tectonic Subsidence

The tectonic subsidence Z_T is related to the total subsidence Z and the deflection A due to the load of sediments by:

$$Z_T = Z - A$$

Assuming local isostatical compensation, the deflection A will be:

$$A = S\left(\frac{\rho s - \rho w}{\rho m - \rho w}\right) + \Delta SL \frac{\rho w}{\rho m - \rho w}$$

(from Watts and Steckler, 1981).

Assuming regional isostasy, the deflection A(x) along the horizontal axis x is obtained by the equation of behavior of a thin elastic plate.

The deflection w(x) due to a point load P, located at x_o is given by:

$$D\frac{d^4 w(x)}{dx^4} + (\rho m - \rho w)g \; w(x) = \rho \delta(x - x_o)$$

(from Timoshenko et al, 1959)

with δ: Dirac function

$$D = \frac{ETe^3}{12(1 - \sigma^2)}$$

The total deflection A(x) is obtained by:

$$A(x) = \sum_{\text{point load}} w(x)$$

the point load P being given by:

$$P(x_o) = \left[S(x_o)\frac{\rho s(x_o) - \rho w}{\rho m - \rho w} + \Delta SL\frac{\rho w}{\rho m - \rho w}\right]\Delta x_o$$

with Δx_o: width of an elementary sedimentary column.

with S   = sediment thickness corrected for compaction (m)
ρs = mean density of sediments (kg cu m)
ρm = mean density of mantle (kg cu m)
ρw = mean density of water (kg cu m)
ΔSL = sea level relative to the present day (m)
g   = average gravity $(m.s^{-2})$
D   = flexural rigidity $(N - m)$
E   = Young's Modulus (Pa)
Te   = elastic thickness (m)
σ = Poisson's ration (no dimension).

APPENDIX 2: THERMAL TRANSFER IN SEDIMENTS[2]

The model including heat transfer by conduction and convection due to the expulsion of water during compaction has been studied in detail by Perrin (1983).

The variation with time of the amount of heat is equal to the heatflow due to conduction and the heat transported by the fluid. The corresponding differential equation is:

$$(\rho c^*)\frac{\delta T}{\delta t} = -(\rho c^* \vec{V}_s + \rho c_e \vec{u})\vec{\text{grad}}\, T + \text{div}(\lambda^* \vec{\text{grad}}\, T)$$

$$\frac{\partial \varnothing}{\partial t} + \text{div}\vec{V}_e = 0$$

$$\frac{\partial \varnothing}{\partial t} + \text{div}(\varnothing - 1)\vec{V}_s = 0$$

$$\vec{u} = \varnothing(\vec{V}_e - \vec{V}_s).$$

The heat capacity of the bulk sedimen ρc^* being:

$$\rho c^* = \rho ce + \rho cs(1 - \varnothing).$$

The thermal conductivity of the bulk sediment * is computed after the semi-empirical formula (see Woodside and Messmer, 1961).

$$\lambda^* = \lambda s\left(\frac{\lambda e}{\lambda s}\right)^{\varnothing}$$

$\vec{u}$ = velocity of filtration of the fluid with respect to the solid matrix $(m.s^{-1})$
V_e = velocity of expulsion of the fluid $(m.s^{-1})$
V_s = velocity of burial of the solid $(m.s^{-1})$
ρce = heat capacity of the fluid $(J\,kg^{-1}\,K^{-1})$.
ρcs = heat capacity of the solid matrix $(J\,kg^{-1}.K^{-1})$
ρe = thermal conductivity of the fluid $(W\,m^{-1}\,°C^{-1})$
ρs = thermal conductivity of the solid matrix $(W\,m^{-1}\,°C^{-1})$
T   = temperature (°K)
t   = time (s)
$\varnothing$ = porosity (no dimension).

The heat flow $\varnothing$ coming out of the basement is computed from a thermal model of lithosphere.

The corresponding equation of conductive heat transfer through the lithosphere is:

$$\frac{\delta T}{\delta t} = K\frac{\partial^2 T}{\partial Z^2} + \frac{A}{\rho c}$$

---

[2]The equations of the thermal model of heat transfer by conduction and convection within the sediments are taken from Combarnous and Bories (1970). The conductive heat transfer theory is developed in Carslaw and Jaeger (1965).

with K = diffusivity of the lithosphere $(m^2.s^{-1})$
ρc = thermal capacity $(J.kg^{-1}.k^{-1})$

The thermal generation in the lithosphere is computed after the empirical formula of Lachenbruch (1965).

$$A = Ao \exp\left(-\frac{Z}{D}\right)$$

Ao = heat generation at the surface of the continental crust (Wm^{-3})
D = parameter related to the decrease of enrichment in the continental crust (around 10 km).

Both thermal models have been solved using a finite difference procedure (Perrin, 1983; Carnahan et al, 1969).

APPENDIX 3: OIL FORMATION AND MIGRATION MODEL[3]

Hydrocarbon Formation

The degradation of organic matter is described as in the Tissot and Espitalie model (1975) by a series of six parallel chemical reactions obeying a kinetic of order 1 and Arrhenius's law:

$$Q = \sum_{k=1}^{6} (\xi_{ko} - \xi_k)$$

$$\frac{d\xi k}{\xi k} = A_k e^{-E_k/RT_{dt}}$$

with Q = quantity of hydrocarbons formed (mg/g C_{org} initial)
ξ_{ko} = initial oil potential relative to reaction k(mg/g C_{org} initial)
ξ_k = residual oil potential relative to reaction k (mg/g C_{org} initial)
A_k = reaction coefficient (s^{-1})
E_k = activation energy (cal/mole)
R = perfect gas constant (R = 2 cal/mole K)
t = time (s)
T = temperature (K)

The values of parameters ξ_{ko}, A_k, and E_k used are fixed for each type of organic matter whose degradation is to be described, on the basis of classification by Tissot et al (1974).

Representation of Fluid Flows (Water and Oil)

The formulation of fluid flows in porous environments is based on Darcy's law adapted to the diphasic case by the use of relative permeabilities:
For water:

$$\vec{V}_e = \frac{-KK_{re}\rho_e g}{\mu_e} \overrightarrow{grad}\, H_e$$

For the hydrocarbon phase, assumed to be unique:

$$\vec{V}_h = \frac{-KK_{rh}\cdot\rho_h g}{\mu_h} \overrightarrow{grad}\, H_h$$

with K = intrinsic permeability (m)
$\vec{V}_e, \vec{V}_h$ = filtration velocity of water and hydrocarbons (m/s)
μ_e, μ_h = dynamic viscosities of water and hydrocarbons (Pa.s)
K_{re}, K_{rh} = relative permeabilities of water and hydrocarbons (no dimension)
ρ_e, ρ_h = densities of water and hydrocarbons (kg/m)
g = gravity acceleration (9.81 m/s)

$$H_e = \frac{P}{\rho_e g} - Z \quad \text{head (water)}$$

$$H_h = \frac{P_h}{\rho_h g} - Z \quad \text{head (hydrocarbons)}$$

P, P_h = pressure of water and oil (Pa)
Z = depth (m).

Characteristic intrinsic permeabilities of the porous environment are calculated according to porosity by the Kozeny-Carman formula, introducing an anisotropy coefficient:

$$K_x = \frac{0 \cdot 2\ \emptyset^3}{(1-\emptyset)^2 S_0^{\ 2}}$$

$$K_y = \lambda K_x$$

with K_x, K_y = horizontal and vertical intrinsic permeabilities (sq m)
$\emptyset$ = porosity (no dimension)
S_0 = specific area of rock matrix (sq m/cu m)
λ = anisotropy coefficient, less than 1 (no dimension).

Once the permeability coefficients for each element of the grid have been calculated, those applicable to each boundary are derived at by interpolation.

Relative permeabilities K_{re} and K_{rh} are assumed to depend on oil saturation by curves similar to those used for modelling fluid movement in reservoirs (Figure 15).

Capillarity Equation

$$P_h - P = P_c$$

P_h = pressure in the oil phase
P = pressure in the water phase
P_c = capillary pressure.

---

[3]The equations used for the model have been taken from Tissot and Espitalie (1975) concerning the oil generation, and from Marle (1972) and Scheidegger (1960) about the fluid flow calculations. We also were inspired from Smith (1971) for the description of compaction.

The value of P_c is computed from a mean radius of access to pores R (in meters) characteristic of each lithology, by the standard formula:

$$P_c = \frac{2\gamma}{R}$$

We selected for γ a value of 30.10^{-3} Newton/n and values of R such that capillary pressure in the clays is approximately 2 MPa.

Representation of Compaction

The equation of effective stress is used to express stresses between the solid matrix and fluids:

$$S_g = \sigma + P$$

with S_g = total stress (P_a)
σ = effective stress (P_a)
P = pressure (P_a)

As we could not do otherwise, we considered only the vertical component of stress, given by the following relation which expresses the weight of overlying sediments:

$$S_g = \int_0^z \rho g dz$$

with z = depth of point considered (m)
ρ = total density of rock (kg/cu m)
g = gravity acceleration.

Total stress S_g is then termed geostatic load. To describe sediment compaction we shall write, like Smith (1971), that effective stress σ is a function of porosity:

$$\sigma = s_k(\emptyset)$$

The index k signifies that the function s_k differs from one lithology to another (Figure 16).

The concept of compaction irreversibility is introduced by linking σ, s_k, and $\emptyset$ as soon as values of $\emptyset$ stop decreasing continuously in a grid unit and tend to increase above the $\emptyset_{min}$ value then reached, using an empirical relation which is easy to calculate. This relation expresses the fact that once values of $\emptyset$ increase, they no longer follow the initial $\sigma = s_k$ curves but, rather, much steeper ones which meet the $\sigma = s_k$ curves at a tangent at the point where the $\emptyset_{min}$ value is reached (Figure 16).

The density for the water (ρ_e) as well as for oil (ρ_h) have been assumed to be constant.

The density and viscosity values used correspond to a low viscosity oil containing dissolved gas:

ρ_h = 750 kg/cu m (density)
μ_h = $ae^{b/T}$ (dynamic viscosity according to Amdrade formula in Latil, 1975).

T is the temperature, a and b are selected so as to obtain a viscosity of 16.10^{-3} Pa.s at 40°C and $1.15.10^{-3}$ Pa.s at 160°C.

Water viscosity according to temperature is calculated by the standard Bingham formula (Latil, 1975).

Algorithm and Boundary Conditions

The geological section under investigation from the beginning of subsidence up to the present time is represented by a finite element grid with conditions at the limits as follows: 1) at the base of the grid, which it is assumed represents an impermeable bedrock affected by subsidence, zero flow and displacement specified; and 2) at the upper boundary of the grid, hydrostatic pressure and overburden by sediments of uniform 2,000 kg/cu m density, vertical displacements free.

The number of grid units in the upward direction increases with evolution, an extra row is added when the depth of the top of the last grid level exceeds the initial thickness of the following level. The number of grid units is constant breadthwise. Using the equations set forth earlier, water and hydrocarbon flow velocities can be calculated at each moment. These velocities are used at each iteration to produce the equations of mass balance of water hydrocarbons the solid matrix for each element of the grid, which in turn enable us to compute the pressures, the porosities, the saturations and the nodes depths through a finite difference method.

APPENDIX 4: THERMODYNAMIC MODEL OF SOLUBILITY OF HC[4]

The thermodynamic equilibrium is expressed by the equation of the fugacities in the gas phase and in the liquid phase, for each component:

$$f_i' = f_i''$$

The fugacities are expressed as follows:

$$f_i' = f_i°(P, T)x'_i\gamma'_i(P, T, s_i)$$

with $f_i°(P, T)$ = fugacity of pure component i at temperature T and pressure P
x'_i = molar percentage of i in phase '
$\gamma'_i(P, T, x_i)$ = activity coefficient of i in phase '.

The activity coefficients γ'_i are computed, introducing the solubility parameters:

$$\log \gamma'_i = \frac{V_i(\delta_i - \overline{\delta'})^2}{RT}$$

with δ_i = solubility parameter of component at P and T
$\overline{\delta'}$ = average solubility parameter in phase '
V_i = co-volume of component i
R = molar gas constant (R = 2 cal/mole K).

Coupled to the mass balance equations, the equations above are sufficient for the determination of the concentrations of each component in both phases for a mixture at given pressure and temperature.

---

[4]The equations of the model were partially taken from Hildebrand (1929).

Geochemical Methods for the Quantitative Evaluation of the Petroleum Potential of Sedimentary Basins

A. E. Kontorovich
SNIGGIMS (Siberian Scientific Research Institute
*   for Geology, Geophysics and Mineral Resources*
Novosibirsk 630081, USSR

This paper gives an overview of the current "state-of-the-art" in geochemical basin evaluation in the USSR.

First, I review the origin of oil and gas accumulations, as related to source-rock type and richness, the degree of maturation of the organic matter, and the changes occurring during migration. The dispersed organic matter in shale or carbonate formations is considered to be the source material for the oil and gas, migration taking place as a solution in formation water (aquabitumoids).

Next, I give a semi-quantitative approach to the calculation of hydrocarbon charge and accumulation efficiency in sedimentary basins, illustrated with examples from Siberian basins. The cornerstone of these methods is determining the bitumoid (extractable organic matter) content of the presumed source beds and its change due to progressing maturation and expulsion. Comparisons between calculated effective expulsion from the source formations and oil and gas field distributions indicate that the richer deposits are concentrated in those parts of the basins where most oil charge was expelled.

Finally, this paper includes a comprehensive list of mostly Soviet literature dealing with geochemical basin evaluation.

INTRODUCTION

In recent years, due to the developments on the energy supply front and the need for projections into the future, the quantitative evaluation of petroleum potential has become of great interest (Moody, 1975; Modelevskiy, 1976; Sickler, 1976). As a consequence, earth scientists have focused attention on methods of petroleum resource evaluation, their reliability, and their accuracy.

In general, quantitative evaluation of petroleum prospects is based either on analog methods or on modeling the processes of petroleum generation and accumulation (Kontorovich, 1976). The conventional distinction between these two approaches has recently become less sharp as each method has tended to borrow from the other. Modern deterministic models, and the critical values of the parameters used in their construction, are often based on well-studied geologic analogs ("calibration set"). On the other hand, modern versions of the analog method, based on mathematical/numerical methods take model notions into account (Kontorovich and Fotiadi, 1976; Bua, 1978; Buyalov and Nalivkin, 1979).

The organic origin of petroleum forms the basis for the models of petroleum accumulation. The geochemical methods of petroleum resource evaluation have made considerable progress in the last two decades. Significant contributions in this field were made in the USSR by Akramkhodzhayev, Vassoyevich, Vyshemirskiy, Kalinko, Karimov, Kartsev, Neruchev, Nesterov, Rogozina, Rodionova, Trofimuk, and Uspenskiy. I have worked extensively on this problem, and the results are presented in a number of papers (Kontorovich, 1970, 1976, 1977, 1978).

This paper focuses on the principles of geochemical petroleum resource evaluation being carried out in the USSR, illustrated by examples from the Siberian petroleum-bearing basins. The basis for these studies is the correlation of the disseminated organic matter (OM) with the oils, based on up-to-date analytical methods. These studies comprise (1) the estimation of OM content; (2) the construction of maps showing organic carbon (OC) and bitumoid content in individual sedimentary strata; (3) the construction of maps showing depositional environment, paleostructure and burial history; and (4) the construction of maps showing katagenetic transformation ("maturity"), generation and migration for oil and gas, oil composition, etc. (Kontorovich, 1970, 1976; Kontorovich et al, 1967, 1971).

More than 60,000 analyses served as a data base for the statistical treatment. The investigations concern sedimentary strata of the Siberian Platform of Upper Proterozoic to Cretaceous age and cover a territory of more than 8 million sq km (3.1 million sq mi). These systematic geochemical studies, combined with extensive geological data, allow the integrated reconstruction of the history of petroleum generation in individual sedimentary basins. Many basins of varying geologic development provided the information for ascertaining which characteristics of petroleum formation are common and which are specific for the individual basins. The main conclusions of these studies are given below, together with some examples. I deal primarily with the work of Soviet scientists who are poorly-known abroad, and the list of references is therefore biased in this respect.

MAIN STAGES OF LITHOGENESIS AND PETROLEUM FORMATION

Comprehensive studies on oil and natural gas (HCG) formation carried out at the molecular and atomic levels, and geological evidence leave little doubt that the source for oil and gas is the organic matter (OM) dispersed in the sediments, and that oil and gas generation is a protracted multi-stage process going hand-in-hand with the phases of lithogenesis.

Definitions and Terminology

For those less familiar with Soviet terminology, I first define terms and notions like "lithogenesis," "katagenesis," "organic matter," "bitumoid," etc. Lithogenesis is the evolution of sediments from their origin through the subsequent transformation during burial. In other words, it starts with weathering and erosion and ends with metamorphism (Strakhov, 1960; Vassoyevich, 1973; Kontorovich and Trofimuk, 1976).

In Soviet petroleum geology, lithogenesis is divided into stages, substages, gradations, and subgradations (Vassoyevich, 1971). The three main stages are: sedimentogenesis (S), diagenesis (D), and katagenesis (K). Katagenesis is subdivided into three substages: protokatagenesis (PK), mesokatagenesis (MK), and apokatagenesis (AK). These are further subdivided into gradations and subgradations (see Figures 14 and 15; see also Karsev et al, 1971; Kontorovich et al, 1974b). Note that the term "diagenesis" as used in the Soviet literature corresponds more or less to "early diagenesis" in the English usage (that is, processes taking place just after sedimentation).

The coalification scale in the USSR can be correlated as follows: coalification rank B corresponds to protokatagenesis ($R_0 = 0.25$ to 0.53); rank D ($= MK_1^1$) and C ($= MK_1^2$) to early mesokatagenesis ($R_0 = 0.53$ to 0.85); rank G ($= MK_2$) to middle mesokatagenesis ($R_0 = 0.85$ to 1.20); and rank K ($= MK_3^1$) and OC ($= MK_3^2$) to late mesokatagenesis ($R_0 = 1.20$ to 2.0). Rank T ($= AK_1$) and PA ($= AK_2$) correspond to apokatagenesis ($R_0 = 2.0$ to 3.5). The R_0 numbers quoted above relate to the reflectance of vitrinite in oil.

The petroleum generating potential of organic matter (OM) disseminated in sediments is controlled by its composition, which is determined by the original living material, its early diagenetic alteration, and the degree of katagenetic transformation. Original organic matter is divided into two distinct classes in the Soviet geological

nomenclature: "humic" OM, genetically associated with higher terrestrial plants, and "sapropelic" OM, mainly associated with plankton and benthos. Formation of the latter is greatly affected by the OM of bacteria formed by the reworking of dead plankton and benthos. However, the terms "humic" and "sapropelic" are not generally accepted, and are often interpreted differently, which is why Vassoyevich and others repeatedly proposed to come to a better defined classification (Vassoyevich, 1973b; Vassoyevich et al, 1974).

According to Vassoyevich, OM can be differentiated chemically into "alinic," "arconic," and "intermediate" classes (Vassoyevich, 1973b). Aliphatic and alicyclic structures dominate in alinic OM, condensed arenic structures dominate in arconic OM. Aquagenic OM is alinic as a rule, but in strata poor in OM it may transform into arconic through (early) diagenetic alterations (Kontorovich et al, 1974a). On the other hand, terragenic OM is arconic on the whole, but some of its components (for example, liptobiolithic coals) are associated with alinic OM. Lastly, the term "bitumoid" is equivalent to the expression, "solvent extractable organic matter;" whereas the term "bitumoid coefficient" is equivalent to the expression,

$$\frac{\text{solvent extractable organic matter}}{\text{total organic matter}}$$

Depositional Environment and Type of Organic Matter

The amount and the chemical composition of the hydrocarbons produced in the course of katagenetic transformation, and consequently the size and composition of the accumulations, are to some extent controlled by the depositional environment and (early) diagenetic alterations of the OM.

The depositional environment controls the original type of OM and its concentration in the sediments (Strakhov, 1960, 1961, 1962; Kontorovich et al, 1967, 1971, 1974b; Gol'bert and Kontorovich, 1978). Aquagenic OM is mainly deposited in marine epicontinental basins. Its concentration varies between tenths of one percent (sometimes as low as hundredths of one percent in carbonates and evaporites) to a few percent or even 15 to 20 percent under starved sediment supply ("euxinic") conditions. Maximum OM concentration in such basins occurs usually in their central, relatively deeper water parts. The deposition of aquagenic OM in such depositional environments started with the origin of life on Earth, in the Proterozoic, and dominated during the early Phanerozoic. Sediments of such marine basins with predominantly planktonic OM are well-studied, particularly the Lenian and Amginian (Lower-Middle Cambrian) of the Siberian Platform (Figure 1), the Lower Frasnian of the East-European Platform, and the Late Jurassic, Berriasian, and Valanginian of West Siberia (Figure 2; Strakhov, 1960, 1961, 1962; Kontorovich et al, 1967, 1971, 1974b; Yevtushenko et al, 1969).

From the Carboniferous on, the proliferation of higher terrestrial plants gave rise to a significant change in the organic matter deposited. Terragenic OM became a prominent type, deposited in continental lacustrine-alluvial

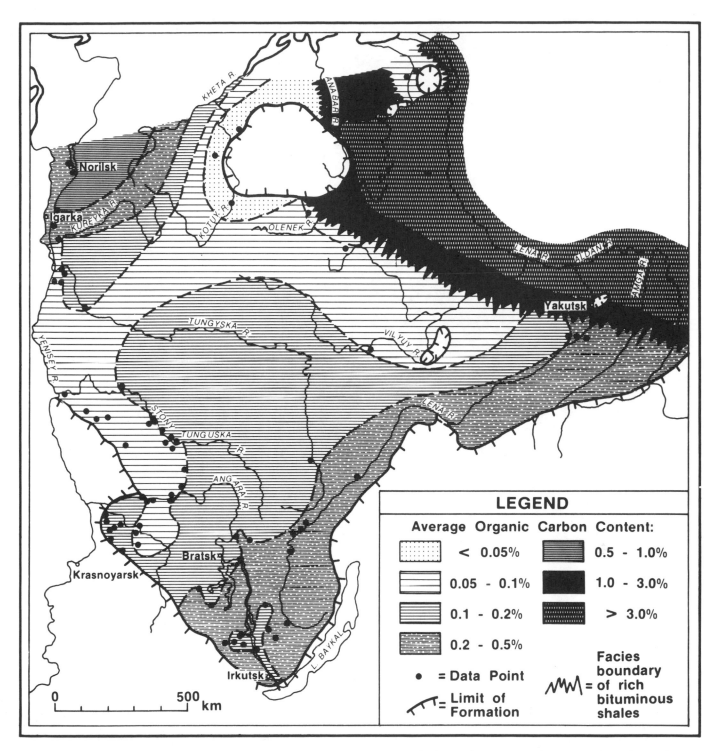

Figure 1. Organic richness map of the Lower-Middle Cambrian formations. Note the facies-change into rich source rocks in the northeast.

and lacustrine-paludal environments under humid climatic conditions. Consequently, from then on concentrated terragenic OM is the main and sometimes the dominant component in the peripheral continental basins of the humid zones. The amount of concentrated OM rapidly decreases from the peripheral to the central parts of these basins. At the same time, the concentration of dispersed OM increases markedly from 0.5 to 1 percent to over 5 percent. Examples of such sedimentary sequences are the coal-bearing deposits of the Carboniferous of the East-European Platform, the upper Carboniferous and Permian of the Siberian Platform, the Lower and Middle Jurassic of West Siberia (Figure 3) and the Turanian Plate, and the Aptian, Albian, and Cenomanian of West Siberia. The deposits of the continental subaqueous facies commonly contain large amounts of aquagenic OM of lacustrine origin.

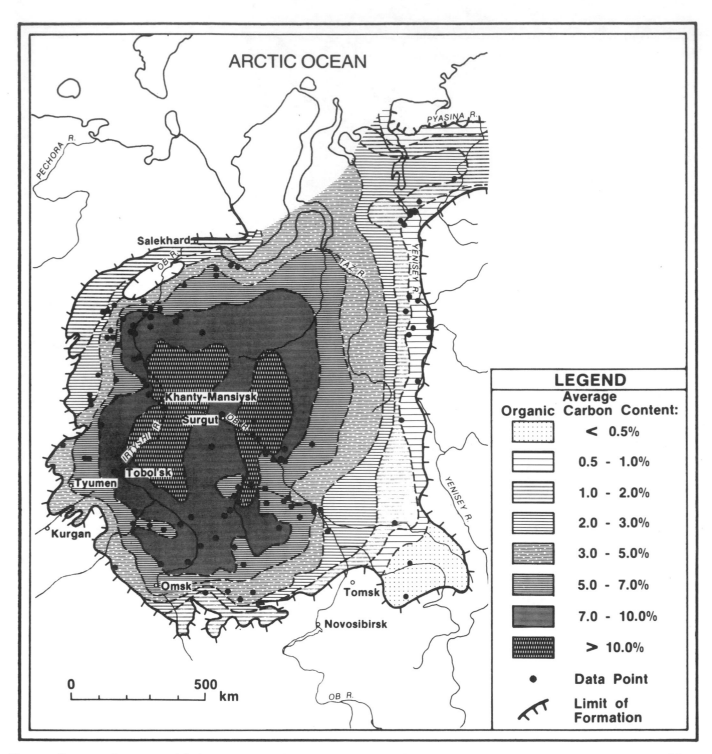

Figure 2. Organic richness map of the Upper Jurassic of Western Siberia. Note the increase in richness toward the basin center, typical for marine environments of deposition.

During arid and semi-arid sedimentogenesis only negligible amounts of OM are accumulated in the continental basins; these sediments are consequently not important for petroleum generation.

Whether oil or gas is predominant strongly depends on the composition of the original type of OM. Oil is primarily present in marine sediments containing aquagenic OM which is katagenetically moderately transformed. Continental coal measures, on the other hand,

predominantly contain gas (Kontorovich, 1970; Kontorovich et al, 1974c). This relationship is generalized in Figure 4. There is also a distinct correlation between the amount of organic matter in sedimentary sequences and the petroleum resources contained in them (Figure 5).

The Effects and Products of Early Diagenesis

The diagenetic stage of lithogenesis has a marked influence on the chemical structure of the OM, and to some

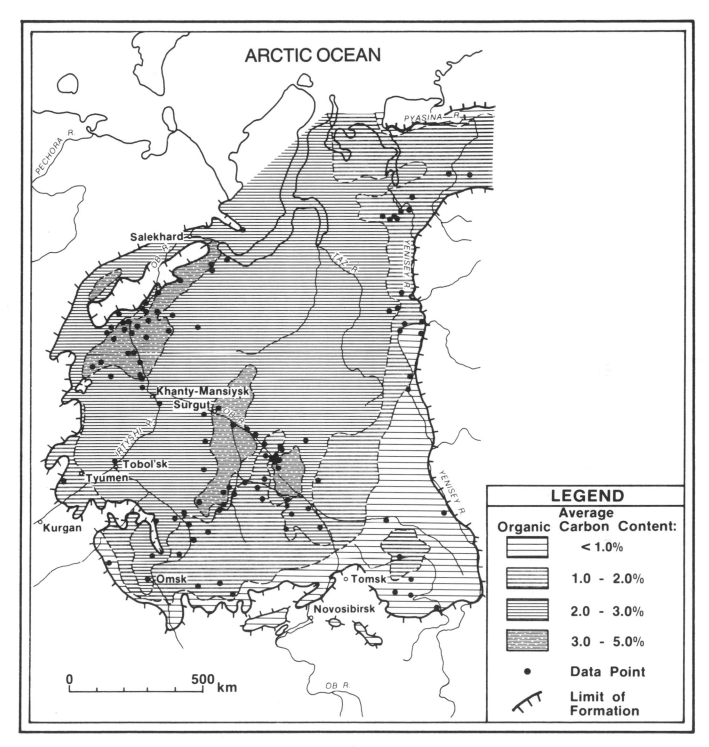

Figure 3. Organic richness map of the Lower-Middle Jurassic. Note the increase in richness toward the basin edge in the west in these predominantly terrestrial formations.

extent it also determines the concentration of OM in sediments until the beginning of katagenesis. Anaerobic biochemical decomposition of OM gives rise to large amounts of gases, including hydrocarbon gases (Sokolov, 1965; Veber, 1966). For a long time, it was believed that the only gas formed during diagenesis was dry methane. Recent studies of Veber, Zorkin, Levshunova, and Chertkova show that the biochemical decomposition of marine sediments produces—along with methane—hydrocarbon gases in the

C_2-C_5 range, normal as well as branched.

Until recently, it was believed that hydrocarbon gases generating during diagenesis diffuse from the sediments into the bottom water and dissipate whereas the accumulation of oil and gas only takes place during katagenesis (Sokolov, 1965). However, it becomes more and more evident that hydrocarbon gases generated during diagenesis may be trapped and form large pools.

The main type of gas pools of biogenic origin seem to be

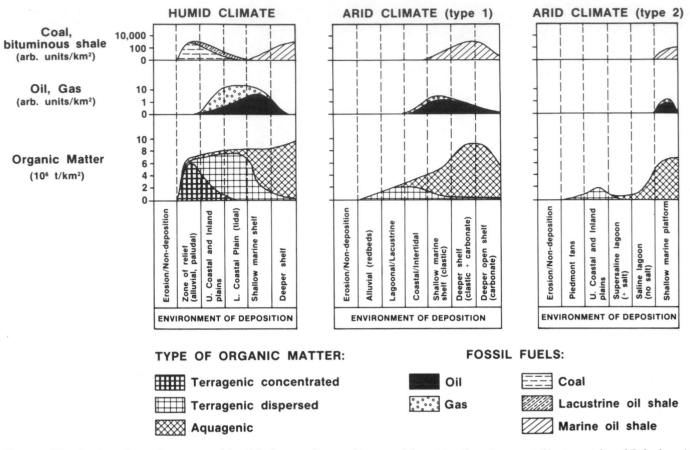

Figure 4. Distribution of organic matter and fossil fuels according to climate and depositional environment (Kontorovich and Polyakova).

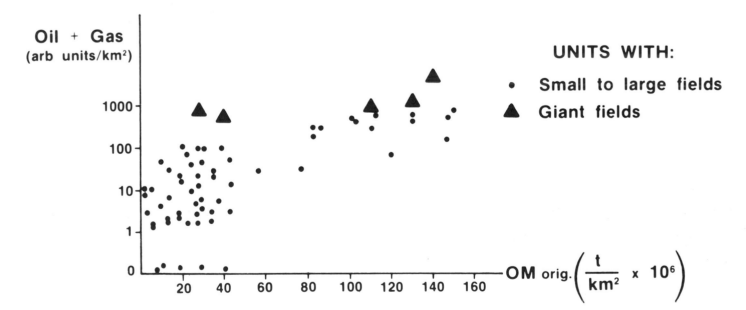

Figure 5. Relation between original organic matter in the sediments and size of hydrocarbon accumulations.

gas-hydrates formed by methane and other hydrocarbon gases, such as carbon dioxide and hydrogen sulfide, in deep-marine and freshwater basins in regions where the bottom temperature is below 4 to 5°C and the water depth exceeds 400 to 500 m (1,312 to 1,640 ft). The possibility of the conservation of hydrocarbon gases in deep-water sediments has been proposed on theoretical grounds by Makagon, Trofimuk, Tsarev, and Chersky (Makagon et al, 1973; Trofimuk et al, 1975), and was later confirmed by experiments.

Subsequent subsidence of hydrate-containing sediments into the zone of katagenesis results in the decomposition of hydrates due to the increase in temperature. The gases so formed will either accumulate in traps on their migration path, or form secondary hydrate deposits in shallow zones, or disseminate (Kontorovich and Trofimuk, 1976). This methane will be enriched in the light isotope of C^{12}; δC^{13} varies between $-50°/oo$ and $-70°/oo$ PDB.

Katagenesis

A further transformation of OM occurs during katagenesis, the main agents being temperature and pressure. It is held by some, including myself, that the rocks enclosing the OM play, therefore, a catalytic role, and consequently the conversion dynamics and the composition of the products depend to a certain degree on the composition of the host rock.

Transformation of organic matter in the zone of katagenesis is most clearly reflected in its compositional changes. The higher the P and T conditions, the higher the carbon content and the lower the hydrogen and especially heterocomponent (NSO) content. Table 1 shows the carbon and hydrogen content of the OM in different sediments of the Siberian region, graded according to their stage of katagenesis as based on vitrinite reflectance data. Also indicated are the depths at which these stages occur.

The composition of bitumoids (= extractable organic matter) at the different substages, gradations, and subgradations of katagenesis is controlled by two processes working in opposite directions (Kontorovich, 1970; Vyshemirskiy et al, 1971; Trofimuk et al, 1973). The first process is the generation and transformation of bitumoids. During proto-, early, and middle mesokatagenesis this process leads to an increase in the bitumoid content of the OM. At the same time, a gradual change in the composition of the bitumoids toward that typical of oil takes place, with the gradual equalization of the relative amounts of odd- and even-numbered n-alkanes, and the formation of hydrocarbons in the gasoline and kerosene range (Figures 6, 7, 8, and 9).

The second process, working in the opposite direction, is the primary migration (expulsion) of bitumoids out of the source rocks. In the organic matter of the source rock this leads to a decreasing concentration of bitumoids, an increase in the content of the less mobile heterocyclic compounds, and the concentration of high-boiling relative to low-boiling hydrocarbons and of naphthenic-aromatic relative to alkanic-naphthenic compounds (Vysemirskiy et al, 1971).

Lithology and temperature control the mode of primary migration to a large extent. I propose, as others do, that primary migration of bitumoids and hydrocarbon gases takes place mainly in true and colloid solution (Kontorovich, 1970; Vyshemirekiy et al, 1971; Trofimuk et al, 1973). The presence of bitumoids in formation water (= aquabitumoids) associated with petroleum-bearing deposits of Western Siberia supports this hypothesis (Kontorovich et al, 1976a). These studies are based on the extraction of the aquabitumoids by chloroform and isobutil alcohol

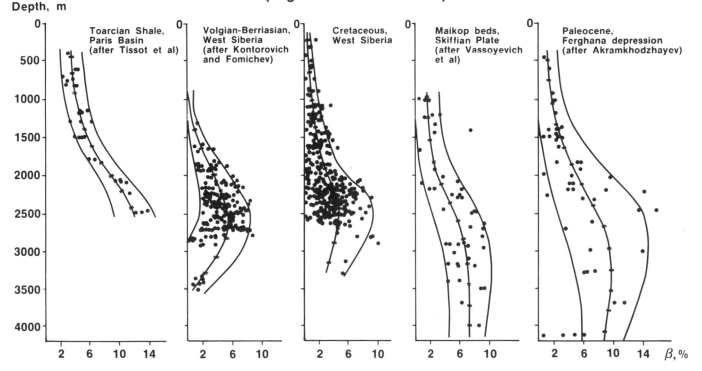

Figure 6. Increase and subsequent decrease in bitumoid (=extractable organic matter) content with increasing maturation; shales and clays.

BITUMOID COEFFICIENT β VERSUS DEPTH OF BURIAL
(Marine Carbonates and Marls)

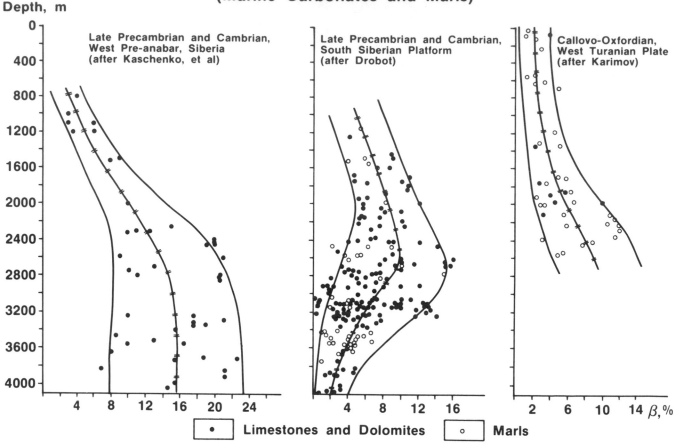

Figure 7. Increase and subsequent decrease in bitumoid (= extractable organic matter) content with increasing maturation; calcareous formations.

successively, and subsequent detailed investigation of the extract (Danilova and Kontorovich, 1977). Water samples from some petroleum-bearing horizons contain large quantities of bitumoids, with a regional average concentration around 20 to 30 mg/l. The composition of the aquabitumoids is as follows: saturated hydrocarbons, 30 to 35 percent; aromatics, 7 to 15 percent; resins, 50 to 60 percent; asphaltenes, 10 to 20 percent. The normal alkanes (mostly C_{15}, C_{19}, C_{21}, C_{25}, C_{27}) constitute 10 to 30 percent of the saturate fraction (Figure 10). Isoalkanes prevail over cycloalkanes, while mono- and bicyclic naphthenes are predominant ring compounds (Figure 11). It is estimated that the formation waters of the entire Western Siberian sedimentary basin contain at present 20 to 25 billion tons of liquid HC, asphaltenes and resins, and about 11 billion tons of C_2-C_5. Thus, at least for this particular petroleum bearing basin, the concept of primary migration of bituminoids with formation water seems to be supported by our data.

Oil and Gas Composition Related to Maturity Stages

The oil generation and expulsion cycle can thus be divided into (1) preparatory maturation of the source rock; (2) the start and gradual increase of oil generation and expulsion; (3) the principal phase of oil generation and

expulsion; (4) decreasing oil generation; and (5) a post-mature stage (see also Figure 19).

The composition of the bitumoids formed and expelled during each of these phases is very characteristic. For example, practically no hydrocarbons in the gasoline and kerosene range are generated during the early oil generation phase (end of PK, beginning of MK_1^1); n-alkane, odd-number preference still prevails and cycloalkanes are much more abundant than isoalkanes. Condensed structures dominate the cycloalkanes (Philippi, 1965; Kontorovich et al, 1967; Tissot and Pelet, 1971; Kontorovich and Danilova, 1973; Kontorovich et al, 1974a). Heavy, naphthenic oils poor in n-alkanes are to be expected as the indigenous products of this early expulsion zone.

In the principal oil generation zone, the OM generates a complex of hydrocarbon and other compounds which are characteristic for normal oils. Here the conditions are most favorable for primary migration, and hence the oil pools are concentrated in this zone. During this principal phase of oil generation, one ton of aquagenic OM expels about 37 kg of oil with the following typical composition: saturated HC, 28 to 29 percent (of which only a few percent are n-alkanes), unsaturated HC, 31 to 32 percent, asphaltenes, etc., making up the rest (Kontorovich et al, 1974a, 1974b). On the other

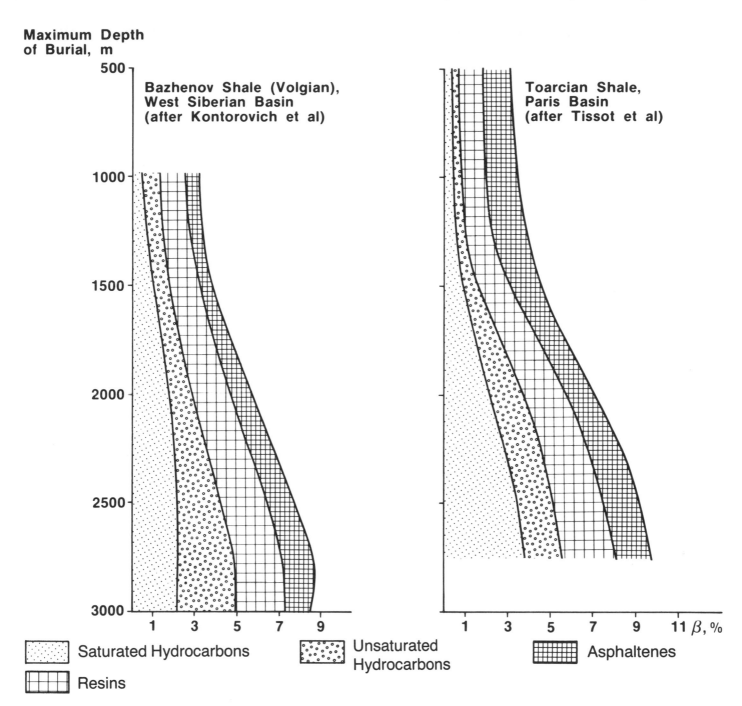

Figure 8. Composition of dispersed aquagenic organic matter versus depth.

hand, one ton of terragenic OM typically produces 10 to 12 kg of paraffinic oil, with 50 percent saturated HC (of which about half are n-alkanes), 40 to 42 percent unsaturated HC and 8 to 9 percent resins and ashaltenes.

During the decreasing oil generation phase, an additional 6 to 7 kg oil is expelled from 1 ton of terragenic OM (20 percent saturates and 60 percent unsaturated compounds). Very small amounts of n-alkanes are expelled during this phase: 300 to 400 g/ton. The oils generated during this phase by aquagenic OM are rich in low-boiling compounds with a predominance of branched alkanes and cyclic compounds.

Considering the above, it follows that it is during the principal phase of oil generation that the expelled oils inherit most of the chemical structure of the lipids of the original aquagenic or terragenic OM. To the contrary, during the phase of early oil generation, expulsion of bitumoids is so difficult that the composition of the oils is to a much larger extent determined by the migration capacity of the bitumoids. During the phase of decreasing oil generation, katagenetic transformations of the OM result again in a weak correlation between the composition of the generated and expelled hydrocarbons and the original OM.

Katagenetic gas generation (as distinct from biogenic gas

LIGHT HYDROCARBON COMPOSITION OF BAZHENOV SHALE, WEST SIBERIA

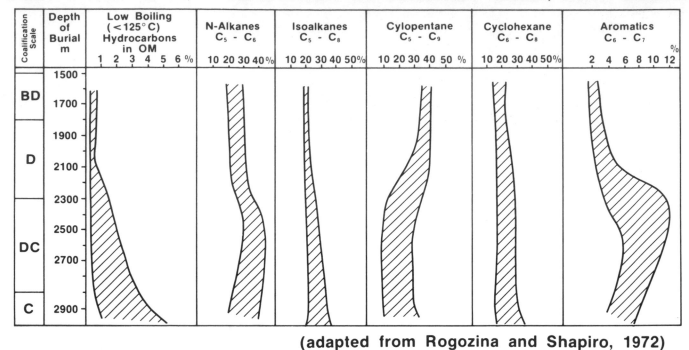

(adapted from Rogozina and Shapiro, 1972)

Figure 9. Variation in the hydrocarbon composition of the Bazhenov shale with depth of burial.

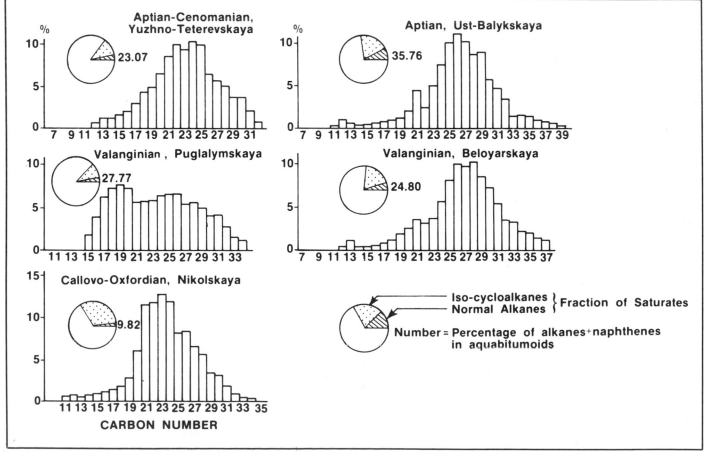

Figure 10. Alkane content in aquabitumoids, West Siberia.

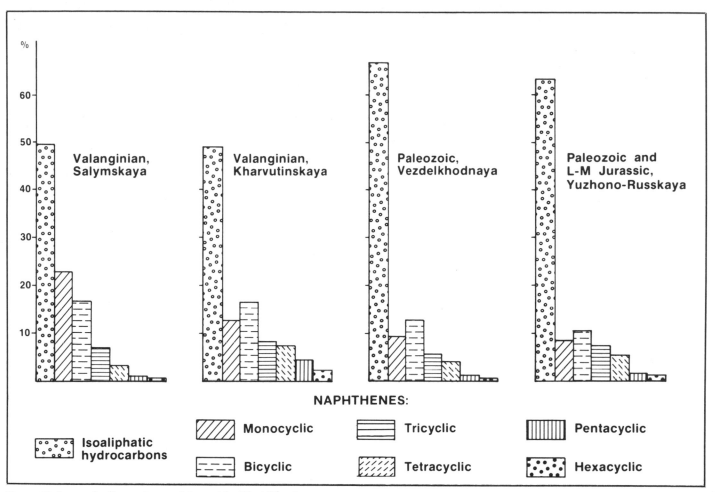

Figure 11. Iso-cycloalkanes in aquabitumoids, West Siberia.

generation) in sediments has two maxima: the upper zone ($PK-MK_1^1$) and the deep zone (MK_3-AK^1) of intense gas generation. The early zone of oil generation coincides with the upper gas generation zone, and the zone of decreasing oil generation corresponds to the deep gas generating zone (Vassoyevich and Uspenskiy, 1954; Akramkhodskayev, 1973; Kontorovich et al, 1973; Neruchev et al, 1973; Regozina et al, 1974; see Figures 12 and 13).

The composition of gases found in these two zones is markedly different. In the shallow zone carbon dioxide dominates while the hydrocarbon gas is mainly presented by methane. The deep zone of intense gas generation at first produces wet hydrocarbon gas with a high condensate content. During apokatagenesis the gas becomes dry, the condensate ratio decreases, and the condensate itself becomes more paraffinic. The typical composition of the gas and condensate of the Valanginian-Hauterivian accumulations of West Siberia is most probably due to their formation in the deep zone of intense gas generation. Isoalkanes, often strongly branched, and cycloalkanes predominate.

The isotopic composition of methane generated during the various stages is not constant: δC^{13} changes from $-50/70^o/_{oo}$ to $-20/30^o/_{oo}$ PDB with increasing maturity (Alexeyev et al, 1973; Prasolov and Lobkov, 1977; Alexeyev et al, 1978).

The above outlined zonation of oil generation is currently widely accepted but there is no general agreement about the zonation of gas generation. Some scientists recognize only the existence of one deep zone of intense gas generation while others overestimate the importance of the upper zone. The various opinions are summarized and presented in Figure 12.

There is a distinct correlation between the intensity of hydrocarbon generation during the various gradations of katagenesis and the localization of oil and gas deposits in petroleum-bearing basins. In Figure 13, the first two columns show the generation intensity of oil and gas. The third and fourth columns give a compilation of Soviet and other data on the vertical distribution of known oil and gas accumulations in the world. As Figure 13 shows, there is a marked correlation between the principal zone of oil accumulation and the zone of early mesokatagenesis, while the upper zone of gas accumulation corresponds to the lower part of protokatagenesis and the upper part of subgradation MK_1^1. The deeper gas accumulation zone corresponds essentially to the middle mesokatagenesis zone. It is very probable, however, that the real maximum of intense gas generation is located lower, at the top of the zone of deep mesokatagenesis.

The two zones of predominant gas accumulation are even more clearly seen in the fifth column which shows the

Table 1. Variation in elemental composition of dispersed organic matter during katagenesis.

Katagenetic Zonation			CARBON %								
			West Siberian Plate			Turnian Plate				Siberian Platform		
			Maximum depth of burial, m	T* Clays	A* Argillites	Maximum depth of burial, m	T Clays	A Argillites	A Carbonate rocks	Maximum depth of burial, m	A Clays, argillites	A Carbonate rocks
PK			<1600	<73.0	<74.0	<1700	<72.0	<71.6	<69.4	<1150	<74.0	<71.0
MK	MK$_1$	MK$_1^1$	1600-2400	73.0-77.4	74.0-78.0	1700-2600	72.0-77.0	71.6-75.2	69.4-73.6	1150-2250	74.0-78.0	71.0-75.0
		MK$_1^2$	2400-3200	77.4-81.8	78.0-82.2	2600-3100	77.0-80.0	75.2-77.8	73.6-76.0	2250-3400	78.0-82.2	75.0-79.8
	MK$_2$		3200-4000	81.8-86.0	82.2-86.2	3100-3600	80.0-85.0	77.8-81.2	76.0-78.4	3400-4500	82.2-86.2	79.8-83.8
	MK$_3$	MK$_3^1$								4500-5300	86.2-89.0	83.8-86.8
		MK$_3^2$								5300-5850	89.0-91.0	86.8-89.1
AK	AK$_1$									5850-6300	91.0-92.5	89.1-90.8
	AK$_2$									6300-6700	92.5-94.0	90.8-92.4
	AK$_3$									>6700	>94.0	>92.4

*(T = mainly terragenic and terragenic OM, A = mainly aquagenic and aquagenic OM)

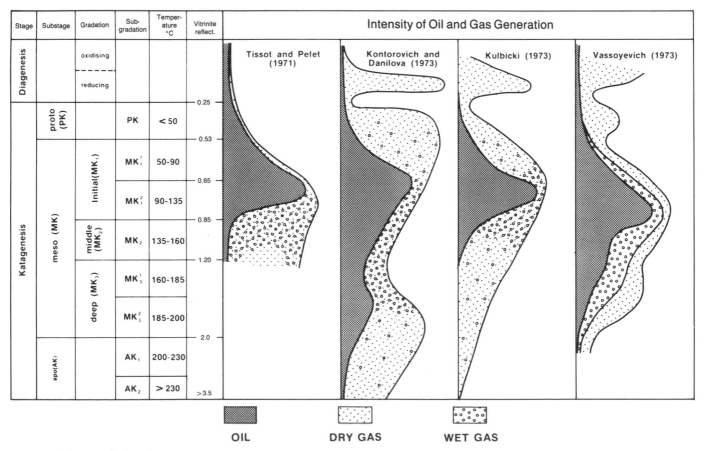

Figure 12. Schemes of oil and gas generation as proposed by various authors.

	HYDROGEN %								
West Siberian Plate			Turnian Plate				Siberian Platform		
Maximum depth of burial, m	T Clays	A Argil-lites	Maximum depth of burial, m	T Clays	A Argil-lites	A Car-bonate rocks	Maximum depth of burial, m	A Clays, argil-lites	A Car-bonate rocks
<1600	>5.4	>7.5	<1700	>4.3	>6.2	>6.3	<1150	>6.7	>6.34
1600-2400	5.4-5.2	7.5-7.4	1700-2600	4.3-4.2	6.2-5.8	6.3-6.2	1150-2250	6.7-5.7	6.3-5.7
2400-3200	5.2-4.8	7.4-7.2	2600-3100	4.2-4.1	5.8-5.5	6.2-6.1	2250-3400	5.7-4.6	5.7-5.0
3200-4000	4.8-4.2	7.2-6.5	3100-3600	4.1-3.9	5.5-5.1	6.1-5.7	3400-4500	4.6-3.5	5.0-4.3
							4500-5300	3.5-2.7	4.3-3.8
							5300-5850	2.7-2.2	3.8-3.5
							5850-6300	2.2-1.8	3.5-3.2
							6300-6700	1.8-1.4	3.2-2.9
							>6700	<1.4	<2.9

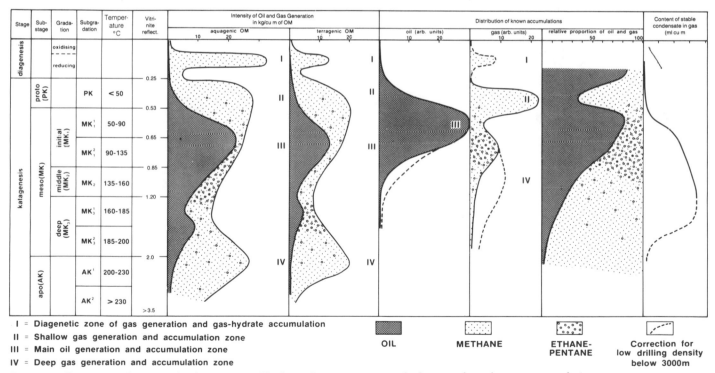

I = Diagenetic zone of gas generation and gas-hydrate accumulation
II = Shallow gas generation and accumulation zone
III = Main oil generation and accumulation zone
IV = Deep gas generation and accumulation zone

OIL   METHANE   ETHANE-PENTANE   Correction for low drilling density below 3000m

Figure 13. Comparison between depth zonation of hydrocarbon generation and oil, gas and condensate accumulations.

relative proportions of oil and gas in the accumulations. The increase in the proportion of oil resources at depths less than 500 to 600 m (1,640 to 1,969 ft) is probably due to unfavorable conditions for the retention of gas at these shallow depths.

Systematic petrographic and chemo-analytical studies of the organic matter make it possible to map maturity zones for individual layers, which are of prime importance for the evaluation of petroleum prospects. Examples are presented in Figures 14 and 15.

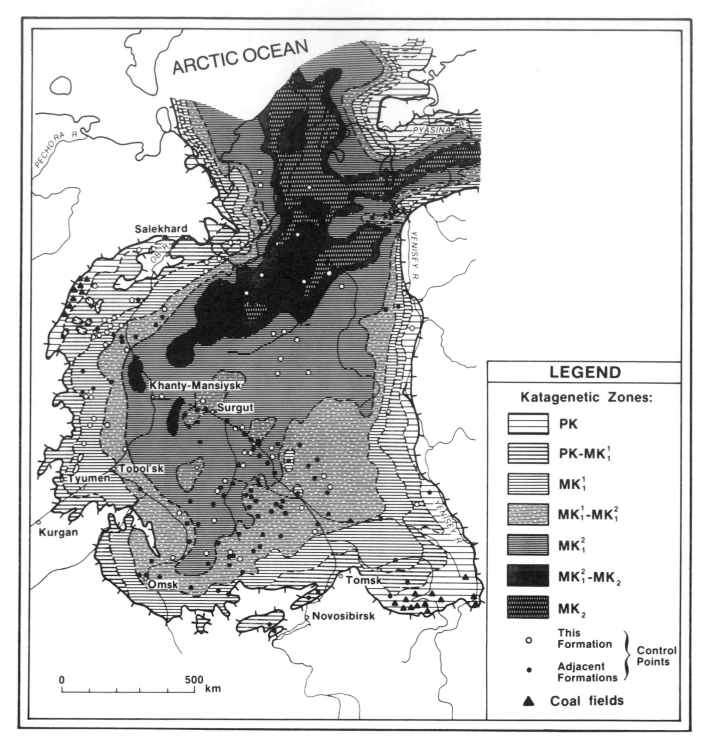

Figure 14. Organic maturity map of the Middle Jurassic of the West Siberian basin (after Kontorovich, Trushkov, and Fomichev).

CLASSIFICATION AND DISTRIBUTION OF GENETIC OIL TYPES

Oil typing and classification has played a prominent role in petroleum geology and geochemistry during the last decades. Already thirty years ago Dobryanskiy (1948) tried to classify oils on a scientific basis. Other publications on oil typing and classification worthy of note are, in chronological order: Uspenskiy and Radchenko (1947),

Radchenko (1965), Starobinets (1966), Botneva (1972), Kuklinkskiy and Pushkina (1974), Kontorovich et al (1974b, 1975, 1976), Stasova (1976), Kontorovich and Stasova (1964, 1977, 1978), Bars (1977), Philippi (1977), and Sabrodina et al (1978).

The theoretical grouping of oils based on criteria fixed a priori is called formal oil classification, and the resulting oil groups are termed class. Exhaustive classifications of this type have been proposed by the author (Kontorovich et al,

1967, 1975), and a comparable classification has been discussed by Vassoyevich and Berger (1968).

Oil Types

The most common classes or groups of classes, similar in hydrocarbon composition, are hereby proposed to be termed oil types. Geochemical data collected in the USSR and elsewhere indicate the existence of four such basic oil types (Table 2; Kontorovich and Stasova, 1978).

Type A (Paraffinic-Waxy)

These oils are of medium to low density, with a low resin content and a high wax content (up to 40 percent). Sulfur, vanadium, and nickel content is low. Paraffinic and predominantly paraffinic oils of this type are enriched in high-boiling n-alkanes exhibiting a maximum at C_{22}-C_{27}. The A-type oils contain C_9-C_{25} isoprenoids in high concentrations (3 to 8 percent). Pristane and phytane are the most significant, pristane being predominant.

Oils of this type are found in sediments of various ages beginning with the Devonian. Usually they occur in terrestrial coal-bearing strata which have reached the subgradations MK_1^2 and MK_2 ($R_0 = 0.65$ to 1.2) of katagenesis. These oils are of Middle Jurassic age in northern and southeastern West Siberia, and of Permo-Triassic, Jurassic, and Cretaceous age in some parts of the Turanian and Skiffian plates.

Oils of comparable composition are known from Permian and Triassic deposits of the Vilyuy hemisyneclise and from Permian deposits of the Kutznetsk coal basin. Commercial oil accumulations of the A type are known from a restricted number of petroliferous basins in the USSR.

Type B (Paraffinic-Low Wax)

Oils of this type are of light to medium density, and poor in waxes. Relatively heavy, highly resinous oils are also found. Ash content and the content of vanadium and nickel are very low; their ratio is close to 1.

Like the A-type oils, the B-type oils are essentially paraffinic (alkanic). The gasoline fraction contains a high proportion of n-alkanes. Their quantity sharply decreases in the kerosene and heavier fractions (that is, unlike A-type oils, high-boiling n-alkanes are insignificant, with a peak concentration around C_{15}-C_{17}). The bulk of the high-boiling fraction is represented by isoalkanes. Isoprenoid content ranges between 2 to 4 percent, and the pristane to phytane ratio is close to 1 (Kontorovich and Stasova, 1977).

B-type oils are widespread in clastic and carbonate-evaporite deposits of the late Precambrian and Lower Cambrian of the Siberian Platform. These oils differ substantially from each other depending on their occurrence in clastic or carbonate reservoirs (Drobot and Presnova, 1975). The carbonate-evaporite associated oils have a higher sulfur content, typically composed of thermally unstable sulfur compounds: low molecular weight mercaptans and sulfides (Obolentsev and Baykova, 1973).

Type C (Naphthenic-Paraffinic)

These oils are characterized by a high density (0.85 to 0.89), significant quantities of resinous components (up to

30 percent) and high concentrations of sulfur (up to 3 percent). The content of vanadium, nickel, and their porphyrin complexes is high, with vanadium predominant. The hydrocarbon composition is predominantly aromatic-naphthenic-paraffinic and naphthenic-paraffinic. The n-alkanes concentration is low, with two maxima: the first at C_7-C_8, the second at C_{17}-C_{19} or more rarely at C_{21}-C_{22}. The distribution of the isoprenoids is comparable to that found in A-type oils, but the pristane to phytane ratio is close to 1, as in the B-type oils.

These C-type oils are relatively rich in thermally stable organic sulfur compounds, mainly sulfides and thiophenes, and on the other hand contain practically no reactive sulfur-bearing components such as mercaptans, elemental sulfur, and hydrogen sulfide.

C-type oils are typically found in marine strata enriched in aquagenic OM in the subgradations M_1^1 and MK_1^2 ($R_0 = 0.53$ to 0.85) of katagenesis, such as in the Volga-Ural and West Siberian provinces. These oils constitute more than 85 percent of the reserves in Western Siberia, where the source rocks are marine shales and marls of Upper Jurassic to Valanginian age.

Type D (Cyclanic-Subnaphthenic)

These oils are characterized by a high density (0.87 to 0.97), the absence of n-alkanes and a low amount of isoalkanes. Isoprenoids such as phytane and pristane are absent or rare. Polycyclic compounds dominate in these oils. Naphthenic-aromatic fractions are relatively abundant, characterized by a significant proportion of naphthenic rings and branched aliphatic chains. Polycyclic naphthenic structures are the primary structural elements of the heterocompounds. The resin, sulfur, vanadium and nickel content varies widely.

Two subtypes of the D-type oils can be distinguished. Low-boiling hydrocarbons are absent from the first subtype (D_1) and present in the second one (D_2) (Table 2). The D_1 subtype often forms the oil leg of oil and gas pools, in which case the gas tends to contain small amounts of naphthene-base condensate.

D-type oils are common in shallow reservoirs, such as the Cenomanian of Western Siberia and South-Mangyshlak, the Lower-Middle Jurassic deposits of Vilyuy hemisyneclise and the Dnieper-Donetsk depression, the post-saline deposits of the Pre-Caspian depression and the Timano-Pechora basin. The tar and heavy oil accumulations in the Athabasca region of Canada appear to be comparable to D-type oils (Demaison, 1977).

Origin of the Different Oil Types

Analysis of the conditions under which oil pools of different types are formed leads to the conclusion that oil type is the result of the interplay of many factors, the most important of which are: the chemical composition of the original organic material, the environment of deposition and early diagenesis of the source rocks, the degree of katagenesis, and the expulsion and migration conditions.

A-, B-, and C-type oils are formed in the principal zone of oil generation under conditions favorable for expulsion, migration, and accumulation of a wide range of components. Consequently, the composition of these oils

Table 2. Chemical composition of different oil types.

Oil type	Oil sub-type	Content % Density (g/cm³)	Sulphur	Paraffin	Asphaltenes + resins	Low-boiling (<125°C) alkanes	n-alkanes	iso-alk	naphth	aromatics	High-boiling (>200°C) alkanes	n-alkanes	iso-alk	naphth	aromatics	Examples
A		0.82 / 0.80-0.85	0.1 / 0-0.2	14 / 7-40	4 / 1-10	70 / 65-90	40 / 30-55	30 / 25-40	25 / 13-30	5 / 2-7	70 / 60-90	32 / 20-45	38 / 30-50	10 / 8-25	20 / 15-25	West Siberian Plate (J₁₋₂), Turanian Plate (J₁₋₂), Skifian Plate (R̄, J), Siberian Platform (P, R̄), Predverkhoyansk marginal foredeep (J)
B		0.83 / 0.80-0.85	0.6 / 0.1-1.3	1.5 / 0.6-2.4	5 / 2-6	80 / 70-90	46 / 28-50	36 / 30-50	18 / 15-20	2 / 1-3	63 / 55-80	8 / 8-15	55 / 30-60	20 / 15-25	17 / 12-25	Siberian Platform (E₃, Rf)
C		0.86 / 0.85-0.89	1.5 / 0.5-3.5	4 / 2-7	15 / 10-30	70 / 60-80	32 / 25-40	38 / 30-45	27 / 20-30	3 / 2-8	40 / 30-60	10 / 6-21	30 / 15-40	25 / 15-35	35 / 20-50	West Siberian Plate (J, K), East European Platform, Volga-Urals region (D, C, P)
D	D₁	0.94 / 0.87-0.90	0.7 / 0.1-1.5	1.1 / 0.5-2.0	20 / 10-40	absent	absent	absent	absent	absent	6 / 4-10	0	6 / 4-10	52 / 40-60	42 / 35-50	West Siberian Plate (K₂), Siberian Platform (J₁), Turanian Plate (J₁₋₂)
D	D₂	0.90 / 0.87-0.92	0.8 / 0.6-1.8	1.0 / 0.5-1.5	15 / 10-30	29 / 20-30	2 / 1-3	27 / 20-30	70 / 60-80	1 / 1-2	15 / 10-20	0	15 / 10-20	35 / 30-50	50 / 49-60	West Siberian Plate (K₂)

reflects closely the chemical structure of the original organic material.

A-type oils are, in general, genetically linked to wax-rich OM of the higher terrestrial plants. B- and C-type oils are genetically associated with planktonic and benthonic OM of marine sediments (Kontorovich et al, 1967, 1974b, 1974c, 1975; Halbouty et al, 1973; Vandebroucke et al, 1975; Albrecht et al, 1976; Durand and Espitalie, 1976; Kontorovich, 1976; Stasova, 1976; Kontorovich and Stasova, 1977, 1978).

The source type is not the only factor controlling the oil types. The transformation of the organic matter due to depositional and early diagenetic processes is also important. In particular, the C-type oils enriched in porphyrins and sulfur are formed in strata accumulating under strongly reducing oxygen-starved conditions in stagnant basins, or where contamination by hydrogen sulfide has taken place. There is reason to believe that the B-type oils are formed in OM-poor marine strata which were laid down in neutral to weakly reducing environments.

Note that significant accumulations of B-type oils are known from the Riphean, Vendian, and Lower Cambrian of the Siberian Platform. Compared to C-type oils, it cannot be excluded that the specific composition of these oils is due to special chemical composition of the ancient phytoplankton. Two points of view exist concerning the origin of D-type oils. As one theory holds it, these oils are "immature" products generated from OM which was not subjected to the principal phase of oil generation. According to the second point of view, these oils are formed in the principal zone of oil generation, and their special composition is due to differentiation during vertical migration and/or selective bacterial destruction of n-alkanes in shallow horizons (Kontorovich et al, 1974b; Stasova, 1977; Radchenko, 1965; Philippi, 1977). Most likely, both mechanisms occur in nature.

The above outlined scheme is, of course, only operative under ideal conditions, which in certain petroleum basins can be modified by, for instance, extensive vertical migration. Oils may undergo significant changes during vertical migration, and the character of these changes varies considerably depending on the mode of primary and secondary migration and the prevailing temperature and pressure.

Oil composition may also be significantly modified during its residence in a trap, perhaps by the difference between rate and composition of supply versus loss, or by the reaction of the oil with host rock or formation water (Kartsev, 1969; Kontorovich et al, 1975; Kontorovich, 1977, 1978). Thus, the enrichment of B-type oils in mercaptans in carbonate strata is apparently the result of their interaction with the reservoir rock.

PETROLEUM ACCUMULATION IN SEDIMENTARY BASINS

The stages of petroleum generation described earlier form a more or less ideal pattern applicable to sedimentary sequences which steadily subside and pass all stages of katagenesis. The real-world processes are, of course, more complicated. A sedimentary basin may be subdivided into a number of plays: porous-permeable units separated by clay, marl, or evaporite seals. These separate plays often are isolated habitats (that is, they have their own particular generation and accumulation conditions) and may be of local or regional extent.

Case 1: Continuous Subsidence

Consider a sedimentary basin A; at time t_o its sedimentary fill contains n regional plays: $a_1, a_2, ..., a_i, ..., a_n$. Consider the evolution of petroleum accumulation in play a_i, which can in a simplified form be described by a rectangular block with height L_i and surface S. At time t the overburden of a_i is h, and a_i contains the amount Q of hydrocarbons. Assume that during the subsidence of a_i over one meter, $q_i(h)$ of hydrocarbons are expelled from the source-rock for each volume unit. The value $q_i(h)$ may be derived for different source-rock lithologies and for different petroliferous basins by either following the methods worked out by Kontorovich et al (1967, 1975, 1976), Neruchev (1962, 1969) and Trofimuk et al (1965, 1976), or using the kinetic model described by Tissot and Pelet (1971). Evidently, $q_i(h)$ depends on sourcerock lithology, richness, type, etc., and varies with h according to the maturity zonation described above.

Let a_i subside to the depth dh over time interval dt. The amount of hydrocarbons expelled from the source rock during this subsidence is:

$$dQ_{exp} = q_i(h)SL_i dh = q_i[h(t)]SL_i \frac{dh}{dt} dt = q_i(t)L_i \frac{dSh}{dt} dt \quad (1)$$

By substitution we can obtain the equivalent equation for charge to the reservoir:

$$dQ_{chg} = \phi(t)V_v(t)dt \quad (2)$$

in which $V_v(t)$, the rate of sedimentation, equals $\frac{dSh}{dt}$, and in which it is assumed that dQ_{chg} is proportional to dQ_{exp}.

Hydrocarbons are not only trapped, they also escape from pools formed at time t. The amount of hydrocarbons lost (disseminated) from pools during interval dt is assumed to be proportional to Q:

$$-dQ_{dis} = \alpha Q dt \quad (3)$$

The value α depends on the lithology of the seals (and other parameters characterizing their quality), the hydrogeologic regime, the amount and type of tectonic activity, etc. During interval dt the net amount accumulated can be described:

$$dQ = \phi(t)V_v(t)dt - \alpha Q dt \quad (4)$$

M.K. Kalinko (1964) was the first to analyze qualitatively the process of petroleum accumulation resulting from the interplay of charge and dissemination in basins with different tectonic histories. However, he did not integrate his concepts with the fundamental maturation theory of petroleum generation. At present, it is possible to construct a better qualitative model based on the recent

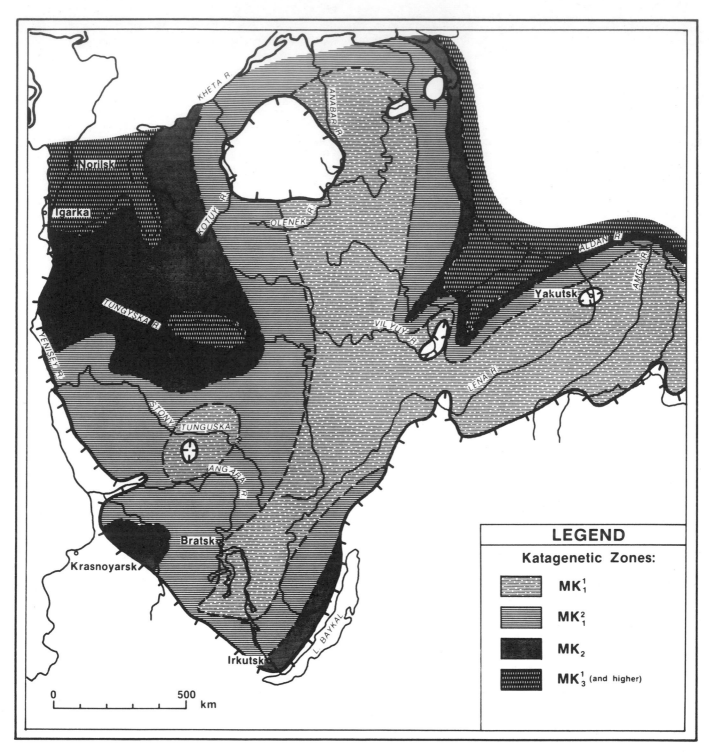

Figure 15. Organic maturity map of the Lower and Middle Cambrian of the Siberian Platform (after Drobot, et al, 1974).

progress in geochemical knowledge. Let a_i continuously subside and from time $t = t_o$, oil starts to accumulate. The amount of oil accumulated in a_i at time t is:

$$Q = e^{-\alpha(t-t_o)} \int_{t_o}^{t} \phi(t) V_v(t) d^{\alpha t} dt \qquad (5)$$

A similar solution of equation (4) can be given for gas. This paper does not further elaborate on equation (5). Using equation (4) and the above developed concepts concerning

the zonation of oil generation I will consider conceptually the evolution of oil accumulation in play (reservoir) a_i.

In Figure 16 the burial history of a_i is shown as well as the generation of oil and gas according to the function $q_i(h)$, assuming $\frac{dh}{dt}$ = constant. The relations dQ_{chg}/dt and dQ_{dis}/dt are also depicted graphically. Given the relation $-dQ_{dis}/dt = \alpha Q$ it is assumed that for gas the value α is twice for oil: $\alpha_g = 2\alpha_0$.

Figure 16 is self-explanatory and shows how the interplay

of charge and dissemination rates of oil and gas lead to net accumulated volumes in the traps. These relations are, of course, schematic and idealized. In reality, the accumulation of hydrocarbons is controlled not only by the rate of generation and loss through seals, but also by the timing of trap formation, by differential entrapment of oil or gas along the migration paths, and by many other factors. Nevertheless, even in this simplified form our model of petroleum accumulation is able to explain, in a rough, qualitative way, the vertical distribution of oil and gas pools as shown on Figure 13. Thus, the bulk of the gas generated in the upper zone of intense gas generation, even if initially accumulated in traps, seems to have escaped prior to the time of principal oil generation. This may explain why gas and condensate pools are comparatively rare in the principal zone of oil accumulation while no gas, generated in the upper zone of gas generation, is found as gas caps or solution gas.

Case 2: Subsidence Interrupted by Uplift

Consider a more complicated case of hydrocarbon accumulation in zone a_i. Let the process take place during time interval (t_o to t_1). Then, at time t_1 the accumulated volume (say of oil) in a_i will be:

$$Q_1 = e^{-\alpha(t_1-t_o)} \int_{t_o}^{t_1} \phi(t)V_v(t)e^{\alpha t}dt$$

At time t_1 the subsidence of the basin has come to a halt or changed into uplift until time t_2, and as a result, generation and charge to the traps has come to a halt. At time t_2, subsidence takes over again, but generation and accumulation in a_i does not recommence until time t_3, when temperature and pressure in a_i exceed those of time t_1. As a result, for time interval (t_1 to t_3), equation (4) can be simplified and becomes:

$$dQ = -\alpha Qdt \qquad (6)$$

During the same period destruction of the accumulations occurs in a_i according to:

$$Q = Q_1 e^{-\alpha(t-t_1)} \qquad (7)$$

and by time t_3 the amount of oil in a_i will be:

$$Q_3 = Q_1 e^{-\alpha(t_3-t_1)}$$

For any time $t > t_3$, the solution of equation (4)—taking accumulation as well as dissemination into account—is:

$$Q = [Q_3 + \int_{t_3}^t \phi(t)V_v(t)e^{\alpha t}dt]e^{-\alpha(t-t_3)} \qquad (8)$$

This situation is illustrated in Figure 17. It shows that with such a tectonic history of the basin, hydrocarbon accumulation takes place in two steps, while during the break in sedimentation partial destruction or degassing of oil accumulations may occur. What influence the break(s) in sedimentation or uplift will have on petroleum accumulation depends on the zone of petroleum generation reached just prior to the break.

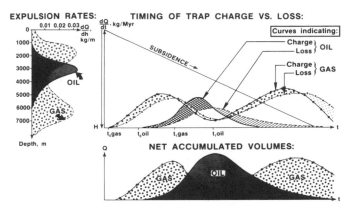

Figure 16. Evolution of hydrocarbon generation and accumulation in a steadily subsiding basin.

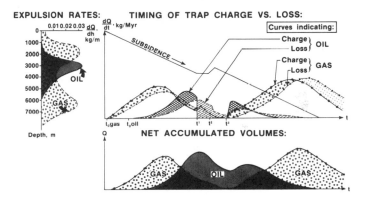

Figure 17. Evolution of hydrocarbon generation and accumulation in a basin where subsidence was interrupted by uplift. Note that during uplift there is a net loss of oil and particularly gas.

Generalized Scheme

We can now set up a generalized scheme for oil and gas accumulation, as shown on Figure 18. For this scheme we have assumed a single play a_i, with its own source rock and reservoir contained by a regional seal. The figure is self-explanatory, and shows how during steady subsidence first gas (enriched in C^{12} isotope) is accumulated, then oil, followed by wet gas and finally dry gas. The accumulations are the net result of charge to the traps and loss from them. Also shown are the postulated changes in the accumulations when uplift or a break in sedimentation occurs. In my view, uplift leads to increased rate of loss which coupled to an arrested generation process results in a rapidly diminishing size of the accumulations.

The early accumulated oil tends to be heavy and enriched in polycyclic alkanes. The oils accumulated in the main oil zone are of a more "normal" naphthenic-paraffinic or aromatic-naphthenic-paraffinic composition. Oils in reservoirs subjected to uplift, and which hence have preferentially lost gas and light components, tend to be relatively heavy and enriched in asphaltic-resinous components. Oils in reservoirs that subside into the deep zone of intense gas generation will be partly destroyed by the increased temperature, with the formation of gas, light hydrocarbons, and insoluble residues (deasphalting of M. Rogers, N. Bailey, and others). As a result, light paraffinic oil and gas-condensate pools will become predominant.

The sketched sequence of events is valid for aquagenic source rocks; similar schemes can be devised for plays containing terragenic or mixed source rock.

QUANTITATIVE PROSPECT EVALUATION BASED ON GEOCHEMICAL DATA

The first attempts at quantitative prospect evaluation based on geochemical ideas were by Arkhangel'skiy in 1927 and by Trask in 1936. They proposed that if Q_{om} is the amount of OM in source rocks, and Q_{oil} is the volume of oil-in-place, then $Q_{oil} = \alpha Q_{om}$, with $\alpha = 0.030$ to 0.035 according to Arkhangel'skiy and 0.022 to 0.150 according to Trask. In 1940, Kudryashova and Starik-Bludov proposed to find out whether a relationship exists between Q_{oil} and the amount of bitumoids dispersed in the source rocks, Q_{bit}. This method has been applied in various modifications up to the late 1950s (Vyshemirskiy et al, 1971; Tissot, 1973). It is still in use today, and is called the "volumetric-genetic" method in Soviet literature.

Bitumoid Expulsion as a Criterion for Effective Generation

Neruchev (1962) pointed out that viewing the unequal charge contribution by source rocks of different lithology and/or burial history, it would be more correct to look for a relationship between oil-in-place and the amount of bitumoids primary migrated (= expelled) from the source rocks:

$$Q_{oil} = KQ_{exp} \qquad (9)$$

The technique proposed by Neruchev to determine Q_{exp} is based on the supposition that the expulsion of bitumoids is of a chromatographic character and that therefore residual bitumoids are richer in heterocyclic compounds and poorer in hydrocarbons than the migrated bitumoids. Developing the concept of dispersed bitumoids, Vassoyevich and later Chernikov and Neruchev proposed to distinguish several classes of these bitumoids. The most important classes are listed below:

1. Syngenetic bitumoid (SB) is essentially unchanged because of insufficient maturity and therefore closely resembles the original OM.
2. Syngenetic residual bitumoid (SRB) has lost part of the mobile components due to expulsion/primary migration and is therefore of a more acidic composition than syngenetic bitumoid.
3. Epigenetic bitumoid (EB) is a migration product and is consequently enriched in the more mobile components.
4. Mixed bitumoid (MB) is a mixture of SB or SRB with EB.

The quantity of expelled and migrated bitumoids depends on the type of OM, the degree of katagenesis, the expulsion conditions as controlled by lithology and structure of the oil-producing strata, etc. In theory, equations (5) or (8) should be more appropriate to predict oil-in-place than equation (9). However, this is not yet possible due to a lack of knowledge on the values of the controlling parameters. Therefore, as in equation (9), the geological parameters are

currently best lumped together in the "accumulation coefficient," K.

Direct methods for recognizing effective source rocks are, in the USSR, generally based on indicators of primary migration of bitumoids, two of which are used most. The first one was introduced by Neruchev in 1965, and independently by Trofimuk and Kontorovich (1965). A clay layer with an evenly distributed OM concentration may be considered as a massive chromatographic column out of which the generated bitumoids migrate with formation water expelled during compaction and with gases formed during katagenesis, whereby the water and gas serve as solvents. The farther we move away from the center of the generating layer, the larger the amount of water or gas which passes through and, consequently, the more the less mobile components begin to migrate (Trofimuk and Kontorovich, 1965; Neruchev and Kovacheva, 1965). Therefore, in such clay layers the content of heterocyclic compounds, arenes, and condensed cyclanes in the bitumoids increases from the center outward.

This theoretically predicted situation was verified experimentally on real-world examples (Neruchev and Kovacheva, 1965; Trofimuk and Kontorovich, 1965; Korotkov, 1966, *in* Zhezhchenko, 1974; Kontorovich et al, 1967; Vyshemirskiy et al, 1971; Tissot and Pelet, 1971; Vandenbroucke, 1971; Kontorovich, 1976).

The second criterion for source-rock recognition makes use of the already mentioned fact that in the katagenetic zone two opposing processes are active: generation and expulsion of bitumoids. As long as the former prevails, the bitumoid coefficient increases. Concomitantly the range of values of this coefficient becomes wider due to the fact that the expulsion conditions vary from place to place within the clay layer. The bitumoid coefficient begins to decrease as soon as primary migration begins to prevail. It should be stressed, however, that these relations are only meaningful when dealing with units of uniform facies, similar type of OM and not too much variation in organic carbon content. As shown by Figures 6, 7, and 8, the actual data on the variation of bitumoid content with burial depth support the postulated relations.

The described criteria for the recognition of primary migration (expulsion) make it possible to solve two important practical problems:

(1) to ascertain the geological and geochemical conditions under which expulsion from source rocks is possible and thus to create the basis for a mapping technique of oil-producing rocks; and,
(2) to elaborate an evaluation technique of the quantity of liquid and gaseous hydrocarbons expelled from source rocks.

Primary migration of bitumoids seems to begin in clays after their burial to a depth of 1,600 to 1,700 m (5,249 to 5,577 ft), at the very beginning to mesokatagenesis ($R_0 = 0.53$), and in carbonates at a depth of 1,900 to 2,000 m (6,234 to 6,562 ft), in the middle of subgradation MK_1^1 of mesokatagenesis ($R_0 \pm 0.60$). Naturally, these depths are only approximate and vary with the geothermal gradient, effective heating time, distribution of OM in the source

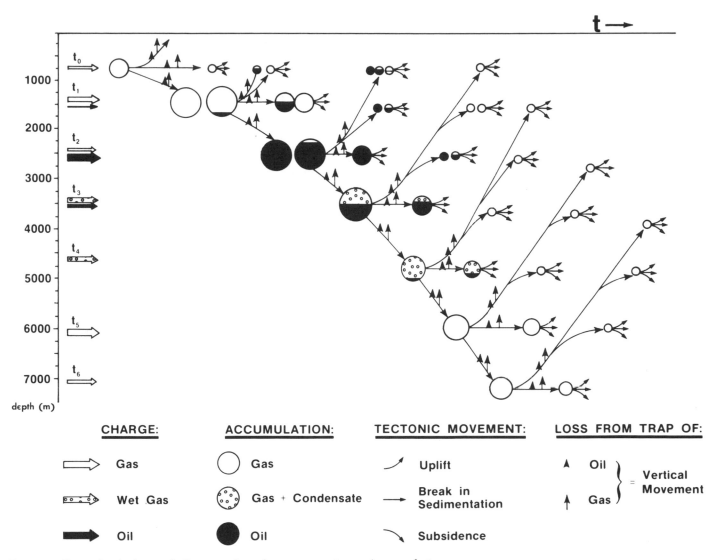

Figure 18. Generalized scheme of oil, gas, and condensate generation and accumulation.

rock, etc. Thus, oil-producing rocks can be mapped based on their maximum depth of burial and especially their degree of maturation.

The generalized patterns of bitumoid coefficient variation with depth is given in Figure 19. The dotted area encloses the observed and extrapolated values of the bitumoid coefficient. The hatched area indicates the hypothetical values of the bitumoid coefficient if the effect of bitumoid loss due to primary migration would be eliminated. A few additional remarks are in order:

1. In drawing the lines for β_{SB}^{max}, β_{SB}^{min}, and $\bar{\beta}_{SB}$, no low-boiling bitumoid fractions were included; and,
2. Correction of the β values for the effect of expulsion could be improved by using the chemical-kinetic model suggested by Tissot and Pelet (1971). Recently Burshtein (1979) has successfully applied a similar model for the Bazhenov formation of Western Siberia.

The upper limit of the zone of incipient oil generation is taken at the depth where the range of β values widens (in

other words, $\beta_{SB}^{max} - \beta_{SB}^{min}$ increases). The upper limit of the principal zone of oil generation is postulated at the depth where $(\bar{\beta}_{SB} - \bar{\beta}_{SRB})$ $= 0.2(\bar{\beta}_{SB} - \bar{\beta}_{SRB})_{max}$, the lower limit where $(\bar{\beta}_{SB} - \bar{\beta}_{SRB})$ $= 0.8(\bar{\beta}_{SB} - \bar{\beta}_{SRB})_{max}$. Finally, the lower limit of the zone of decreasing oil generation is taken at the depth where $\bar{\beta}_{SB} - \bar{\beta}_{SRB} = $ constant.

Effective Oil Generation From a Shale Source Rock

When mapping the areal distribution of source rocks, one of the principal criteria should be the quantity of expelled bitumoids. Reconsider the clay layer discussed before. According to our model, if primary migration out of this layer took place, the composition of the residual bitumoids should change systematically from its center to any adjacent reservoir. The closer to the reservoir, the more residual in composition the bitumoid (that is, the richer it should be in heterocompounds and the poorer in hydrocarbons).

Consider a clay layer a_i with clay density ρ and bitumoid content b_i. The bitumoid composition parameters are V_o in SB, V_{exp} in EB, and $V_{res} = f(Z)$ in SRB.

Then, the amount of bitumoids expelled from the differential of the volume: $dv = dxdydz$ within a layer (see also equation 9) is:

$$\Delta q = \rho b_i(x, y, z)\frac{V_o - V_{res}}{V_{exp} - V_o}\,dxdydz$$

If the dimensions of the layer are X_i, Y_i, L_i then for the whole layer we may write:

$$q(X_i, Y_i, L_i) = \frac{\rho}{V_{exp} - V_o}\iiint_{v_i} b_i(x, y, z)[V_o - V_{res}(z)]dxdydz \quad (10)$$

For a complex geologic body with n layers, the total amount of bitumoids expelled is:

$$Q = n\int_0^\infty\int_0^\infty\int_0^\infty q(X, Y, L)\phi(X, Y, L)dxdydz \quad (11)$$

A.A. Trofimuk and I determined that the equation $V_{res} = f(Z)$ for heterocomponents (NSO_s) in bitumoids is:

$$N + S + O = \frac{\alpha}{\beta + h} + \gamma \quad (12)$$

where h is the distance from the point in the clay layer to the nearest reservoir, and α, β, and γ are constants determined

from empirical data by the least-squares method. The values of these parameters as determined by A.S. Fomichev and myself for several regions of West Siberia are given in Table 3. Table 4 shows the classification of syngenetic bitumoid based on this principle.

The data on bitumoids, obtained by routine extraction methods, are used for the evaluation of b, as well as parameters α, β, and γ. The gasoline and kerosene fractions are, however, lost in the analytical process. That is why Q_{exp}, determined on the basis of these equations, must be corrected for these low-boiling fractions. This is usually done by comparison with their content in crude oils of the region.

The above outlined technique can be used for the evaluation of the amount of bitumoid expelled from shale formations or argillaceous units in sand-shale sequences, and for the mapping of the spatial variations in the oil-generating potential of source rocks in such sequences. The area of effective source rocks is divided by a square grid, and all parameters necessary for the evaluation of the

Table 3. Values of the parameters α, β, γ in the equation $(NSO) = \frac{\alpha}{\beta + h} + \gamma$ describing the heterocompound content in bitumoids in a clay layer in relation to the distance to the nearest reservoir, for various regions in West Western Siberia.

Area	Age		α	β	γ
Igrim-Narykar	J_{1-2}	MK_1^1	34.00	1.76	6.07
Sargat	J_{1-2}	PK_3-MK_1^1	19.85	2.75	6.82
Surgut	J_{1-2}	MK_1	21.05	2.74	3.87
Novyy Vasyugan	J_3vlg	PK_3-MK_1^1	49.73	8.86	5.08
Shaim	J_3vlg	PK_3	75.60	12.60	6.40
Surgut-Nizhnevartovsk	K_1v	PK_3-MK_1^1	13.08	3.84	6.49

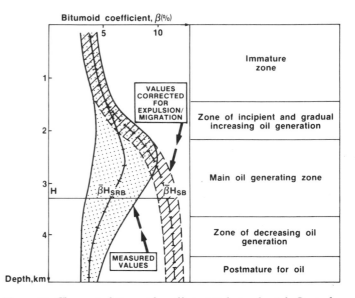

Figure 19. Change in bitumoid coefficient β during burial. Central lines indicate mean values. Compare with Figures 6 and 7.

Table 4. Elemental composition of syngenetic bitumoid after correction for the effect of expulsion.

Type of organic matter	Environment of deposition	Geochemical facies	Gradation, subgradation of katagenesis*	Elemental composition of autochthonous syngenetic bitumoid, %				
				C	H	N	S	O
Essentially terragenic	Shallow-water marine and continental	Pyritic-sideritic and sideritic	PK	77-80	8-10	0.3-0.6	1.0-0.8	13.7-8.6
			MK_1^1	80-82	8-10	0.3-0.6	0.8-0.5	10.9-6.9
			MK_1^2	82-84	8-10	0.3-0.6	0.5-0.3	9.2-5.1
			MK_2	84-86	8-10	0.3-0.6	0.3-0.2	7.4-3.2
Aquagenic	Deep-water marine	Pyritic and essentially pyritic	PK	78-81	9-11	0.3-0.6	3.0-2.0	9.7-5.4
			MK_1^1	81-83	9-11	0.3-0.6	2.0-1.0	7.7-4.4
			MK_1^2	83-85	9-11	0.3-0.6	1.0-0.5	6.7-2.9
			MK_2	85-87	9-11	0.3-0.6	0.5-0.3	5.2-1.1

*For correlation with vitrinite reflectance scale see Figures 12 and 13.

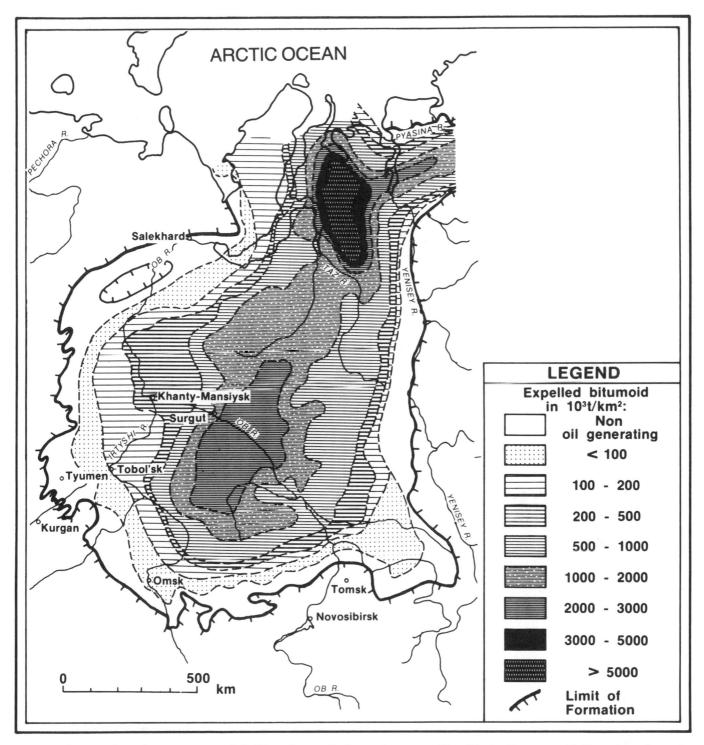

Figure 20. Effective oil-generating zonation of the Upper Jurassic Bazhenov formation, West Siberia. Compare with Figure 21 and note that the northern area, though less organic rich, is a more effective charge area due to higher maturity.

amount of expelled bitumoids are determined at the grid intersections expressed either as t/sq km (q_s) or t/cu km (q_v). The total amount of expelled bitumoids is then obtained by summation.

The calculation of the amount of bitumoid expelled at a given grid point from the Bashenov shale formation of West Siberia is presented as an example. For this point, L = 18 m, ρ = 2.4 t/cu m, b = 0.8 percent, α = 49.7, β = 8.86, and

γ = 5.1. Assume that V_{exp} = 2.0 and that the correction for light fraction loss is 1.5. Then for an area of 1 sq km, and using equation (11), we obtain:

$$q_s = \frac{2 \times 49.7 \times 0.8 \times 10^4 \times 2.4}{5.1 - 2.0} 1.5 \frac{1}{0.43} \lg\left(1 + \frac{18}{2 \times 8.86}\right)$$

$$= 640 \times 10^3 \text{t/sq km}$$

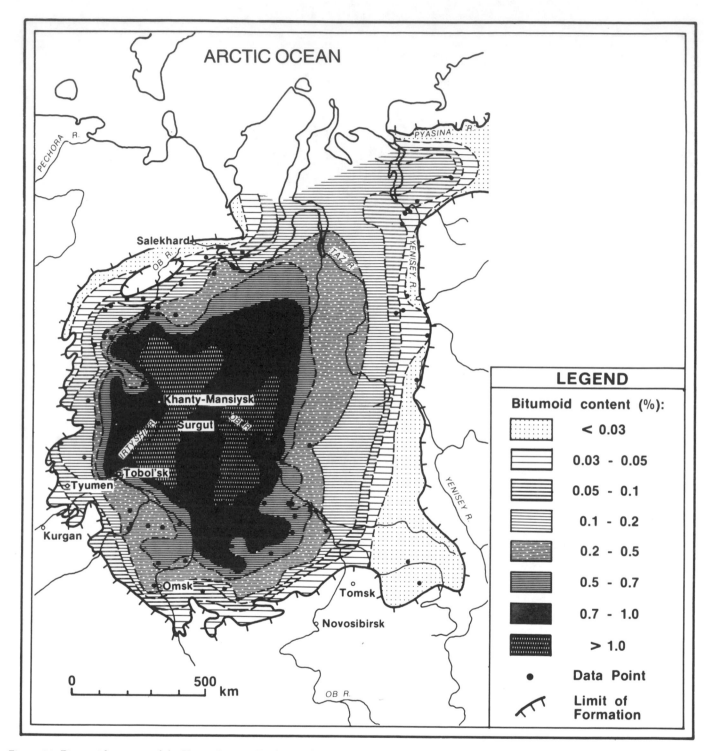

Figure 21. Bitumoid content of the Upper Jurassic Bazhenov formation, West Siberia.

Using this technique, maps of the oil-generating potential for all formations of Western Siberia were constructed (Kontorovich et al, 1967, 1971). As an example, a map of the bitumoid concentration in Volgian and Berriasian sediments and a map of the extension of effective source rocks in this interval are given as Figures 21 and 20 respectively. When analyzing these maps in 1967, it was noted that the principal regions of oil accumulation in the West Siberian Basin appear to be confined to the areas with the most intense bitumoid expulsion from Volgian and Berriasian sediments.

Effective Oil Generation From Non-Shale Source Rocks

The above method of source rock mapping is not always feasible in practice, however, because of the lack of information on the regional variations in bitumoid composition and richness in the clay layers. Moreover, this method cannot be applied to sand-silt and carbonate

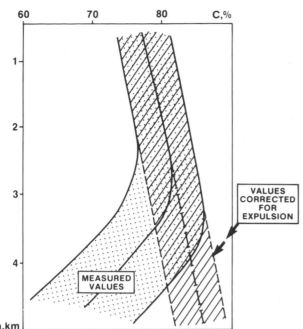

Figure 22. Change in carbon content of bitumoids during burial.

formations, which need a different approach now to be discussed.

If we know the thickness (L) of an oil-generating formation, its density, SRB content, and the coefficient of bitumoid expulsion (K) within a certain area (S), then:

$$Q_{exp} = (0.01K/1 - K)\rho LSb. \tag{13}$$

From (13), we obtain equations for the amount of expelled bitumoids per area, respectively, and volume of effective source rock (in t/sq km and t/cu km; L in meters):

$$q_s = (K/1 - K)\rho Lb \times 10^4 \tag{14}$$

$$q_v = (K/1 - K)\rho b \times 10^7 \tag{15}$$

To determine the average coefficient of bitumoid expulsion at a given point, see Figure 19. At depth H = 3,300 m, $\bar{\beta}_{H(SRB)}$ = 6 percent, and $\bar{\beta}_{H(SB)}$ = 12.5 percent. A given mass of OM, M, contains $0.01\bar{\beta}_{H(SRB)} \times M$ bitumoid at this depth, but if no primary migration took place its content would have been $0.01\bar{\beta}_{H(SB)} \times M$. Thus, the average coefficient of primary migration is

$$K = \frac{\bar{\beta}_{H(SB)} - \bar{\beta}_{H(SRB)}}{\bar{\beta}_{H(SB)}} \tag{16}$$

which in this example equals 0.52.

The map of the oil-generating strata of Lenian and Amginian age (Cambrian) of the Siberian Platform may serve as an example (Figure 24). On a given point on the Anabar anteclise, we read off the following values: L = 350 m, ρ = 2.72 t/cu m, b = 0.022 percent, K = 0.60. then, according to equations (14) and (15),

$$q_s = \frac{0.6}{0.4} \times 2.72 \times 350 \times 0.022 \times 10^4 = 0.31 \times 10^6 t/sq\ km$$

$$q_v = \frac{0.6}{0.4} \times 2.72 \times 0.022 \times 10^7 = 0.9 \times 10^6 t/cu\ km$$

The coefficient of primary migration may also be evaluated from the compositional parameters of the bitumoids, such as the carbon content (Figure 22). The combined data indicate that the coefficient of bitumoid expulsion from source rocks varies widely and that, therefore, no a priori average coefficient of expulsion can be given.

Effective source rock mapping based on a different principle, the evaluation of the amount of generated bitumoids, was proposed in 1971 by Tissot and Pelet. The shortcoming of this technique is that it does not take into account bitumoids which might have been expelled from source rocks.

Gas Generation

Less attention is generally paid to estimating the generated gas volumes, and consequently to the quantitative prediction of gas reserves. In principle, this problem could be solved in two ways. The first is to construct chemical-kinetic models according to the scheme proposed by Tissot, but for gas this is not yet possible. The second is based on material balance calculations. Uspenskiy (1954) worked out the theoretical basis for that approach, which was applied by Kontorovich and Rogozina (1967).

Concluding Remarks

The mapping techniques of hydrocarbon generating potential outlined above are widely used in the USSR. They have been applied to the Mesozoic formations of the West Siberian Basin, the Turanian and Skiffian plates, the late Precambrian, Paleozoic, and Mesozoic of the Siberian Platform, the Paleozoic of the Dnepr-Donets trough, and the late Precambrian and lower Paleozoic deposits of the Moscow synclise.

The next step is to select typical petroleum-generating drainage areas ("kitchens"). For these areas, the amount of expelled liquid and gaseous hydrocarbons can be calculated and the oil- and gas-in-place resources are rather accurately known. Through statistical correlation, the coefficient of accumulation for oil and gas can then be determined. Based on a sufficiently large set of typical (standard) accumulation coefficients determined this way, the expected reserves for other areas can be evaluated. Figure 25 shows that even for single fields a rather strong correlation seems to exist between reserves and amount of calculated expelled oil.

It should be stressed that estimating standard accumulation coefficients and evaluating total oil and gas resources are very complicated tasks, requiring a profound geological and geochemical knowledge of the hydrocarbon habitat of the study area. For example, there are three alternative hypotheses on the origin of the Cenomanian gas pools in the northern part of Western Siberia. According to the first hypothesis, this gas was generated in the coal-bearing deposits of the Lower Cretaceous and Cenomanian. According to the second hypothesis, the gas

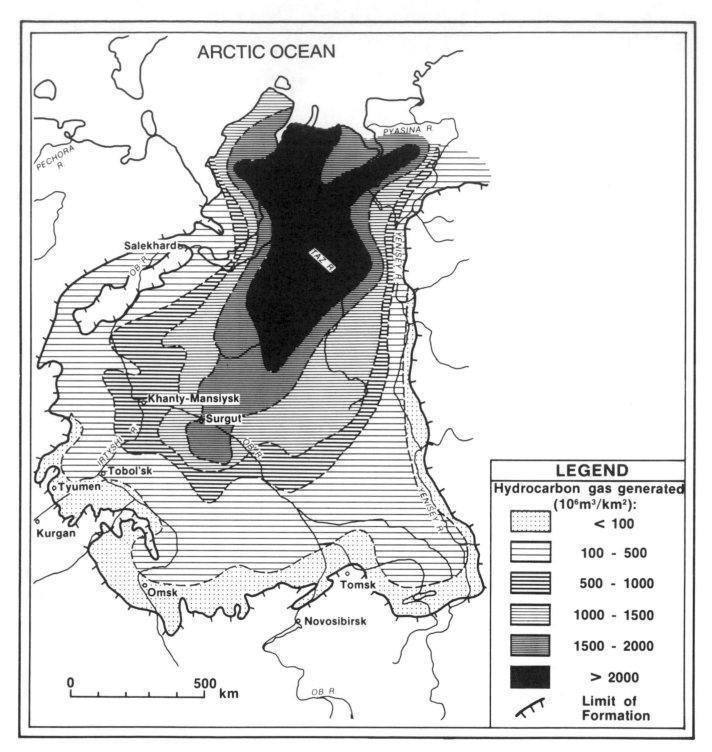

Figure 23. Effective gas-generating zonation of the Lower-Middle Jurassic of West Siberia. Giant gas accumulations are located in the highly effective charge area in the North.

migrated from the Jurassic. In the third hypothesis, it is of mixed Jurassic-Cretaceous origin. Of course, depending on the working hypothesis adopted, the accumulation coefficients are estimated differently. Only detailed typing studies based on isotopes make it possible to establish genetic relations between source horizons and oil and gas accumulations, and to decide which alternative is probably correct.

It is evident that geochemical methods for the quantitative evaluation of the petroleum potential of sedimentary basins are powerful tools. At the same time, the importance of these geochemical methods for prospect evaluation is often overestimated, and the limitations inherent to an actively developing branch of science, where much remains to be resolved, are not always sufficiently realized.

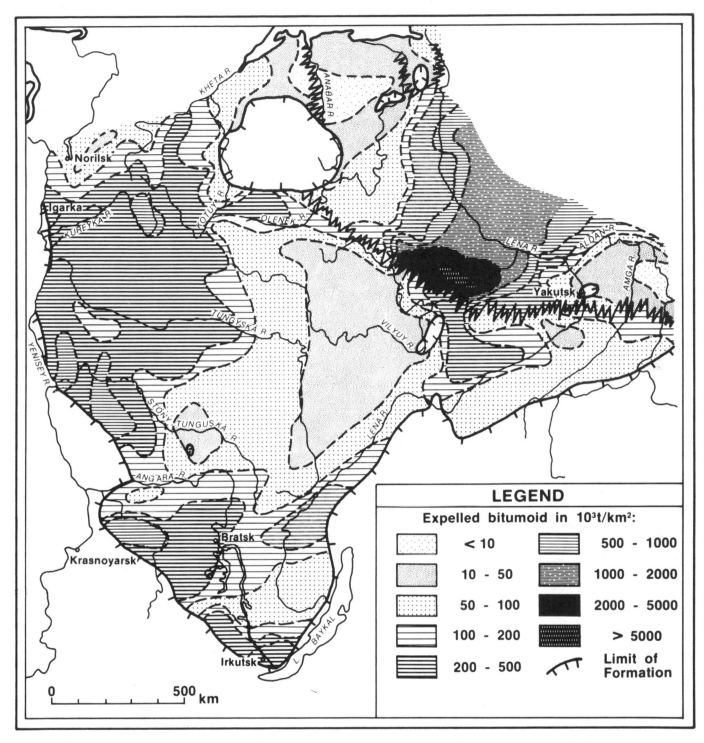

Figure 24. Effective oil-generating zonation of the Lower-Middle Cambrian of the Siberian Platform. Compare with Figure 1, and note the influence of the low maturity across the central north-south arch on the effective charge potential.

ACKNOWLEDGMENTS

I am indebted to G.J. Demaison for his suggestion to write this paper. Critical reviews of this paper by N.B. Vassoyevich, R.J. Murris, N.P. Laverov, V.V. Semenovich and F.S. Surkov are gratefully acknowledged. Also, I am grateful to L.I. Vinogradova, O.A. Drapp, Ye.A. Kontorovich, L.I. Levina, T.I. Tolmacheva, and S.J. Shaulina for their help in preparing the manuscript for printing.

REFERENCES CITED
(English translation of Russian titles only)

Akramkhodzhayev, A.M., 1973, Organic matter; the main source of oil and gas (in the light of new laboratory data):

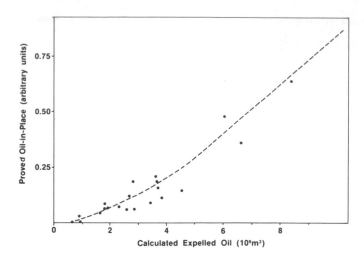

Figure 25. Oil-in-place in accumulations versus calculated expelled oil from their respective drainage areas; Mesozoic fields of the Sredne-Ob region, West Siberia.

Tashkent, FAN, 203 p.

Alexeyev, F.A., et al, 1973, Isotopic composition of carbon in gases of biochemical origin: Moscow, Nedra, 88 p.

——— , et al, 1978, Methane: Moscow, Nedra, 310 p.

Alpern, B., 1976, Fluorescence et réflectance de la matière organique dispersée et évolution des sédiments: Bulletin Centres de Recherches, Pau-SNPA, v. 10, p. 201-220.

Anonymous, 1975, Exploration risks in finding oil and gas: Shale Shaker, v. 26, p. 57-58.

——— , 1976, Quantitative methods of predicting petroleum prospects: Minsk, Be1NIGRI, 256 p.

——— , 1976, Principles of geological petroleum zonation applied to the prediction of the petroleum content of the subsurface: Moscow, Nedra, 301 p.

Biederman, E.W., 1965, Crude oil composition—a clue to migration: World Oil, v. 161, n. 7, p. 78-82.

Bogorodskaya, L.I., 1976, Oxygen-containing functional groups of sapropelic disseminated organic matter, *in* Geochemical criteria for the evaluation of Mesozoic and Paleozoic oil and gas prospects in Siberia: Novosibirsk, SNIIGG & MS Trudy, v. 231, p. 26-36.

Bordovskiy, O.K., 1967, Processes of organic matter accumulation and transformation in marine and oceanic sediments, *in* Origin of oil and gas: Moscow, Nedra, p. 22-32.

Botneva, T.A., 1972, Cyclic pattern in the processes of petroleum generation: Moscow, Nedra, 256 p.

Bua, K., 1978, Application of the notion of petroleum-bearing zone and of the analysis of similarity for the evaluation of resources, *in* Evaluation methods for the prognosis of oil and gas deposits: Moscow, Nedra, p. 104-107.

Burshtein, L.M., 1979, Geochemical-kinetic simulation of bitumoid generation in the Bazhenov formation of the West Siberian petroleum-bearing basin, *in* Evaluation methods of oil and gas source potential in sediments: Moscow, Moscow University Seminar, Abstracts of Reports, p. 117-118.

Burst, Y.F., 1969, Diagenesis of Gulf Coast clayey sediments and its possible relation to petroleum migration: AAPG Bulletin, v. 53, p. 73-93.

Buyalov, N.I., and V.D. Nalivkin, eds., 1979, Evaluation methods of petroleum potential: Moscow, Nedra, 332 p.

Cordell, R.Y., 1972, Depth of oil origin and primary migration: a review and critique: AAPG Bulletin, v. 56, p. 2029-2067.

Danilova, V.P., and A.E. Kontorovich, 1977, Rational system of methods for analyzing water-dissolved organic matter, *in* Disseminated organic matter in rocks and methods for their study: Novosibirsk, Proceedings, p. 108-119.

Demaison, G.J., 1977, Tar sands and super giant oil fields: AAPG Bulletin, v. 61, p. 1950-1961.

Dobryanskiy, A.F., 1948, Oil geochemistry: Leningrad, Gostoptekhizdat, 476 p.

Drobot, D.I., and R.N. Presnova, 1975, Sulfur-organic compounds in Vendian and Lower Cambrian oils from the Siberian Platform: Geologiya Nafti i Gaza, n. 1, p. 47-52.

——— , et al, 1974, Geochemical evaluation criteria for the petroleum potential of Precambrian and Lower Cambrian deposits of the southern Siberian Platform: Moscow, VNIGNI Trudy, v. 146, 160 p.

Dunton, M.L., and J.M. Hunt, 1962, Distribution of low molecular weight hydrocarbons in recent and ancient sediments: AAPG Bulletin, v. 46, p. 2246-2248.

Durand, B., 1975, Indices optiques, potential pétrolier et histoire thermique des sédiments, *in* Pétrogr. matière org. sédiments, relat. paléotemp. et potent. pétrol.: Paris, p. 205-215.

——— and J. Espitalié, 1971, Formation et évolution des hydrocarbures de C_1 à C_{15} et des gaz permanents dans les argiles du Toarcien du bassin de Paris, *in* Advances in Organic Geochemistry: Pergamon Press, p. 455-468.

Gol'bert, A.V., and A.E. Kontorovich, eds., 1978, Paleobiofacies of Volgian and Neocomian petroleum-bearing deposits of the West Siberian Plate: Moscow, SNIIGG & IMS Trudy, v. 348, 87 p.

Grunau, H.R., and U. Gruner, 1978, Source rocks and the origin of natural gas in the Far East: Journal of Petroleum Geology, v. 1, n. 2, p. 3-56.

Halbouty, M.T., et al, 1973, Factors controlling the formation of giant oil and gas fields; classification of basins, *in* Geology of giant oil and gas fields: Moscow, Mir, p. 410-431 (AAPG Memoir 14, Russian translation).

Holder, G.D., et al, 1976, Hydrate formation in subsurface environments: AAPG Bulletin, v. 60, p. 981-994.

Kalinko, M.K. 1964, Principal regularities in the distribution of petroleum in the earth's crust: Moscow, Nedra, 207 p.

Kartsev, A.A., 1969, Principles of petroleum geochemistry: Moscow, Nedra, 272 p.

——— , et al, 1971, Principal stages of oil and gas formation, *in* Developments in obtaining knowledge on the generation, migration and accumulation of oil and gas, and appropriate methods of petroleum prospect evaluation: Moscow, Vneshtorgizdat, p. 3-31.

Kontorovich, A.E., 1968, A scheme for the determination of oil-producing deposits: Academic Proceedings, USSR Doklady, v. 179, n. 3, p. 675-677.

——— , 1970, Theoretical principles of a volumetric-genetic method for the evaluation of potential petroleum resources: Novosibirsk, SNIIGG & MS Trudy, v. 95, p. 4-51.

——, 1976, Geochemical methods for the quantitative evaluation of petroleum potential: Moscow, Nedra, 250 p.

——, 1977, Historical approach to the quantitative evaluation of petroleum potential, *in* Principal geological and geophysical problems of Siberia: Novosibirsk, SNIIGG & MS Trudy, v. 250, p. 46-57.

——, 1978, Genetic principles for the separate evaluation of oil and gas, *in* Sediment-migrational theory of oil and gas generation: Moscow, Proceedings, p. 189-204.

——, and O.F. Stasova, 1964, On the geochemistry of oils from the West Siberian plains: Geologiya i Geofizika, n. 2, p. 13-24.

——, and ——, 1977, Types of oils in nature and regularities of their occurrence in the stratigraphic sequence: Moscow, 8th International Congress on Organic Geochemistry, Abstract of Reports, v. 1, p. 173-175.

——, and ——, 1977a, Geochemistry of Jurassic and Paleozoic oils of the southeastern West Siberian Plate, *in* Problems of geology and petroleum prospect evaluation in the Jurassic of the West Siberian Plate: Novosibirsk, SNIIGG & MS Trudy, v. 255, p. 46-62.

——, and ——, 1978, Types of oil in the sedimentary cover of the Earth: Geologiya i Geofizika, n. 8, p. 3-13.

——, and E.A. Rogozina, 1967, Limits of hydrocarbon gas formation in Mesozoic deposits of the West Siberian lowlands: Novosibirsk, SNIIGG & MS Trudy, v. 65, p. 13-25.

——, and S.G. Neruchev, 1971, Katagenesis of dispersed organic matter and petroleum formation, *in* Problems of oil and gas exploration in Siberia: Novosibirsk, Proceedings, p. 51-69.

——, and V.P. Danilova, 1973, Oil and gas generation in coal-bearing sedimentary strata, *in* New data for the geology and petroleum prospect evaluation of the Siberian platform: Novosibirsk, SNIIGG & MS Trudy, v. 167, p. 73-82.

——, and A.A. Trofimuk, 1973, Methods to study the evolution of oil and gas pools: Geologiya Nefti i Gaza, n. 7, p. 18-24.

——, and ——, 1976, Lithogenesis and oil/gas formation, *in* Fossil fuels, geological and geochemical problems of oil and bituminous rocks: Moscow, Proceedings, 25th International Geological Congress, p. 19-36.

——, and E.E. Fotiadi, 1976, Principles of quantitative prediction and exploration of oil and gas pools, *in* Fossil fuels, geological and geochemical problems of oil and bituminous rocks: Moscow, Proceedings, 25th International Geological Congress, p. 63-75.

——, et al, 1967, Petroleum-generating strata and conditions of oil formation in Mesozoic deposits of the West Siberian Lowlands: Leningrad, Nedra, 224 p.

——, et al, 1971, Geochemistry of Mesozoic deposits in the petroleum bearing basins of Siberia: Novosibirsk, SNIIGG & MS Trudy, v. 118, 86 p.

——, et al, 1973, The change of chemical composition of humic organic matter and its paramagnetic properties in the zone of katagenesis: Academiya Proceedings, USSR Doklady, v. 209, n. 6, p. 1431-1434.

——, et al, 1974a, Anaerobic transformations of organic matter in ancient marine sediments: Academiya Proceedings, USSR Izvestiya, Series in geology, n. 9, p. 112-123.

——, et al, 1974b, Organic geochemistry of Mesozoic petroleum-bearing deposits in Siberia: Moscow, Nedra, 192 p.

——, et al, 1974c, Limits and characteristics of the process of petroleum generation and accumulation in the zone of katagenesis, *in* Lithology and geochemistry of oil and gas provinces of Siberia: Novosibirsk, SNIIGG & MS Trudy, v. 193, p. 5-12.

——, et al, 1975, Petroleum geology of Western Siberia: Moscow, Nedra, 680 p.

——, et al, 1976, Geochemistry of bitumoids in subsurface waters of the West Siberian petroleum-bearing basin, *in* Geochemical criteria for the evaluation of the oil and gas prospects of the Mesozoic and Paleozoic of Siberia: Novosibirsk, SNIIGG & MS Trudy, v. 231, p. 3-25.

Kuklinskiy, A.Y., and R.A. Pushkina, 1974, Degree of branching of paraffin chains in naphthenes of high-boiling point oil fractions: Neftekhimia, v. 14, n. 4, p. 514.

Larskaya, E.S., ed., 1974, Influence of lithogenetic processes on the formation and productivity of petroleum source beds: Moscow, VNIGRI Trudy, v. 158, 195 p.

Makagon, Y.F., et al, 1973, The possibility of natural gas formation in gas-hydrate pools of the bottom zones of seas and oceans: Geologiya i Geofizika, n. 4, p. 3-6.

Modelevskiy, M.S., 1976, Oil and gas resources of foreign countries: Geologiya Nefti i Gaza, n. 8, p. 56-69.

Moody, J.D., 1975, An estimate of the world's recoverable crude oil resources: London, Proceedings, 9th World Petroleum Congress, v. 3, p. 11-20.

Neruchev, S.G., 1962, Source formations and migration of oil: Leningrad, Gostoptekhizdat, 224 p.

——, 1969, Source formations and migration of oil: Leningrad, Nedra, 240 p.

——, and I.S. Kovacheva, 1965, The influence of geologic conditions on the oil yield of source rocks: Academiya Proceedings, USSR Doklady, v. 162, p. 913-914.

——, et al, 1973a, Transformation of dispersed sapropelic organic matter at the stage of sediment diagenesis: Nauk, USSR Doklady, v. 212, n. 4, p. 972-975.

——, et al, 1973b, The main gas-phase as a stage in the katagenetic evolution of dispersed sapropelic organic matter: Geologiya i Geofizika, n. 10, p. 14-16.

——, et al, 1976, Paramagnetism of organic matter as an indication of the oil and gas generation process: Geologiya Nefti i Gaza, n. 10, p. 49-55.

Obolentsev, R.D., and A.Y. Baykova, 1973, Organo-sulfuric compounds of oils of the Volga-Urals region and Siberia: Moscow, Proceedings, 263 p.

Perry, E.A., and J. Hower, 1972, Late-stage dehydration in deeply buried pelitic sediments: AAPG Bulletin, v. 56, p. 2013-2021.

Philippi, G.T., 1965, On the depth, time and mechanism of petroleum generation: Geochimica et Cosmochimica Acta, v. 29, p. 1021-1049.

——, 1977, On the depth, time and mechanism of origin of the heavy to medium-gravity naphthenic crude oils:

Geochemica et Cosmochimica Acta, v. 41, p. 33-52.

Prasolov, E.M., and V.A. Lobkov, 1977, On the conditions of methane formation and migration (according to their isotopic carbon composition): Geokhimia, n. 1, p. 122-134.

Radchenko, O.A., 1965, Geochemical regularities in the world distribution of petroleum-bearing regions: Leningrad, Nedra, 314 p.

Rogozina, E.A., et al, 1974, On the place and conditions for the occurrence of the main gas-generating phase in the course of sediment burial: Academiya Proceedings, USSR Izvestiya, Seriya Geologicheskaya, n. 9, p. 124-132.

Romankevich, E.A., 1979, Advances in the study of organic matter in seas and oceans, *in* Accumulation and deposition of sediments: Moscow, Proceedings, p. 7-17.

Ronov, A.B., 1958, Organic carbon in sediments (in connection with hydrocarbon source rocks): Geokhimia, n. 5, p. 409-423.

Sheriff, R.E., 1974, Seismic detection of hydrocarbons—the underlying physical principles, *in* Houston, 5th Annual Offshore Technology Conference: Preprints, v. 2, p. 637-649.

Sickler, R.A., 1976, A survey of petroleum resources in the world outside centrally planned economics: Laxenburg, Austria, First IIASA Conference of Energy Resources, p. 204-206.

Sokolov, V.A., 1965, Processes of petroleum formation and migration: Moscow, Nedra, 276 p.

Starobinets, I.S., 1966, Geochemistry of petroleum in Central Asia: Moscow, Nedra, 292 p.

Starobinets, I.S., 1974, Geological-geochemical peculiarities of gas condensates: Leningrad, Nedra, 151 p.

Stasova, O.F., 1976, On the problem of oil generation in continental coal-bearing strata, *in* Geochemical criteria for hydrocarbon prospect evaluation in Mesozoic and Paleozoic formations of Siberia: Novosibirsk, SNIIGG & MS Trudy, v. 231, p. 111-118.

—— , 1977, Mesozoic oil types in Western Siberia, *in* Dispersed organic matter in organic-rich deposits and methods for their study: Novosibirsk, Proceedings, p. 83-88.

Strakhov, N.M., 1960, Basics of the theory of lithogenesis, part 1: Moscow, AN USSR, 212 p.

—— , 1961, Basics of the theory of lithogenesis, part 2: Moscow, AN USSR, 574 p.

—— , 1962, Basics of the theory of lithogenesis, part 3: Moscow, AN USSR, 550 p.

Tissot, B., 1973, Vers l'évaluation quantitative du pétrole formé dans les bassins sédimentaires: Revue de l'Association français des technicieus du Pétrole, n. 222, p. 27-31.

—— , and R. Pelet, 1971, New data on the origin and migration of oil, their mathematical modeling and use in exploration, *in* Recent advances in the understanding of hydrocarbon generation, migration and accumulation and in the development of evaluation methods of oil and gas prospects: Moscow, Vneshtorgizdat, p. 75-97.

Trofimuk, A.A., and A.E. Kontorovich, 1965, Some problems of the theory of the organic origin of oil and of the determination of source rocks: Geologiya i Geofizika, n. 12, p. 3-14.

—— , and S.G. Neruchev, eds., 1976, Hydrocarbon generation during sediment lithogenesis: Novosibirsk, Proceedings, 198 p.

—— , et al, eds., 1972, Distribution and conditions of formation of petroleum pools in Mesozoic rocks of the West Siberian Lowlands: Novosibirsk, SNIIGG & MS Trudy, v. 131, 312 p.

—— , et al, 1973, Progress in the theory of the organic origin of oil, *in* Actual geological and geochemical problems of fossil fuels: Moscow, Proceedings, p. 32-42.

—— , et al, 1975, Resources of biogenic methane in the World Ocean: Academiya Proceedings, USSR Doklady, v. 225, n. 4, p. 936-939.

Uspenskiy, V.A., 1954, Material balance of the processes occurring during the metamorphism of coal seams: Academiya Proceedings, USSR Izvestiya, Seriya Geologicheskaya, n. 1, p. 94-101.

—— , and O.A. Radchenko, 1947, On the problem of the origin of oil types: Leningrad, Gostoptekhizdat, 60 p.

—— , et al, 1964, Principles of the genetic classification of bitumen: Leningrad, Nedra, 267 p.

Vandenbroucke, M., 1971, Etude de la migration primaire: variation de composition des extraits de roche à un passage roche mère/reservoir, *in* Advances in organic geochemistry: Pergamon Press, p. 547-566.

Vassoyevich, N.B., 1958, Oil formation in terrigenous rocks (with reference to the Chokrak-Karaganda beds of the Terski foredeep), *in* Questions on the origin of oil: Leningrad, Gostoptekhizdat, p. 9-220.

—— , 1967, Theory of the sedimentary-migratory origin of oil (historical outline and current status): Academiya Proceedings, USSR Isvestiya, Seriya Geologicheskaya, n. 11, p. 135-156.

—— , 1971, On the terminology used in the study of organic matter in recent and fossil sediments, *in* Organic matter in sediments and rocks: Moscow, Proceedings, p. 218-238.

—— , 1973a, Lithogenesis, in the Great Soviet Encyclopedia, v. 14: Moscow, p. 559.

—— , 1973b, Principal regularities characterizing the organic matter in recent and fossil sediments, *in* The nature of organic matter and fossil deposits: Moscow, Proceedings, p. 11-59.

—— , and V.A. Uspenskiy, 1954, Petroleum geology, *in* Field guide to the geology of petroleum, v. 2: Leningrad, Gostoptekhizdat, p. 152-293.

—— , and G.A. Amosov, 1967, Geological and geochemical evidence for oil generation by organisms, *in* Origin of oil and gas: Moscow, Nedra, p. 5-22.

—— , and M.G. Berger, 1968, On the nomenclature of oils and their fractions according to their hydrocarbon composition: Geologiya Nefti i Gaza, n. 12, p. 38-41.

—— , et al, 1969, Principal phase of oil formation: Moscow University Vestnik, Seriya, 4, n. 6, p. 3-27.

Veber, V.V., 1966, Sedimentary facies favorable for oil formation: Moscow, Nedra, 274 p.

Vilks, G., and M.A. Rashid, 1977, Methane in the sediments of a subarctic continental shelf: Geoscience Canada, v. 4, n. 4, p. 191-197.

Vyshemirskiy, V.S., 1963, Geological conditions for coal and oil metamorphism: Saratov. University, 377 p.

—— , et al, 1971, Migration of dispersed bitumoids:

Novosibirsk, Nauka, 167 p.

Waples, D.W., 1979, Simple method for oil source bed evaluation: AAPG Bulletin, v. 63, p. 239-248.

Welte, D.H., 1964, Uber die Beziehungen zwischen Erdölen und Erdölmuttergesteinen: Erdöl und Kohle, Erdgas und Petrochemie, v. 17, p. 417-429.

Yevtushenko, V.M., et al, 1969, Main geochemical and lithological features of Cambrian deposits of Cuonamian type on the Siberian Platform, *in* Lithology and resources of alluvial platform deposits of Siberia: Novosibirsk, SNIIGG & MS Trudy, v. 98, p. 72-76.

Zabrodina, M.N., et al, 1978, Chemical types of petroleums and the transformation of petroleum in nature: Neftekhimia, v. 18, n. 2, p. 280-290.

Zhizhchenko, V.P., 1974, Paleogeographical methods in petroleum-bearing regions: Moscow, Nedra, 375 p.

New Albany Shale Group (Devonian-Mississippian) Source Rocks and Hydrocarbon Generation in the Illinois Basin

Mary H. Barrows
Illinois State Geological Survey
Champaign, Illinois

Robert M. Cluff
*Consulting Geologist**
Denver, Colorado

The New Albany Shale Group (Devonian-Mississippian) has long been suspected of being one of the major petroleum source beds in the Illinois Basin because of its high organic content and its stratigraphic relationship to many of the producing horizons. Previous stratigraphic and lithologic studies of the shales show that the New Albany was deposited in a deep-water, stratified anoxic basin that was conducive to the preservation of abundant, high-quality, organic matter in the sediments.

Detailed geochemical and microscopic analyses reported in this study were undertaken on drill cutting and core samples from 238 wells in Illinois, southwestern Indiana, and western Kentucky in an attempt to more clearly define the source rock potential of the New Albany and to identify areas where generation most likely occurred. The sampling pattern was chosen to represent all major stratigraphic intervals of the New Albany Shale over the entire basin.

Results of this study show that the organic matter types are closely controlled by the depositional environment of the shales and can be largely predicted by facies analysis. Laminated, black, anoxic shales contain high amounts (2.5 to 9 percent total organic content [TOC]) of sapropelic, Type II kerogen. Predominant organic components of this facies are amorphous organic matter, alginite (mainly *Tasmanites*), and minor amounts of vitrinite, semifusinite, and inertinite. In contrast, the bioturbated, greenish-gray, dysaerobic shales contain only modest amounts of humic, Type III organic matter (1 to 2 percent TOC). The laminated black shale facies is considered an excellent liquid hydrocarbon source, whereas the bioturbated gray shale facies is unlikely to have generated significant quantities of any hydrocarbons.

Vitrinite reflectance and liptinite fluorescence reveal a regionally consistent pattern of increasing maturation southward toward the area of greatest paleoburial depths and possibly higher heat flow. Highest maturation levels for the New Albany Shale are found in southeastern Illinois and adjacent western Kentucky: they are within the uppermost oil window and approach the wet gas or gas condensate generation zone. New Albany levels of maturity found across most of the central Illinois Basin are within the principal oil generation zone and are consistent with the oily, generally undersaturated condition of Illinois petroleum reservoirs. Furthermore, the geographic distribution of both favorable organic facies and maturation in the New Albany is also consistent with a geologic model, which implies that this stratigraphic unit has sourced most of the basin's oil and gas fields. All Mississippian and Pennsylvanian fields, which account for more than 90 percent of the basin's reserves, are probably charged with New Albany oil that may be locally diluted with hydrocarbons from younger, minor source beds. Many, if not all, of the Devonian and Silurian pools are also associated with nearby New Albany source rocks that most likely charged these stratigraphically deeper reservoirs. Only a few small fields in the Ordovician "Trenton" Limestone follow a pattern that is inconsistent with a New Albany source and contain geochemically distinct crudes.

INTRODUCTION

Oil has been commercially produced from the Illinois Basin for nearly 100 years. It is estimated that during that time more than 3.2 billion barrels of oil have been produced, primarily from Devonian and younger strata (Jacob Van Den Berg, personal communication). Despite this long history of petroleum exploration and production, relatively few attempts have been made to identify and evaluate the probable source beds for oil and gas in the basin. This paper presents the results of a recently completed study of the New Albany Shale Group (Devonian-Mississippian), which is thought to have played a major role in petroleum generation throughout the Illinois Basin.

---

*Formerly with the Illinois State Geological Survey.

111

This study was initiated in 1976 at the request of the United States Department of Energy (DOE) as one of the several cooperative studies of Mississippian-age and Devonian-age black shales throughout the northeastern United States. These studies were undertaken to evaluate the potential of the shales as sources of commercial hydrocarbons, particularly natural gas. Devonian shale strata in the Appalachian Basin from eastern Tennessee to southern New York have yielded commercial quantities of natural gas from fractured reservoirs since the early 1800s. The Big Sandy field in eastern Kentucky has produced more than 2 Tcf of gas from these shales (Hunter and Young, 1953).

The Illinois State Geological Survey studied the New Albany Shale Group in Illinois to evaluate geologic and geochemical factors thought to play a role in gas accumulation (Bergstrom, Shimp, and Cluff, 1980; Cluff, Reinbold, and Lineback, 1981). Our objectives in this study were to characterize the types of organic matter in the shale by petrographic methods and to relate variations in organic matter to the stratigraphy, paleogeography, and paleoenvironment. We also attempted to map the degree of thermal maturity of the organic matter in the shales, chiefly by means of vitrinite reflectance analysis. Our work relied primarily on reflected light microscopy of acid-insoluble (kerogen) residues of New Albany Shale.

Lastly, we attempted to evaluate the overall role of the New Albany as a source bed in the Illinois Basin, and to identify areas where hydrocarbon generation is most likely to have occurred within the New Albany Shale Group. Because our study focused on only one of several potential source beds in the Paleozoic section, we have not attempted to account for all of the oil and gas fields in the basin nor to evaluate possible contributions of other source rocks to petroleum accumulations now thought to be sourced primarily from the New Albany Shale Group. Until detailed studies of other potential source beds in the stratigraphic column are completed, and until studies employing modern geochemical tests to correlate probable source rocks with associated petroleum accumulations are completed, many questions concerning the relationships between petroleum source rocks and accumulations in the Illinois Basin will remain unanswered.

REGIONAL GEOLOGY

The Illinois Basin is a broad, relatively shallow intracratonic basin covering an area of about 60,000 sq mi (155,400 sq km) in southern Illinois, southwestern Indiana, and western Kentucky (Figure 1). The sedimentary fill reaches a maximum thickness of about 14,000 ft (4,267 m) near the center of the Fairfield Basin (a major sub-basin) in southern Illinois, and probably exceeds 20,000 ft (6,096 m) in portions of the Moorman Syncline of western Kentucky. This sedimentary fill consists of Paleozoic rocks ranging in age from Middle Cambrian (or possibly older) to Pennsylvanian; a thin veneer of Pleistocene overlies much of the northern part of the basin, and in the south a slight overlap of Cretaceous and Tertiary rocks occurs at the northern end of the Mississippi Embayment. The basin is bounded by several arches and domes as shown in Figure 1.

The structure of the Illinois Basin is the complex product of a long history of tectonism (Swann, 1967; Atherton, 1971; and Willman et al, 1975). Apparently, subsidence of the Illinois Basin began in Cambrian time and continued intermittently throughout the Paleozoic. The greatest depth-of-burial of basin sediments probably occurred during the late Pennsylvanian or Permian (Damberger, 1971). During the Paleozoic, arches defined the western, northern, and eastern flanks of the basin, but it was open to the south. The Du Quoin Monocline and La Salle Anticlinal Belt had sharply defined the western and eastern margins (respectively) of the Fairfield Basin by the close of the Mississippian. There is evidence that these structures also influenced earlier sedimentation (Buschbach and Atherton, 1979; Cluff, Reinbold, and Lineback, 1981). Although sedimentation probably continued into Permian time, any Permian or post-Permian sediments have since been eroded with the exception of about 3,000 ft (460 m) of presumed Permian strata recently found in a graben in western Kentucky (Kehn, Beard, and Williamson, 1979). Cretaceous sediments are found in extreme southern Illinois at the northern limit of the Mississippi Embayment. These two exceptions notwithstanding, the lack of a post-Pennsylvanian sedimentary record makes it very difficult to establish the tectonic history of the basin from the Permian to the present. Uplift of the Pascola Arch closed the basin on the south sometime between the Pennsylvanian and Late Cretaceous (Buschbach and Atherton, 1979). Movement along the Shawneetown-Rough Creek fault zone (Krausse and Treworgy, 1979) and emplacement of numerous igneous intrusive bodies in southern Illinois (Clegg and Bradbury, 1956) are also thought to be post-Pennsylvanian, probably Late Permian age.

Ultimate oil production from Paleozoic rocks of the Illinois Basin is estimated at approximately 4 billion barrels (Swann and Bell, 1958; Bond et al, 1971). Most production to date is from lenticular sandstone reservoirs in Chesterian (Upper Mississippian) and Pennsylvanian strata (Figure 2). Most of the major petroleum accumulations are associated with several large anticlinal structures, although numerous smaller stratigraphic oil and gas fields have also been discovered. Significant oil reserves also occur in carbonate reservoirs of Valmeyeran (Middle Mississippian), Devonian, and Silurian age (Figure 2). A minor amount of oil has been produced from Champlainian (Middle Ordovician) "Trenton" Limestone.

ILLINOIS BASIN SOURCE ROCKS (PREVIOUS STUDIES)

Previous evaluation of potential petroleum source strata in the Illinois Basin focused only on the quantity of organic matter (total organic carbon) in various stratigraphic intervals. Organic-bearing sediments proposed as possible source strata (Bond et al, 1971) include: (1) Pennsylvanian coals and associated marine black shales, which have total organic carbon values as high as 30 percent; (2) Chesterian dark gray shales (mostly prodeltaic deposits), which have organic carbon values ranging between 1 and 1.5 percent; (3) Valmeyeran fine-grained limestones, including probable tidal flat, lagoonal, and deep-water deposits; (4)

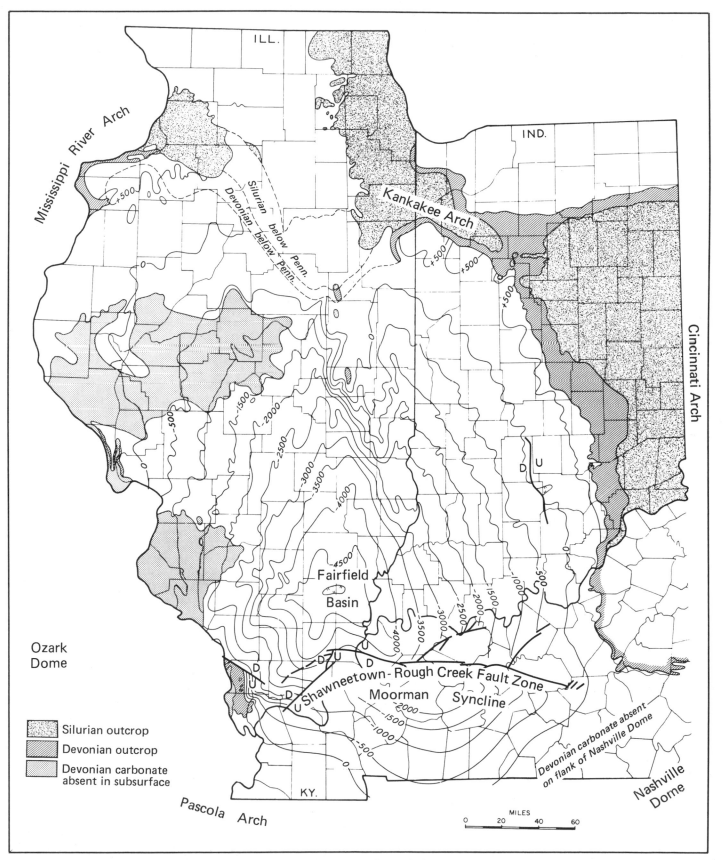

Figure 1. Geologic structure of the Illinois Basin on the base of the New Albany Shale Group. In the stippled areas of western Illinois, the New Albany Shale rests directly upon Silurian or, in a few places, Ordovician carbonates (after Swann, 1967; Stevenson, Whiting, and Cluff, 1981).

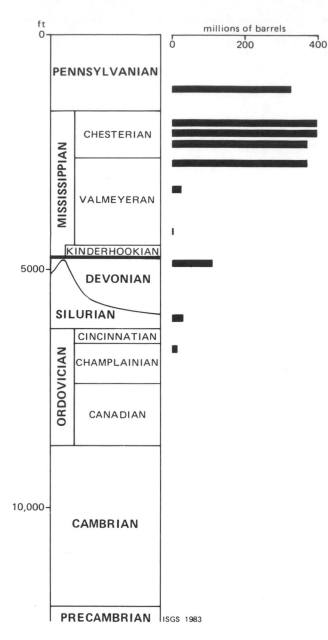

Figure 2. Geologic column for the Illinois Basin and stratigraphic distribution of cumulative oil production to 1955 (after Swann and Bell, 1958). Although many millions of barrels of reserves have been produced since the original publication of this illustration, the relative importance of the various pay zones has remained the same. Stratigraphic position of the New Albany Shale shown by heavy line at top of Devonian.

Kinderhookian-Middle Devonian marine black shales of the New Albany Group; (5) Lower and Middle Devonian fine-grained limestones; (6) Cincinnatian (Upper Ordovician) dark shales of the Maquoketa Group, which have organic carbon contents ranging from 1.25 to greater than 4 percent; (7) Cambrian and Ordovician carbonates of the Knox Megagroup; and (8) Cambrian shale and shaly carbonates of the Eau Claire Formation. Recent unpublished work suggests that some parts of the Middle Ordovician Plattville Group are also possible source rocks.

Of the units cited by Bond et al (1971) as potential source rocks on the basis of organic carbon content alone, the New Albany Shale Group is the thickest, most widespread, continuous, organic-rich unit in the Illinois Basin. Because of its proximity to oil reservoirs in the underlying Devonian and Silurian carbonates, geologists have long suspected that it played a major role in petroleum generation across much of the basin.

Stevenson and Dickerson (1969) examined approximately 350 New Albany Shale samples throughout Illinois. They found organic carbon contents ranging from less than 1 percent to more than 9 percent.

Lineback (1970) reported organic carbon contents as high as 20 percent by weight in some beds of the New Albany in Indiana. However, Stevenson and Dickerson (1969) were unable to demonstrate any relationship between high organic carbon contents in the New Albany Shale Group and the locations of oil reservoirs in the underlying carbonate and sandstone reservoirs of the Hunton Megagroup (Silurian-Devonian), which were believed to have been charged with oil from the New Albany Shale Group. One possible reason why no relationship was found between organic carbon content and proven reservoirs is that the New Albany Shale contains sufficient organic matter to be a potential source rock throughout its extent. Surprisingly, little work had been published until the present study to describe the organic matter qualities (kerogen types) in the New Albany or to identify those mature areas where petroleum generation is likely to have occurred. Hence, the early attempts at correlating organic carbon distributions, alone, with regional petroleum occurrence may not have been fruitful since regional maturation levels and kerogen types were not considered.

STRATIGRAPHY AND DEPOSITIONAL ENVIRONMENTS OF THE NEW ALBANY SHALE

The New Albany Shale Group and equivalent strata (Devonian-Mississippian) occur widely in the subsurface of the Illinois Basin. The shale is thickest (more than 460 ft, or 140 m) in Hardin County, Illinois, and adjacent western Kentucky (Figure 3), and thins from this southern depocenter in a roughly concentric pattern. The New Albany Shale Group is also thick in western Illinois (Figure 3).

Investigations of new Albany stratigraphy, lithology, and structure in Illinois, Indiana, and western Kentucky, respectively, were published by Cluff, Reinbold, and Lineback (1981); Lineback (1970); Hasenmueller and Woodard (1981); and Schwalb and Norris (1980). These studies indicated a regional facies pattern in which laminated, black organic-rich shales are dominant in the central area of the southern depocenter (Figure 3). From this depocenter, the black shales thin and grade laterally into bioturbated, gray and greenish-gray, organic-lean shales toward the northern and western margins of the shale basin. Gray shale facies are dominant in the area of relatively thick shales in western Illinois. These regional trends are illustrated in the stratigraphic cross section (Figure 4) and by a map showing gross thickness of radioactive "black" shale (Figure 5). Because of its high

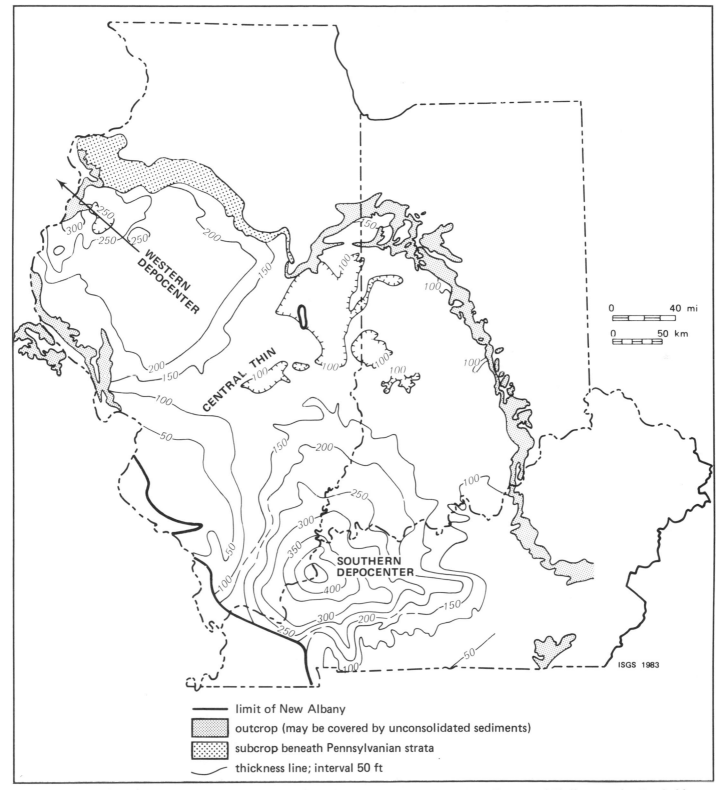

Figure 3. Total thickness of the New Albany Shale Group in the Illinois Basin (from Bergstrom, Shimp, and Cluff, 1980; after Reinbold, 1978; Bassett and Hasenmueller, 1978; and Schwalb and Potter, 1978).

organic content, the black shale facies tends to have high concentrations of radioactive elements such as uranium and thorium. These concentrations cause "hot" zones on natural gamma-ray logs—in some cases 200 to 400 API units above normal shale background.

The stratigraphic nomenclature of the New Shale Group is complex and beyond the scope of this paper. Each of the stratigraphic units is, for the most part, based on a single

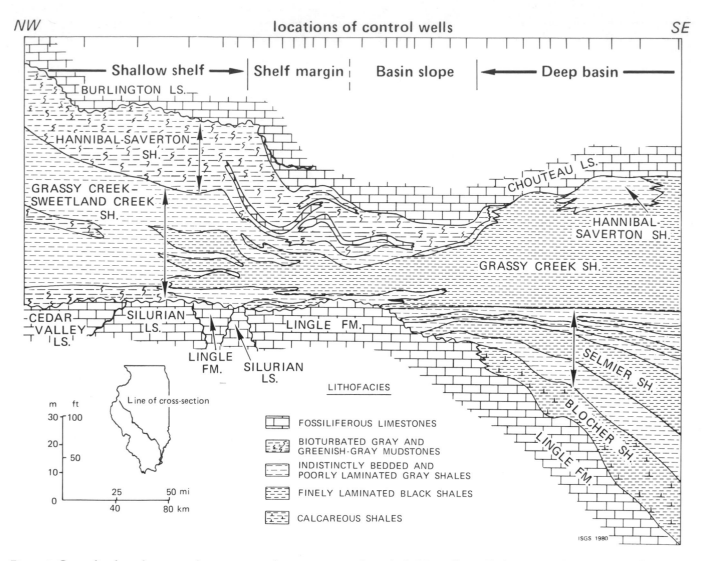

NW   locations of control wells   SE

Figure 4. Generalized northwest-southeast stratigraphic cross section through the New Albany Shale group in Illinois (modified from Reinbold, 1978) illustrating broad lateral facies changes from shelf to slope to basin environments (after Cluff, Reinbold, and Lineback, 1981). The finely-laminated black shales are the richest in total organic carbon.

dominant shale lithology that is both laterally and vertically persistent. Thickness maps of the many identified shale units have been published in the references cited previously. Interpretation of these thickness and facies patterns, in conjunction with studies of cores and outcrops, suggests that the New Albany was deposited in a relatively deep-water, stratified, largely anoxic basin centered in southeastern Illinois and western Kentucky (Cluff, 1980; Cluff, Reinbold, and Lineback, 1981).

As stated previously, the Illinois Basin was open to the south, probably until the Mesozoic when the Pascola Arch rose and closed it off. Most of the present-day anticlinal folds did not influence the deposition of the New Albany Shale; only the LaSalle Anticlinal Belt and Du Quoin Monocline appear to have had sufficient relief to affect Devonian shale deposition. The deepest part of the basin probably was considerably south of its present location—presumably near the area of greatest shale thickness (Figure 3) or even farther south, in western Kentucky. New Albany paleogeography, as well as it can be

defined by gradual facies changes in the shale, is shown in Figure 6.

Within the deep basin area, anoxic conditions generally prevailed throughout New Albany deposition. The lack of oxygen precluded the existence of burrowing macrofauna and restricted the microbial flora to anaerobes. Abundant organic matter was preserved in the sediment, resulting in thick accumulations of black shale. In contrast, the shallow shelf region of west-central Illinois was better oxygenated, probably because of the moderate mixing of superficial water by wind, waves, currents, and thermally driven convection. Burrowing infauna and aerobic bacteria prospered and almost completely destroyed most organic matter before it could be buried. The environment was not fully oxygenated, however, as calcified invertebrates were almost totally absent. This environmental condition corresponds to the "dysaerobic" zone of Rhoads and Morse (1971).

The basin slope region was a transitional environment that was at times largely anoxic, and at other times partly

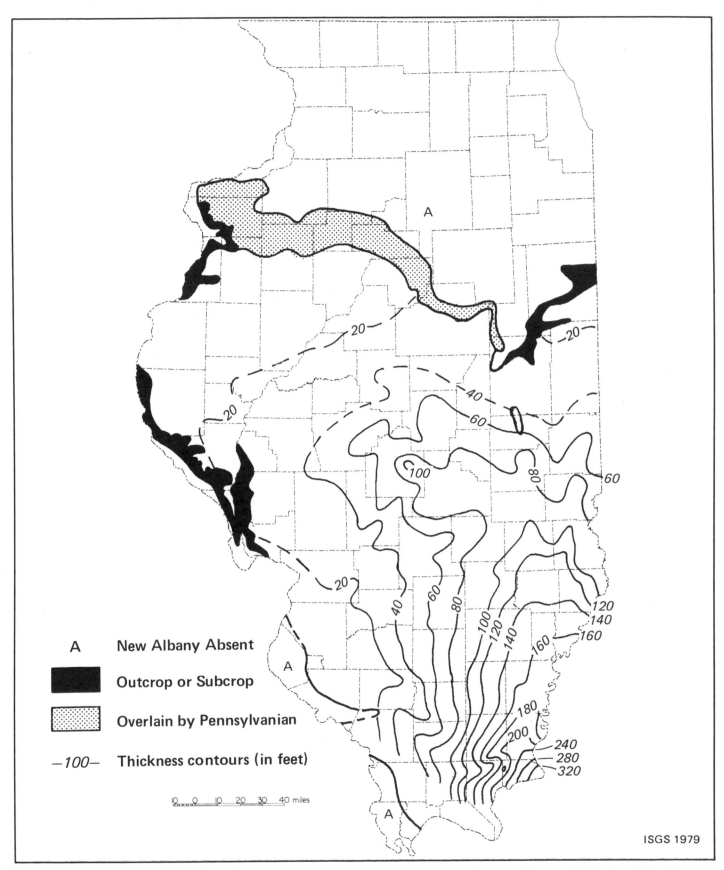

Figure 5. Cumulative thickness of radioactive shale within the New Albany Shale Group in Illinois. Radioactive shale is here defined as having a gamma ray log value 60 API units above normal shale base line (from Cluff, Reinbold, and Lineback, 1981). Data are not available for Indiana or western Kentucky due to lack of gamma ray logs.

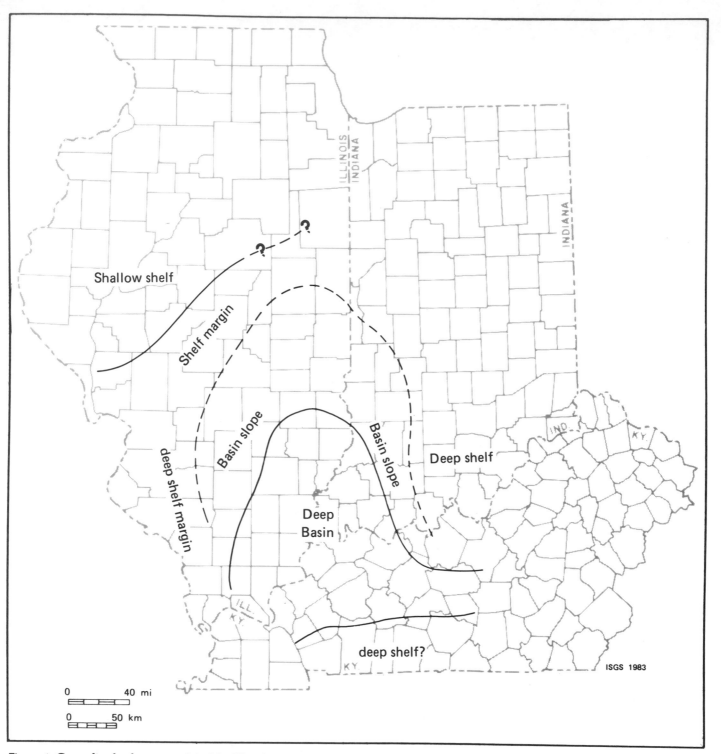

Figure 6. Generalized paleogeography of the Illinois Basin during New Albany deposition (from Lineback, 1980).

dysaerobic. The interface between these two stratified zones apparently fluctuated within the water column, and as a result swept repeatedly across the slope area of central Illinois and adjacent Indiana. Thus, the two major facies belts interfinger and grade into one another in this area (Figure 4). The amount of bathymetric relief across the basin slope is difficult to determine. If one uses the analogy of modern anoxic basins as proposed by Byers (1977), the

dysaerobic shelf area was probably under not less than 50 m (150 ft) of water, and the fully anoxic zone under at least nearly 150 m (nearly 500 ft) of water.

The paleogeographic and depositional interpretation presented above is critical to understanding the distribution of hydrocarbon source facies in the New Albany. As is shown in the following sections, the types of organic matter found in the shales and their state of preservation are

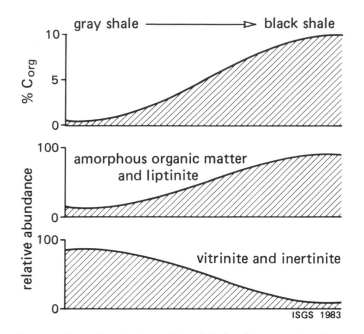

Figure 7. Generalized relationships of shale color, organic carbon content, and organic matter type for the New Albany Shale Group.

intimately associated with the depositional conditions and patterns. Once the depositional and stratigraphic framework is understood, the petroleum source potential can be confidently projected into areas for which little or no information is available.

RESULTS OF GEOCHEMICAL AND MICROSCOPIC INVESTIGATIONS OF NEW ALBANY SHALE[1]

Quantity of Organic Matter

Previous studies by Stevenson and Dickerson (1969) and Lineback (1970), and geochemical data obtained for this study (Frost, *in* Bergstrom, Shimp, and Cluff, 1980) suggest that sufficient organic matter has been preserved within at least some of the beds of the New Albany Shale Group to qualify them as potential source beds for hydrocarbons throughout its present extent.

Most of the samples analyzed for this study had organic carbon values greater than the critical lower limit range cited by Tissot and Welte (1978). Of 371 core and drill-cutting samples of shale, 319 samples had organic carbon contents greater than 1.0 percent (up to 15.6 percent), 40 samples had values between 0.50 and 1.0 percent, and only 12 samples had less than 0.50 percent organic carbon. Frost (*in* Bergstrom, Shimp, and Cluff, 1980) reports that the lowest organic carbon values are generally found in the uppermost shale units, the Hannibal and the Saverton shales (Figure 4). Organic carbon values are generally moderate in the northwest part of the basin for both the Grassy Creek (average 2.79 ± 0.72 percent) and Sweetland Creek shales (1.66 ± 1.45 percent). In all other parts of the basin the Grassy Creek and Sweetland Creek shales have high to very high organic carbon contents—averaging from 5 to 9 percent in the Grassy Creek Shale and from 2.5 to 6.5 percent in the Sweetland Creek

Shale (including the Selmier Shale in the southeast). The lowermost shale unit, the Blocher Shale, has high organic carbon values averaging from 4 to 9 percent. The limited number of analyses done in this study for samples in Indiana and Kentucky indicates that organic carbon values in these areas are comparable to or higher than values obtained for Illinois shales (Frost, *in* Bergstrom, Shimp, and Cluff, 1980). In general, the black shales have organic carbon contents averaging from 2.5 to 9 weight percent; greenish-gray shales average 1 to 2 weight percent organic carbon. The relationship between shale color and organic carbon is illustrated in Figure 7.

High organic content of the shale generally correlates directly with high natural gamma-ray intensities recorded on borehole radioactivity logs and with high gas release values for core samples (Cluff and Dickerson, 1982; Lineback, 1980).

Types of Organic Matter

The types of insoluble organic matter (kerogen) deposited and preserved in a sedimentary rock subsequently influence the amount and composition of hydrocarbons that can be generated from that rock (Tissot and Welte, 1978; Hunt, 1979). To evaluate the quality of insoluble organic matter in a potential source bed, both geochemical and microscopic methods are commonly used. In our study, both reflected white and fluorescent light microscopy were the main methods used to determine organic composition (see Annex I for details of the methods used) although some pyrolysis and kerogen elemental analysis measurements were also considered.

Our microscopic analyses of New Albany Shale samples indicate that the occurrence and abundance of the various types of organic matter are sedimentary facies dependent. The greenish-gray bioturbated shale facies contain little organic matter, and often acid-maceration treatment of 20 to 30 grams of shale does not yield sufficient material to permit characterization of the types of organic matter present. Only vitrinite and inertinite, the humic group macerals, are observed in the concentrated organic matter pellets of these shales. Some isolated patches of undissolved mineral matter with small, unconnected stringers of amorphous organic matter are also observed within the pellets of the greenish-gray shales (Figure 7). This predominance of humic material in the greenish-gray shales corresponds to the Type III kerogen of Tissot and Welte (1978). This kerogen assemblage is interpreted to result from selective preservation of only those organic constituents most resistant to destruction by benthic invertebrates (detritus feeders) and aerobic bacteria in the moderately-oxygenated gray shale environment.

The brownish-black and black laminated shale facies deposited under anoxic water is rich in amorphous organic matter that is lower in reflectance than vitrinites in the same sample and which often fluoresces a weak, dark brown. It usually accounts for 90 to 95 percent of the organic matter, while recognizable macerals (liptinite, vitrinite, inertinite)

---

[1]The methods used for evaluating the quantity, quality, and maturation of organic matter are described in Annex 1 at the end of this paper.

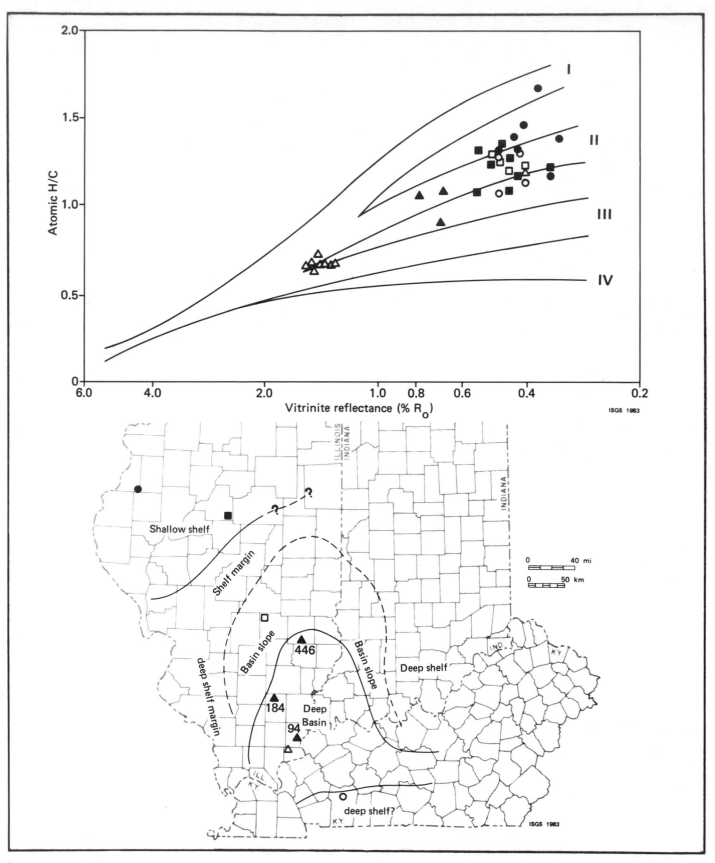

Figure 8. Atomic H/C versus percent vitrinite reflectance for New Albany Shale Group core and drill cutting samples (core data from Zielinski, 1977a, 1977b; Zielinski and Nance, 1979; drill cutting data are Illinois State Geological Survey data; kerogen type boundaries after Dow, 1977). Index map shows location of sampled cores and paleogeographic provinces for the New Albany Shale.

account for only 5 to 10 percent of the concentrates (Figure 7). The amorphous matter in the laminated black shale facies is believed to be derived primarily from marine algae, although very small unrecognizable fragments of humic material may also be included in our estimates of the amounts of amorphous organic matter. The alginite *Tasmanites* is the most commonly recognized liptinite maceral, and it is even easily identified in thin sections, core samples, and drill cuttings due to its large size and distinctive shape. Its abundance appears to vary randomly. Vitrinite is generally more common than inertinite except in southeastern Illinois, where the inertinite maceral semifusinite is commonly observed. This mixed assemblage of predominantly marine, with some nonmarine, organic matter constituents in the brownish-black and black shales is typical of Type II kerogen, as defined by Tissot and Welte (1978). This kerogen assemblage is interpreted to be well-preserved, locally-derived marine planktonic organic matter.

A limited number of published elemental analyses of New Albany organic matter support our microscopic interpretation of kerogen types. As part of the cooperative DOE study, Mound Facility collected and analyzed several samples of New Albany Shale from cores in each of the major paleogeographic provinces of the Illinois Basin. For the samples selected, they determined the ash-free atomic weight percent of organic carbon, hydrogen, nitrogen, and oxygen. Vitrinite reflectance analyses were also run on splits of the same samples. These data, compiled from several reports, are summarized in Figure 8 in which the atomic H/C ratios are plotted against the vitrinite reflectance level of the sample (Zielinski, 1977a, 1977b; Zielinski and Nance, 1979). Note that most of the data plots within the region are considered typical of Type II kerogens; for samples with vitrinite reflectances greater than 1.00 percent, the chemical data does not clearly pinpoint the kerogen type.

Although in polished blocks of shale the amorphous organic matter surrounds and is not clearly distinguishable from the inorganic matrix, the continuous network-like nature of this material is clearly apparent in the acid concentrated organic residues from most areas of Illinois. In southeastern Illinois where vitrinite reflectance levels range from 0.90 to greater than 1.35 percent, however, the continuous network of amorphous material is rare, and in its place a non-fluorescent, lighter gray mass of unidentifiable fragments is more commonly observed (Figure 9a). Surrounding this region is an area in which vitrinite reflectance values range from 0.70 to 0.90 percent and where a mixture of the network and the unidentifiable fragments are observed within the same sample. This change in the nature of the amorphous organic matter is interpreted to be the result of increasing maturation, and probably is a direct indicator of areas where the amorphous organic matter has been chemically and physically altered by hydrocarbon generation.

Dark- to medium-gray uniform materials, generally of lower reflectance than vitrinite within the same sample, were observed as fillings in fusinite pores (Barrows, Cluff, and Harvey, 1979). No fluorescence of the filling materials was observed. Dark- to medium-gray angular particles of lower reflectance than vitrinite within the same sample were

also observed. These particles appear to have been fillings of inorganic material dissolved by the acid maceration process. No fluorescence of these angular particles was observed. The fillings and the angular particles may be solid hydrocarbons (Castano, personal communication). Others refer to similar particles as exsudatinite (Stach et al, 1982), pyrobitumen (Alpern, 1970), or bitumen (Robert, 1973). Such particles are thought to be residues produced during maturation (Stach et al, 1982), and their presence is strong evidence of liquid hydrocarbon generation. The distribution of these two types of particles is shown in Figure 9b.

Organic Maturation

A vitrinite reflectance map for the New Albany Shale Group (Figure 10) shows distinct patterns that indicate trends of maturation of the organic matter within the shales. These maturation trends can be related to regional tectonics and paleoburial depths. Three areas of the New Albany Shale exhibit reflectances significantly greater than 0.6 percent $\overline{R}_o$. The northernmost of these is centered in Cumberland and Jasper counties, Illinois, between the Clay City Anticlinal Belt to the west and the La Salle Anticlinal Belt to the east. The sediments in this southward plunging syncline have been buried by an additional 1,000 to 2,000 ft (305 to 610 m) compared to equivalent strata immediately to the east and west. Therefore, this area of the New Albany has been subjected to slightly elevated burial temperatures since the mid-Paleozoic.

The second area of increased reflectance is represented by a narrow finger of the 0.7 percent $\overline{R}_o$ contour extending northward into Wayne and Hamilton counties, Illinois (Figure 10). This area is the present day area of maximum burial of the New Albany Shale (Figure 1) and, like the area discussed previously, has been subjected to slightly elevated burial temperatures.

The highest reflectances in the New Albany Shale are in extreme southern Illinois, predominantly south of the Rough Creek Lineament. The maximum reflectances obtained exceed 1.2 percent $\overline{R}_o$ and were observed in Hardin County and vicinity. Although the New Albany is not buried to great depths in this region, southern Illinois is characterized by extensive faulting, mineralization, and scattered small igneous intrusions. Damberger (1971, 1974) suggested that this area may have been intruded by a deep-seated igneous pluton sometime after the Pennsylvanian, resulting in significantly increased heat flow and maturation of organic matter in the overlying sedimentary strata. Alternatively, stratigraphic evidence indicates that most of the Devonian, Mississippian, and Pennsylvanian strata thicken progressively southward, and it is likely that the area of greatest paleoburial depth was significantly south of the present day structural center of the basin (Figure 1).

Similar patterns of maturation based on seam moisture and calorific value were observed in the Pennsylvanian Herrin (No. 6) Coal Member of the Carbondale Formation (Damberger, 1971, 1974). The mean maximum vitrinite reflectance values for Pennsylvanian coals vary in a random pattern in the north and central part of Illinois. In these regions values range from 0.45 to 0.60 percent, with most values between 0.50 and 0.57. In the southeastern part of

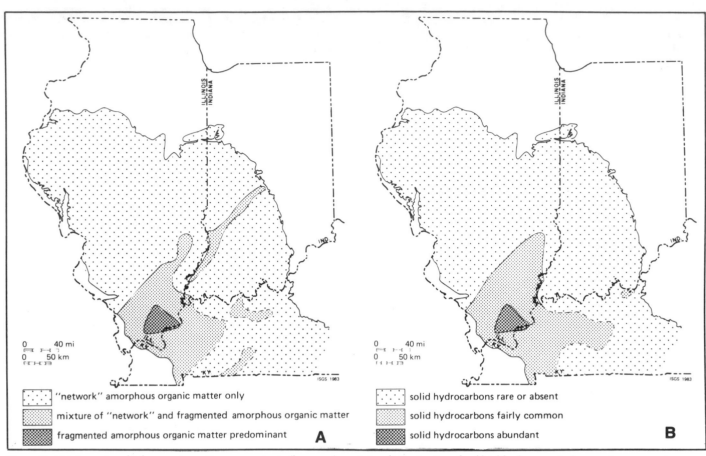

"network" amorphous organic matter only

mixture of "network" and fragmented amorphous organic matter

fragmented amorphous organic matter predominant   **A**

solid hydrocarbons rare or absent

solid hydrocarbons fairly common

solid hydrocarbons abundant   **B**

Figure 9. (A) Areas in which organic matter concentrate (kerogen) contains distinct network of highly-fragmented amorphous organic matter; (B) Areas in which organic matter concentrate (kerogen) contains solid hydrocarbons occurring as fillings or angular particles (after Barrows, Cluff, and Harvey, *in* Bergstrom, Shimp, and Cluff, 1980). Kentucky data are preliminary.

the state, mean maximum reflectance values increase in a pattern similar to the maturation pattern of the New Albany Shale; in this region, values range from 0.65 to 1.10 percent (Hansman et al, 1978). No Pennsylvanian coal reflectance values are available for the region where reflectance exceeds 1.20 percent in the New Albany Shale.

Vitrinite reflectance generally increases with burial depth. However, the reflectance of vitrinite in the New Albany Shale Group is only slightly higher than, or about the same as, the vitrinite reflectance of the overlying Pennsylvanian coals in many parts of the basin, despite the fact that the New Albany Shale occurs some 1,200 to 3,800 ft (370 to 1,160 m) deeper. Bostock and Foster (1975) suggested 0.012 percent R_o/100 ft (.04 percent R_o/100 m) as a reasonable reflectance gradient for the Illinois Basin. Using this gradient and the known vitrinite reflectance values for the Pennsylvanian coals, reflectance values for the New Albany Shale would be predicted to range from 0.64 to 1.33 percent. Measured reflectance values for the New Albany Shale in the areas where data are also available for the Pennsylvanian coals range from 0.44 to 1.21 percent, or about 10 to 20 percent less than expected values. More recent data indicates that Paleozoic reflectance gradients range from 0.002 to 0.030 percent R_o/100 ft (0.0056 to 0.098 percent R_o/100 m) depending on location in the basin and the stratigraphic intervals studied (Barrows, Cluff, and

Harvey *in* Bergstrom, Shimp, and Cluff, 1980; Barrows, 1981). Our preliminary data show no consistent variations or trends, and the discrepancies between Pennsylvanian coal and New Albany Shale samples have not been resolved.

There are several possible explanations for the observed, but lower than expected, reflectance values in the New Albany Shale. Part of the reflectance fallback may be accounted for by the random orientation method used to measure reflectance in dispersed organic matter, versus the maximum reflectance method used for coal samples. In addition, it is possible that the vitrinite measured in the New Albany Shale could be a different type from the vitrinite in Pennsylvanian coals. Tissot and Welte (1978) suggest that even within closely associated samples from the same rock type, variations in reflectance occur that may result from differences in vitrinite types and organic matter preservation. Other types of organic matter present could also affect vitrinite reflectance values. Hutton and Cook (1980) demonstrated that vitrinite reflectance values are significantly lowered by the presence of alginite, an abundant organic component in the New Albany Shale. Teichmueller suggested that at intermediate reflectance levels (0.50 to 0.70 percent R_o) exudatinite or liquid petroleum may impregnate the fine pores in vitrinite and thereby lower its reflectance (Stach et al, 1982). It is also

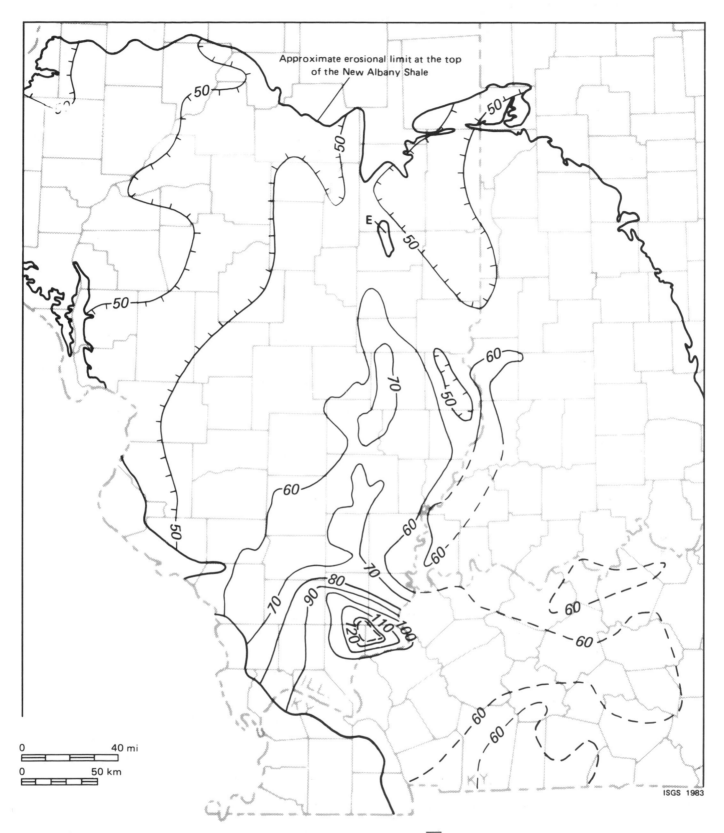

Figure 10. Mean random reflectance of vitrinite for New Albany Shale samples (% $\overline{R_o}$ × 100) (after Barrows, Cluff, and Harvey, *in* Bergstrom, Shimp and Cluff, 1980; Barrows, Potter, and Dillon, in preparation). Kentucky data are preliminary.

possible that vitrinite in different rock types might be at different maturation levels because of variations in the heat conductivity of the surrounding rocks (Stach et al, 1982). The fact that vitrinite reflectance values for the New Albany Shale are lower than might be predicted, based on Pennsylvanian coalification gradients, is important because it suggests that significant petroleum generation could have occurred at reflectance levels that are lower than would otherwise be expected for the main zone of oil generation.

In recent years it has been recognized that oil is generated within a relatively narrow maturation range (the "oil window"), and that natural gas is formed over a much broader range. For Type II kerogens, the range of vitrinite reflectances corresponding to the oil generation window is between 0.5 and 1.3 percent reflectance (Figure 11). Peak oil generation and expulsion occur at maturation levels corresponding to reflectance greater than 0.6 to 0.7 percent R_o (Hood, Gutjahr, and Heacock, 1975; Dow, 1977). The maturation level referred to as the immature stage corresponds to vitrinite reflectance values less than 0.50 percent.

Recent work (van Gijzel, 1967; Ottenjahn, Teichmuller, and Wolf, 1975; Robert, 1979) indicates that the fluorescence of liptinites may also be used as a guide to the rank of coals and to the maturity of organic matter dispersed in sedimentary rocks as well as aid in maceral identification (Cook and Kantsler, 1980). Tissot and Welte (1978) indicate that with increasing maturation, fluorescence intensity decreases and fluorescence colors move from the yellow toward the red. Fluorescence generally disappears at vitrinite reflectance values greater than about 1.0 percent. Although visual fluorescence studies are not as precise as vitrinite reflectance studies, the patterns observed (Figure 12) support the patterns of increasing reflectance and maturity illustrated in Figure 10. The fluorescence intensity is also useful for discriminating reflectance patterns that result from increasing thermal maturity from those that arise from other phenomena. For example, the slight increase in vitrinite reflectance values along the northern erosional edge of the New Albany in east central Illinois is probably due to early Pennsylvanian oxidation of organic matter. The fluorescence map does not show any anomaly in this region.

Pyrolysis of Shales

Dickerson and Chou (*in* Bergstrom, Shimp, and Cluff, 1980) reported the results of controlled stepwise pyrolysis-gas chromatography analyses for several selected samples of New Albany Shale from Illinois and western Kentucky. In general, they found that heating the shale up to 350°C released the highly volatile C_1 to C_8 hydrocarbons and medium volatile C_9 to C_{12} hydrocarbons that are trapped in the shale matrix. Samples from a core in Wayne County, Illinois, near the center of the Fairfield Basin (Figure 1) and in an area where vitrinite reflectance values exceed 0.70 percent (Figure 10), yielded C_9 to C_{12} hydrocarbons in greater proportion than C_1 to C_8 hydrocarbons. Samples from cores in other, lower maturity areas yielded the C_1 to C_8 hydrocarbons in greater proportions. Further heating of the shale samples to 550°C and above resulted in the evolution of large amounts of C_1

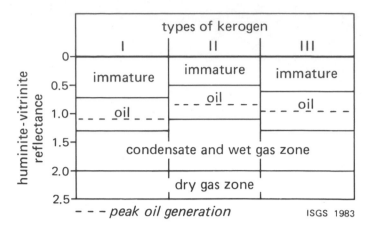

Figure 11. Approximate boundaries of the oil and gas generation zones in terms of vitrinite reflectance for kerogen types. Boundaries may change slightly due to mixing of various kerogen types and/or variations in time-temperature relationships (after Tissot and Welte, 1978).

to C_8 gases, as well as moderate amounts of heavier hydrocarbons. The hydrocarbons derived at these high pyrolysis temperatures were formed by the breakdown of the non-volatile organic matter in the shale. Gas yields varied linearly with organic carbon content, up to a maximum of about 30 standard cubic feet of gas/cubic foot of shale.

The results of their pyrolysis analyses show that: (1) considerable amounts of volatile hydrocarbons, presumably formed by natural maturation and generation, are trapped in the shale matrix; (2) with increasing thermal maturity the proportion of heavier hydrocarbons (C_{9+}) increases; and (3) over most of the basin the New Albany Shale has considerable generative potential remaining. These results agree with the conclusions derived from petrographic studies as summarized above.

SUMMARY

Area of Hydrocarbon Generation in the New Albany Shale

The previously described patterns of major depositional environments, shale facies, organic matter types, and maturation levels in the New Albany can be combined to give a fairly complete picture of where and in what relative proportions hydrocarbons have been generated. Evidence suggesting that extreme southern Illinois and parts of western Kentucky were areas of intense oil generation (Figure 13) includes:

1. Predominance of black shales with organic carbon contents generally over 2.0 weight percent which have an aggregate thickness of 100 to 350 ft (30.5 to 107 m) (Figure 5).
2. Predominance of Type II, oil-prone kerogen in the black shale facies (Figure 7).
3. Thermal maturations within the oil generation window for Type II kerogen (Figure 11) as shown by vitrinite reflectance values > 0.60 percent $\overline{R_o}$ (Figure 10) and alginite fluorescence within the orange or higher range (Figure 12).
4. Presence of solid hydrocarbons as pore and fissure

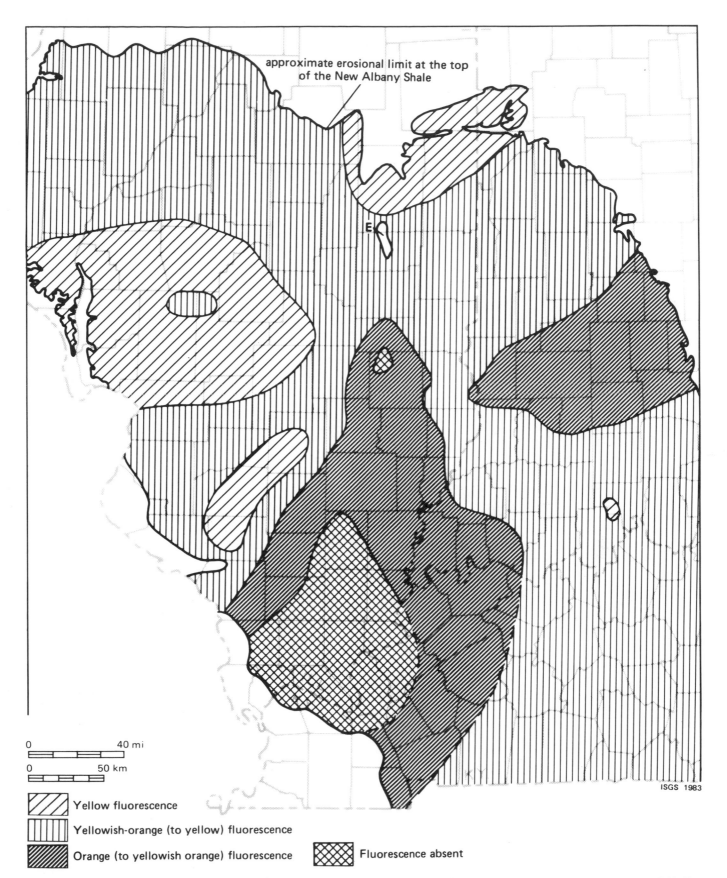

approximate erosional limit at the top of the New Albany Shale

0 40 mi

0 50 km

ISGS 1983

Yellow fluorescence

Yellowish-orange (to yellow) fluorescence

Orange (to yellowish orange) fluorescence

Fluorescence absent

Figure 12. Color of liptinite fluorescence in the New Albany Shale (after Barrows, Cluff, and Harvey, *in* Bergstrom, Shimp, and Cluff, 1980). Kentucky data are preliminary.

126   *Barrows and Cluff*

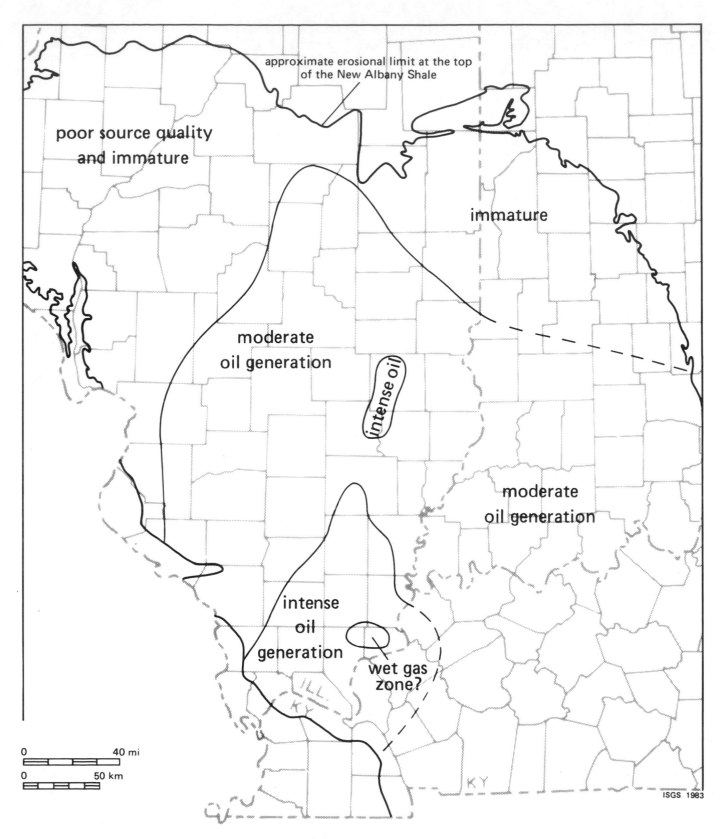

Figure 13. Generalized areas of oil and gas generation within the New Albany Shale.

Table 1. Calculation of time-temperature index of maturity (TTI) for the New Albany Shale, using method of Waples (1980).

Assumptions: (1) burial to 5,000 ft (1,524 m) by end of the Pennsylvanian with negligible erosion since then; (2) constant geothermal gradient from Mississippian to present.

Temperature Interval	r^n	Duration (M.Y.)	Interval TTI	Sum of TTI	Estimated $\overline{R}_o$
20-30°C	2^{-8}	25	0.097	0.097	<0.30
30-40°C	2^{-7}	40	0.312	0.409	0.40
40-50°C	2^{-6}	280	4.375	4.784	0.52

fillings within the shale matrix and extensive alteration of amorphous kerogen from a continuous network to a fragmented, particulate appearance (Figure 9).

Within the region of intense oil generation shown on Figure 13 is a small area where vitrinite reflectance values exceed 1.1 percent $\overline{R}_o$, the approximate "floor" of oil generation. At these levels of maturity wet gas and condensate are the major hydrocarbons formed by breakdown of kerogen (Harwood, 1977). The wet gas zone appears to be restricted to parts of Saline, Hardin, and Pope Counties in southeastern Illinois. Although data are scarce for western Kentucky, it is possible that similarly high levels of maturity occur in the deeper parts of the Moorman Syncline in western Kentucky. Many of the significant gas fields in the Illinois Basin have been found in the eastern part of the Moorman Syncline, suggesting gas generation to the west and subsequent migration into shallower traps updip. Virtually all the known remaining gas in the basin occurs as solution gas within oil reservoirs and a few gas-capped oil fields. Meents (1981) published analyses of natural gas from various sources in Illinois, including 412 gas wells and 166 oil wells. His compilation shows that most of the gas in southern Illinois is moderately wet, which is consistent with its having been co-generated with liquid petroleum.

Surrounding the region of intense oil generation is a broad area—encompassing much of southwestern and central Illinois, southwestern Indiana, and western Kentucky—where oil generation may have been less intense but still significant (Figure 13). The New Albany is generally thinner in this region and contains a smaller proportion of organic-rich shales, but still contains at least some beds with good source quality. Reflectance and fluorescence analysis indicate lower levels of maturity ($\overline{R}_o$ =0.50 to 0.60 percent; yellow to yellowish-orange fluorescence), but values that are still within at least the initial stages of oil generation (Figure 11). Gas analyses from wells in this region (Meents, 1981) indicate largely dry gas with only traces of ethane and heavier hydrocarbons.

It is probable that no significant hydrocarbon generation occurred in the New Albany Shale across most of western Illinois or along the northern fringes of the New Albany in Illinois and Indiana. Stratigraphic studies show that in these areas only a very small part of the shale section contains appreciable organic matter, and even in those beds it is in low concentrations. Reflectance and fluorescence studies indicate the organic matter has not yet reached the point of oil generation—probably because of insufficient burial.

Paleoburial Depths and Geothermal Gradient

The present geothermal gradient of the Illinois Basin ranges from approximately 1.2 to 2.0°F/100 ft, depending on location, and the subsurface temperature of the New Albany in the central area of the basin is now 110 to 120°F (43° to 49°C). Given that the New Albany was buried to at least its present depth by the close of the Pennsylvanian (a period of about 65 million years), and assuming that the present geothermal gradient is representative of the paleogradient from Permian to the present (a period of about 280 million years), then by using Lopatin's method as outlined by Waples (1980) we can predict the thermal maturity that the New Albany should have attained. The cumulative "time-temperature index" resulting from this simplified burial history is about 4.8 (Table 1), which corresponds to a vitrinite reflectance level of near 0.52 percent $\overline{R}_o$. This predicted value is significantly less than the observed values of 0.70 percent $\overline{R}_o$ and higher in the center of the Illinois Basin, and also falls short of the level of maturation required for petroleum generation (estimated to be at a time-temperature index of 15).

Damberger (1971, 1974) noted that the rank of Pennsylvanian coals in Illinois cannot be accounted for by the present burial depth and temperature, and either an additional 5,000 ft (1,524 m) of burial or a significant increase in the geothermal gradient at some time in the past is required to achieve the known degree of coalification. Considering the wide variation of reflectance gradients and the paucity of data on the post-Pennsylvanian geologic history of the basin, we feel most conclusions regarding paleoburial depths or paleogeothermal gradients are premature. Nevertheless, the available data indicate that the New Albany was once buried much deeper than at present, with an unknown amount of sedimentary cover having been removed by post-Pennsylvanian erosion. Geologic evidence cited previously also suggests that a deep-seated igneous pluton was emplaced beneath southern Illinois during or shortly after the Permian. Emplacement of a large igneous body would have resulted in increased heat flow and higher geothermal gradients for an undetermined length of time. Whatever caused the increased maturation of the New Albany over that which is accounted for by its present burial temperatures, it is improbable that any active generation is taking place today given the shallow burial depth and low temperatures.

Importance of New Albany-Sourced Oil in the Illinois Basin

Maps showing the location and approximate areal extent of oil production from Pennsylvanian, Chesterian, Valmeyeran, Devonian, Silurian, and Trenton reservoirs in the Illinois Basin are presented in Figures 14 to 19. Comparison of these maps with the map showing likely areas of oil and gas generation from the New Albany Shale (Figure 13) is quite revealing.

Most of the basin's Valmeyeran oil (Figure 16) occurs within or near the area of intense oil generation in the New Albany Shale shown on Figure 13. Most of the remaining accumulations in the basin occur in Chesterian (Figure 15) and Pennsylvanian (Figure 17) strata, primarily in areas immediately surrounding the area of intense generation and almost entirely within the area of at least moderate generation.

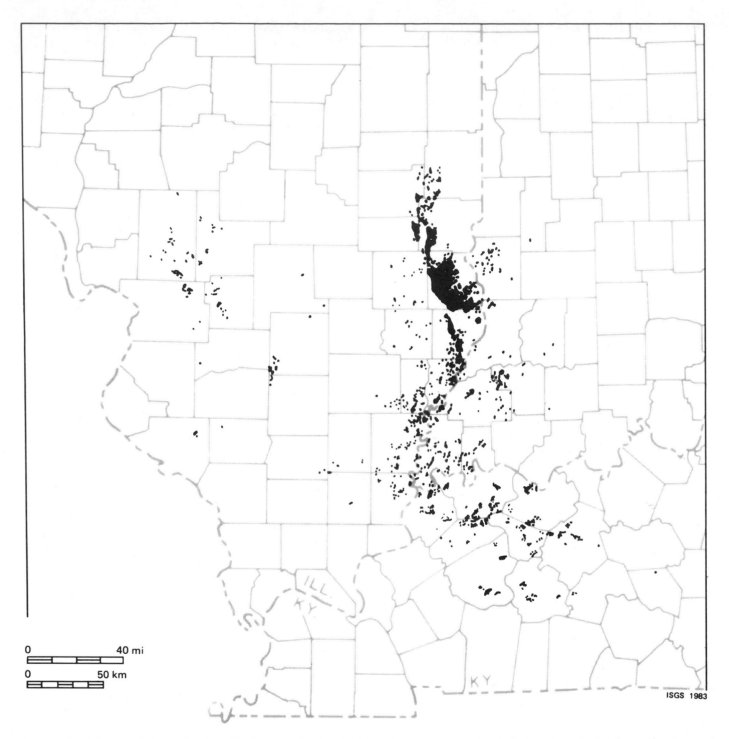

Figure 14. Distribution of Pennsylvanian petroleum production. Note that for the pay maps illustrated in Figures 14-19, the productive areas shown for Illinois and Indiana are areas within a field with actual production in the zone indicated. For Kentucky, such detailed information is not readily available; therefore, if production is reported for that pay zone the *entire field* is shown as having production (Schwalb, Wilson, and Sutton, 1971). Compare Figures 10 and 13 for relation of oil producing areas to probable oil generation in the New Albany Shale.

Ninety percent or more of the basin's known oil reserves could, therefore, be accounted for by oil that was generated within the New Albany Shale and subsequently migrated into traps in younger strata. As indicated by isoreflectance contours superimposed over a stratigraphic section across the basin (Figure 20), all of the Mississippian and

Pennsylvanian fields overlie or are near the area of greatest generation from New Albany Shale source beds.

Five of the largest oil fields in the basin—Louden, Salem, New Harmony Consolidated, Lawrence County, and Main Consolidated—lie near but outside the area of most intense generation. Two very large fields—Clay City Consolidated

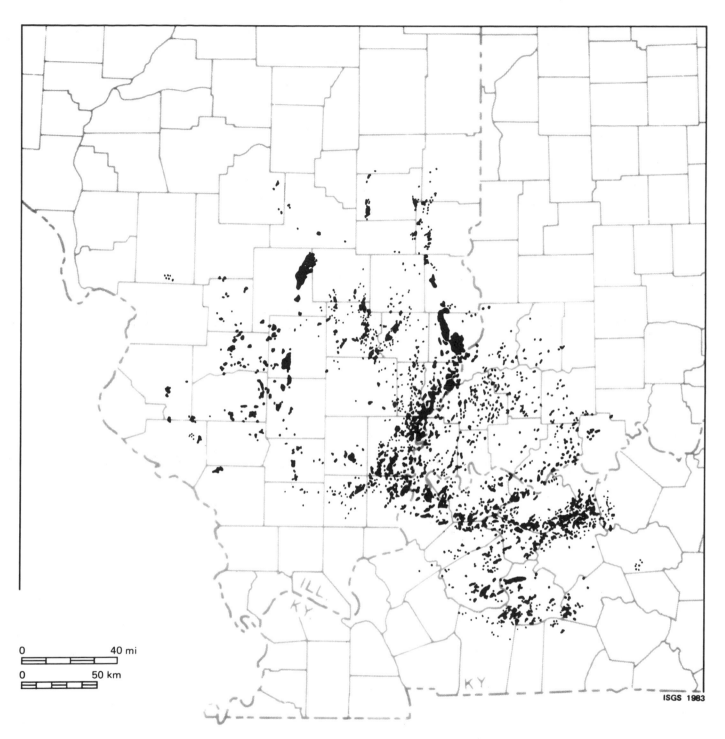

Figure 15. Distribution of Chesterian petroleum production.

and Dale Consolidated—lie within or adjacent to the area of intense generation. This suggests two important generalizations: (1) at the time of generation and early migration large structures probably gather the hydrocarbons generated in broad areas surrounding them, thereby ultimately forming major accumulations; (2) even at maturity levels less than the peak of oil generation, very large (100 MMbbl+) accumulations can be formed if rocks of sufficient source quality and thickness exist nearby, and if sufficient time has elapsed.

Based on comparisons between the known producing areas and the areas of probable generation in the New Albany Shale, it appears that the bulk of the Mississippian and Pennsylvanian oils in the Illinois Basin were derived from the New Albany, with perhaps some minor dilution by thin source beds within the various younger carbonates and shales. Although the relative contribution of younger source beds to the basins reserves cannot be quantitatively assessed at present, all of the younger strata are at lower maturity levels than the New Albany and none approach it in total

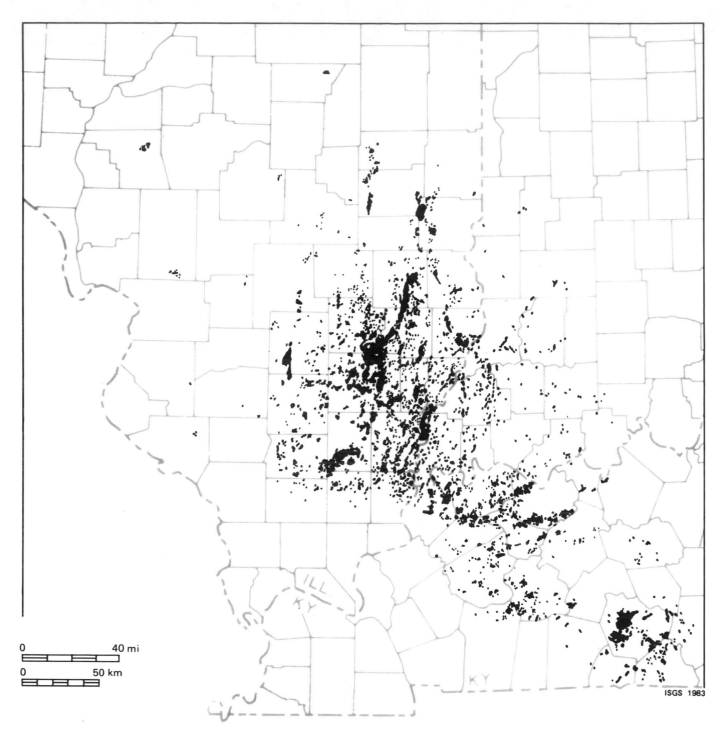

Figure 16. Distribution of Valmeyeran petroleum production.

bulk volume of potential source material. Most of the Upper Mississippian and Pennsylvanian strata were deposited in deltaic and shallow-marine, oxygenated environments. The quality of the organic matter in these sediments is not believed to be as rich a source material as the Type II kerogen in the New Albany. This problem is presently being studied.

Unfortunately, only a few chemical analyses of Illinois Basin crude oils have been published and the question of crude oil-source rock correlation has not been studied in depth. Rees, Henline, and Bell (1943) showed that most Mississippian and Pennsylvanian oils in Illinois are very similar. Their analyses of the gravity, sulfur content, and fractional distillation products of Illinois crudes all fall within a narrow range that is generally typical of crudes derived from Type II kerogen (Figure 21). Some Pennsylvanian age crudes vary from this mean trend and may represent bacterially degraded crudes, or perhaps a different source rock altogether. The close grouping of crude compositions shown in Figure 21 is strong evidence for a

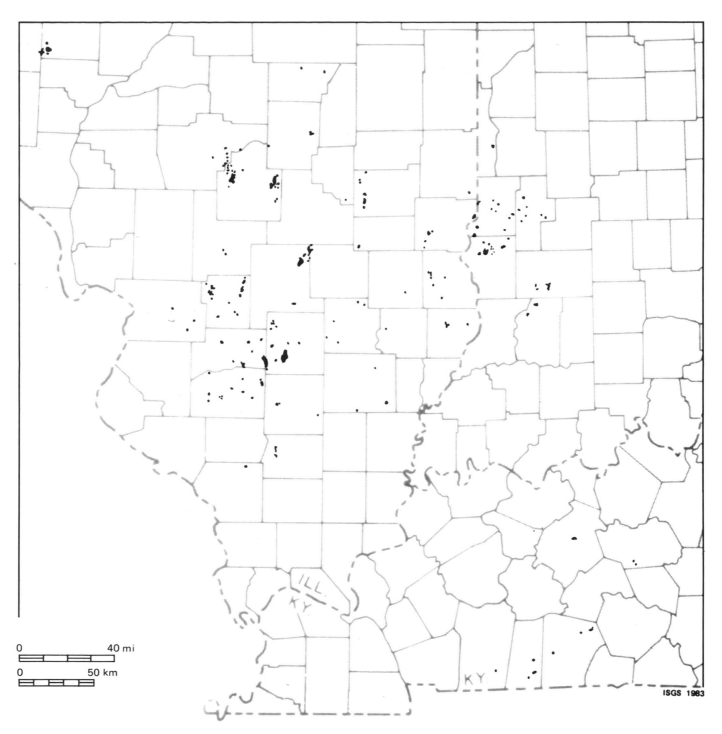

Figure 17. Distribution of Devonian petroleum production.

common source, or at least sources with very similar
organic matter types. If multiple source beds with different
kerogen types had played significant roles in the formation
of petroleum accumulations in the basin, then such a low
variation in composition would be very difficult to explain.

Devonian and Silurian oil production (Figures 17 and 18)
occurs mostly toward the fringes of the basin, away from
the areas of most oil generation in the New Albany Shale.
The basin cross section (Figure 20) shows that although
these fields occur in units stratigraphically below the New

Albany, they all are found structurally higher than mature
New Albany source rocks. The deepest production yet
established in the Illinois Basin (5,400 ft or 1,646 m) is from
the Devonian Dutch Creek Sandstone Member at Mill
Shoals field in White County, Illinois. Mature New Albany
Shale is found structurally deeper in the synclines on either
side of this field. The distribution of Devonian and Silurian
fields is largely controlled by the occurrence of
reservoir-quality porosity—which is generally found
toward the margins of the basin. These reservoirs could

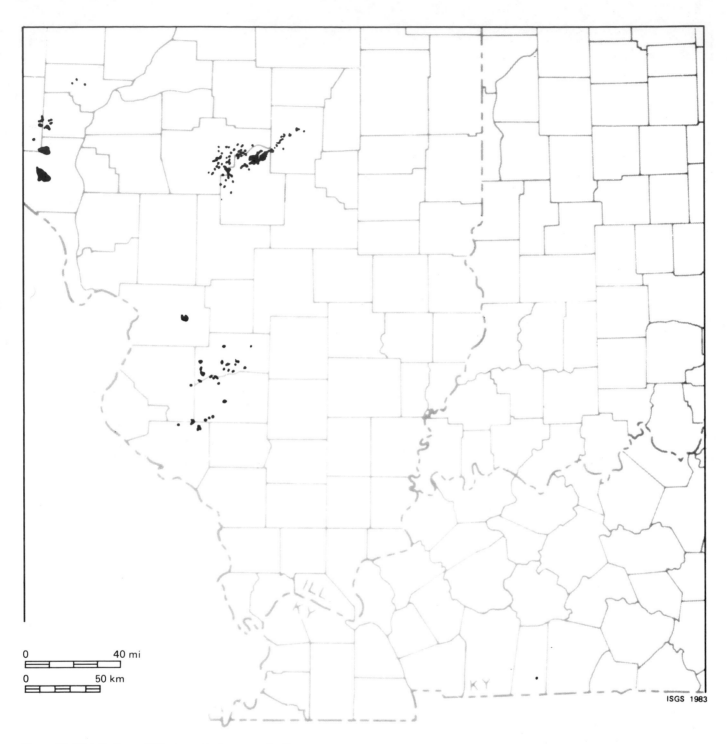

Figure 18. Distribution of Silurian petroleum production.

have been charged with New Albany oil that migrated fairly long distances from structurally deeper, mature areas. One problem with this interpretation, however, is the absence of continuous, porous carrier beds to provide a migration pathway.

Alternatively, the Devonian and Silurian reservoirs may have been locally charged from the overlying New Albany Shale or organic-bearing carbonate beds within the Devonian-Silurian succession itself, or from the deeper Maquoketa Shale Group (Ordovician). This raises the

question of how much oil could have been generated at the relatively low levels of thermal maturity (that is, around $\overline{R}_o = 0.5$) found in these areas. Considering the high organic content of the New Albany, it appears likely that oil fields up to several million barrels in size can be formed in regions that would conventionally be thought of as immature.

The distribution of known "Trenton" pools is also worth noting. Like most of the Devonian-Silurian pools, they are confined to the margins of the basin, where dolomitization has created reservoir-quality porosity in what is regionally a

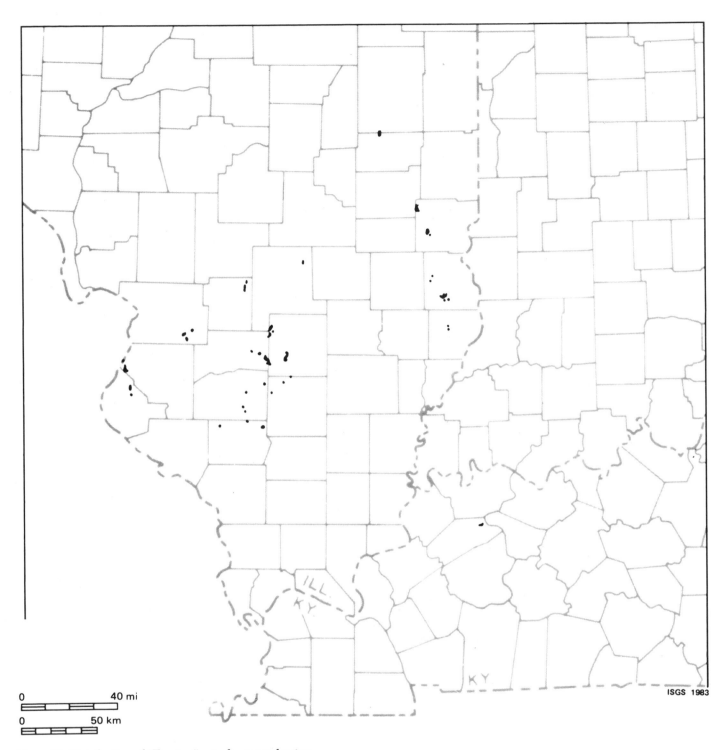

Figure 19. Distribution of "Trenton" petroleum production.

thick sequence of dense limestones (Figure 19). Thus, they pose a similar set of problems and potential answers. Oils from "Trenton" pools differ from those of other oil fields in the Illinois Basin, however, in that they are highly paraffinic (Rees, Henline, and Bell, 1943), much like Ordovician oils in other basins of the world. This geochemical evidence suggests an entirely different source rock for the "Trenton" accumulations—possibly the organic-rich shales in the overlying Maquoketa (Figure 20).

CONCLUSIONS

Based on determinations of the quantity, quality, and maturation of organic matter, it is likely that extreme southern Illinois and far western Kentucky were areas of intense oil generation from the New Albany. Petroleum generation from the New Albany may also have been significant in parts of central and southwestern Illinois, southwestern Indiana, and western Kentucky. Comparison

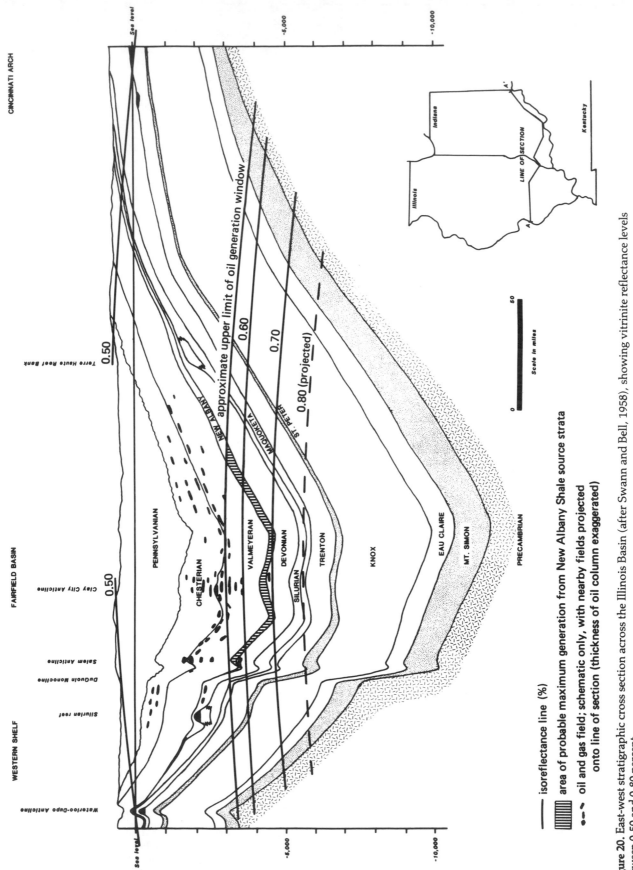

Figure 20. East-west stratigraphic cross section across the Illinois Basin (after Swann and Bell, 1958), showing vitrinite reflectance levels between 0.50 and 0.80 percent.

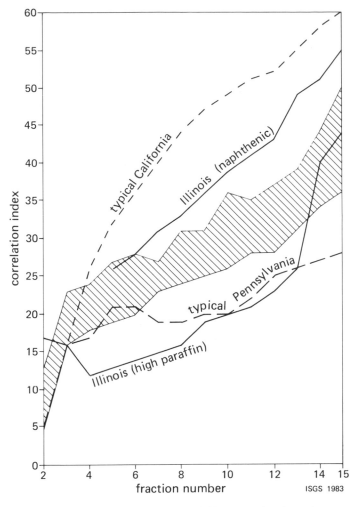

Figure 21. Chemical characteristics of Illinois crude oils as shown by correlation index numbers. Shaded area includes 70 percent of all Illinois crude oils analyzed, representing all productive horizons and most significant fields (from Rees, Henline, and Bell, 1943).

of the areas of likely petroleum generation outlined from this study with the locations of known oil-bearing reservoirs suggests that the New Albany may account for 90 percent or more of the basin's known oil accumulations.

ACKNOWLEDGMENTS

We are grateful for the cooperation of the Indiana and Kentucky geological surveys. Gerry Carpenter and Nancy Hasenmuller from the Indiana Geological Survey provided the Indiana data for Figures 14, 15 and 16. Richard Leininger and Nancy Hasenmuller provided New Albany Shale samples from Indiana for analyses. Paul E. Potter from the University of Cincinnati and Patrick Gooding from the Kentucky Geological Survey supplied the New Albany Shale samples from Kentucky.

ANNEX 1: RESEARCH METHODS

Procedures

Petrographic data discussed in this report were obtained from samples from 238 drill holes. No samples were

included from areas where the New Albany crops out or from areas where the shale has been eroded and is overlain by Pennsylvanian or Pleistocene sediments.

We sampled 12 New Albany Shale cores in Illinois, one in Indiana, and two in Kentucky to relate variations in organic richness, types of organic matter, and maturation to depth, stratigraphic position, and depositional facies. The remainder of the samples studied were obtained from hand-picked, washed, drill cuttings from the files of the Illinois, Indiana, and Kentucky geological surveys (Bergstrom, Shimp, and Cluff, 1980).

Analysis of Total Organic Carbon

Total organic carbon contents of 230 core samples and 160 drill-cutting samples were determined, using a combination of combustion and wet-chemical methods (Frost, *in* Bergstrom, Shimp, and Cluff, 1980).

Microscopic Analyses

The types of insoluble organic matter (kerogen) deposited and preserved in a sedimentary rock subsequently influence the amount and composition of hydrocarbons that can be generated from that rock (Tissot and Welte, 1978; Hunt, 1979). To evaluate the quality of insoluble organic matter in a potential source bed, both geochemical and microscopic methods are commonly used. Geochemical methods have achieved wide acceptance because they are considered to be objective, accurate, and reproducible. One of the main disadvantages with geochemical techniques, however, is that only the bulk chemistry of the organic material can be obtained, and determinations of the relative mixture of individual components of the kerogen are not possible. Another disadvantage is that some geochemical techniques cannot differentiate between those effects due to kerogen type and those due to maturation (Jones and Edison, 1978).

In contrast, microscopic methods are generally perceived to be less precise and more subjective than geochemical analyses. They are nonetheless valuable because: (1) you can assess both organic quality and maturation in a single analysis; (2) you can detect errors in geochemical analyses; (3) you can visually determine contamination of the sample by caving or drilling additives; and (4) you can commonly detect the presence of reworked organic matter.

Powell, Creaney, and Snowdon (1982) report that there is often a marked discrepancy between the interpretation of source rock potential based on chemical methods and those based on microscopic methods, and that the lack of agreement arises from the inability to consistently distinguish between hydrogen-rich and hydrogen-poor amorphous organic matter using either transmitted or reflected light microscopy. They suggest that reflected light microscopy, in conjunction with fluorescent light microscopy, offers more promise as a petrologic tool for accurate source rock assessment than does transmitted light microscopy alone.

In our study, both reflected white and fluorescent light microscopy were used to determine organic composition. As in previous studies of this type by other researchers, we had considerable difficulty in identifying and classifying certain organic constituents, and particularly the

amorphous organic matter fraction. Although a number of
classification systems for dispersed organic matter have
been proposed for reflected light studies (Teichmuller and
Ottenjahn, 1977; Alpern and Cheymol, 1978; Robert, 1979,
1981; Bostick, 1979; Alpern, 1980; Combaz, 1980; and
Cook and Kantsler, 1980), a uniform system has not yet
been adopted by the International Committee for Coal
Petrology (ICCP). Most of the proposed systems classify the
figured organic matter constituents along traditionally
established ICCP Maceral Groups (liptinite, vitrinite, and
inertinite), but there is little agreement on how to classify
the amorphous organic matter commonly observed in
fine-grained source rocks (ICCP Commission II, 1981).
Many workers lump the amorphous organic matter in the
liptinite maceral group; however, we chose to identify the
amorphous organic matter as a separate organic matter
type, because the origin and oil-generating potential of
amorphous organic matter is not easily determined
microscopically (Powell, Creaney, and Snowden, 1982).
We, therefore, recognized two basic types of organic
matter: (1) the figured organic matter, which includes
macerals from the liptinite, vitrinite, and inertinite group
(Stach et al, 1982); and (2) unfigured or amorphous organic
matter.

Most of the petrographic data were obtained on organic
matter (kerogen) that we isolated and concentrated from its
inorganic matrix using a combination of HCl, HF, and
heavy liquid separation techniques (Gray, 1965). The
isolated kerogen was mounted in epoxy and polished prior
to examination. Details of the procedures are outlined in a
paper by Barrows, Cluff, and Harvey (1979). We also
prepared several polished blocks of shale from core samples
to observe the distribution of organic matter within the
rock. A few vitrain samples (mostly *Callixylon* stems and
fragments) obtained from core bedding-plane surfaces were
also mounted in epoxy without chemical treatment,
polished, and examined. Each kerogen pellet was evaluated
to determine types of organic matter present and,
concurrently, organic maturity of the sample.

Types of Organic Matter

Visual estimates of the occurrence and abundance of
organic matter types (liptinite, vitrinite, inertinite, and
amorphous organic matter) were made under reflected
white light. Particle size, organic matter preservation, and
any unusual particles observed were also recorded. Due to
the tendency of some liptinites to blend into the background
of the mounting media under reflected light (Kantsler,
1980), our estimates of the relative abundance of liptinites
were checked by fluorescence analysis. Observation under
ultraviolet excitation is a useful tool in determining the
relative abundance of liptinites because at moderately low
levels of maturity (R_o 1.00 percent), liptinites fluoresce
various shades and intensities of yellow, orange, and brown
depending on the organic preservation and maturity of the
sample.

Organic Maturation

We determined the level of organic maturity of each
sample using vitrinite reflectance and visual fluorescence
analyses. We calculated the mean random reflectance of

vitrinite and the standard deviation of the readings and
recorded a reflectance class histogram for each sample.

Samples that had fewer than 20 readings or that had a
standard deviation greater than 0.15 were considered
unrepresentative and were excluded from the final
compilation of the data. We took random rather than
maximum vitrinite reflectance measurements because it was
impractical to keep very fine vitrinite particles centered
under the measuring spot during the stage rotation
necessary to obtain maximum values. The color and relative
intensity of fluorescence of the liptinite fraction of the
samples were estimated visually as another measure of the
maturity of the organic matter. Because *Tasmanites* (an
alginite) was the most easily and consistently recognized
liptinitic material in our samples, the final evaluation of the
fluorescence data was made on the fluorescence of only the
Tasmanites. Details for the procedures for these microscopic
analyses are given in Barrows, Cluff, and Harvey (1979).

Pyrolysis Analyses

Geochemical studies of the organic matter in the New
Albany Shale were not performed specifically for this study.
Several other investigators at the Illinois Geological Survey
and at other institutions, however, reported the results of
various geochemical analyses that were performed as part of
the overall DOE sponsored study of the New Albany Shale
Group. Dickerson and Chou (*in* Bergstrom, Shimp, and
Cluff, 1980) reported on the stepwise pyrolysis of selected
shale samples from across the entire Illinois Basin. The
pyrolysis temperature was incrementally increased from 60
to 750°C. Between 250 and 450°C, volatile hydrocarbons
trapped in the shale matrix are evolved from the sample in
proportions that vary with organic content and location of
the sample. At 550°C and above, fragmentation of the
non-volatile organic matter was initiated. Stepwise
pyrolysis is similar to conventional Rock-Eval pyrolysis
(summarized in Tissot and Welte, 1978) in that the
evolution of volatiles at low pyrolysis temperatures
corresponds to the Rock-Eval "S" peak, and the generation
of hydrocarbons from kerogen at high pyrolysis
temperatures corresponds to the Rock Eval "S_2" peak.

REFERENCES CITED

Alpern, B., 1970, Classification petrographique des
  constituants organiques fossiles des roches sedimentaires:
  Revue de L'institut Francais du Petrole et Annales des
  combustibles Liquides, v. 25, n. 11, p. 1233-1266.
———, 1980, Petrographie du kerogene, *in* B. Durand, ed.,
  Kerogen insoluble organic matter from sedimentary
  rocks: Paris, Editions Technip, p. 339-384.
——— and D. Cheymol, 1978, Reflectance et fluorescence
  des organoclastes du Toarchien du Bassin de Paris en
  fonction du la profondeur et de la temperature: Revue de
  l'Institut Francais du Petrole, v. 33, n. 4, p. 515-535.
Atherton, E., 1971, Tectonic development of the eastern
  interior region of the United States: Illinois State
  Geological Survey, Illinois Petroleum 96, p. 29-43.
Barrows, M.H., 1981, Petrographic studies of sedimentary
  organic matter in the Illinois Basin: Lexington, Kentucky,
  North American Coal Petrographers' Meeting Abstracts.

——, R.M. Cluff, and R.D. Harvey, 1979, Petrology and maturation of dispersed organic matter in the New Albany Shale Group of the Illinois Basin *in* Proceedings, Third Eastern Gas Shales Symposium: U.S. Department of Energy METC/SP-79/6, p. 58-114.

——, P.E. Potter, and J.W. Dillion, in preparation, Vitrinite reflectance map of the New Albany Shale (Devonian and Mississippian): University of Kentucky, Kentucky Geological Survey.

Bassett, J.L., and N.R. Hasenmuller, 1978, The New Albany Shale and correlative strata in Indiana, *in* Proceedings, First Eastern Gas Shales Symposium: U.S. Department of Energy MERC/SP-77/5, p. 183-194.

Bergstrom, R.E., N.F. Shimp, and R.M. Cluff, 1980, Geologic and geochemical studies of the New Albany Shale Group (Devonian-Mississippian) in Illinois: Illinois State Geological Survey Final Report to U.S. Department of Energy, Contract DE-AC21-76ET12142, 183 p.

Bond, D.C., et al, 1971, Possible future petroleum potential of region 9—Illinois Basin, Cincinnati Arch, and northern Mississippian Embayment, *in* I.H. Cram, ed., Future petroleum provinces of the United States—their geology and potential: AAPG Memoir 15, v. 2, p. 1165-1218.

Bostick, N.H., 1979, Microscopic measurements of the level of catagenesis of solid organic matter in sedimentary rocks to aide exploration for petroleum and to determine former burial temperatures—a review: Society of Economic Paleontologists and Mineralogists Special Publication No. 26, p. 17-43.

——, and J.N. Foster, 1975, Comparison of vitrinite reflectance in coal seams and in kerogen of sandstones, shales, and limestones in the same part of a sedimentary section, *in* B. Alpern, ed., Petrographie de la matiere organique des sediments, relations avec la paleotemperature et le potential petrolie: Paris, Centre National de al Recherche Scientifique, p. 13-25.

Buschbach, T.C., and E. Atherton, 1979, History of the structural uplift of the southern margin of the Illinois Basin, *in* J.E. Palmer and R.R. Dutcher, eds., Depositional and structural history of the Pennsylvanian System of the Illinois Basin, part 2; invited papers: Illinois State Geological Survey Guidebook Series 15a, p. 112-115.

Byers, C.W., 1977, Biofacies patterns in euxinic basins; a general model, *in* H.E. Cook and P. Enos, eds., Deep-water carbonate environments: Society of Economic Paleontologists and Mineralogists Special Publication No. 25, p. 5-17.

Clegg, K.E., and J.C. Bradbury, 1956, Igneous intrusive rocks in Illinois and their economic significance: Illinois State Geological Survey, Report of Investigation, 19 p.

Cluff, R.M., 1980, Paleoenvironment of the New Albany Shale Group (Devonian-Mississippian) of Illinois: Journal of Sedimentary Petrology, v. 50, p. 767-780.

——, and D.R. Dickerson, 1982, Natural gas potential of the New Albany Shale Group (Devonian-Mississippian) in southeastern Illinois: Society of Petroleum Engineers Journal, v. 22, n. 2, p. 291-300.

——, M.L. Reinbold, and J.A. Lineback, 1981, The New Albany Shale Group of Illinois: Illinois State Geological Survey Circular 518, 83 p.

Combaz, A., 1980, Les kerogenes vus au microscope, *in* B. Durand, ed., Kerogen insoluble organic matter from sedimentary rocks: Paris, Editions Technip, p. 55-112.

Cook, A.C., and A.J. Kantsler, eds., 1980, Oil shale petrology workshop: Keiraville Kopiers, 100 p.

Damberger, H.H., 1971, Coalification pattern of the Illinois Basin: Economic Geology, v. 66, p. 488-494.

——, 1974, Coalification patterns of Pennsylvanian coal basins of the Eastern United States, *in* Carbonaceous materials as indicators of metamorphism: Geological Society of America Special Paper 153, p. 53-74.

Dow, W.G., 1977, Kerogen studies and geological interpretations: Journal of Geochemical Exploration, v. 7, p. 79-99.

Gray, J., 1965, Extraction techniques, *in* B. Kummel and D. Ramp, eds., Handbook of paleontological techniques: San Francisco, W.H. Freeman and Company, p. 530-587.

Hansman, M.H., et al, 1978, Organic maturity of the New Albany Shale Group (Devonian-Mississippian) in Illinois: Geological Society of America Abstracts with Programs, v. 10, n. 7, p. 416.

Harwood, R.J., 1977, Oil and gas generation by laboratory pyrolysis of kerogen: AAPG Bulletin, v. 61, p. 2082-2102.

Hasenmuller, N.R., and G.S. Woodward, 1981, Studies of the New Albany Shale (Devonian and Mississippian) and equivalent strata in Indiana: Indiana Geological Survey Contract Report to U.S. Department of Energy, Contract DE-AC-21-76MC05204, 100 p.

Hood, A., C.C.M. Gutjahr, and R.L. Heacock, 1975, Organic metamorphism and the generation of petroleum: AAPG Bulletin, v. 59, p. 986-996.

Hunter, C.D., and D.M. Young, 1953, Relationship of natural gas occurrence and production in eastern Kentucky (Big Sandy Gas Field) to joints and fractures in Devonian bituminous shale: AAPG Bulletin, v. 37, p. 282-299.

Hunt, J.M., 1979, Petroleum geochemistry and geology: San Francisco, W.H. Freeman and Company, 617 p.

Hutton, A.C., and A.C. Cook, 1980, Influence of alginite on the reflectance of vitrinite from Joadja, NSW, and some other coals and oil shales containing alginite: Fuel, v. 59, p. 711-714.

International Committee for Coal Petrology Commission II, 1981, Minutes of April 8 meeting: Pau, France, International Committee for coal Petrology Commission II, 9 p.

Jones, R.W., and T.A. Edison, 1978, Microscopic observations of kerogen related to geochemical parameters with emphasis on thermal maturation, *in* D.F. Oltz, ed., Symposium in geochemistry: Los Angeles, Pacific Section Society of Economic Paleontologists and Mineralogists, p. 1-12.

Kantsler, A.J., 1980, Aspects of organic petrology with particular reference to exinite group of macerals, *in* A.C. Cook and A.J. Kantsler, eds., Oil Shale Petrology Workshop, Wollongong: Keiraville Kopiers, p. 16-41.

Kehn, T.M., J.G. Beard, and A.D. Williamson, 1979, The Mauzy Formation, a new stratigraphic unit of Permian age in western Kentucky: U.S. Geological Survey Open

File Report.

Krausse, H.-F., and C.G. Treworgy, 1979, Major structures of the southern part of the Illinois Basin, *in* J.E. Palmer and R.R. Dutcher, eds., Depositional and structural history of the Pennsylvanian System of the Illinois Basin, part 2; invited papers: Illinois State Geological Survey Guidebook Series 15a, p. 115-120.

Lineback, J.A., 1970, Stratigraphy of the New Albany Shale in Indiana: Indiana Geological Survey, Bulletin 44, 73 p.

――――, 1980, Coordination study of the Devonian black shale in the Illinois Basin; Illinois, Indiana, and Western Kentucky: Illinois State Geological Survey Contract/Grant Report 1981-1 to U.S. Department of Energy, Contract DE-AS21-78MC08214, 36 p.

Meents, W.F., 1981, Analyses of natural gas in Illinois: Illinois State Geological Survey, Illinois Petroleum 122, 64 p.

Ottenjahn, K., M. Teichmuller, and M. Wolf, 1975, Spectral fluorescence measurements of sporinites in reflected light and their applicability for coalification studies, *in* Petrographie de la matiere organique des sediments, relations avec la paleotemperature et le potentiel petrolie: Paris, p. 49-65.

Powell, T.G., S. Creaney, and L.R. Snowdon, 1982, Limitations of use of organic petrographic techniques for identification of petroleum source rocks: AAPG Bulletin, v. 66, p. 430-435.

Rees, O.W., P.W. Henline, and A.H. Bell, 1943, Chemical characteristics of Illinois crude oils with a discussion of their geologic occurrence: Illinois State Geological Survey Report of Investigation No. 88, 128 p.

Reinbold, M.L., 1978, Stratigraphic relationships of the New Albany Shale Group (Devonian-Mississippian) in Illinois: Morgantown, WV, Preprints, Second Annual Eastern Gas Shales Symposium, U.S. Department of Energy, METC/SP-70/6, v. 1, p. 443-454.

Rhoads, D.C., and J.W. Morse, 1971, Evolutionary and ecologic significance of oxygen-deficient marine basins: Lethaia, v. 4, p. 413-428.

Robert, P., 1973, Analyse microscopique des charbons et des bitumes disperses dans les roches et mesure de leur pourvoir reflecteur, *in* Advances in organic geochemistry: Paris, Editions Technip, p. 549-569.

――――, 1979, Classification des matieres organiques en fluorescence application aux roches-meres petroliers: Pau, France, Bulletin des Centres de Recherches Exploration-Production Elf-Aquitaine, v. 3, n. 1, p. 223-263.

――――, 1981, Classification of organic matter by means of fluorescence; application to hydrocarbon source rocks: International Journal of Coal Geology, v. 1, p. 101-137.

Schwalb, H.R., and R.L. Norris, 1980, Studies of the New Albany Shale in western Kentucky—final report: U.S. Department of Energy METC/SP-8215-1, 55 p.

――――, and P.E. Potter, 1978, Structure and isopach map of the New Albany-Chattanooga-Ohio Shale (Devonian and Mississippian) in Kentucky (western sheet): Kentucky Geological Survey X, 1 sheet.

――――, E.N. Wilson, and D.G. Sutton, 1971, Oil and gas map of Kentucky, sheet 2, west-central part: Kentucky Geological Survey, University of Kentucky, Series X, 1 sheet.

Stach, E., et al, 1982, Stach's textbook of coal petrology, 3rd revised edition: Berlin, Gebruder Borntraeger, 428 p.

Stevenson, D.L., and D.R. Dickerson, 1969, Organic geochemistry of the New Albany Shale in Illinois: Illinois State Geological Survey, Illinois Petroleum 90, 11 p.

――――, L.L. Whiting, and R.M. Cluff, 1981, Geologic structure of the base of the New Albany Shale Group in Illinois: Illinois State Geological Survey, Illinois Petroleum 121, 2 p.

Swann, D.H., 1967, A summary geologic history of the Illinois Basin, *in* Geology and petroleum production of the Illinois Basin: Illinois and Indiana-Kentucky Geological Societies, Symposium Volume, p. 3-21.

――――, and A.H. Bell, 1958, Habitat of oil in the Illinois Basin, *in* Habitat of oil: AAPG Special Volume, p. 447-472.

Teichmuller, M., and K. Ottenjann, 1977, Liptinite und lipoid stoffe in einem erdolmuttergestein: Erdol U. Kohle, v. 30, p. 387-398.

Tissot, B.P., and D.H. Welte, 1978, Petroleum formation and occurrence, a new approach to oil and gas exploration: Berlin, Springer-Verlag, 538 p.

van Gijzel, P., 1967, Autofluorescence of fossil pollen and spores with special reference to age determination and coalification: Leidse Geologische Mededelingen, v. 40, p. 263-317.

Waples, D.W., 1980, Time and temperature in petroleum formation; application of Lopatin's method to petroleum exploration: AAPG Bulletin, v. 64, p. 916-926.

Willman, H.B., et al, 1975, Handbook of Illinois stratigraphy: Illinois State Geological Survey Bulletin 95, 261 p.

Zielinski, R.E., 1977a, Physical and chemical characterization of Devonian gas shale, quarterly status report, April 1-June 30: Mound Facility Report to the U.S. Department of Energy, Contract No. EY-76-C-040053.

――――, 1977b, Physical and chemical characterization of Devonian gas shale, quarterly status report, October 1-December 31: Mound Facility Report to the U.S. Department of Energy, Contract No. EY-76-C-040053, 210 p.

――――, and S.W. Nance, 1979, Physical and chemical characterization of Devonian gas shale, quarterly status report, April 1-June 30: Mound Facility Report to the U.S. Department of Energy, Contract No. EY-76-C-040053, 505 p.

This article is reprinted from the *AAPG Bulletin*, v. 62, p. 98-120.

Organic Geochemistry, Incipient Metamorphism, and Oil Generation in Black Shale Members of Phosphoria Formation, Western Interior United States

George E. Claypool
Alonza H. Love
Edwin K. Maughan
U.S. Geological Survey
Denver, Colorado

Nearly the whole range of organic metamorphism is reflected in the composition of sedimentary organic matter in shale members of the Permian Phosphoria Formation of western Wyoming and adjacent states. The different degrees of thermal maturity are recognizable on the basis of the amount, composition, and molecular nature of the extractable organic matter, and the color and elemental composition of the kerogen.

Organic matter in black shale members of the Phosphoria exhibits maximum conversion to petroleum hydrocarbons at temperatures corresponding to a 2.5 to 4.5-km range of burial depth. At shallow depths, Phosphoria sedimentary rocks are rich in extractable organic matter of asphaltic composition, and contain an immature hydrocarbon assemblage which is unlike mature petroleum. At greater depths there is extreme depletion of extractable organic matter and loss of hydrocarbons by thermal destruction and expulsion.

Oil which is believed to have been derived from the Phosphoria black shales also appears to be limited regionally to areas in which the reservoir rocks were buried to depths of from 2.5 to 4.5 km at the end of the Cretaceous. This suggests that there is an effective "window" for Paleozoic oil in the Cordilleran region.

If all of the oil in Paleozoic reservoirs in central Wyoming were derived from the Phosphoria black shales, and if hydrocarbon generation and migration efficiencies were on the order of 10 to 20 percent each, then much of the oil would have been generated and migrated from as far away as eastern Idaho prior to metamorphism, because of insufficient quantities of organic matter in the immediate area of the producing reservoirs.

INTRODUCTION

The Phosphoria Formation is believed to be the source of much of the oil that has been found in upper Paleozoic rocks in the northern and central Rocky Mountain region. Bowen (1918) and Condit (1919) first recognized oil shale in the Phosphoria. Cheney and Sheldon (1959) speculated that these phosphatic shale beds were sources of petroleum found in the Park City Formation in southwestern Wyoming and northern Utah. Stone (1967) extended this concept to include most of the oil in Paleozoic reservoirs in the Big Horn basin. Sheldon (1967) suggested that oil in upper Paleozoic rocks originated in shales of the Phosphoria and migrated from eastern Idaho and western Wyoming into central and eastern Wyoming in response to accumulating overburden and eastward-progressing tectonic forces.

Brongersma-Sanders (1948) and McKelvey (1959) also pointed out the association of phosphorite and oil. They attributed this association to the high organic productivity caused by upwelling marine waters. Powell et al (1975)

studied the organic geochemistry of black pelletal phosphorites in an effort to clarify the nature of the phosphorite/oil association. They found that although sedimentary phosphorite is not exceptionally rich in organic matter, a substantial part of it is soluble in organic solvents and has a predominantly asphaltic composition. They also suggested that phosphate-rich rocks could yield a heavy, nitrogen-rich oil at an early stage of diagenesis. Our study complements the work of Powell et al (1975) in that we have applied similar analytic techniques to the black phosphatic shale which was deposited in strata adjacent to the phosphorite beds and probably represents a deeper water facies. The black shale of the Phosphoria Formation in the region shown in Figure 1 is exceptionally rich in organic matter and is volumetrically extensive. The shale, rather than the phosphorite, probably is the better petroleum source rock.

GEOLOGIC SETTING

Stratigraphic Relations

The Meade Peak and Retort Phosphatic Shale Members

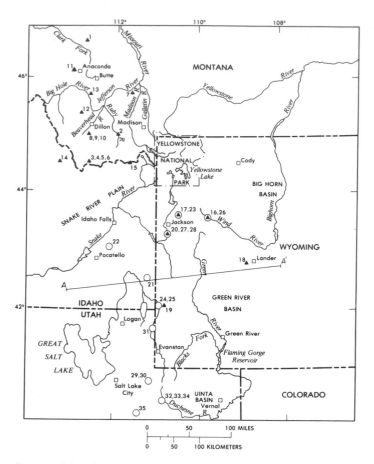

Figure 1. Map showing samples, localities, and geographic areas in Utah, Idaho, Wyoming, and Montana referred to in text; approximate location of cross-section *AA'* . Circles are Meade Peak localities and triangles are Retort localities.

of the Phosphoria Formation are parts of the complex stratigraphy of the Permian rocks of eastern Idaho and adjacent areas in Utah, Wyoming, and Montana. The stratigraphy and nomenclature of the Phosphoria and related rocks were summarized by Cheney (1957), McKelvey et al (1959), Sheldon (1963), Cressman and Swanson (1964), and McKee et al (1967). The stratigraphic relations and nomenclature used in this report were adapted from the report of McKelvey et al (1959) and are shown in Figure 2.

The members of the Phosphoria are composed of the dark-gray shale, phosphorite, and chert of marine origin. These beds thin eastward from Idaho into western Wyoming and intertongue with the predominantly carbonate rock of the members of the Park City Formation in central Wyoming. The Phosphoria and Park City also intertongue with Shedhorn Sandstone in southwestern Montana and northwestern Wyoming, a relation not shown in Figure 2.

Each of the rock types within the Phosphoria Formation was described by McKelvey et al (1959, p. 5) as part of marine transgressive and regressive cycles. Two complete cycles are present in western Wyoming (Figure 2) in the sequence of carbonate rock, chert, phosphorite, carbonaceous shale and then the reverse back to carbonate

rock. The lateral or vertical sequence may be incomplete at any location and only part of a cycle may be present. The carbonaceous shale beds in the Meade Peak and Retort Members represent the maximum transgression reached in each of the cycles. The facies relations and the paleogeographic setting of the Meade Peak Member and equivalent strata at the time of the earlier of the two transgressions were shown by Sheldon et al (1967). Similar facies and paleogeography of the Retort Member at the time of the later of the two transgressions were shown by Maughan (1966).

Structural Setting

The Phosphoria Formation was deposited on the edge of the continental shelf in a bight of the Cordilleran seaway. This large, open bay lay east of the late Paleozoic mobile belt adjacent to a bend in the geosynclinal axis from a northeasterly to a northwesterly trend (Roberts and Thomasson, 1964).

The two transgressive-regressive cycles recorded in the sedimentary sequence of the Phosphoria probably occurred in response to large movements of the Sonoran orogeny farther west in the mobile belt. Northward tilting of the entire four-state region between the time of deposition of the Meade Peak and the time of deposition of the Retort is indicated by a northward shift of both southern and northern shorelines and a corresponding shift of the center of maximum average organic carbon deposition in these two members (cf. Maughan, 1975, Figures 4 and 5). The later Permian rocks also record regional subsidence, which was greater on the west near the geosyncline and continued through the Mesozoic.

Overthrusting occurred in the western part of the area during Jurassic time, and subsequent crustal shortening brought outer parts of the bay against more shoreward areas as progressively younger thrust movements affected areas farther east. At least 60 km of crustal shortening was calculated by Crittenden (1961) in northern Utah, and possibly as much as 100 km is postulated in Idaho and Wyoming.

Depths of Burial

Maximum burial of the Permian strata throughout most of the region is assumed to have occurred at the end of the Cretaceous Period, although important exceptions to this generalization are recognized. Subsequent to deposition, Permian strata were buried by increments to increasingly greater depths through the Mesozoic Era. Additions to the sedimentary cover above the Permian frequently were followed by nondeposition, or even by erosion, during this era; but there was a slow net accumulation of sedimentary cover in the Triassic and Jurassic Periods. In contrast, thick Cretaceous sediments, especially those of Late Cretaceous age, were deposited rapidly, and they generally provide most of the post-Permian sediment accumulation. Within the Tertiary intermontane basins, sedimentary loading of the Permian strata continued from the Mesozoic into the Cenozoic Era, and maximum burial may not have been reached until about middle Eocene at some places. However, such scattered, additional, post-Cretaceous burial has not been incorporated in our calculations because most

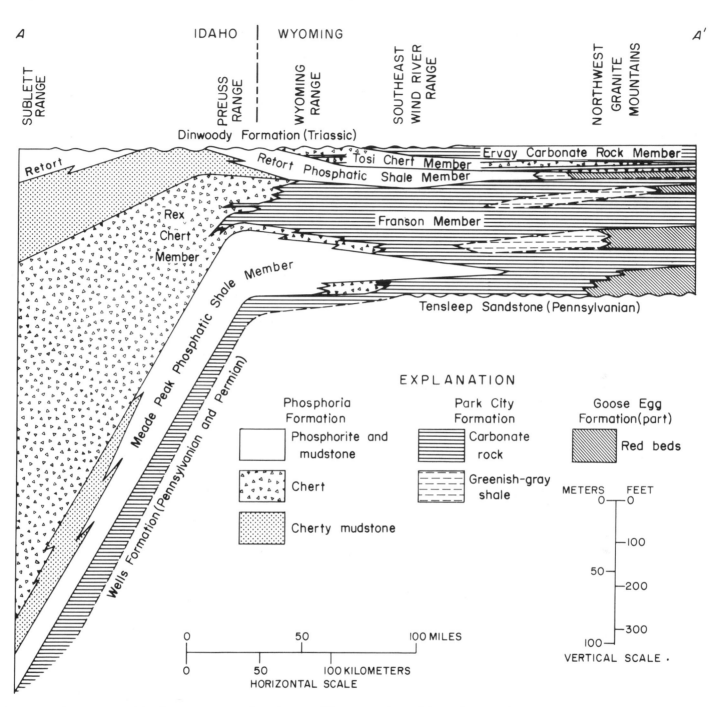

Figure 2. Cross-section *AA'* showing stratigraphic relations of Permian rocks in Idaho and Wyoming. Approximate location of cross section shown on Figure 1.

of our Phosphoria samples are from the flanks of the intermontane basins where Cenozoic sedimentation was negligible.

Generalized thicknesses of Mesozoic rocks are shown on Figure 3, which was compiled from the sources listed in Table 1. The approximate thickness of Mesozoic strata at each sample locality (Table 1) was obtained from measured sections or from cross sections in each area. In addition to these thicknesses, the isopachs shown on Figure 3 for central Wyoming, northwestern Colorado, and northeastern Utah, and in part for those in eastern Idaho and western Montana,

are composites derived from regional thickness maps of the Triassic (McKee et al, 1959), the Jurassic (McKee et al, 1956), and the Cretaceous (McGookey et al, 1972). Additionally, some interpretations of regional thicknesses of Cretaceous units in southwestern Montana were derived from McMannis (1965).

Post-Permian strata were thinner on the continental shelf and thicker toward the Cordilleran geosyncline. This cover is markedly thicker west of a hinge line which separated the tectonically active geosynclinal trough from the more stable shelf and craton. The hinge, across which the strata of all

Table 1. Sample localities and thickness of overlying Mesozoic sedimentary rocks (inferred maximum depth of burial).

Sample No.	Place	Section, Township, and Range	Mesozoic Thickness (Inferred Maximum Depth of Burial in Meters)	Thickness Reference
		Retort Phosphatic Shale Member, southwestern Montana		
1.	Anderson Mine, Garnet Range	19-10N-10W	5,000	Gwinn, 1961
2.	Canyon Camp, Ruby River	18-9S-3W	2,600	Hadley, 1969
3.	Middle Fork, Little Sheep Creek	4-15S-9W	2,050	Ryder and Scholten, 1973; *E.T. Ruppel, 1976
4.	Middle Fork, Little Sheep Creek	4-15S-9W	2,050	Ryder and Scholten, 1973; *E.T. Ruppel, 1976
5.	Middle Fork, Little Sheep Creek	4-15S-9W	2,050	Ryder and Scholten, 1973; *E.T. Ruppel, 1976
6.	Middle Ford, Little Sheep Creek	4-15S-9W	2,050	Ryder and Scholten, 1973; *E.T. Ruppel, 1976
7.	Sappington Cn., Jefferson River	26-1N-2W	1,700-2,500	Berry, 1943
8.	Retort Mtn., Blacktail Mtns.	23-9S-9W	<1,500	Scholten and others, 1955; Klepper, 1950; Lowell, 1965
9.	Retort Mtn., Blacktail Mtns.	23-9S-9W	<1,500	Scholten and others, 1955; Klepper, 1950; Lowell, 1965
10.	Retort Mts., Blacktail Mtns.	23-9S-9W	<1,500	Scholten and others, 19755; Klepper, 1950; Lowell, 1965
11.	Anaconda Quarry, Anaconda Wm. Spgs.	25-5N-12W	3,500	Emmons and Calkins, 1915
12.	Kelly Gulch, Pioneer Mtns.	3-6S-11W	<1,500	*E.T. Ruppel, 1976
13.	Garrison Wm. Spgs. Creek, Garnet Range	19-10N-9W	1,600	Fraser and Waldrop, 1972
		Retort Phosphatic Shale Member, eastern Idaho		
14.	Hawley Creek, Beaverhead Range	36-16N-27E	?	—
15.	Taylor Creek, Centennial Mtns.	14-14N-40E	3,000	*I.J. Witkind, 1976; Honkala, 1953
		Retort Phosphatic Shale Member, western Wyoming		
16.	Stoney Pt. on Wind River	24-42N-108W	3,750	Keefer, 1957
17.	Gros Ventre Slide	6-42N-43E	4,400	Love and others, 1973
18.	Baldwin Creek, Wind River Range	18-33N-100W	2,300	Rohrer, 1973
19.	Cokeville Hydro. Plant, Tump Range	35-25N-118W	6,000	Rubey, 1973 a, b
20.	Astoria Hot Spgs., Snake River Cn.	32-39N-116W	4,500	Schroeder, 1974
		Meade Peak Phosphatic Shale Member, eastern Idaho		
21.	Georgetown Cn., Preuss Range	19-10S-45E	8,300	Cressman, 1964
22.	Gay Mine	15-4S-37E	5,500-6,100	*D.E. Trimble, 1976
		Meade Peak Phosphatic Shale Member, western Wyoming		
23.	Gros Ventre Slide	5-42N-43E	4,500	Love and others, 1973
24.	Rocky Point, Cokeville	4-24N-119W	7,000	-estimated-
25.	Rocky Point, Cokeville	4-24N-119W	7,000	-estimated-
26.	Stony Point on Wind River	24-42N-108W	3,800	Keefer, 1957
27.	Astoria Hot Spgs., Snake River Cn.	32-39N-116W	4,600	Schroeder, 1974
28.	Astoria Hot Spgs., Snake River Cn.	32-39N-116W	4,600	Schroeder, 1974
		Meade Peak Phosphatic Shale Member, northeastern Utah		
29.	Franson (Pinon) Cn., Uinta Mtns.	14-1S-6E	6,800	Hintze, 1973, p. 130
30.	Franson (Pinon) Cn., Uinta Mtns.	14-1S-6E	6,800	Hintze, 1973, p. 130
31.	Benjamin Mine, Crawford Mtns.	18-11N-8E	8,500	Hintze, 1973, p. 124
32.	W. Fork, Duschene River, Uinta Mtns.	25-1N-9W	3,500	Hintze, 1973, p. 130, 144
33.	W. Fork, Duschene River, Uinta Mtns.	25-1N-9W	3,500	Hintze, 1973, p. 130, 144
34.	W. Fork, Duschene River, Uinta Mtns.	25-1N-9W	3,500	Hintze, 1973, p. 130, 144
35.	Rt. Fork, Hobble Creek, Wasatch Range	19-7S-5E	3,300	Hintze, 1973, p. 138, 140

*Oral communication.

three Mesozoic systems thicken, is approximately the same as the hinge across which the Permian strata are shown to thicken on Figure 2.

The thick wedge of Mesozoic strata west of the hinge along the foreland overthrust belt probably was deposited in a trough, and the wedge seems to be bounded by generally thinner Mesozoic strata, especially Cretaceous rocks, farther west. Thinner Mesozoic beds are known in the vicinity of sample locality 35, which is west of the Charleston overthrust fault in Utah; in the vicinity of locality 22, which is west of the Bannock overthrust fault in Idaho; and in the locality of samples 3-6, which is west of

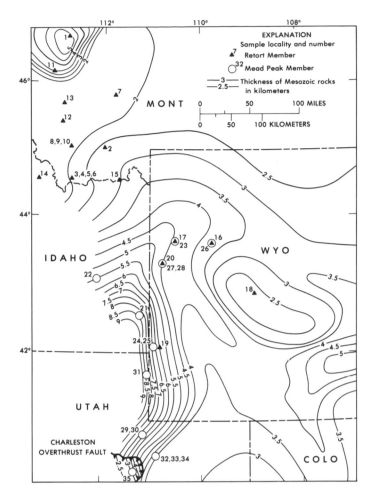

Figure 3. Generalized thicknesses of Mesozoic rocks in western Wyoming and parts of adjacent states.

the Tendoy thrust fault in southwest Montana. Increasing tectonic activity during development of the Laramide orogeny probably created anticlinal uplifts and upthrusts at places in the western parts of the region where Mesozoic sediments, and especially Cretaceous sediments were not deposited, were thinly deposited, or were eroded. The result is that burial of the Permian strata lessened within short distances to the east. In contrast to these areas of relatively shallow burial are the thick accumulations in the nearby foreland deep; some areas have sediment thicknesses which exceed 10 km in the vicinity of the point where the Utah, Idaho, and Wyoming borders join.

Sediment accumulation after the end of the Cretaceous continued locally within intermontane basins that remained after the Laramide orogeny. These additional local accumulations do not significantly affect our interpretations and conclusions about the depth of burial of the Permian rocks, because most of our samples were obtained from the margins of these basins where Tertiary deposition was minimal or where erosion and uncovering of the Permian strata have been dominant since the end of the Cretaceous. One possible exception may be the Permian strata at Little Sheep Creek in the Tendoy Mountains (locality of samples 3-6, Figure 1) where maximum burial by younger sediments possibly included as much as 3,050 m of the Upper

Cretaceous, Paleocene, and Eocene Beaverhead Formation (Ryder and Scholten, 1973) in addition to about 2,050 m of Mesozoic sediments. Deep burial at this locality is not indicated by the geochemical nature of the organic matter in these samples as described elsewhere in this report. The thick Beaverhead Formation was deposited, in part, syntectonically with movements of the Tendoy fault (Ryder and Scholten, 1973), and most of these conglomerates were confined to the east of the fault, which lies 2 km east of the sample locality on Little Sheep Creek. A few small, thin patches of conglomerate are present west of the fault, and it seems probable that the Tendoy overthrust block, which includes the sample locality, was principally an area of uncovering rather than of additional accumulation in Late Cretaceous and early Tertiary times. Similarly, in Utah, the Upper Cretaceous Price River Formation and thick lower Tertiary sedimentary rocks could have buried deeply the Permian strata at sample locality 35; however, aspect of the organic matter suggests that burial there was not deep although thick Cretaceous to lower Tertiary sedimentary rocks occur a few kilometers east of this locality on Hobble Creek.

The total thickness of Mesozoic rocks could not be determined from measurements of the strata in vicinity of sample localities 14, 24, 25, and 31. At these places, appreciable parts of the Mesozoic strata are absent and maximum depths of burial are uncertain. Estimated depths are derived from the isopach trends which suggest the probable Mesozoic thicknesses at these localities.

SAMPLES

All but one of the samples were obtained from surface localities given in Table 1. Most of the samples are from relatively fresh exposures. An attempt was made to obtain unweathered samples. To determine effects of weathering, certain key localities were resampled, and at one locality a shallow core was taken to obtain fresher material that was used in a comparative study (Clayton and Swetland, 1976). In all cases the analyses of replicate samples confirmed the analyses of the potentially more weathered rock previously sampled. The comparisons indicate that although the organic geochemical composition of our samples has been changed slightly by weathering, similar to the effects demonstrated by Leythaeuser (1973), the weathering is believed not to be a significant factor in the interpretation of our results.

ANALYTIC PROCEDURES

Pulverized samples (< 100 mesh) were extracted with chloroform or benzene-methanol azeotrope in a Soxhlet apparatus for 20 to 24 hours. The filtered extract solution or an aliquot was evaporated under nitrogen to an arbitrarily defined solvent-free point, and the weight of the total extract used to calculate the bitumen concentration. The bitumen isolated was chromatographed on silica gel, eluting successively with heptane, benzene, and benzene-methanol (1:1) to collect the saturated hydrocarbon, aromatic hydrocarbon, and nonhydrocarbon fractions, respectively.

The saturated hydrocarbon fractions were analyzed

EXPLANATION

further by gas chromatography on 2-m × 3-mm packed columns (OV-1 5 percent on Chromosorb P), temperature programmed from 100 to 320°C at 10°C per minute.

Insoluble organic matter (kerogen) was concentrated from the solvent-extracted rock samples by successive hydrochloric-hydrofluoric-hydrochloric acid treatment, followed by zinc bromide density separation of insoluble heavy minerals. The kerogen concentrates were mounted on glass slides and examined under the microscope for color of organic particles, according to a general color scale similar to that of Staplin (1969). Elemental analysis (C, H, N) was performed on kerogen concentrates using a Perkin-Elmer

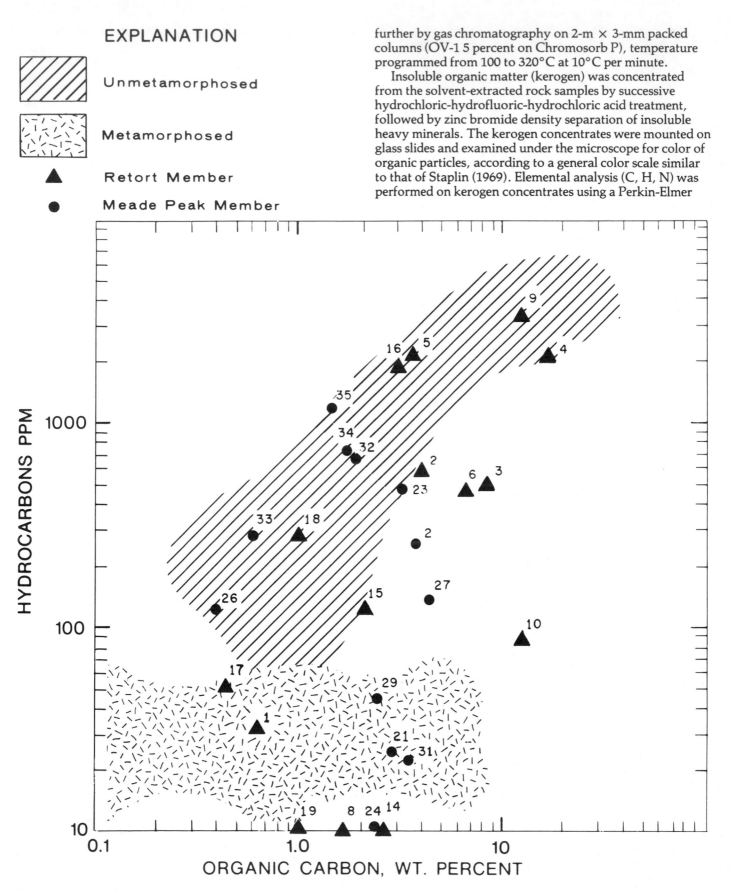

Figure 4. Hydrocarbon concentration versus organic carbon content in unmetamorphosed and metamorphosed sediments of Phosphoria Formation.

214D CHN analyzer. The analytical laboratories of the U.S. Geological Survey at Lakewood, Colorado, determined organic carbon values. Total carbon was determined with the Leco induction furnace, mineral carbon was determined gasometrically, and organic carbon was calculated as the difference between the two values.

RESULTS

The analytic results are summarized in Table 2. In addition to organic carbon in weight percent, we report the concentration of bitumen in parts per million, the total hydrocarbons (sum of the heptane and benzene eluates from silica get) in parts per million relative to the dry weight of the original rock sample. Also computed are the saturated-to-aromatic hydrocarbon ratios and the hydrocarbon-to-organic carbon ratios for each sample. For selected samples we have estimated and reported kerogen color, according to a general yellow to light brown to orange brown to brown to dark brown to black scale, and have determined atomic H/C ratios of the kerogen concentrates.

In the following sections we first present results of hydrocarbon and organic carbon determinations for the Meade Peak and Retort Members separately. The results of kerogen analyses and gas chromatography of saturated

Table 2. Organic geochemical analyses of black shales of the Phosphoria Formation.

Sample	Field No.	Organic Carbon Wt. %	CHCl₃ Bitumen ppm	Total Hydrocarbons ppm	Sat./Arom.	Hydc./Org. C. %	Kerogen Concentrate Color	Atomic H/C
Retort Phosphatic Shale Member, southwestern Montana								
1.	73M46H	0.6	75*	32	5.4	0.50	dk. brn.	1.1 (?)
2.	73M42K	4.0	1,870*	578	0.4	1.45	org. brn.	1.0
3.	73M44B	8.4	6,790*	480	0.04	0.56	—	—
4.	73M44J	19.1	16,830	2,080	0.1	1.09	brn.	1.2
5.	73M44G	3.9	18,860	2,350	0.06	6.04	—	—
6.	73M30AG	6.3	2,140	485	0.2	0.8	org.-lt. brn.	1.0
7.	74M22BE	0.8	—	—	—	—	lt. brn.	0.7
8.	74M29BE	1.6	358	11	0.1	0.07	—	—
9.	73M29FI	11.9	13,650	3,320	0.2	2.79	org.-brn.	1.2
10.	73M29JL	13.3	12,030	85	0.02	0.06	—	—
11.	73M45C	1.2	—	—	—	—	dk. brn.-blk.	—
12.	73M41C	0.5	—	—	—	—	dk. brn.-blk.	—
13.	74M27	—	—	—	—	—	dk. brn.-blk.	—
Retort Phosphatic Shale Member, eastern Idaho								
14.	74M33AK	2.8	28	9	2.0	0.03	dk. brn.	0.41
15.	74M37AP	2.3	4,290	105	0.5	0.5	—	—
Retort Phosphatic Shale Member, western Wyoming								
16.	73M34D	2.9	5,380*	1,920	1.0	6.6	org. brn.	1.2
17.	73M32E	0.5	119*	53	7.8	1.1	org. brn-dk. brn.	1.0
18.	74M10CJ	1.1	571	274	2.7	2.5	—	—
19.	74M42BE	1.0	227	10	1.5	0.1	dk. brn.	0.6
20.	75PS16	2.2	21	5	1.5	0.02	—	—
Meade Peak Phosphatic Shale Member, eastern Idaho								
21.	73M21L	2.9	80*	27	5.8	0.09	brn.	0.4
22.	73M25AR	3.9	208	183	5.8	0.5	—	—
Meade Peak Phosphatic Shale Member, western Wyoming								
23.	73M31G	3.2	880*	561	1.4	1.8	dk. brn.	0.6
24.	73M43AD	2.5	31	11	2.7	0.04	dk. brn.	0.4
25.	75PS14	0.9	22	6	2.0	0.07	—	—
26.	73M34A	0.4	423*	128	0.5	3.2	brn.	1.2
27.	73M33G	4.2	345*	126	3.3	0.3	blk.	0.4
28.	75PS15	1.2	38	3	2.0	0.03	—	—
Meade Peak Phosphatic Shale Member, northeastern Utah								
29.	73M11D	2.7	124*	46	10.5	0.2	dk. brn.	0.4
30.	75PS4	0.9	27	9	8.0	0.1	—	—
31.	73M17J	3.5	91*	24	5.0	0.07	dk. brn.	0.4
32.	74M2BK	1.9	1,100	675	1.5	3.6	brn.	1.0
33.	75PS7	0.6	704	265	0.8	4.4	—	—
34.	75PS8	1.9	1,500	800	0.4	4.2	lt. brn.	1.1
35.	74M4AH	1.6	1,990	1,280	1.7	8.0	lt. brn.	1.0

* = 3:2 benzene:methanol extraction

— = no determination

hydrocarbons are given in the following sections, for the Phosphoria samples as a group.

These results are interpreted within the context of three general facies of organic metamorphism—immature, mature, and metamorphosed—on the basis primarily of chemical composition of sedimentary organic matter.

Baker and Claypool (1970) and Evans and Staplin (1971) discussed the recognition of these facies. More recently, Hood et al (1975) have described a detailed 20-interval classification scheme for level of organic metamorphism (LOM) based primarily on coal rank and the reflectance of vitrinite. Earlier, Vassoyevich et al (1967) had extended the Russian concept of lithogenesis to provide a classification scheme based on processes which account for compositional differences reflected in the coal-rank series. All of these classification schemes reflect the same general process, that is, the thermochemical transformation of sedimentary organic matter. The analytic methods employed in our study do not permit use of the more detailed classification schemes.

Hydrocarbons and Organic Carbon

Meade Peak Phosphatic Shale Member—There is a wide variation in hydrocarbon content (11 to 1,280 ppm) even though the organic carbon contents of Meade Peak samples are relatively constant (14 out of 15 samples in the range from 0.9 to 4.2 percent). The organic carbon content of the Meade Peak Member is above the average of shales in general (Hunt, 1972). However, the contents of extractable hydrocarbons of certain samples are well below average, both on an absolute basis and relative to the organic carbon content. This is shown in Figure 4, a log-log plot of hydrocarbon versus organic carbon. Also shown on Figure 4 for comparison with Meade Peak samples are patterned fields indicating normal unmetamorphosed and metamorphosed sediments, based on analyses of contact and regionally metamorphosed mud rocks by Baker and Claypool (1970). The terms metamorphosed and unmetamorphosed are used in the standard geologic sense, in that the organic geochemical characteristics of Phosphoria samples are being compared with those of true, low-grade metamorphic rocks. Unmetamorphosed sediments generally show a positive correlation similar to that established for the rocks of the Pennsylvanian Cherokee Group from southeastern Kansas and correlatives in northeastern Oklahoma (Baker, 1962). Incipiently metamorphosed samples are consistently lower in extractable hydrocarbons at all levels of organic content. Six of the Mead Peak samples fall in the unmetamorphosed field, four in the field suggestive of incipient metamorphism, and two in intermediate positions in Figure 4.

Hydrocarbon composition, as expressed by the saturated-to-aromatic hydrocarbon ratio, is plotted in Figure 5 versus hydrocarbon-to-organic carbon ratio in percent for Meade Peak samples. Unmetamorphosed sediments generally exhibit saturated-to-aromatic hydrocarbon ratios less than 2 and a wide range of hydrocarbon-to-organic carbon ratios. The inferred effect of incipient metamorphism is preferential loss of aromatic hydrocarbons relative to saturated hydrocarbons. This effect, combined with the absolute loss of hydrocarbons at

all levels of organic carbon content, defines a metamorphic trend for extractable hydrocarbons in sedimentary rocks (Baker and Claypool, 1970). Incipiently metamorphosed rocks are characterized by saturated-to-aromatic hydrocarbon ratios greater than 2, and hydrocarbon-to-organic carbon ratios of much less than 1 percent. The same Meade Peak samples which were classified as unmetamorphosed in Figure 4, also fall in the unmetamorphosed field of Figure 5, whereas the remaining samples show the effects of incipient metamorphism.

Retort Phosphatic Shale Member—Samples of the Retort Member are more variable than the Meade Peak. Organic carbon ranges from 0.5 to 16 percent, whereas extractable heavy hydrocarbons range from 10 to 3,320 ppm. Extractable bitumen is extremely variable ranging from 75 to 30,000 ppm. Moreover, the hydrocarbons as a proportions of the bitumen are consistently less for the Retort (3 to 48 percent; average 22 percent) than for the Meade Peak (34 to 88 percent; average 48 percent). Compared with the Meade Peak samples the Retort samples are, in general, higher in organic carbon, much higher in total extractable bitumen, but only moderately higher in hydrocarbons. These differences are illustrated in Figure 4, where extractable heavy hydrocarbons are plotted versus organic carbon. Among most samples of the Retort there is a better general relation between hydrocarbons and organic carbon, within the range of unmetamorphosed rocks. Four samples of the Retort Member (localities 1, 8, 14, 19) have hydrocarbon contents which are low enough to suggest incipient metamorphism. When saturated-to-aromatic hydrocarbon ratio is plotted against the hydrocarbon-to-organic carbon ratio on Figure 5, two of the same Retort samples are in the incipiently metamorphosed field, with the third (locality 19) just below it with saturated-to-aromatic hydrocarbon ratio of 1.5. The rest of the Retort samples plotted on Figure 5 also have unusual organic compositions. Many of these samples are clustered in the lower left-hand corner, having saturated-to-aromatic hydrocarbon ratios less than 0.5 and hydrocarbon-to-organic carbon ratios less than 0.8 percent. These characteristics, together with the low proportion of hydrocarbons in the bitumen, are typical of thermally immature sediments. However, even the Retort samples which have hydrocarbon-to-organic carbon ratios above 1 percent generally have saturated-to-aromatic hydrocarbon ratios of 1 or less and a high proportion of asphaltic compounds in the bitumen. It is probable that these are also primary characteristics of organic matter deposited in phosphate-rich sediments, as was suggested by Barbat (1967) and confirmed by Powell et al (1975).

Samples from two localities (17 and 18) have unusually high saturated-to-aromatic hydrocarbon ratios (7.8 and 2.7) and have hydrocarbon-to-organic carbon ratios greater than 1 (1.1 and 2.5). These samples exhibit effects of weathering, which causes loss of both hydrocarbons and organic carbon (Leythaeuser, 1973), with preferential loss of aromatic hydrocarbons (Clayton and Swetland, 1976). Thus, to some extent, the effects of weathering mimic the effects of incipient metamorphism, as is reflected in the composition of the extractable organic matter. This illustrates (1) the hazard of using surface materials for

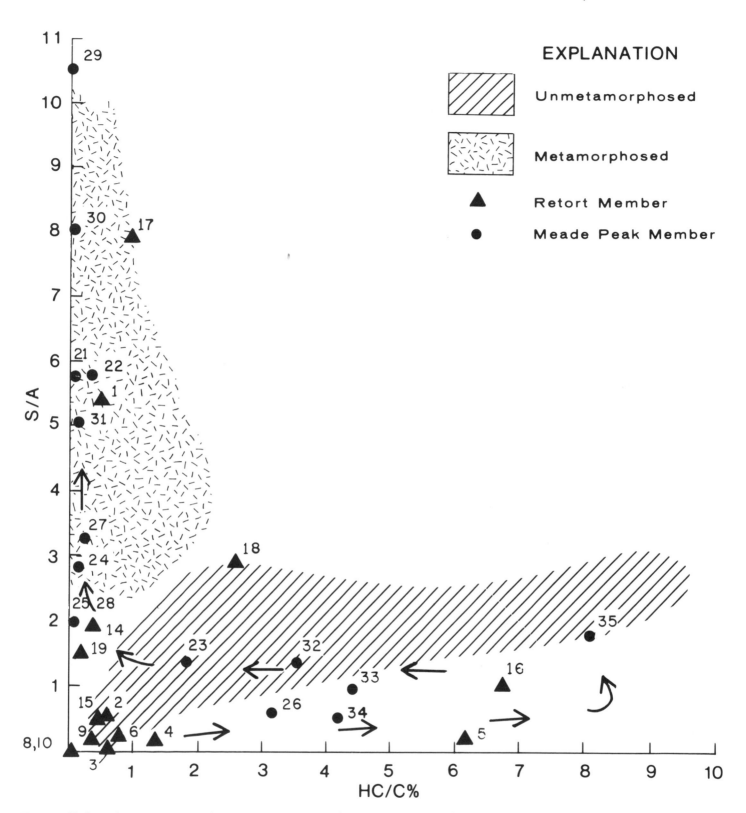

Figure 5. Hydrocarbon-to-organic carbon ratio versus saturated-to-aromatic hydrocarbon ratio. Effect of metamorphism on hydrocarbon concentration and relative proportions of hydrocarbon types in Phosphoria Formation. Arrows indicate suggested path of thermal alteration.

organic geochemical studies, and (2) the danger of relying only on the composition of extractable organic matter for interpretation of temperature history.

The suggested evolution of the composition of extractable organic matter in the Phosphoria black shales as a function of thermal alteration is along the path indicated

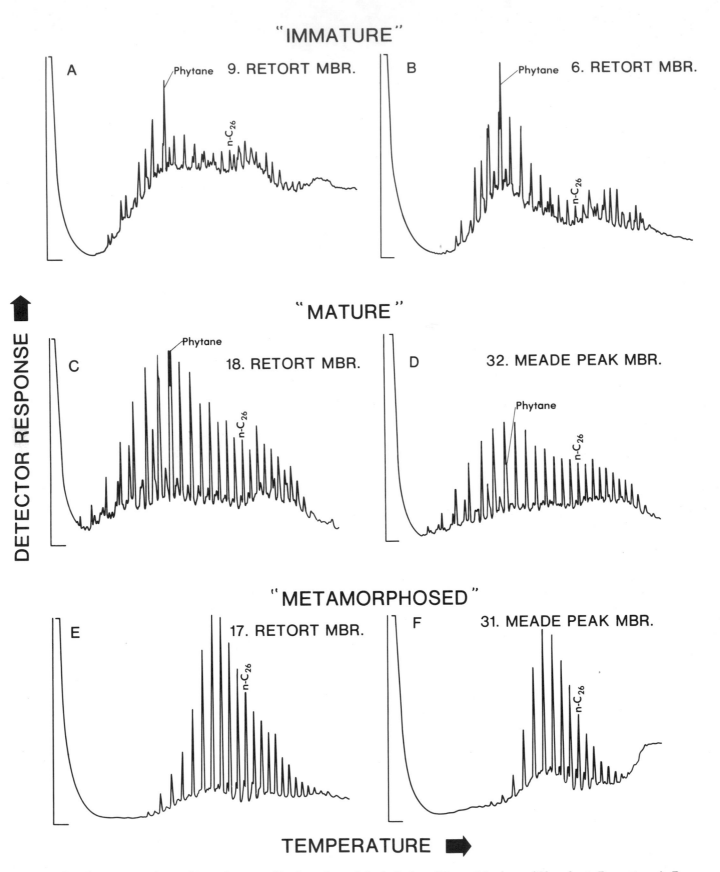

Figure 6. Gas chromatographic analyses of saturated hydrocarbons, Meade Peak and Retort Members of Phosphoria Formation. **A, B,** are "immature" Retort samples from localities 9 and 6. **C,** and **D,** are "mature" samples of Retort (locality 18) and Meade Peak (locality 32). **E, F,** are "metamorphosed" samples from localities 17 and 32. See Figure 1 and Table 1 for sample localities.

by arrows in Figure 5.

Kerogen

Kerogen was concentrated from selected samples of the Meade Peak and the Retort Members of the Phosphoria Formation. Colors and atomic H/C ratios of the concentrates are listed in Table 2. All but one of the Meade Peak samples analyzed had atomic H/C ratios of less than 1 and kerogen colors of dark brown to black. All but three Retort samples analyzed had yellow to dark-brown kerogen colors, and all but one had atomic H/C ratios greater than 1.

The kerogen composition is generally consistent with the interpretation of organic metamorphic facies based on extractable hydrocarbons. All of the samples which were classified as immature on the basis of extractables (i.e., had hydrocarbon-to-organic carbon less than 1 percent, saturated-to-aromatic hydrocarbon ratio less than 0.5, and low percentages of hydrocarbons in the bitumen) had atomic H/C ratios greater than 1.0 and yellow to brown kerogen colors. Most of the Phosporia samples classified as mature on the basis of extractables (hydrocarbon-to-organic carbon greater than 1 percent) had atomic H/C ratios in the range of 1 to 0.6 and brown to dark-brown kerogen colors. Incipiently metamorphosed Phosphoria samples (hydrocarbon-to-organic carbon less than 0.2 percent, saturated-to-aromatic hydrocarbon ratios greater than 2) generally had atomic H/C ratios of 0.4 and very dark brown to black kerogen colors.

Gas Chromatographic Analyses of Saturated Hydrocarbons

Selected chromatograms of saturated hydrocarbons extracted from the Phosphoria Formation are displayed in Figure 6. All of the samples were run under the same instrumental conditions, although size of sample injected and attenuation of detector response have been adjusted in some cases to produce a comparable display for each sample. No attempt has been made to quantify peak areas, and tentative identifications based on relative retention times are shown for index purposes only.

There appears to be continuous variation among saturated hydrocarbon assemblages of the Phosphoria Formation. Samples classified as thermally immature, on the basis of kerogen and extractable organic composition, have saturated hydrocarbon mixtures in which normal alkanes are subordinate to the presumed isoprenoids, pristane and phytane, and in which the unresolved, complex, branches—cyclic, saturated hydrocarbon mixture (i.e., the hump) predominates over the resolved peaks. Chromatograms for samples from localities 9 and 6 shown in Figure 6A and 6B are typical of the "immature" Retort samples.

Samples which are thermally mature—as indicated by hydrocarbon-to-organic carbon ratios, kerogen colors, and atomic hydrogen-to-carbon ratios—have saturated hydrocarbon chromatograms in which the normal alkanes are comparable or slightly predominant over pristane and phytane, and in which the resolved compounds in general are more predominant. Samples from localities 18 and 32, shown in Figure 6C and 6D, are examples of "mature" saturated hydrocarbon assemblages.

Incipiently metamorphosed samples are characterized by saturated hydrocarbon mixtures with an apparently narrower boiling point range, and commonly by a predominance of normal alkanes. Examples are shown in Figure 6E and 6F.

We believe that the characteristics of the saturated hydrocarbon mixtures, as shown by the gas chromatograms, are a function of presumed temperature history and no stratigraphy. That is, the Meade Peak and the Retort Members at the same apparent degree of thermal transformation have the same general characteristics. This interpretation is consistent with a uniform initial organic composition for both members.

DISCUSSION

The two black shale members of the Phosphoria Formation, the Mead Peak and the Retort, were sampled and selected for analysis in such a manner that they constitute two nearly distinct geographic groups (Figure 1). Although the original organic composition of the two shale members were qualitatively very similar, there are present-day differences owing to regional variations in temperature history. Other factors being equal, temperature history of the sediments is controlled primarily by maximum depth of burial and proximity to igneous heat sources. Maximum depth of burial for most of the black shale facies of the Phosphoria was at the end of the Cretaceous, and can be estimated from the total thickness of Mesozoic sedimentary rocks (Figure 3) overlying Permian rocks in this region.

The hydrocarbon-to-organic carbon ratios are plotted in Figure 7 against the thicknesses of overlying Mesozoic sedimentary rocks, taken as the inferred maximum depth of burial. A maximum in the normalized hydrocarbon content is indicated for samples buried to inferred maximum depths of about 3 km (10,000 ft). Hydrocarbon-to-organic carbon values fall off very abruptly for samples believed to have been buried deeper than 3 km, and below 5 km the extractable hydrocarbons have an incipiently metamorphosed character. Kerogen concentrates of deeply buried (> 5 km) samples have uniformly dark-brown to black colors and atomic hydrogen-to-carbon ratios from 0.40 to 0.6.

The depth of the maximum in the curve of hydrocarbon-to-organic carbon versus depth, shown in Figure 7, is not known precisely. The sample from locality 35 is plotted at an inferred maximum burial depth of 3.3 km; however, this depth estimate is less certain for this sample because of its location in the Charleston overthrust block. Among the phosphoria samples analyzed, this sample appears to have had the optimum temperature history for liquid-hydrocarbon generation.

The curve shown in Figure 7 is identical in form to curves published by Hood and Castaño (1974) and Albrecht and Ourisson (1969). By analogy, we interpret that organic matter in the Phosphoria Formation has been "overcooked" in the general region of the Idaho-Wyoming thrust belt during burial to depths of as much as 9 km or more. In contrast, the Retort Member of the Phosphoria in southwestern Montana never has been buried so deeply and its organic matter is relatively immature, never having been subjected to temperatures sufficient for a significant degree of hydrocarbon generation. One sample of the Retort Member of southwestern Montana (locality 5) has a hydrocarbon-to-organic carbon ratio of 6 percent. By itself this might suggest deeper than the inferred burial of this locality, however our other geochemical criteria (kerogen color, H/C ratio, hydrocarbon proportion of bitumen, gas chromatography of saturated hydrocarbons) indicate that the organic matter in this sample is also immature.

The Phosphoria Formation east and south of the basin foredeep on the stable shelf in Wyoming and northern Utah retains the chemical imprint of the optimum depth of burial for liquid-hydrocarbon generation.

Samples from the Astoria Hot Springs locality (20, 27, 28) are shown on Figure 7 as darkened symbols, because they apparently have been subjected to higher temperatures than is otherwise indicated by their inferred depth of burial. Other samples for which the apparent degree of metamorphism is inconsistent with the inferred burial history are discussed in the next section.

Temperature Effects Related to Igneous Intrusions

Four samples of the Retort Member have low hydrocarbon-to-organic carbon ratios and high saturated-to-aromatic ratios suggesting that they had histories of high temperatures (Figure 5). Two of these samples (17, 19) are from the Idaho-Wyoming thrust belt, and they probably have been subjected to burial depths greater than 4 and 8 km, respectively, as was suggested in the previous section. The other two samples (1, 14) are from the region of batholithic intrusions shown on Figure 8. It is very likely that the composition of the organic matter in these samples reflects incipient contact or regional thermal metamorphism rather than burial metamorphism, although the chemical effects are the same.

Kerogens of nine Retort samples from southwestern Montana were examined. The location of altered kerogen samples is shown on Figure 8, and is consistent with a regional gradient of incipient metamorphism across southwest Montana related to proximity to the batholiths.

A strikingly similar gradient in C^{13}/C^{12} of organic carbon in the Phosphoria formation was reported by Rooney (1956). His results are shown in Figure 8. The contours are based on Rooney's original interpretation, but have been adjusted to even increments of δC^{13}_{PDB} which were calculated from the absolute abundances reported by Rooney, using the C^{13}/C^{12} ratio for PDB given by Craig (1957). The 15 per mill change in the carbon isotopic composition of the organic matter in Phosphoria sediments seen on Figure 8 also may reflect a relation between gradient of contact and regional metamorphism across southwestern Montana. Similar but smaller isotopic effects also have been observed by Barker and Friedman (1969), Baker and Claypool (1970), and McKirdy and Powell (1974). The latter also showed a relation between δC^{13} and atomic H/C ratio for carbonaceous matter in Precambrian sedimentary rocks.

Rooney's interpretation was that the change in isotopic composition reflected environmental variations relative to an ancient shoreline. However, isotopic changes of this magnitude generally are not observed in modern nearshore marine environments (Sackett and Thompson, 1963), especially within the limits of environmental variation suggested by the lithologic uniformity of Phosphoria sediments.

Relevance to Petroleum Occurrence

The geochemistry of organic matter in the Phosphoria Formation, as shown by analyses of outcrop samples, reflects primarily the rock temperature at the maximum depth of burial. Permian rocks in the overthrust belt apparently have been subjected to temperatures sufficient to cause loss or destruction of previously existing liquid hydrocarbons, and to produce a high degree of carbonization of the kerogen (low atomic H/C). The Phosphoria and equivalents deposited on the stable shelf on the east retain the chemical imprint of optimum temperatures for maximum conversion of organic matter to liquid hydrocarbons. The Retort Member of the Phosphoria Formation in southwestern Montana is low in hydrocarbons relative to organic matter; the type of hydrocarbons present and the low degree of carbonization of the kerogen (high atomic H/C) suggest that this region has not been subjected to temperatures required for a significant generation of hydrocarbons.

The occurrence of oil in upper Paleozoic rocks in central Wyoming can be evaluated in terms of organic metamorphic facies of the Phosphoria source rock. The regional distribution of organic-rich rocks which contain chemically recognizable petroleum hydrocarbons (i.e. mature source rocks) has been advanced as a primary factor controlling the distribution of oil within Lower Cretaceous Mowry Shale and equivalent rocks (Nixon, 1973) and the Pennsylvanian Tyler Formation (Dow, 1974; Williams, 1974) in the northwestern interior of the United States. In both of these studied areas, the oils are within or immediately adjacent to regions in which the organic-rich rocks have been buried to depths in excess of 2.1 km (7,000 ft). In a similar manner, Phosphoria-derived oils appear to be limited to areas where Permian rocks were buried more than 2.1 km but less than about 4.9 km at the end of the Cretaceous. In the Phosphoria, this regional coincidence of mature source beds and oil occurrence may not be the primary controlling factor, because conduit and reservoir rocks of Permian age also are generally coincident with petroleum occurrence. However, the underlying sandstone of Pennsylvanian age, such as the Tensleep Sandstone which contains most of the Phosphoria-derived oil reserves in central Wyoming, is present throughout the region and is barren by between 2 and 5 km of sediment at the end of the Cretaceous. This supports the requirements for the proper thermal history analogous to that of the Mowry and the Tyler source-rock sections.

In western Wyoming and eastern Idaho the Phosphoria black shales apparently have been subject to temperatures sufficient to cause a drastic depletion of indigenous liquid hydrocarbons. High temperatures probably also caused the destruction or loss of reservoired liquid hydrocarbons in this region, if originally present.

Likewise, no Phosphoria-derived petroleum occurrences are known in the area of southwestern Montana where the

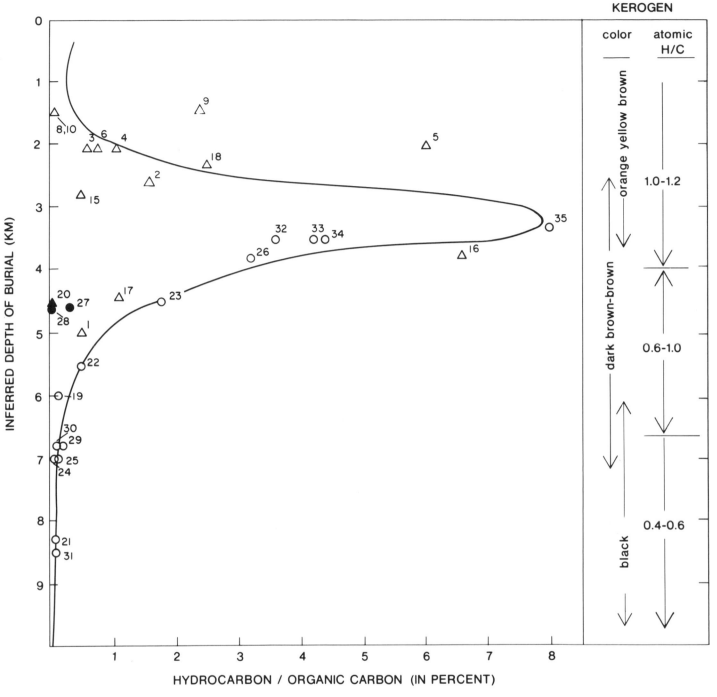

Figure 7. Hydrocarbon-to-organic carbon ratio and composition of kerogen versus inferred maximum depth of burial (thickness of overlying Mesozoic rocks).

organic matter in the Retort Member has characteristics indicating a history of low temperature. Whether the lack of commercial petroleum is due to absence of suitable conduit and reservoir beds or to a temperature history insufficient to bring about generation of petroleum hydrocarbons cannot be determined definitively on the basis of our results.

Powell et al (1975) suggested that heavy, asphaltic oil could be generated from relatively immature phosphate-rich source sediments. Our analyses confirm that thermally immature Phosphoria sediments contain abundant

extractable organic matter, although they are relatively deficient in petroleumlike hydrocarbons. Stone's (1967) theory for the source and accumulation of oil in the Big Horn basin invokes early migration (Early Jurassic) of immature oil (low API gravity and high sulfur content) from Phosphoria sediments. The composition of this oil is assumed to have been modified subsequently owing to increased burial depths and temperatures during and since the Laramide orogeny. Orr (1974) has presented a well-documented case for high-temperature alteration

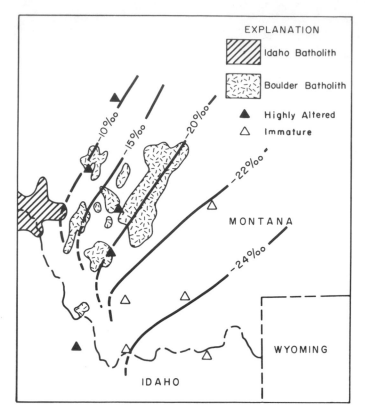

Figure 8. Effect of proximity to Boulder and Idaho batholiths on stable carbon isotope ratio (δC^{13}_{PDB}) and color alteration of kerogen in Retort Phosphatic Shale Member of Phosphoria Formation. Stable carbon isotope interpretation based on Rooney (1956).

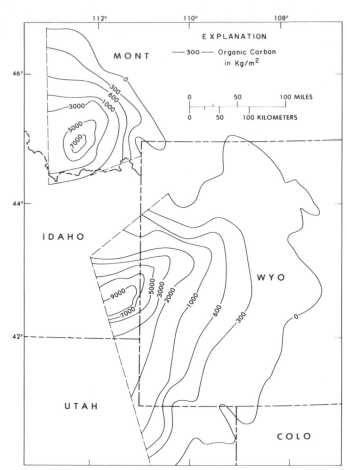

Figure 9. Distribution of organic carbon in black shale facies of Phosphoria Formation (kg/sq m).

(thermal oxidation) of oils in the Big Horn basin. However, the original nature of the Phosphoria-derived oils is still an open question because of the possibility that low-temperature alteration processes, chiefly biodegradation (Winters and Williams, 1969) and water washing, may have transformed the original oil into the low-API-gravity, asphaltic, high-sulfur oil characteristic of shallow (<1.8 km) Big Horn basin reservoirs subjected to recharge by fresh, oxygenated surface water (Todd, 1963).

If it could be demonstrated that low-temperature alteration (biodegradation and water washing) was a major factor in determining the quality of Phosphoria-derived oil, it would considerably weaken the case for early migration of heavy, asphaltic oil from Phosphoria black shales. Moreover, it would greatly strengthen the analogy of similar oil-generating requirements in the Phosphoria, the Mowry Shale (Nixon, 1973), and the Paleozoic source rocks of the Williston basin (Dow, 1974), that is, temperatures equivalent to those prevailing at depths of burial in excess of 2.1 km at the end of the Cretaceous Period in the northern Rocky Mountain region.

Quantitative Evaluation of Oil Potential

Consideration of thermal history in conjunction with regional distribution of organic matter may provide an objective basis for evaluating potential petroleum resources. In addition, the questions of volumetric sufficiency of

Phosphoria black shales as oil sources and local versus long-distance migration, can be evaluated in a more quantitative way using these considerations.

On the basis of the amount and distribution of organic carbon, and on the regional temperature history of the sediments indicated by the chemical composition of the organic matter, the total quantity of liquid hydrocarbons generated in the Phosphoria Formation can be estimated. An estimate also can be made of the quantity of migrated hydrocarbons believed, on the basis of the geologic and geochemical evidence, to have been derived from Phosphoria source rocks. By comparing these quantities we can estimate degrees of hydrocarbon expulsion, migration, and accumulation which are consistent with different volumes of effective source rocks and associated distances of migration.

The regional distribution of organic carbon in Phosphoria sediments was determined by Maughan (1975). Figure 9 was developed from maps by Maughan (1975) showing the thickness and organic carbon content of the Meade Peak and the Retort Members of the Phosphoria Formation. These maps were combined to estimate the regional distribution of organic carbon in the Phosphoria Formation, by multiplying thickness times organic carbon content times average density to give carbon content per

Table 3. Summary of hydrocarbons from black shale members of Phosphoria.

A. Disseminated Hydrocarbon Inventory, Phosphoria Fm.

By Region, Based on Degree of Maturity	Organic Carbon (10^9 Metric Tons)		Observed Heavy Hydrocarbon Content (10^9 Metric Tons)		Estimate of Heavy Hydrocarbons Generated (Q) @ 20% Conversion (10^9 Metric Tons)	
1. Southwestern Montana	62		0.31		0	(immature)
2. Ida., Wyo., Utah, Colo.	132		1.44		30.7	
By depth of burial:						
a. >5 km		81		0.08		19.0
b. 3.5 to 5 km		34		0.68		7.8
c. 2 to 3.5 km		17		0.68		3.9
Totals	194		1.75		30.7	

B. Migrated Hydrocarbons Presumed Derived from Phosphoria Fm.
 1. Cumulative production plus proven reserves (Q_{R_e}) — 0.6×10^9 metric tons
 2. Migrated heavy hydrocarbons as original oil in place, economic and subeconomic occurrences plus oil lost by secondary alteration processes (Q_E) — 2.4×10^9 metric tons

unit area, in this case kilograms of organic carbon per square meter. An average density of 2.8×10^3 kg/cu m was used, rather than trying to adjust for density differences caused by variation of composition or depth of burial.

We suggest that this distribution of the quantity of organic carbon in black shales is largely a primary depositional characteristic, and is modified only slightly by subsequent diagenetic processes. Therefore, the organic carbon contours shown in Figure 9 approximate the distribution of organic matter initially available for conversion to petroleum hydrocarbons.

The total quantity of organic carbon in Phosphoria Formation black shales can be estimated by integration across the carbon isopleths in Figure 9. This estimate is summarized in Table 3, for the Phosphoria Formation as a whole, and for certain regional subdivisions. The total amount of organic carbon in black shale of the Phosphoria Formation is 194×10^9 MT, within the boundaries indicated in Figure 9. About 62×10^9 MT of carbon is contained in thermally immature organic matter in southwestern Montana, which for purposes of this comparison is assumed not to have contributed to petroleum accumulation. This leaves 132×10^9 MT of carbon in thermally mature and postmature Phosphoria black shale in parts of Wyoming, Idaho, and Utah.

The total quantity and regional distribution of organic carbon summarized in Table 3 provide base figures for the estimate of petroleum hydrocarbons still present and originally generated in the black shale members of the Phosphoria Formation. Heavy hydrocarbons still present are estimated by applying the average observed hydrocarbon-to-organic carbon ratio in particular regions, to the total amount of organic carbon in those regions. For the immature Retort Member in southwestern Montana, the average hydrocarbon-to-organic carbon ratio is 0.5 percent. In the rest of the Phosphoria, in regions defined by inferred maximum depth of burial at the end of the Cretaceous, the average hydrocarbon-to-organic carbon ratios are: between 2 and 3.5 km, 4 percent; between 3.5 and 5 km, 2 percent; and greater than 5 km, 0.1 percent. Multiplying these factors times the quantity of organic carbon in each region

gives an estimate of the extractable heavy hydrocarbons present in black shale beds of the Phosphoria Formation. As summarized in the second column of Table 3, a total of 1.75 $\times 10^9$ MT of extractable heavy hydrocarbons is estimated to be present in Phosphoria black shale. Of this total disseminated hydrocarbon inventory, about 18 percent, or 0.31×10^9 MT are a non-petroleumlike, immature assemblage of hydrocarbons in the areas where the Phosphoria apparently never has been buried deeply enough to generate petroleum hydrocarbons. Therefore, about 1.44×10^9 MT of heavy hydrocarbons are present in the thermally mature and postmature black shale of the Phosphoria Formation.

For comparison, our estimated quantity of migrated heavy hydrocarbons, as economic and subeconomic original oil in place in Paleozoic reservoir rocks of central Wyoming, northwestern Colorado, and northeastern Utah, is shown as 2.4×10^9 MT at the bottom of Table 3. This estimate was arrived at in the following manner.

1. Cumulative production and proved reserves (6×10^9 bbl) were estimated for Paleozoic reservoirs in the Big Horn basin, Wind River basin, Green River basin, Casper arch area, plus Elk Springs, Rangely, and Ashley Valley fields. On the basis of evidence presented by Barbat (1967), Stone (1967), and Sheldon (1967), we assume that this oil is derived from the black shales of the Phosphoria Formation.

2. Original oil in place (12×10^9 bbl) was estimated by doubling cumulative production plus proved reserves. This factor was arrived at by consideration of (a) the estimated proportion of oil in place ultimately recoverable by waterflood production, and (b) the extent of waterflood production in these fields as of 1974. This 12×10^9 bbl agrees exactly with the estimate by Curry (1971) of cumulative production plus total future oil in place in Paleozoic reservoirs in the Big Horn, Wind River, and Green River basins, if oil is 80 percent of total production, as it is in NPC Region 4 (North Rocky Mountains) as a whole.

3. Subeconomic occurrences of migrated hydrocarbons, plus oil lost owing to the combined effects of seepage, exhumation, inspissation, biodegradation, water washing,

thermal cracking, and thermal oxidation were estimated to be roughly equal to the original oil in place above (12×10^9 bbl), thus resulting in a further doubling of the estimate for total original oil in place (24×10^9 bbl). This is a rough estimate based on (a) the frequency of oil shows in wells drilled in central Wyoming (Partridge, 1958) and other Paleozoic oil provinces, and (b) the apparent extent of secondary-alteration processes, assuming a relative enrichment of sulfur content due to hydrocarbon loss, compared with an assumed primary oil containing 0.5 to 0.8 percent sulfur.

4. Assuming an average gravity of 30° API and 70 percent heavy-hydrocarbon content, 24×10^9 bbl of oil is equivalent to 2.4×10^9 MT of heavy hydrocarbons.

The 2.4×10^9 MT of heavy hydrocarbons estimated to have migrated from Phosphoria black shale as oil is completely out of balance with the 1.44×10^9 MT of heavy hydrocarbons remaining in those same rocks. This discrepancy apparently is caused by heavy hydrocarbon loss and/or destruction in the part of the Phosphoria Formation in Idaho and western Wyoming that contains most of the organic matter and which is inferred to have been buried deeper than 5 km at the end of the Cretaceous. During an earlier time (i.e., Early Cretaceous) these sediments must have passed through the zone of optimum temperature for hydrocarbon generation. If expulsion and eastward migration also occurred during that time, some heavy petroleum hydrocarbons would have escaped burial and destruction. Therefore, the estimated quantity of migrated heavy hydrocarbons probably should be compared with the amount of heavy hydrocarbons originally generated, rather than with the amount now remaining disseminated in the rocks.

The amount of hydrocarbons originally generated can be estimated by applying some maximum conversation ratio to the total quantity of organic carbon in mature and postmature black shale. The observed maximum hydrocarbon-to-organic-carbon ratio is 8 percent for the Meade Peak Member and 6.6 percent for the Retort Member of the Phosphoria Formation (Table 2). This estimate is a lower limit for hydrocarbon generation, for it is based *only* on extractable nonvolatile hydrocarbons *remaining* in outcrop samples. Other estimates of hydrocarbon-generating capacities are provided by Fisher assays and pyrolysis yields. Condit (1919) reported an average yield of 84 l of oil/MT for the thermally immature Retort Member in southwestern Montana. In our study, rock samples with 6 percent organic carbon gave a pyrolysis response equivalent to a 4 percent oil yield (i.e., 100 g of rock contains 6 g organic carbon and yields 4 g of "oil" upon pyrolysis). The "oil" or pyrolysis product is about one-third heavy hydrocarbons (the remaining two-thirds being nonhydrocarbons and volatile light hydrocarbons). This provides an estimated conversion ratio of organic carbon to heavy hydrocarbons of 22 percent. If we apply an average conversion ratio of 20 percent to the quantities of organic carbon in Table 3 (after restoring an amount of carbon equivalent to the hydrocarbons presumed lost or destroyed), an estimate is derived for the total amount of heavy hydrocarbons originally generated and available for migration (tabulated in the third column of Table 3).

We are now in a position to compare the estimated quantity of Phosphoria-derived migrated heavy hydrocarbons (2.4×10^9 MT) with the estimated total amount of heavy hydrocarbons generated in black shale of the Phosphoria Formation (30.7×10^9 MT). About 8 percent of the heavy hydrocarbons generated was expelled, and about 2 percent accumulated as *recoverable* petroleum, if the Phosphoria black shale over the entire region shown in Figure 9 served as effective source rocks. This supposition requires migration distances on the order of 400 km. If burial depths of at least 2 km were required for the generation of petroleum hydrocarbons from Phosphoria black shale, then the part of the Phosphoria west of the 5-km isopach for Mesozoic rocks did not generate oil until the Early Cretaceous—or 125 million years ago (Figure 10; Sheldon, 1967). Migration from this region would have to have been completed by the onset of the Laramide orogeny about 65 million years ago, when uplift of Precambrian mountain ranges created the present structural basins. This leaves a period of 60 million years for the extensive lateral migration to have taken place. J.M. Hunt (according to McDowell, 1975) stated that oil moves in the direction of decreasing hydrostatic pressure at a rate of 2.5 to 7.6 cm/year, or 25 to 76 km/million years. At this rate the required migration distance easily could be traveled in the allowable time, provided suitable conduits for migration existed.

However, the requirement for 2 km of overburden—before the principal phase of hydrocarbon generation could take place—was not achieved until Early Cretaceous, which might increase the likelihood that potential conduit beds of Pennsylvanian and Permian age were made ineffective by cementation. Fox et al (1975) concluded that the earliest cementation of the Tensleep Sandstone was during or shortly after the Laramide orogeny. A more localized oil source still may seem preferable from this standpoint, but it is harder to justify in light of the smaller quantities of hydrocarbons available if migration distances were shorter. If we limit the region of effective Phosphoria source rocks to that part of the formation which has been buried between 2 and 3.5 km, then a degree of expulsion of 60 percent is required to account for the 2.4×10^9 MT of Phosphoria-derived migrated heavy hydrocarbons from the estimated 3.9×10^9 MT generated. This estimate seems extremely high, although McDowell (1975) believed that when source and reservoir are interlayered, source rocks commonly may expel more than 50 percent of the oil they generate.

The estimated quantities of heavy hydrocarbons generated, expelled, and accumulated as recoverable oil are summarized in Figure 10, a box diagram after the style of McDowell (1975), specifically for the Phosphoria Formation. The combinations of migration distance, quantity of hydrocarbons in effective source rocks, and degree of expulsion required to account for the estimated quantity of Phosphoria-derived migrated hydrocarbons are summarized in Table 4. It is not possible to place closer limits on efficiencies of hydrocarbon expulsion and migration in the Phosphoria Formation unless effective volumes of source sediments are defined more adequately by geologic constraints.

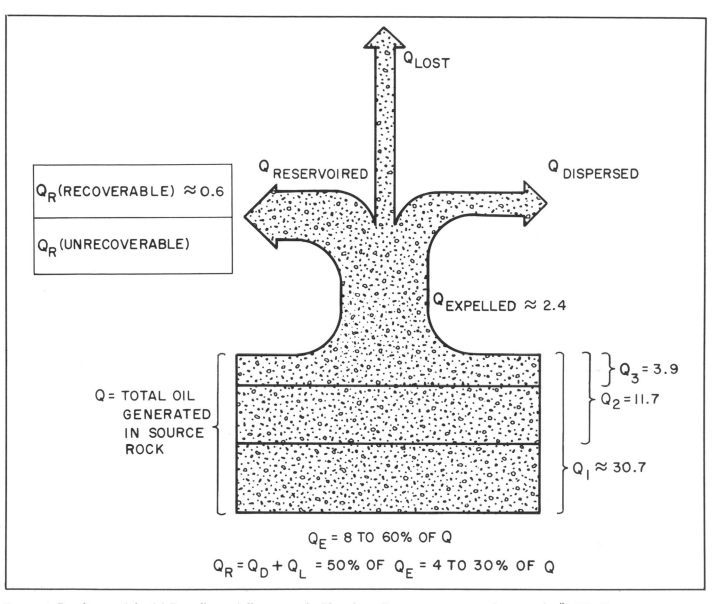

Figure 10. Box diagram (after McDowell, 1975) illustrating, for Phosphoria Formation, quantities (in units of 10^9 MT) of heavy hydrocarbons generated, expelled, dispersed, lost, and reservoired. Q = total oil generated; Q_E = expelled; Q_D = dispersed; Q_L = lost; Q_R = reservoired.

Table 4. Permissible combinations required for estimated Phosphoria-derived migrated hydrocarbons.

If Migration Distances were (km)	Quantity of Hydrocarbons in Effective Source Rocks (MT)	Required Degree of Hydrocarbon Expulsion from Source Rocks (%)
400	30.7×10^9	8
175-250	11.7×10^9	20
100	3.9×10^9	60

There also may be reason to question the validity of our assumption that *only* the organic-rich black shale members served as effective source rocks. Argillaceous carbonates of the Park City Formation may have contributed to petroleum accumulation, thus significantly lessening the requirements for migration distance and/or expulsion efficiency.

In addition, burial of these source rocks to requisite depths for hydrocarbon generation probably continued to be reached locally in the Tertiary intermontane basins after the end of the Cretaceous. The generation, thermal destruction, and losses to outcrops of petroleum in these areas probably have affected the total hydrocarbon inventory.

SUMMARY AND CONCLUSIONS

The present organic geochemical variability of the black shale facies of the Phosphoria Formation is primarily a function of temperature history, as has been inferred from maximum depths of burial and/or proximity to igneous heat sources. This variability of the organic matter is reflected consistently in the chemical composition of both extractable and nonextractable organic fractions of outcrop samples. Three general facies of organic metamorphism are recognizable: (1) thermally immature sediments characterized by low extractable heavy (C_{12+}) hydrocarbon yields relative to organic carbon and chloroform-soluble bitumen contents, with yellow to brown colors and atomic hydrogen-to-carbon ratios greater than 1.1 in the insoluble organic matter (kerogen); (2) thermally mature sediments with extractable hydrocarbon contents in excess of 1 percent of the total organic carbon, kerogen of brown to dark-brown color and atomic hydrogen-to-carbon ratios in the range from 1.1 to 0.6; (3) incipiently metamorphosed sediments with extractable hydrocarbon contents of less than 0.2 percent of the total organic carbon and brownish-black to black kerogen with atomic hydrogen-to-carbon ratios in the range of 0.6 to 0.4. In addition, each metamorphic facies is characterized by a distinctive type of saturated hydrocarbon distribution, as determined by packed-column gas chromatography.

The temperature history of the Phosphoria Formation is related to thickness of overlying Mesozoic rocks (maximum depth of burial) and/or distance from margins of the Idaho and boulder batholiths. Thus, the immature facies of the Phosphoria is confined to southwestern Montana at distances greater than about 50 km from the margins of batholiths where burial depths probably never exceeded 2 km and regional igneous activity is absent. The incipient metamorphic organic facies is localized along the western Wyoming-Idaho-Utah boundary, which corresponds to the foredeep of the Mesozoic Cordilleran geosyncline, and in southwestern Montana and central Idaho at or near the contacts with the Boulder and Idaho batholiths. The thermally mature organic facies is located on the platform east of the Cordilleran geosyncline across central Wyoming, and in (allochthonous) thrust sheets in eastern Utah and east-central Idaho.

Petroleum deposits, which are believed to be derived from the black shale facies of the Phosphoria, occur exclusively in the region of thermally mature organic facies, predominantly in laterally adjacent Phosphoria carbonate or underlying Permian and Pennsylvanian reservoirs.

The black shale facies of the Phosphoria contains about 194×10^9 MT of organic carbon. This amount is taken as an index of the organic matter originally deposited. We calculate that about 30×10^9 MT of heavy (C_{12+}) petroleum hydrocarbons may have been generated from the organic matter in thermally mature and incipiently metamorphosed parts of the Phosphoria Formation. We have excluded 62×10^9 MT of organic carbon in thermally immature sediments in southwestern Montana from this calculation, and assumed a 20 percent conversion of organic matter to petroleum heavy hydrocarbons. Only about 1.4×10^9 MT of heavy hydrocarbons remain disseminated in these Phosphoria black shales, primarily because hydrocarbons were lost as a consequence of deep burial. Petroleum heavy hydrocarbons generated and expelled from Phosphoria black shale, are estimated at 2.4×10^9 MT. This oil can be accounted for by combined expulsion and migration efficiencies of 8 percent, if migration distances of as much as

400 km are permitted. If the area of effective Phosphoria source rocks is limited to the region of thermally mature, unmetamorphosed organic facies, then migration distances would be less than 100 km, but expulsion and migration efficiencies of 60 percent would be required. Because this is extremely high, either long-distance migration must be invoked, or rocks other than the Phosphoria black shale members made a significant contribution to Paleozoic oil in central Wyoming, northwestern Colorado, and northeastern Utah.

ACKNOWLEDGMENTS

This report is based on analytic work by Jimmie Bell, Jerry L. Clayton, Tom G. Ging, John M. Patterson, and Paul J. Swetland. Clayton and Swetland also contributed independent field and laboratory studies on the reproducibility of regional geochemical variations and the geochemical effects of weathering.

REFERENCES CITED

Albrecht, P., and G. Ourisson, 1969, Diagénèse des hydrocarbures saturés dan une série sédimentaire épaisse (Douala, Cameroun): Geochimica et Cosmochimica Acta, v. 33, p. 138-142.

Baker, D.R., 1962, Organic geochemistry of Cherokee Group in southeastern Kansas and northeastern Oklahoma: AAPG Bulletin, v. 46, p. 1621-1642.

—— and G.E. Claypool, 1970, Effects of incipient metamorphism on organic matter in mudrock: AAPG Bulletin, v. 54, p. 456-468.

Barbat, W.N., 1967, Crude-oil correlations and their role in exploration: AAPG Bulletin, v. 51, p. 1255-1292.

Barker, F. and I. Fredman, 1969, Carbon isotopes in pelites of the Precambrian Uncompahgre Formation, Needle Mountains, Colorado: Geological Society of America Bulletin, v. 80, p. 1403-1408.

Berry, G.W., 1943, Stratigraphy and structure at Three Forks: Geological Society of America Bulletin, v. 54, p.1-30.

Bowen, C.F., 1918, Phosphatic oil shale near Dell and Dillon, Beaverhead County, Montana: U.S. Geological Survey Bulletin, 661-I, p. 315-328.

Brongersma-Sanders, M. 1948, The importance of upwelling water to vertebrate paleontology and oil geology: Verhandelingen der Koninklijke Nederlandse Akademie van Wetenschappen, Afdeeling Natuurkunde, Tweede Sec., v. 45, no. 4, 112 p.

Cheney, T.M., 1957, Phosphate in Utah and an analysis of the stratigraphy of the Par City and Phosphoria Formations, Utah: Utah Geological and Mineralogical Survey Bulletin 59, 54 p.

—— and R.P. Sheldon, 1959, Permian stratigraphy and oil potential, Wyoming and Utah, *in* Intermountain Association of Petroleum Geologists 10th Annual Field Conference: Intermountain Association of Petroleum Geologists, p.90-100.

Clayton, J.L., and P.J. Swetland, 1976, Subaerial weathering of sedimentary organic matter (abs.): Geological Society of America Abstracts with Programs, v. 8, p. 815.

Condit, D.D., 1919, Oil shale in western Montana, southeastern Idaho, and adjacent parts of Wyoming and Utah: U.S. Geological Survey Bulletin 711-B, p. 15-40.

Craig, H., 1957, Isotopic standards for carbon and oxygen and correction factors for mass-spectrometric analysis of carbon dioxide: Geochimica et Cosmochimica Acta, v. 12, p. 133-149.

Cressman, E.R., 1964, Geology of the Georgetown Canyon-Snowdrift Mountain area, southeastern Idaho: U.S. Geological Survey Bulletin 1153, 105 p.

—— and R.W. Swanson, 1964, Stratigraphy and petrology of the Permian rocks of southwestern Montana: U.S. Geological Survey Professional Paper 313-C, p. 275-569.

Crittenden, M.D., Jr., 1961, Magnitude of thrust faulting in northern Utah, *in* Geological Survey research 1961: U.S. Geological Survey Professional Paper 424-d, p. D128-D131.

Curry, W.H., 1971, Summary of possible future petroleum potential, Region 4, Northern Rocky Mountains, *in* Future petroleum provinces of the United States—their geology and potential: AAPG Memoir 15, p. 538-546.

Dow, W.G., 1974, Application of oil-correlation and source-rock data to exploration in Williston basin: AAPG Bulletin, v. 58, p. 1253-1262.

Emmons, W.H., and F.C. Calkins, 1915, Description of the Philipsburg quadrangle, Montana: U.S. Geological Survey Geology Atlas Folio 196, 25 p.

Evans, C.R. and F.L. Staplin, 1971, Regional facies of organic metamorphism, *in* Geochemical exploration: Proceedings, 3d International Geochemical Exploration Symposium, Canadian Institute of Mining and Metallurgy Spec. Vol. II, p. 517-520.

Fox, J.E., et al, 1975, Porosity variation in the Tensleep and its equivalent, the Weber Sandstone, western Wyoming—a log and petrographic analysis, *in* Symposium on deep drilling frontiers of the central Rocky Mountains: Rocky Mountain Association of Geologists, p. 185-216.

Fraser, G.C., and H.A. Waldrop, 1972, Geologic map of the Wise River quadrangle, Silver Bow and Beaverhead Counties, Montana: U.S. Geological Survey Geologic Quadrangle Map GQ-988, scale 1:24,000.

Gwinn, V.E., 1961, Geology of the Drummond area, central-western Montana: Montana Bureau Mines and Geology Special Publication 21 (Geol. Map 4), scale about 1 in. to 1 mi.

Hadley, J.B., 1969, Geologic map of the Varney quadrangle, Madison County, Montana: U.S. Geological Survey Geologic Quadrangle Map GQ-814, scale 1:62,500.

Hintze, L.F., 1973, Geologic history of Utah: Brigham Young University Geology Studies, v. 20, pt. 3, 181 p.

Honkala, F.S., 1953, Preliminary report on geology of Centennial Range, Montana-Idaho, phosphate deposits: U.S. Geological Survey, Open-File Report 196.

Hood, A., and J.R. Castaño, 1974, Organic metamorphism—its relationship to petroleum generation and application to studies of authigenic minerals: Coordinating Comittee for Offshore Prospecting Technical Bulletin, v. 8, p. 85-118.

——, C.C.M. Gutjahr, and R.L. Heacock, 1975, Organic metamorphism and the generation of petroleum: AAPG Bulletin, v. 59, p. 986-996.

Hunt, J.M., 1972, Distribution of carbon in crust of earth: AAPG Bulletin, v. 56, p. 2273-2277.

Keefer, W.R., 1957, Geology of the Du Noir area, Fremont County, Wyoming: U.S. Geological Survey Professional Paper 294E, p. 155-221.

Klepper, M.R., 1950, A geologic reconnaissance of parts of Beaverhead and Madison Counties, Montana: U.S. Geological Survey Bulletin 969-C, p. 55-85.

Leythaeuser, D., 1973, Effects of weathering on organic matter in shales: Geochimica et Cosmochimica Acta, v. 3, p. 1183-1194.

Love, J.D., et al, 1973, Geologic block diagram and tectonic history of the Teton region, Wyoming-Idaho: U.S. Geological Survey Miscellaneous Geologic Investigations Map I-730.

Lowell, W.R., 1965, Geologic map of the Bannack-Brayling area, Beaverhead County, Montana: U.S. Geological Survey Miscellaneous Geologic Investigations Map I-433, scale 1:31,680.

Maughan, E.K., 1966, Environment of deposition of Permian salt in the Williston and Alliance basins, *in* 2d Symposium on salt, v. 1, Geology, geochemistry and mining: Northern Ohio Geological Society, p. 35-47.

——, 1975, Organic carbon in shale beds of the Permian Phosphoria Formation of eastern Idaho and adjacent states—a summary report: Wyoming Geological Association 27th Annual Field Conference, p. 107-115.

McDowell, A.N., 1975, What are the problems in estimating the oil potential of a basin?: Oil and Gas Journal, v. 73, no. 23, p. 85-90.

McGookey, D.P., et al, 1972, Cretaceous system, *in* W.W. Mallory, ed., Geologic atlas of the Rocky Mountain region: Rocky Mountain Association of Geologists, p. 190-228.

McKee, E.D., et al, 1956, Paleotectonic maps of the Jurassic system: U.S. Geological Survey Investigations Map I-175, 6 p.

—— et al, 1959, Paleotectonic maps of the Triassic system: U.S. Geological Survey Miscellaneous Geologic Investigations Map I-300, 33 p.

——, et al, 1967, Paleotectonic maps of the Permian System: U.S. Geological Survey Miscellaneous Geologic Investigations Map I-450, scale (plate 1-8) 1:5,000,000, 164 p.

McKelvey, V.E., 1959, Relation of upwelling marine waters to phosphorite and oil (abs.): Geological Society of America Bulletin, v. 70, p. 783-1784.

——, et al, 1959, The Phosphoria, Park City and Shedhorn Formations in the western phosphate field: U.S. Geological Survey Professional Paper 313-A, 47 p.

McKirdy, D.M. and T.G. Powell, 1974, Metamorphic alteration of carbon isotopic composition in ancient sedimentary organic matter—new evidence from Australia and South Africa: Geology, v. 2, p. 591-595.

McMannis, W.J., 1965, Resumé of depositional and structural history of western Montana: AAPG Bulletin, v. 49, p. 1801-1823.

Nixon, R.P., 1973, Oil source beds in Cretaceous Mowry

Shale of northwestern interior United States: AAPG Bulletin, v. 57, p. 136-161.

Orr, W.L., 1974, Changes in sulfur content and isotopic ratios of sulfur during petroleum maturation—study of Big Horn basin Paleozoic oils: AAPG Bulletin, v. 58, p. 2295-2318.

Partridge, J.F., Jr., 1958, Oil occurrence in Permian, Pennsylvanian, and Mississippian rocks, Big Horn basin, Wyoming, *in* L.G. Weeks, ed., Habitat of oil: AAPG, p. 293-306.

Powell, T.G., P.J. Cook, and D.M. McKirdy, 1975, Organic geochemistry of phosphorites—relevance to petroleum genesis: AAPG Bulletin, v. 59, p. 618-632.

Roberts, R.J., and M.R. Thomasson, 1964, Comparison of late Paleozoic tectonic history of northern Nevada and central Idaho, *in* Short papers in geology and hydrology: U.S. Geological Survey Professional Paper 475-D, p. D1-D6.

Rohrer, W.L., 1973, Geologic map of the Phosphate Reserve in the Lander area, Fremont, Wyoming: U.S. Geological Survey Mineral Investigations Field Studies Map MF-305, scale 1:24,000.

Rooney, L.F., 1956, Organic carbon in Phosphoria Formation: AAPG Bulletin, v. 40, p. 2267-2271.

Rubey, W.W., 1973a, Geologic map of the Afton quadrangle and part of Big Piney quadrangle, Lincoln and Sublette Counties, Wyoming: U.S. Geological Survey Miscellaneous Geologic Investigations Map I-686, Scale 1:62,500.

———, 1973b, New Cretaceous formations in the western Wyoming thrust belt, *in* Contributions to stratigraphy: U.S. Geological Survey Bulletin 1372-I, p. I1-135.

Ryder, R.T., and R. Sholten, 1973, Syntectonic conglomerates in southwestern Montana—their nature, origin and tectonic significance: Geological Society of America Bulletin, v. 84, p. 773-796.

Sackett, W.M., and R.R. Thompson, 1963, Isotopic organic carbon composition of recent continental derived clastic sediments of eastern Gulf Coast, Gulf of Mexico: AAPG Bulletin, v. 47, p. 525-528.

Scholten, R., K.A. Keenmon and W.O. Kupsch, 1955, Geology of the Lima region, southwestern Montana and adjacent Idaho: Geological Society of America Bulletin, v. 66, p. 345-403.

Schroeder, M.L., 1974, Geologic map of the Camp Davis quadrangle, Teton County, Wyoming: U.S. Geological Survey Geologic Quadrangle Map GQ-1160, scale 1:24,000.

Sheldon, R.P., 1963, Physical stratigraphy and mineral resources of Permian rocks in western Wyoming: U.S. Geological Survey Professional Paper 313-B, 273 p.

———, 1967, Long-distance migration of oil in Wyoming: Mountain Geologist, v. 4, p. 53-65.

———, 1972, Phosphate deposition seaward of barrier islands at edge of Phosphoria sea in northwestern Wyoming (abs.): AAPG Bulletin, v. 56, p. 653.

———, E.K. Maughan, and E.R. Cressman, 1967, Environment of Wyoming and adjacent states— interval B, *in* Paleotectonic maps of the Permian System: U.S. Geological Survey Mineral Investigations Map I-450, p. 48-54.

Staplin, F.L., 1969, Sedimentary organic matter, organic metamorphism and oil and gas occurrence: Canadian Petroleum Geology Bulletin, v. 17, p. 47-66.

Stones, D.S., 1967, Theory of Paleozoic oil and gas accumulation in Big Horn basin, Wyoming: AAPG Bulletin, v. 51, p. 2056-2114.

Todd, T.W., 1963, Post-depositional history of Tensleep Sandstone (Pennsylvanian), Big Horn basin, Wyoming: AAPG Bulletin, v. 47, p. 599-616.

Vassoyevich, N.B., et al, 1967, Hydrocarbons in the sedimentary mantle of the earth: Proceedings, 7th World Petroleum Congress, Mexico City, v. 2, p. 37-46.

Williams, J.A., 1974, Characterization of oil types in Williston basin: AAPG Bulletin, v. 58, p. 1243-1252.

Winters, J.C., and J.A. Williams, 1969, Microbiological alteration of crude oils in the reservoir: New York, American Chemical Society, Petroleum Chemistry Division, p. E22-E31.

Reprinted from the Williston Basin Symposium, 1978, The Montana Geological Society, 24th Annual Conference, Billings, Montana.

Petroleum Geology of the Bakken Formation Williston Basin, North Dakota and Montana

Fred F. Meissner
Filon Exploration Corporation
Denver, Colorado

The Bakken Formation is relatively thin and is found (relatively) deep in the Williston basin. In spite of its insignificant volume (compared to the total sedimentary section) its organic-rich shales are documented source rocks. The generation/migration scheme postulated for the Bakken (source rock) and its associated oil-productive reservoirs are believed to be somewhat universal in units and basins throughout the world.

INTRODUCTION

The Bakken Formation is a relatively thin unit and is limited in areal distribution to the deeper part of the Williston Basin (Figure 1). In spite of its insignificant volume when compared with that of the total sedimentary section, the unit is undoubtably one of the most important when considered in relation to the presence of oil and gas. Organic-rich shales in the Bakken have been documented as excellent petroleum source-rocks (Dow, 1974; Williams, 1974) and are believed to have generated the tremendously large volumes of oil found in reservoirs somewhat distantly located above and below the unit. Production has been established within the Bakken itself, and considerable remaining exploration potential may exist within the elusive fracture-type reservoirs which characterize the unit. Since the Bakken is relatively isolated by seemingly impervious overlying and underlying lithologies and is the only source-rock within several thousand feet of vertical stratigraphic section, studies of its hydrocarbon-generation (maturity) pattern, associated physical changes and fluid pressure phenomena, and its relation to known reservoirs and accumulations may be of value in deciphering mechanisms and routes of migration and in adding a factor of predictability to the overall science of petroleum geology.

STRATIGRAPHY OF THE BAKKEN FORMATION AND ASSOCIATED UNITS

The Bakken Formation (Figure 2) is the relatively thin basal unit of a thick sequence of predominantly carbonate rocks (the overlying Madison Group) deposited during a major cycle of onlap-offlap sedimentation which began in uppermost Devonian (?) - lowermost Mississippian time (Bakken transgression) and extended to upper middle-Mississippian (Meramec) time (Charles regression). This onlap-offlap "depositional cycle" has been termed the "Tamaroa sequence" by Wheeler (1963). It can be identified as a rock unit surrounding most of the North American craton and is associated with a major amount of production on a continental scale. The basal Tamaroa transgressive unit equivalent to the Bakken can be recognized as a "black shale unit" known variously as the Exshaw/Banff in the Alberta Basin/Northern Rocky Mountains; the Pilot in the Cordilleran area; the "Lower Mississippian Black Shale" in the Permian Basin; the Woodford in the Anadarko Basin/Arbuckle Mountains; the Chattanooga in the eastern Mid-Continent/Southern Appalachian Basin; the Antrim in the Michigan Basin and the New Albany in the northern Appalachian Basin. Although immediate lateral correlatives of the Bakken are present in outcrops of central and western Montana and adjacent Alberta, the unit as formally defined (Nordquist, 1953) is restricted to the subsurface of the Williston Basin in eastern Montana/western North Dakota in the United States (Figures 1 and 3) and southern Saskatchewan/southwestern Manitoba in Canada.

The Bakken Formation unconformably overlies the Upper Devonian Three Forks Formation (Figure 2). The Three Forks (Peale, 1893; Sandbert and Hammond, 1958) averages about 150 ft in thickness, and consists primarily of interbedded yellowish-gray, greenish-gray, orange and red siltstones and shales, generally highly dolomitic. A few of the silty zones contain conglomeratic dolomite clasts and pebbles. Thinly-layered anhydrite beds occur near the base. A sandy zone up to 10 ft thick occurs somewhat erratically at the top of the formation, just beneath the lower Bakken

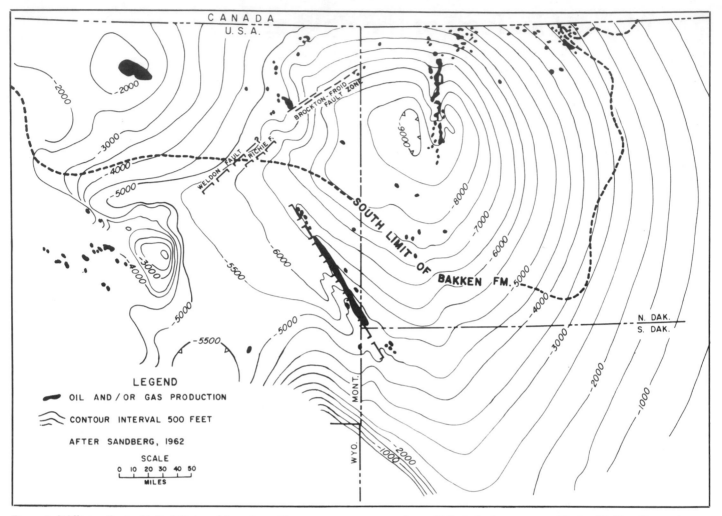

Figure 1. Williston Basin, United States of America with structure contours on base of Mississippian strata and limit of Bakken Formation.

Shale. At Antelope field this sandy zone is productive and is informally referred to as the "Sanish Sand."

At its type locality near the center of the Williston Basin (Figures 3 and 4), the Bakken may be divided into three members:

1. an Upper Shale Member
2. a Middle Siltstone Member
3. a Lower Shale Member

The total formation ranges in thickness from a maximum of 140 ft near the center of the Basin to a subsurface "0" limit on the eastern, southern and southwestern flanks (Figure 3). The three members can be correlated regionally, with most thickness changes taking place in the Middle Siltstone Member. Regional correlations indicate that the Bakken overlaps the underlying Devonian unconformity, with each succeedingly younger member overlapping the older member toward the depositional zero limit of the unit (Cross Section, Figure 4).

The Upper and Lower Shale Members of the Bakken have apparently identical lithologies throughout most of their areal extent and consist of hard brittle often waxy-looking black shale which has a very dark brown color when examined as cuttings with a microscope under strong light. Thin-sections of the shales show them to be composed mostly of indistinct organic material with lesser amounts of clay, silt, and dolomite grains. The Lower Bakken Shale appears to become less-organic and more-clayey, -silty, and -dolomitic near its zero limit, particularly on the western flank of the Basin.

The lithology of the Middle Siltstone Member of the Bakken varies somewhat unpredictably from a light-to-medium-gray very-dolomitic fine-grained siltstone to a very-silty fine-crystalline dolomite. Dark carbonaceous mottles and partings are common. The unit is often faintly laminated and occasionally contains fine-scale cross-bedding.

The Bakken is conformably overlain by the Lodgepole Formation—the basal unit of the Madison Group (Figure 2). The lithology of the lowermost Lodgepole, just above the Bakken, generally consists of dark gray dense lime mudstones interbedded with a few dark gray calcareous shales.

RESERVOIR PROPERTIES OF THE BAKKEN FORMATION AND ASSOCIATED UNITS

Effective matrix porosities and permeabilities of the Bakken Formation, the underlying Three Forks and the

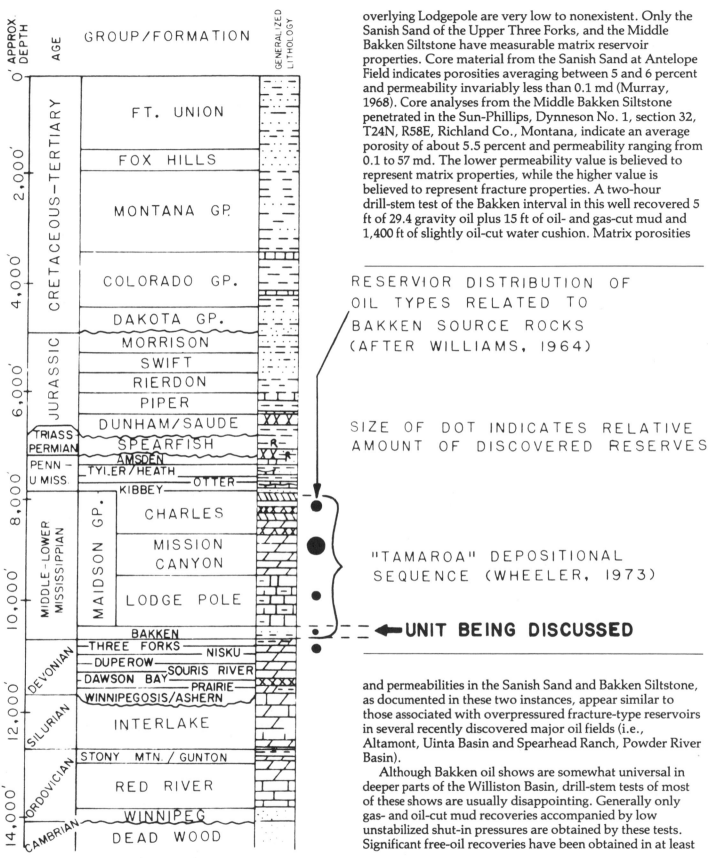

overlying Lodgepole are very low to nonexistent. Only the Sanish Sand of the Upper Three Forks, and the Middle Bakken Siltstone have measurable matrix reservoir properties. Core material from the Sanish Sand at Antelope Field indicates porosities averaging between 5 and 6 percent and permeability invariably less than 0.1 md (Murray, 1968). Core analyses from the Middle Bakken Siltstone penetrated in the Sun-Phillips, Dynneson No. 1, section 32, T24N, R58E, Richland Co., Montana, indicate an average porosity of about 5.5 percent and permeability ranging from 0.1 to 57 md. The lower permeability value is believed to represent matrix properties, while the higher value is believed to represent fracture properties. A two-hour drill-stem test of the Bakken interval in this well recovered 5 ft of 29.4 gravity oil plus 15 ft of oil- and gas-cut mud and 1,400 ft of slightly oil-cut water cushion. Matrix porosities

RESERVIOR DISTRIBUTION OF OIL TYPES RELATED TO BAKKEN SOURCE ROCKS (AFTER WILLIAMS, 1964)

SIZE OF DOT INDICATES RELATIVE AMOUNT OF DISCOVERED RESERVES

"TAMAROA" DEPOSITIONAL SEQUENCE (WHEELER, 1973)

← UNIT BEING DISCUSSED

and permeabilities in the Sanish Sand and Bakken Siltstone, as documented in these two instances, appear similar to those associated with overpressured fracture-type reservoirs in several recently discovered major oil fields (i.e., Altamont, Uinta Basin and Spearhead Ranch, Powder River Basin).

Although Bakken oil shows are somewhat universal in deeper parts of the Williston Basin, drill-stem tests of most of these shows are usually disappointing. Generally only gas- and oil-cut mud recoveries accompanied by low unstabilized shut-in pressures are obtained by these tests. Significant free-oil recoveries have been obtained in at least

Figure 2. Generalized stratigraphic column, central Williston Basin, showing the location of the Bakken Formation and the distribution of oil types related to Bakken source-rocks.

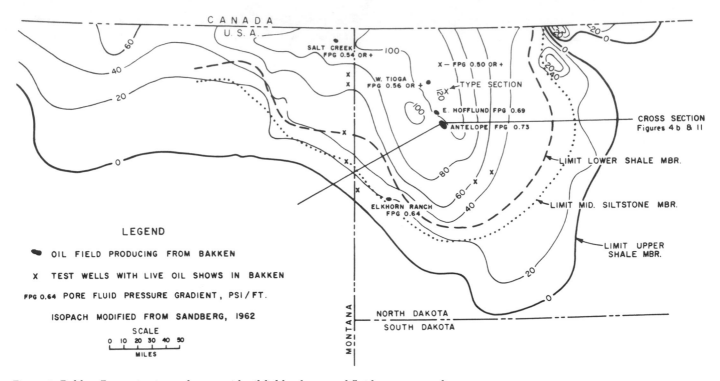

Figure 3. Bakken Formation isopach map with oil fields, shows and fluid pressure gradients.

8 widely-spaced wildcat wells in which subsequent completions were not attempted. Bakken production has been established from five fields in the U.S. portions of the Basin (Figure 3). Four of these fields are small accumulations; the fifth (Antelope field) has about 44 wells which have produced some 10 million bbl of oil and 29 billion cu ft of gas. A significant fact regarding fluid found in the Bakken is that no formation water is ever recovered during drill-stem tests or initial well completions. Small water cuts have been observed in advanced stages of depletion at Antelope field; however, this water is probably extremely fine-pore capillary water, and is not suggestive of water encroachment or the existence of an oil-water contact. This leads to the conclusion that hydrocarbons are essentially the only movable fluid found in the Bakken.

Most Bakken oil wells have productive rates far in excess of those theoretically possible from the low to nonexistent matrix-reservoir properties present in the interval. Murray has made strong argument for tensional (extensional) fracturing being the major cause of reservoir development in the Bakken/Sanish zone at Antelope field. This argument appears to be equally valid wherever oil DST recoveries or actual production has been established in the Bakken. In fact, it would appear that the Bakken will produce oil and gas—and these fluids only—wherever it is found in a fractured state within the deeper part of the Basin.

Although fracturing is apparently of major importance in establishing a viable Bakken reservoir, it appears that the marginal amount of matrix porosity observed in the Middle Bakken Silt and the Sanish Sand are a strong contributing factor in establishing storage volume. Most fracture reservoirs which have been extensively studied have bulk-fracture porosities in the neighborhood of only one-half percent (39 bbl/acre-ft total reservoir volume) or

less, even though permeabilities may range to hundreds or thousands of millidarcies. The fact that Bakken/Sanish wells at Antelope field have produced as much as 900,000 bbl of oil per 160 acres from an approximately 120-ft thick gross interval (or 46.4 bbls/acre-ft actual recovered volume) at initial rates of as much as 1,400 bbl per day would seem to indicate that 1) the major contribution to storage volume must come from marginal matrix porosities with high oil saturations in the Middle Bakken Silt and Sanish Sand zones, while 2) the major contribution to satisfactory production rates must come from a pervasive fracture system which renders the section relatively permeable.

Another factor of interest in evaluating Bakken reservoir properties is that all stabilized formation-fluid pressure measurements from wildcat or field wells in which formation fluid has been recovered indicate the Bakken to be anomalously overpressured (Figure 3). Several unstabilized measurements also indicate overpressure of an unknown maximum magnitude. Documented fluid-pressure gradients are as high as 0.73 psi/ft (Antelope field). These abnormally high pressures are evidently discretely confined to the Bakken/Sanish interval as shown by pressure gradient/depth profile at Antelope field, where overlying and underlying reservoir zones, separated from the Bakken by tight strata, are actually found to be abnormally overpressured (Figure 5). The reason for the anomalous Bakken fluid overpressure will be discussed in a later section.

GEOCHEMICAL (SOURCE-ROCK) PROPERTIES OF THE BAKKEN FORMATION

According to modern theory, most oil and gas is generated from rocks which 1) have certain critical organic

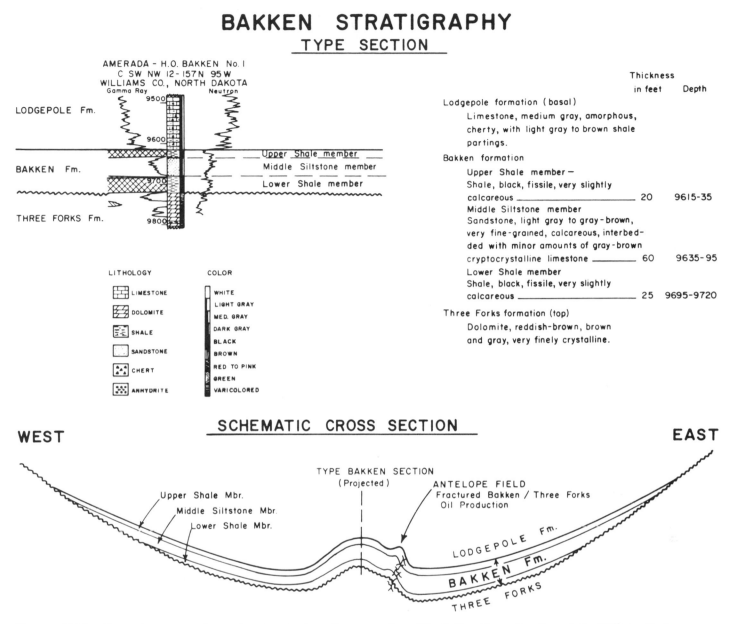

BAKKEN STRATIGRAPHY
TYPE SECTION

AMERADA - H.O. BAKKEN No. I
C SW NW 12 - 157N 95W
WILLIAMS CO., NORTH DAKOTA
Gamma Ray   Neutron

LODGEPOLE Fm.

BAKKEN Fm.
Upper Shale member
Middle Siltstone member
Lower Shale member

THREE FORKS Fm.

LITHOLOGY   COLOR

LIMESTONE   WHITE
DOLOMITE   LIGHT GRAY
SHALE   MED. GRAY
SANDSTONE   DARK GRAY
CHERT   BLACK
ANHYDRITE   BROWN
                   RED TO PINK
                   GREEN
                   VARICOLORED

Thickness
in feet   Depth

Lodgepole formation (basal)
    Limestone, medium gray, amorphous,
    cherty, with light gray to brown shale
    partings.
Bakken formation
    Upper Shale member—
    Shale, black, fissile, very slightly
    calcareous _____ 20   9615-35
    Middle Siltstone member
    Sandstone, light gray to gray-brown,
    very fine-grained, calcareous, interbed-
    ded with minor amounts of gray-brown
    cryptocrystalline limestone _____ 60   9635-95
    Lower Shale member
    Shale, black, fissile, very slightly
    calcareous _____ 25   9695-9720
Three Forks formation (top)
    Dolomite, reddish-brown, brown
    and gray, very finely crystalline.

SCHEMATIC CROSS SECTION

WEST   EAST

Upper Shale Mbr.
Middle Siltstone Mbr.
Lower Shale Mbr.

TYPE BAKKEN SECTION
(Projected)

ANTELOPE FIELD
Fractured Bakken / Three Forks
Oil Production

LODGEPOLE Fm.
BAKKEN Fm.
THREE FORKS

Figure 4. Bakken Formation type section and east-west schematic cross-section of the Bakken Formation through the Williston Basin (cross-section located in Figure 3).

content and which 2) have undergone sufficient thermal alteration (metamorphism) associated with time and burial-temperature to have cracked chemical bonds between complex primary organic materials and released hydrocarbons (Philippi, 1965; Welte, 1965; Landes, 1967; Momper, 1972; LaPlante, 1974). The high organic content of the Upper and Lower Bakken Shales as observed in thin-section studies, together with the fact that small chips of the rock visually generate significant amounts of oil and gas when heated in the test tube (Trask and Patnode, 1942, page 62), indicate that they are hydrocarbon source-rocks. Murray (1968), noting the "petroliferous" nature of the shales, their universal association with shows in sample cuttings and certain diagnostic petrophysical properties in their electrical and mechanical log character, was one of the first to speculate on the importance of the Bakken interval as a hydrocarbon source.

Recently, both Dow (1974) and Williams (1974) have expanded knowledge on the source-rock properties of the Bakken. Williams analyzed 26 samples of Bakken Shale and found them to contain from 0.65 to 10.33 weight percent organic carbon, with an average of 3.84 percent. On the basis of sophisticated geochemical work on hydrocarbon extracts from the Bakken, he further related the Bakken source rock to the distinctive type of oil found and produced from the underlying Nisku Formation, the overlying Madison group and the Bakken/Sanish interval itself (Figure 2). Dow's work presented a series of areal geochemical maps of the Bakken and its related oil types. From these maps he was able to speculate on migration paths, accumulation patterns, etc. He estimates that Bakken source rocks have generated in the neighborhood of 10

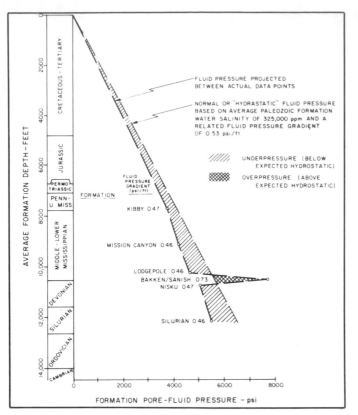

Figure 5. Reservoir fluid pressure vs. depth, Antelope field, McKenzie County, North Dakota.

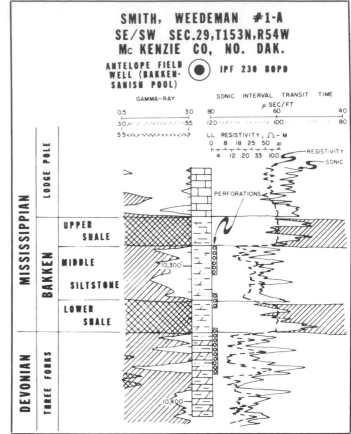

Figure 6. Typical electrical and mechanical log behavior, Bakken and adjacent formations.

billion bbl of oil in the deeper portions of the basin, where they are buried to a postulated "maturity" depth greater than 7,000 ft. Considering Dow's evidence and arguments, it seems reasonable that although a large part of the hydrocarbons generated in the Bakken have migrated vertically downward and upward to underlying and overlying Nisku and Madison reservoirs, such migration would probably not have taken place until after all possible matrix-and fracture-reservoir volume in the Bakken had been effectively charged. This would account for the widespread occurrence of Bakken shows and the observation that hydrocarbons are the only mobile fluid phase found within the unit in the deeper part of the basin.

PETROPHYSICAL PROPERTIES OF THE BAKKEN FORMATION AND THEIR RELATION TO SOURCE-ROCK PRINCIPLES AND RESERVOIR OVERPRESSURING

Murray (1968) noted the peculiar behavior of the Bakken Shale Members are expressed on gamma-ray, sonic, neutron and resistivity logs and was one of the first to relate their petrophysical character to source-rock and overpressured reservoir properties. Using well known log-interpretation principles, the author has expanded considerably on Murray's observations and has utilized resistivity and sonic logs to map 1) source rock maturity and 2) formation fluid overpressure.

The Upper and Lower Bakken Shale Members are regionally characterized by 1) anomalously high gamma-ray

radioactivity, 2) anomalously low, but highly variable sonic velocity (high transit time) and 3) either very-high or very-low resistivity. Typical log character of these parameters is shown in Figure 6. The unusually low sound-velocity (high transit-time) indicated for the Bakken shales is believed to be due for the most part to their high content of low-velocity organic material, although additional affects are also contributing factors, as will be discussed shortly.

Most normal shale units are characterized by low electrical resistivity because of 1) the basic conductivity of most shale clays and 2) their relatively high porosity, filled with conductive water. Logs through the Bakken Formation at depths less than about 6,500 ft are characterized by these "normal" low shale resistivities; however, logs from deeper penetrations indicate anomalously high resistivities. A resistivity vs. depth plot for data obtained in 32 widely-separated control wells is shown in Figure 7 (index map for these points is shown in Figure 10). Murray (1968), concluded that the anomalously-high "essentially infinite" resistivity characterizing the Bakken Shale Members throughout the central Williston Basin was a natural consequence of the fact that they were hydrocarbon-saturated source-rocks. In view of this, the relatively rapid depth-related change from low to high resistivity shown in Figure 7 is believed to represent the onset of "maturity" of hydrocarbon generation and

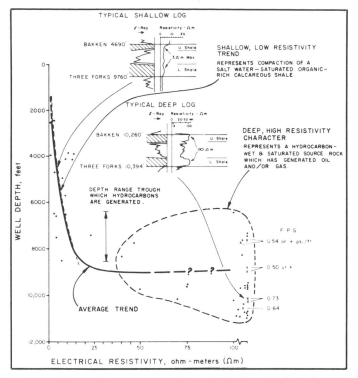

Figure 7. Electrical resistivity vs. depth for Bakken Shale with "trend" and source-rock "maturity" interpretations indicated.

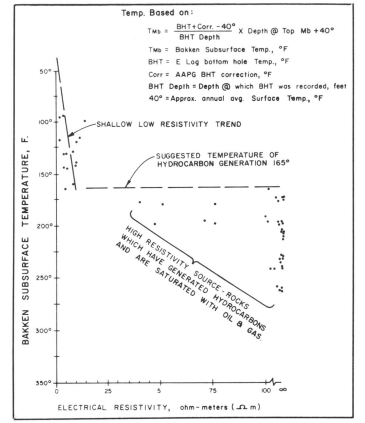

Figure 8. Electrical resistivity vs. subsurface temperature for Bakken Shale, with "trend" and source-rock "maturity" interpretations indicated.

consequent replacement of high-conductive pore water in the organic shales with nonconductive hydrocarbons. Numerous investigators (i.e., Philippi, 1968; Tissot et al, 1971; Nixon, 1973; Claypool et al, 1978) have documented the dramatic increase in extractable heavy hydrocarbons from source-rocks in other basins which have gone through the hydrocarbon-generation ("maturity") threshold. Plots of extractable-hydrocarbon to residual organic-carbon ratio vs. depth prepared by these investigators (i.e., Philippi, 1968, Figure 5, page 33 and Figure 7, page 37) have a form similar to the resistivity vs. depth plot of Figure 7. Although no information is available on the extractable hydrocarbon content vs. depth relation for the Bakken Formation, it is anticipated that such data would show changes in hydrocarbon content which would be correlative with changes in resistivity. If this logic is correct, the resistivity effect might be taken as a measure of the equilibrium saturation required to achieve a continuous hydrocarbon-phase expulsion mechanism from the rock.

As shown in Figure 7, the depth range through which the shallow low-resistivity/depth trend (which follows a normal shale compaction trend) changes to the high-resistivity character indicates maturity occurs from about 6,200 to 8,200 ft. The existence of this rather wide depth range suggests that the actual depth of hydrocarbon generations is not uniform throughout the basin; e.g., that a unique surface which may represent "maturity" is nonplanar and/or non-horizontal. Since the point at which hydrocarbon generation starts is actually more directly related to temperature than depth, temperatures for the formation resistivities plotted in Figure 7 were estimated from bottom-hole temperature data and a resistivity vs. temperature plot was prepared (Figure 8). This plot shows

an extremely abrupt change from low-resistivity immature Bakken source rocks to abnormally high-resistivity rocks at about 160°F. This temperature is about that of the "critical temperature" for oil generation in Devonian-Mississippian-age rocks according to recent work (Connan, 1974) involving the calibration of thermochemical first order-reaction-rate theory to the concept of hydrocarbon generation (Figure 9). This is an observation that strongly supports the idea that the change in resistivity of the Bakken source rocks is uniquely related to the onset of hydrocarbon generation and the replacement of conductive pore-water with thermally generated nonconductive hydrocarbons. Figure 10 is a map of Bakken Shale Member electrical resistivities and formation temperatures showing 1) the areas of "mature" high- resistivity and "immature" low-resistivity source rocks and 2) formation temperatures related to these maturity/resistivity realms. Figure 11 is a cross section which schematically depicts relations between depth, source rock maturity, resistivity and temperature in the Bakken.

It is a well-known fact that anomalous fluid pressures affect the electrical resistivity of rocks (Hottman and Johnson, 1965; MacGregor, 1965); however, the effects of maturity and resulting hydrocarbon saturation which cause high resistivities in the Bakken shales are in the opposite direction of low resistivities caused by high fluid pressures and far overshadow them in their total effect. Realizing that sonic velocities are also influenced by formation-fluid

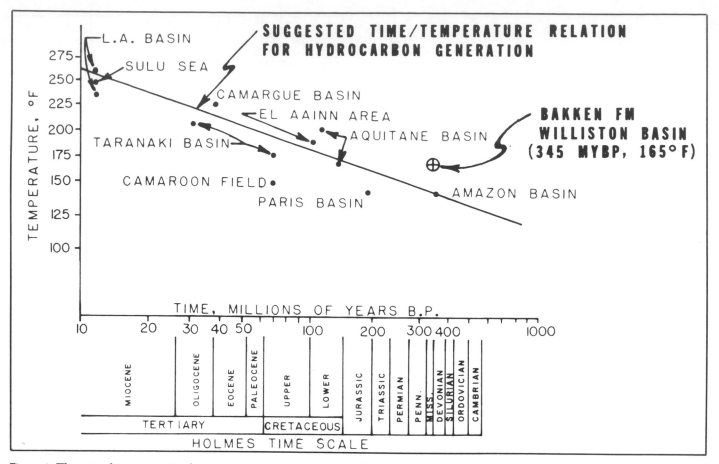

Figure 9. Theoretical time vs. critical temperature required to generate hydrocarbons from source rocks (after Connan, 1974) and its relation to interpreted "maturity" in the Bakken Formation.

pressure, an attempt was made to map the extent of known abnormally-high fluid pressures in the Bakken through the use of sonic-transit-time vs. depth plots as proposed by Hottman and Johnson. As shown in Figure 12a (which was prepared from sonic log data in the same 32 wells utilized in construction of Figures 7, 8, and 10), two trends—one "shallow," and one "deep"—can be discerned in the transit time/velocity vs. depth plot. The "shallow trend" shows little scatter and depicts a normal slight increase of velocity with depth related to the compaction process. The "deep trend" is characterized by considerable scatter and, although within this trend velocities increase with depth as above, the average velocities of this trend are much lower than those predicted by a downward projection of the shallow trend. The shallow trend is further distinguished from the deep trend on the basis of the "low" and "high" resistivity categories separating water-saturated "immature" and hydrocarbon-saturated "mature" source rocks as described in the preceding paragraph. The fact that the deep trend is known to contain overpressured reservoirs probably accounts for both the general shift to lower velocities and the apparent scatter in the actual values. Because sound velocities in oil are less than those in water, replacement of water by oil in the pore spaces of a source-rock which has become mature leads to a proportional diminishment of source velocity in the overall rock (Poh-hsi Pan and DeBremaecker, 1970). This effect is shown schematically in

the velocity/transit time vs. depth plot of Figure 13. The effect of abnormal pressure in lowering the velocity of sound in the rock is shown for a series of Gulf Coast shales in Figure 14a. The magnitude of abnormality in the velocity of the overpressured shales with respect to that of a normal shale at the same depth is a direct relation to the amount of abnormal fluid pressure. If a series of control points reflecting known pressure conditions in a given rock unit can be obtained, a calibration can be derived and unknown fluid pressures can be determined from sonic-log data (Figure 14b). The transit time/velocity vs. depth relation interpreted to show the effects of both oil saturation (maturity) and the magnitude of overpressuring in the Bakken Shale Members is shown in Figure 12b. This interpretation has been extended to the plan map of Figure 15. This map depicts the extent and magnitude of overpressuring in the Upper and Lower Bakken Shale Members. By inferrence, the same pressure properties are present in the immediately adjacent Bakken Silt and Sanish Sand (if present).

A comparison between the areas of Bakken overpressuring and the areas of source rock maturity as shown in Figures 10 and 15 shows that they are essentially identical and suggests that there is a direct relationship between hydrocarbon generation and the occurrence of abnormally high fluid pressure. This phenomenon has been observed elsewhere (i.e., the Uinta and Powder River

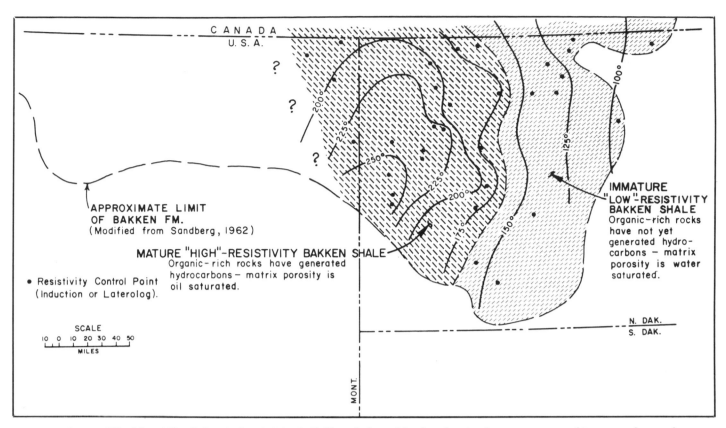

Figure 10. Areas of "high" and "low" electrical resistivity in Bakken shales, with subsurface isotherm contours and interpreted area of source-rock "maturity".

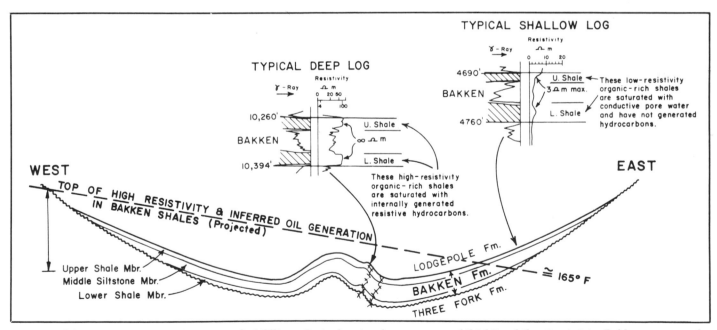

Figure 11. Schematic east-west section across the Williston Basin showing the occurrence of "high" and "low" resistivity Bakken source-rock shales and their interpreted relationship to hydrocarbon generation (cross-section located in Figure 3).

Basins) and the reason for it is of some importance. Formation-fluid overpressures may be caused by a number of processes (see Houston Geological Society, Abnormal Subsurface Pressure Study Group Report for an excellent resumé). Two processes that seem to be important in the phenomenon being discussed here are 1) undercompaction and 2) metamorphic phase change. It must be realized that in most cases involving undercompaction, fluid overpressuring is a manifestation of excess fluid or excess porosity in a rock when compared to normal fluid pressure

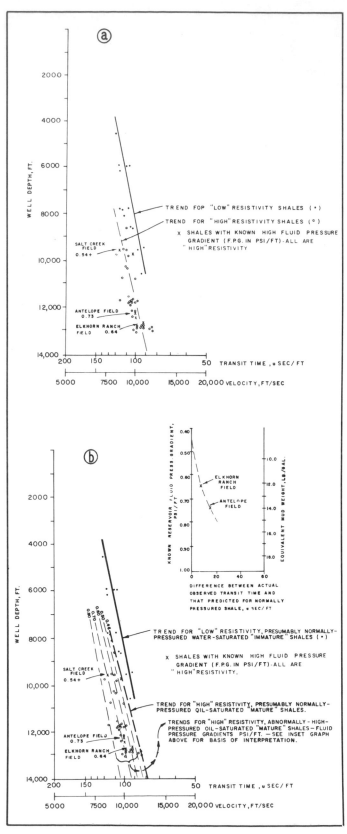

Figure 12. Sound velocity and transit-time vs. depth for Bakken Shale. a) Data points, trends for "high" (mature) and "low" (immature) resistivity shales, and known occurrence of pore-fluid overpressure. b) Fluid-saturation species and interpreted pore-fluid pressure gradients.

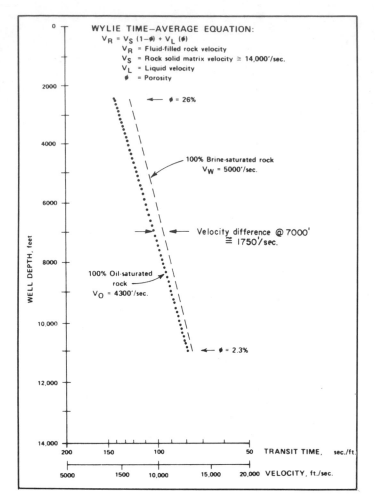

Figure 13. Theoretical effects of pore-fluid species on sound velocity in porous rocks based on Wylie time-average equation.

(e.g., "hydrostatic" conditions) and porosity at the same depth. The well-known fluid overpressures in deep shales and sands of the Gulf Coast are related to water as the excess fluid, and their origin is most widely attributed to rapid depositional rates not in equilibrium with the normal compaction process. In the case of overpressuring related to hydrocarbon generation, hydrocarbons appear to be the only excess fluid. The author believes that abnormal pressures found associated with mature Bakken source rocks are basically caused by:

1) the inhibited structural collapse of the rock framework as overburden-supporting solid organic material (estimated to be at least 25 volume percent of the rock) is converted to non-overburden supporting hydrocarbon pore fluid (e.g., oil and/or gas); and
2) the increased volume occupied by metamorphosed organic residue plus generated hydrocarbon fluids above those occupied by the unaltered organic material.

Anomalous pressures in the Bakken are believed to be maintained by the combination of large hydrocarbon volumes generated at high rates and the relative isolation of the Bakken by extremely tight rocks in the underlying Three Forks and overlying Lodgepole Formations.

THEORY AND CONTROLS FOR LOCALIZING RESERVOIR FRACTURING IN THE BAKKEN

The occurrence of fracture-reservoirs in close association with mature oil-generating source rocks and abnormally-high reservoir-fluid pressures, as found at all fields producing from the Bakken formation in the Williston Basin, (and some fields elsewhere in the world, as the Uinta and Powder River Basins) suggests a direct cause-and-effect relationship. This relationship is strongly supported by currently accepted basic failure-theory for porous isotropic homogeneous brittle-elastic rocks. This theory (Secor, 1965, and others) is schematically depicted according to standard graphical Mohr's circle and failure envelope representation in Figure 16a. Shown is a "failure envelope" superimposed on a coordinate system, upon which it is possible to represent any possible stress field characterized by principal and normal stressed (plotted along or parallel to the horizontal abcissa) and shear stressed (plotted along or parallel to the vertical ordinate). Any particular stress field within the coordinate system is represented by a "Mohr's stress circle" which intersects the abcissa at values of the maximum (normally vertical with reference to the earth's surface) and minimum (normally horizontal) principle stresses (e.g., S_1 and S_2 respectively). The shape and dimensions of the failure envelope characterize certain physical properties of the rock, such an tensile strength (intersection of the envelope with the abcissa and normally a negative value), shear strength (intersection of the envelope with the ordinate, with both negative and positive value intersections), and the degree of brittle-elasticity of the material (as measured by the amount of similarity between shape and dimensions of the Griffith and Mohr-Coulomb-Navier failure theory). Stress circles within the concave portion of the failure envelope represent structurally stable non-fracture conditions. Stress circles which, because of an enlarging or laterally shifting stress field circle or a decreasing envelope size, become tangent to the failure envelope represent structurally unstable conditions and indicate fracture failure of the rock (Figure 16b). If the point of tangency between the stress circle and the failure envelope is on the positive (+) or compressive side of the principle or normal stress coordinate origin, the failure will be in the form of a shear fracture (fault) and will not intrinsically be associated with the formation of reservoir-making porosity and permeability. If the point of tangency is on the negative (−) or tensile side of the origin, the failure will be in the form of an open extension fracture and will be intrinsically associated with the creation of porosity and permeability and hence the creation of fracture-reservoir conditions. Geometric relations between the failure envelope and the stress field coordinate system are such that large stress circles (e.g., large differential stress fields with large differences between maximum and minimum stress values) with compressive principal stresses are required for normal shear fracture failure; smaller stress circles, with at least one of the principal stresses in a negative (−) tensile sense are required for tensile (extension) fracture failure.

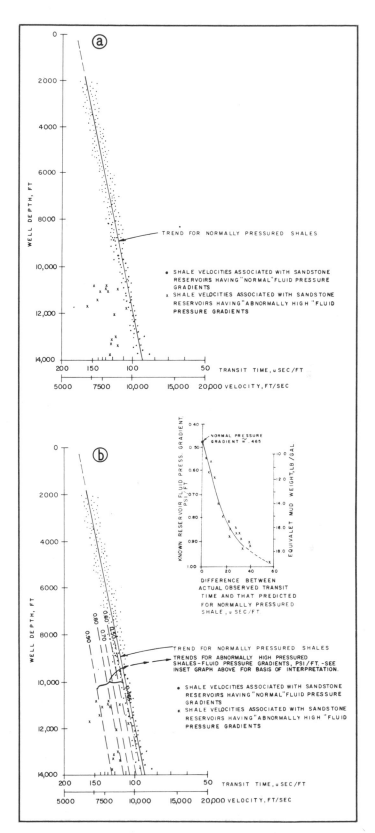

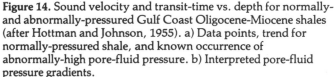

Figure 14. Sound velocity and transit-time vs. depth for normally- and abnormally-pressured Gulf Coast Oligocene-Miocene shales (after Hottman and Johnson, 1955). a) Data points, trend for normally-pressured shale, and known occurrence of abnormally-high pore-fluid pressure. b) Interpreted pore-fluid pressure gradients.

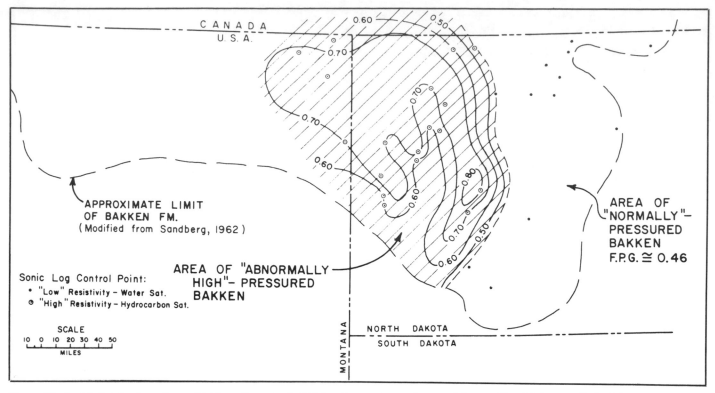

Figure 15. Aerial distribution of pore-fluid gradients in the Bakken Formation and areas of "normal" and "abnormally-high" pressure. Note coincidence of "low" electrical resistivity with "normal" pressure and "high" resistivity with "abnormally-high" pressure. Compare area of abnormally-high pressure with area of interpreted source-rock maturity indicated in Figure 10.

The stress fields represented by Mohr's stress circles are actually of two types in porous fluid-filled materials:

1) a "total stress" field, wherein the stress circle is defined by the "total" or externally applied maximum and minimum principal stresses characterized by S_1 and S_2 in Figure 17a; and

2) an "effective stress" field, wherein the stress circle is defined by the internal or "effective" maximum and minimum principle stresses characterized by σ_1, σ_2 in Figure 17a.

According to Hubert and Rubey (1959) the type of stress field which actually causes porous rock failure is that related to internal "effective stresses" rather than externally applied or "total stresses." "Total stress" (S) and "effective stress" (o) are related by the pressure (p) of fluid filling the pores of the rock according to the following equation:

$$\sigma = S - p$$

As shown in Figures 17a and 17b, the position of an effective stress circle with respect to the failure envelope is strongly influenced by any fluid pressure existing within the pores of a rock. The general effect of introducing or increasing pore-fluid pressure in a rock is to shift a stress circle both toward the failure envelope and toward a more tensile direction. Shifting of the stress field due to increasing pore-fluid pressure may cause either closed shear fracture failure (faulting) or open tensile fracture failure. The shifting or large stress field circles which become tangent to the compressive portion of the failure envelope leads to shear failure, while the shifting of critically small circles which

become tangent to the tensile portion leads to tension (extension) fracturing. An enlarging stress field affecting a thick column of sedimentary rocks may cause failure of different types in individual beds depending on 1) the brittle-elastic or plastic nature of each individual bed, 2) bed strength, or 3) bed pore-fluid pressure. An individual overpressured bed may fail in open tension fracturing, usually at an early, low differential stress condition, while overlying and underlying normally-pressured beds fail by faulting at a later and somewhat larger stress condition as shown in Figure 18.

From the preceding discussion of basic fracture theory, it can be seen that two basic parameters essentially control the type of open tension fractures that create reservoir conditions:

1) fluid overpressuring; and

2) differential stress

Critical fluid overpressures leading to fracture failure can be caused by hydrocarbon generation as discussed in the previous section. Critical differential stress fields can be caused by a number of processes, including:

1) regional tectonic forces;

2) burial and uplift;

3) diagenetic processes (i.e., stylolitization); and

4) secondary "bending" and "stretching" associated with local tectonic features.

Local "bending," which created a critical differential stress field existing in conjunction with regional fluid over-pressuring due to hydrocarbon generation, is believed to have created the fracture reservoir in the Bakken/Sanish interval at Antelope field, McKenzie County, North

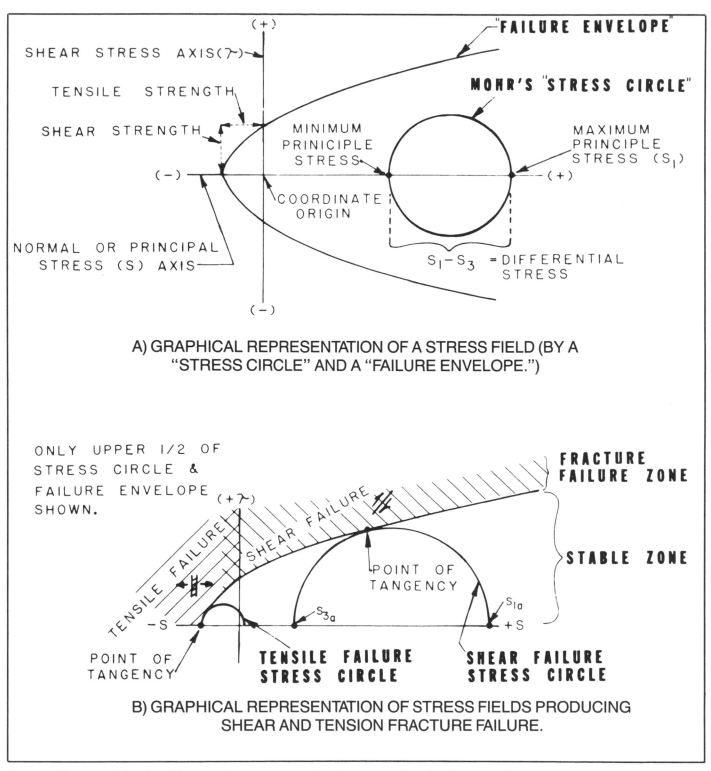

A) GRAPHICAL REPRESENTATION OF A STRESS FIELD (BY A "STRESS CIRCLE" AND A "FAILURE ENVELOPE.")

B) GRAPHICAL REPRESENTATION OF STRESS FIELDS PRODUCING SHEAR AND TENSION FRACTURE FAILURE.

Figure 16. Basic principles of rock fracture theory as illustrated by "failure envelopes" and "stress circles".

Dakota. Murray (1968) and Finch (1969) have written excellent papers describing Antelope field.

The author concurs with the basic data and conclusions of Murray and Finch concerning Antelope field, with the exception that Finch ascribes fluid overpressuring in the Bakken to hydrocarbon generation rather than to tectonic compression. Further, he believes that the overpressure

described by the two investigators is an absolutely essential element in forming the reservoir fracture system, as described in preceding paragraphs. It should be noted that the geometrical factors described as "radius of curvature" and "curvature" are a quantitative measure of the process term referred to as "bending" in this report. Certain additional observations concerning this field can be made.

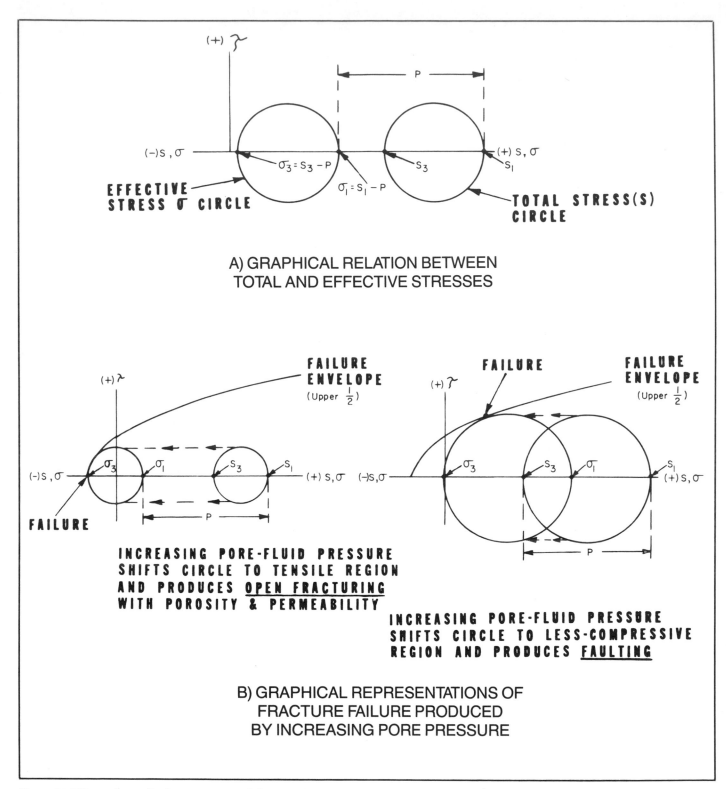

Figure 17. Effects of pore-fluid pressure on rock fracture theory.

The basic structural configuration of Antelope field as proposed by Finch (Figure 19) and Murray (Figure 20) consists of a northwest-southeast trending closed anticline with additional non-closing plunge to the southeast. The structure is strongly asymmetric, with a gentle southwest flank and a much steeper northeast flank. The overall structural configuration of the antelope field is believed to represent a classic "drape fold" (Stearns, 1971) overlying a vertically uplifted basement block which is faulted at depth beneath the steep limb of the anticline (Figure 21). As shown by Finch (Figure 19), "bending" in the shallower, normally-pressures beds over the drape fold has lead to

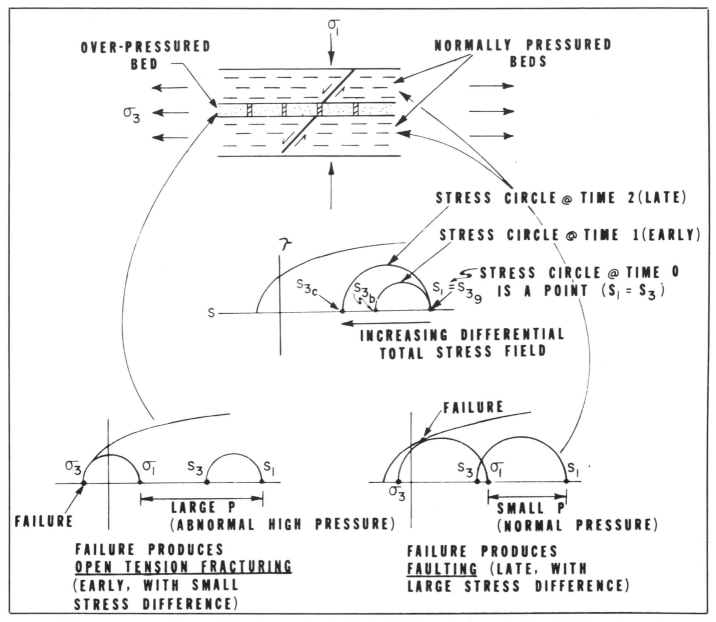

Figure 18. Schematic section and fracture-failure diagrams showing that both faulting and open tension (extension) fracturing may be produced in a series of sedimentary beds with a rising stress field affecting units having different pore-fluid pressure.

stress conditions producing shear-fracture failure as manifested in a series of compensating "keystone-type" normal faults. In contrast to shallower beds, "bending" (or "curvature") in the deeper overpressured Bakkan/Sanish section has led to stress conditions producing open tensile fracture failure, as manifested in the production "fairway" mapped by Murray (Figure 20). The overall effect of "bending" in producing the two types of secondary fracture failure over the Antelope structure is schematically shown in Figure 22. Note the similarity between the secondary faulting and tension fracturing pattern shown in this diagram to that depicted in Figure 18.

CONCLUSIONS AND DISCUSSION OF IMPLICATIONS

In the preceding sections an attempt was made to

describe the petrophysical behavior the Upper and Lower Bakken Shale Members and to speculate on the reasons for this behavior. Arguments were presented which strongly suggest that the unusual petrophysical behavior is a direct reflection of the source-rock properties of these units. The conclusions drawn from these speculations and arguments—while being of a very preliminary nature and obviously requiring further investigation—have very important implications regarding 1) petrophysical behavior in general, 2) oil generation and expulsion mechanisms, 3) the identification of "maturity" in source-rocks, and 4) contributions to exploration plays.

An interesting comparison may be drawn between the petrophysical behavior of reservoir rocks into which oil has migrated and accumulated, and source-rocks from which oil has evidently been generated and expelled. Petrophysical

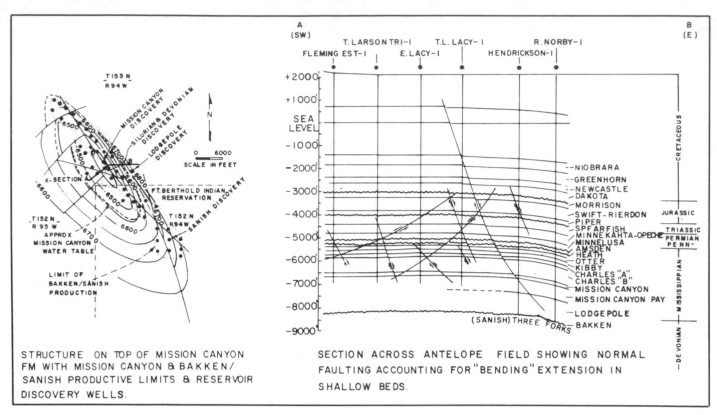

Figure 19. Antelope field, McKenzie County, North Dakota. Structure-contour map and cross-section interpretation after Finch (1969). Note presence of faulting (shear fracturing) at shallow depths, presumably associated with normal pore-fluid pressures.

techniques involving well log measurements of such properties as sonic-transit time, electrical resistivity and density have routinely been used to evaluate such parameters as lithology, porosity, hydrocarbon saturation, fluid pore pressure, etc.; however, most of these measurements have been applied to quantify states-of-condition in coarse-grained reservoir rocks. It seems only logical that similar behavior would manifest itself in the lithology (e.g., organic content), porosity, fluid saturation, pore pressure, etc., of fine-grained source rocks and thus be of considerable effort to an exploration concept in which a consideration of source-rocks is important.

The petrophysical behavior of the Bakken and its speculative interpretation implies that generated oil is expelled from the rock as a continuous porosity-saturating phase through an oil-wet rock matrix. The inferred mechanism of expulsion appears related to a mechanical-potential energy imbalance caused by anomalous fluid pressure introduced by compactional processes related to rock failure incurred as 1) a result of increasing effective stress due to increasing depth of burial, and also 2) by the conversion of solid overburden-supporting kerogen to liquid, nonoverburden-supporting, mobile and expellable oil. An additional supporting expulsion mechanism could be related to pressure increases produced by the greater volumes occupied by metamorphosed organic material (residual kerogen plus generated hydrocarbons) over those occupied by the original kerogen. A conceptual scheme illustrating the proposed generation, saturation, compaction and

expulsion process is shown in Figure 23.

The petrophysical behavior noted in the Bakken may not be characteristic of all source-rock lithologies; however, some similarities may exist. Analyses suggest the Bakken to be unusually rich in organic carbon and to also contain a considerable amount of inorganic carbonate material. The high weight-percentage of organic carbon in the Bakken may be indicative of extremely large volume percentages of organic material in the rock—perhaps as much as 25 to 50 percent. This could mean that the majority of mineral matter forming the walls and throat interconnections of the porosity network in the rock consists of preferentially oil-wet organic material. Further, carbonate mineral grains may also be oil-wet due to the absorption of organic compounds on their surfaces during deposition in an extremely organic-rich environment.

Petrographic examination of Bakken thin sections shows that organic material is well-distributed throughout the rock and that it is not concentrated into obvious laminations, as is characteristic of many other source-rocks. This may have a great effect on average rock wetability and equilibrium-hydrocarbon saturation, in that less-organic-rich rocks may be mostly water-wet and any hydrocarbon generated or expelled within the porosity network may be confined to a low-saturation insular phase. Similarly, if a continuous network or system of laminations composed or organic-oil-wet material exists in the otherwise water-wet rock, hydrocarbon saturation and expulsion paths may be limited to only a small portion of the total rock volume. The effects of hydrocarbon saturation in these types of rocks

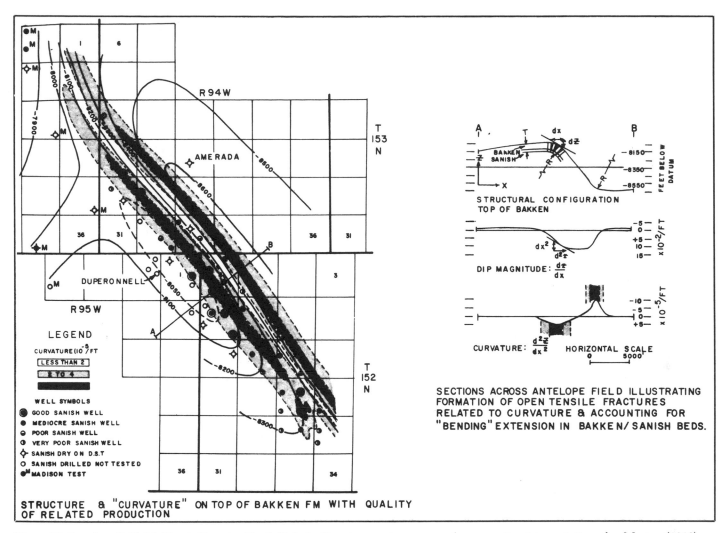

Figure 20. Antelope field, McKenzie County, North Dakota. Structure-contour map and cross-section interpretation after Murray (1968). Note presence of open tensile fracturing associated with production and "curvature" in the highly over-pressured Bakken Formation.

may produce petrophysical changes which are either too subtle to detect or are of a much less obvious magnitude than that typified by the Bakken. However, if these changes are detectable in other source-rocks, a wide field of application to the mapping of mature source-rocks based on existing well control seems possible.

The effect of fluid saturation and pore pressure on the sonic properties of the Bakken shale may make seismic techniques amenable to the delineation of "mature" and geopressured areas.

Speculative conclusions regarding the relationship between source-rock maturity, hydrocarbon generation, geopressuring and fracturing suggest an opportunity in exploration for unrecognized and unlooked-for "unconventional" accumulations of potentially very large regional extent. A logical place to look for oil would naturally be within the source-rock body itself, wherever the generation process has led to the temporary or permanent geopressuring of fluid in the rock and the subsequent creation of economically significant porosity and permeability in a system of extension fractures. Unexplored areas of fractured Bakken are believed to exist

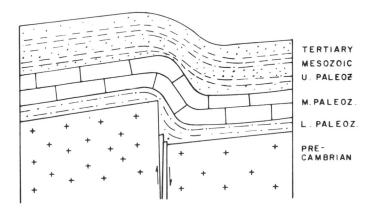

Figure 21. Schematic cross section of a typical Rocky Mountain "drape fold" (after Stearns, 1971). Note similarity of overall structure to that at Antelope field.

in the Williston Basin. The concept may be equally as valid for similar actively generating overpressured source-rocks in other stratigraphic sections and basins.

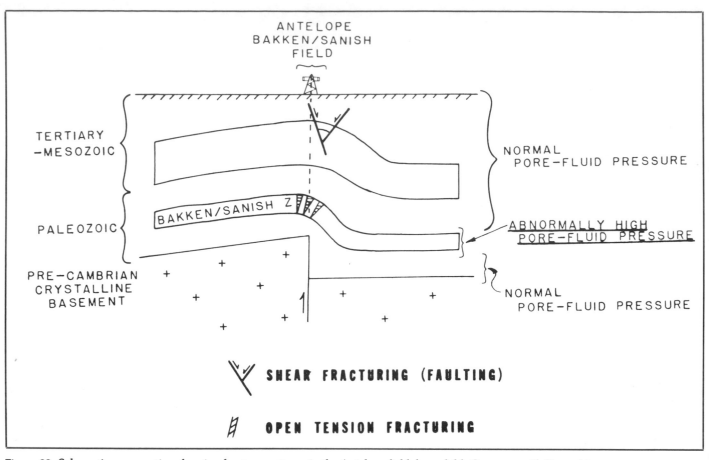

Figure 22. Schematic cross section showing fracture patterns in the Antelope field drape fold. Compare with Figure 18.

Occurrences of abnormally high pore-fluid pressure, where hydrocarbons are the only moveable phase present and which may be uniquely related to an actively-generating and well-defined source-rock unit, have been noted in other sedimentary rocks and areas of the world. The author has noted such occurrences in the Powder River, Wind River, Green River, Uinta, Paradox, Anadarko, Delaware and San Joaquin Basins.

The relation of oil generation in Bakken source-rock shales to the creation of fluid overpressure and the subsequent formation of vertical open tension fractures may provide a key for discerning migration and accumulation patterns of oils thought to originate from the Bakken. Oils found in extracts of Bakken shale, within fractured Bakken reservoirs and within more conventional reservoirs some distance above the Bakken in the Mission Canyon and some distance below the Bakken in the Nisku are apparently identical and can be presumed to have the same origin (Williams, 1974; Dow, 1974; Figure 2 this paper). Further, the Bakken is the only source-rock which has been identified for a distance of several hundred feet above or below the reservoir sections. Considering the general impermeability of "tight" rocks in the Lodgepole Formation separating the Bakken source from a Mission Canyon reservoir or of similar "tightness" in rocks of the Three Forks separating the Bakken from a Nisku reservoir, the question may be raised as to how upward and downward vertical migration takes place outward from the Bakken through "tight" intervening

rocks to reservoirs in which large volumes of related oil have been found. Dow proposed vertical fractures through the Lodgepole and localized along the Nesson anticlinal axis as being responsible for upward migration into the Mission Canyon and believed lateral migration across faults was necessary to charge the Nisku (Dow, 1974, pages 1257-1259 and Figure 12). The author believes vertical fracture paths are much more extensive and are essentially related to the area of oil-generation-caused fluid-overpressure within the area of Bakken source-rock maturity.

The schematic cross-section shown in Figure 24 depicts a scheme of hydrocarbon generation, migration and accumulation relating Bakken source rocks to the distribution of known types of oil accumulations. Salient features of the scheme are summarized as follows:

1) Hydrocarbon source-rock bodies comprising the upper and lower Bakken Shale units are mature in the deeper portions of the Williston Basin where they have been exposed to present-day maximum burial-related temperatures of about 165°F or more.
2) Hydrocarbon generation within the zone of maturity has caused the creation of a zone of abnormally high fluid pressure within the Bakken and closely adjacent beds.
3) The abnormally high fluid pressures have caused the creation of vertical fractures in the adjacent confining beds comprising the Lodgepole and Three Forks Formations. Fracturing appears to be preferentially

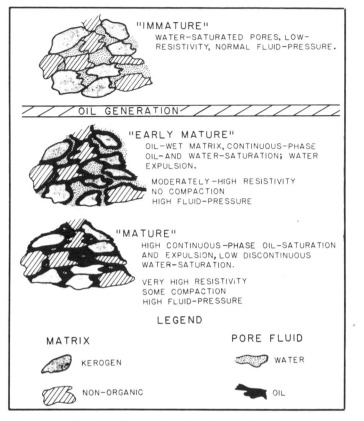

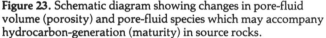

Figure 23. Schematic diagram showing changes in pore-fluid volume (porosity) and pore-fluid species which may accompany hydrocarbon-generation (maturity) in source rocks.

upward towards the Mission Canyon regionally-extensive matrix-porosity interval on the eastern flank of the Basin. The Lodgepole on this flank of the Basin must have a lower "fracture gradient" than the Three Forks. On the western flank of the Basin, where the Lodgepole becomes siliceous and thicker, fracturing evidently occurs downward through the Three Forks to the Nisku. Fracturing within the Bakken itself—both shale and siltstone members—also takes place in areas where Bakken is highly stressed, as in the zone of strong "bending" or "curvature" at Antelope field.

4) Outward migration of oil created within the Bakken overpressure "cell" takes place through the spontaneously generated fracture system. Upward vertical migration takes place through Lodgepole fractures and Mission Canyon matrix porosity until the first evaporite unit of the "Charles facies" is encountered. After encountering the Charles "seal" migration occurs laterally updip until structural and stratigraphic traps are encountered and accumulation takes place. Trap types present in the Mission Canyon include those controlled by a) subcrop truncation b) pinchout of porous tidal flat facies into tight sabka evaporite facies c) basement structural traps d) single- and multiple-stage salt solution traps and e) combinations of any of the preceding conditions. Downward vertical migration to matrix porosity in the Nisku takes place where fractures are formed through the Three Forks. Lateral updip migration beneath the Three Forks "seal" then takes place until a trap is

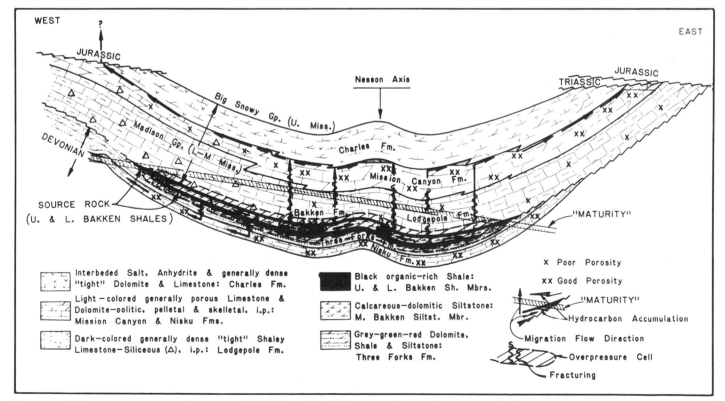

Figure 24. Schematic east-west section across the Williston Basin showing source-rock maturity, fluid over-pressure, fracture, migration and hydrocarbon accumulation patterns in the Bakken Formation and adjacent units.

encountered. Nisku accumulations appear to be
confined to "closed" basement or salt-solution
anticlinal features. Based on the known occurrence of
Bakken-type oil in a few small fields located primarily
on the west flank of the Basin, fracturing and
downward migration through the Three Forks must be
relatively inefficient when compared to upward
fracturing and migration and is essentially limited to
the area where the Lodgepole is thick and siliceous and
the Mission Canyon has regionally poorer
matrix-porosity development.

5) Fractures through the Lodgepole and Three Forks
which "confine" the fluid-overpressure cell within the
Bakken are believed to "breathe" and act as a safety
valve to the development of extremely high
fluid-overpressures developed in the Bakken as a result
of hydrocarbon generation. Once fracturing occurs at
some critical high pressure, generated hydrocarbons
bleed-off from the Bakken to the more-normally
pressured Mission Canyon or Nisku and relieve
pressure within the Bakken. When fluid pressures falls
below that of the *minimum principal* tectonic *stress*
(e.g., the *effective stress* becomes compressive instead
of tensile) the fractures close. The fractures may open
again when more oil is generated and the fluid pressure
is again raised to a critical value again or the whole
system may merely equilibrate to a pressure sufficient
to maintain a steady-state of oil generation and
outward migration through barely open fractures. The
fact that open fractures are present in the
overpressured Bakken/Sanish reservoir at Antelope
field seems to imply substantial elastic contrasts
between the properties of the Bakken/Sanish
lithologies and those of the confining Lodgepole
Formation which separates it from a productive
Mission Canyon matrix reservoir. Fluid pressure
gradients mapped in the Bakken Formation and based
on a sonic-behavior relation (Figure 15) imply that a
fluid pressure ("frac") gradient of about .60 to .80
psi/ft is required to open fractures in confining units
and allow outward migration from overpressured and
currently hydrocarbon-generating Bakken
source-rocks.

The generation/migration scheme postulated for the
Bakken source-rock unit and its associated oil-productive
reservoir units is believed to be somewhat universal in other
units and basins throughout the world. For example —oil in
Upper Cretaceous sandstone reservoirs of the Powder River
Basin in Wyoming are separated from overpressured and
actively generating source-rocks in the Niobrara Formation
by a thousand feet of so of intervening Pierre shales. A
similar scheme of vertical migration through fractured
overpressures shale may explain patterns of generation and
accumulation on the lower Gulf Coast, where mature source
rocks charging Middle and Upper Tertiary reservoirs must
be extremely deep and lie within or at the base of an
extremely thick geopressured shale section.

REFERENCES

Anderson, E.M., 1951, The dynamics of faulting: Edinburg, Oliver and Boyd, 206 p.
Connan, J., 1974, Time-temperature relation in oil genesis: AAPG Bulletin, v. 58, n. 12, p. 2516-2521.
Claypool, G.E., A.H. Love, and E.K. Maughan, 1978, Organic geochemistry, incipient metamorphism, and oil generation in black shale members of Phosphoria formation, western interior United States: AAPG Bulletin, v. 62, p. 98-120.
Dow, W.G., 1974, Application of oil-correlation and source-rock data to exploration in the Williston Basin: AAPG Bulletin, v. 58, n. 7, p. 1253-1262.
Finch, W.C., 1969, Abnormal pressure in the Antelope field, North Dakota: Journal of Petroleum Technology, July, p. 821-826.
Hottman, C.E., and R.K. Johnson, 1965, Estimation of formation pressures from log-derived shale properties: American Institute of Mining and Metallurgy for Petroleum Engineering Transactions, v. 234, p. 717-722.
Houston Geological Society, 1971, Abnormal subsurface pressure—A study group report 1969-1971: 92 p.
Hubbert, M.K., 1951, Mechanical basis for certain familiar geological structures: Geological Society of America Bulletin, v. 62, p. 355-372.
——, and W.W. Rubey, 1959, Role of fluid pressure in mechanics of overthrust faulting: Geological Society of America Bulletin, v. 70, n. 1. p. 115-166.
Landes, K.K., 1967, Eometamorphism and oil and gas in time and space: AAPG Bulletin, v. 51, n. 6, p. 828-841.
LaPlante, R.E., 1974, Hydrocarbon generation in Gulf Coast Tertiary sediments: AAPG Bulletin, v. 58, n. 7, p. 1281-1289.
MacGregor, J.R., 1965, Quantitative determination of reservoir pressures from conductivity log: AAPG Bulletin, v. 49, n. 9, p. 1502-1511.
Momper, V.A., 1972, Evaluating source beds for petroleum (abs.): AAPG Bulletin, v. 56, n. 3, p. 640.
Murray, G.H., Jr., 1968, Quantitative fracture study-Sanish pool, McKenzie County, North Dakota: AAPG Bulletin, v. 52, n. 1, p. 57-65.
Nixon, R.P., 1973, Oil source beds in cretaceous Mowry shale of northwestern interior United States: AAPG Bulletin, v. 57, p. 136-161.
Nordquist, J.W., 1953, Mississippian stratigraphy of Northern Montana: Billings Geological Society Guidebook, 4th Annual Field Conference, p. 68-82.
North Dakota Geological Survey, 1974, Production statistics and engineering data, oil in North Dakota, second half of 1973: 262 p.
Pan, Poh-hsi and J.Cl. de Bremaeker, 1970, Direct location of oil and gas by the seismic reflection method: Geophysical Prospecting; v. 18 (Dec.), p. 712-727.
Peale, A.C., 1893, Paleozoic section in the vicinity of Three Forks, Montana: U.S. Geological Survey Bulletin 110, 56 p.
Philippi, G.T., 1965, On the depth, time, and mechanism of

petroleum generation: Geochemica et. Cosmochim Acta, v. 29, p. 1021-1049.

———, 1968, Essentials of petroleum formation process are organic source and a subsurface temperature controlled chemical reaction mechanism, *in* P.A. Schenk and I. Havanaar, eds., Advances in organic geochemistry: Oxford, Pergamon, p. 25-46.

Price, N.J., 1966, Fault and point development in brittle rock: Oxford, Pergamon Press, 176 p.

Sandberg, C.A., and C.R. Hammond, 1958, Devonian system in Williston Basin and central Montana: AAPG Bulletin, v. 42, n. 10, p. 2293-2334.

Secor, D.T., Jr., 1965, Role of fluid pressure in jointing: American Journal of Science, v. 263, p. 633-646.

Stearns, D.W., 1971, Mechanisms of drape folding in the Wyoming province: Wyoming Geological Association Guidebook, 23rd Annual Field Conference, p. 125-143.

Tissot, B., et al, 1971, Origin and evolution of hydrocarbons in early Toaracian shales, Paris Basin, France: AAPG Bulletin, v. 55, p. 2177-2193.

Trask, P.D., and W. Patnode, 1942, Source beds of petroleum: AAPG Special Publication, 556 p., see especially p. 62.

Welte, D.H., 1965, Relationship between petroleum and source rock: AAPG Bulletin, v. 49, n. 12, p. 2246-2268.

Wheeler, H.E., 1963, Post-Sauk and Pre-Absaroka Paleozoic stratigraphic patterns in North America: AAPG Bulletin, v. 46, n. 8, p. 1497-1526.

Williams, J.A., 1974, Characterization of oil types in the Williston Basin: AAPG Bulletin, v. 58, n. 7, p. 1242-1252.

Geochemical Exploration in the Powder River Basin

James A. Momper
Consultant
Tulsa, Oklahoma

Jack A. Williams
Amoco Production Company
Tulsa, Oklahoma

Geochemical and geological data were used to identify effective source rocks and oil-types, and to determine stratigraphic sequences and areas that are prospective for crude oil and thermal hydrocarbon gas. The source rock volumes and generation-expulsion performance data for each effective source sequence provided the basis for calculating quantities of expelled oil and gas. These quantities readily account for discovered in-place reservoir oil of more than 7 billion barrels and relatively minor amounts of gas, mainly associated.

Lower and Upper Cretaceous source beds expelled most of the indigenous oil. These oils are chemically similar, regardless of their source. Lower Cretaceous Mowry Shale and Upper Cretaceous Niobrara and Carlile formations expelled most of the discovered oil. Oil expulsion from Cretaceous source rocks began during the early Tertiary and continued through much of Miocene time as the expulsion fronts moved up section and updip. Laramide structure controlled directions of migration of Cretaceous oil.

The second major type of oil is nonindigenous to the Powder River Basin and is correlated to the remote Upper Permian Phosphoria Formation source area centered in southeastern Idaho. This oil entered northeastern Wyoming during Late Jurassic time, before the Powder River Basin formed, through carrier beds of Pennsylvanian and Permian age. Phosphoria-type oil is preserved in four separate parts of the basin, primarily in sandstone reservoirs of Early Permian age in the Minnelusa and Tensleep formations.

A minor oil-type found in the southeastern part of the Powder River Basin was expelled from relatively thin, local shales of Pennsylvanian age.

Several giant fields with more than 100 million barrels of recoverable oil and major oil fields of at least 50 million barrels are located on structural positives around the periphery of the Powder River Basin. These salients served as gathering areas to concentrate migrating oil. Other large fields are in stratigraphic traps oriented parallel with structural strike on the eastern flank, this orientation permitting large accumulations to form from a big drainage area in downdip source rocks.

Meteoric water, aerobic bacteria, distillation, and thermal cracking are affecting the quality of preserved oil. Two types of bacterial alteration are common. Much of the gas generated with oil has escaped or dispersed. Oxygenated recharge waters appear to be degrading organic matter in Cretaceous source rocks around the basin perimeter. Both chemical and physical properties of rocks and fluids proved to be useful in defining prospective areas for the various types of oil.

INTRODUCTION

The Powder River Basin in northeastern Wyoming and southeastern Montana is an excellent petroleum province for testing and applying various geochemical techniques and concepts because of its diverse geology and the intensive exploration in the basin for the last 30 years.

A large number of oil, rock, water, and gas samples were analyzed for this study. The 235 oil samples were collected from wells and seeps in widely distributed geographic locations from all of the important reservoirs of the basin and from most of the minor reservoirs. At least 40 discrete

productive reservoirs have been discovered in all systems from the Mississippian to the Tertiary inclusively. The majority of the more than 500 rock samples analyzed (cores, cuttings, and outcrops) were from the Cretaceous System. In addition, Bureau of Mines oil sample analytical data and other published geochemical information pertaining mainly to the Cretaceous Mowry Shale were used.

Typical of the several Bureau of Mines studies are those by Wenger and Reid (1958), Biggs and Espach (1960), Wenger and Ball (1961), and Coleman et al (1978). Papers by Hunt (1953), Curtis (1958), Strickland (1958), and McIver (1962) described crude oil and its occurrences in the

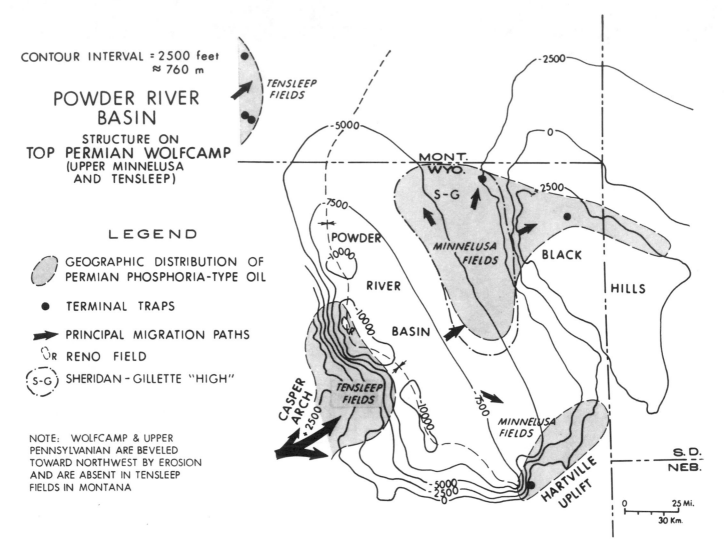

CONTOUR INTERVAL = 2500 feet
≈ 760 m

POWDER RIVER
BASIN
STRUCTURE ON
TOP PERMIAN WOLFCAMP
(UPPER MINNELUSA
AND TENSLEEP)

L E G E N D

GEOGRAPHIC DISTRIBUTION OF
PERMIAN PHOSPHORIA-TYPE OIL

● TERMINAL TRAPS

➤ PRINCIPAL MIGRATION PATHS

⌣R RENO FIELD

(S-G) SHERIDAN - GILLETTE "HIGH"

NOTE: WOLFCAMP & UPPER
PENNSYLVANIAN ARE BEVELED
TOWARD NORTHWEST BY EROSION
AND ARE ABSENT IN TENSLEEP
FIELDS IN MONTANA

Figure 1. Distribution of Phosphoria-type oil in four areas of the Powder River Basin. Migration preceded evolution of the structural basin. Gaps along migration paths in the basin resulted, in part, from thermal cracking of oil to gas. Isolated shows are present east-southeast of the terminal trap on the northern margin of the Black Hills in South Dakota.

Powder River Basin. Hedberg (1968) discussed the significance of high-wax oils.

Published information on source rocks in the Powder River Basin is quite limited. Four papers on the Mowry Shale (Early Cretaceous) have been published: Schrayer and Zarella (1963, 1966, 1968) and Nixon (1973). A cursory study of the geochemistry of some Upper Cretaceous shales by Merewether and Claypool was published in 1980.

OIL TYPES

Three basic oil types occur in the basin—a non-indigenous Permian Phosphoria-type (Figure 1), a Cretaceous-type, and a Pennsylvanian-type of minor importance. Phosphoria-type oil was generated in a remote source area external to the basin and migrated into the area during Late Jurassic time. The Powder River Basin formed during Late Cretaceous-early Tertiary time. Phosphoria-type oil occurs in reservoirs of Mississippian through Jurassic age in and around the basin. Expulsion of

indigenous Cretaceous and Pennsylvanian oils occurred after the basin formed.

Cretaceous-type oil was derived from multiple source beds in both the Lower and Upper Cretaceous (Figure 2), resulting in two subtypes with only subtle differences in chemical characteristics. The Lower Cretaceous (LK) subtype is principally Mowry oil (Figure 3). Most of the Upper Cretaceous (UK) subtype was derived from the Niobrara and Carlile formations (Figure 4). The two subtypes commonly are commingled and are therefore difficult to distinguish. Identifications of the LK- and UK-subtypes are further complicated by several alteration processes that have affected oil to varied degrees in many of the fields. Cretaceous-type oil is the most abundant and widespread in the basin, having been discovered in reservoirs as old as Pennsylvanian and as young as early Tertiary, and in all but the northernmost quarter of the basin.

Oils expelled from Upper Cretaceous source beds are mainly confined to the southwestern part of the basin and

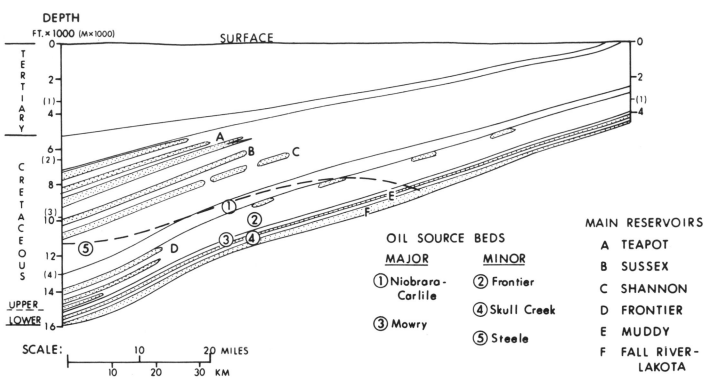

Figure 2. Stratigraphic relationships of oil source beds and main reservoirs within Cretaceous System in the Powder River Basin. Argillaceous rocks with significant oil-generating capability below heavy dashed line expelled oil, and are effective oil source beds. Line of section (southwest-northeast) is on Figure 3.

its margins. Only a few small scattered pools of UK-subtype oil occur updip east of the effective source area (Figure 4). The limiting factor for more entrapment in that updip area appears to be a lack of reservoir beds. Some Upper Cretaceous reservoirs out in the basin are interpreted to contain LK-subtype oil, suggesting that faults or fractures may have permitted transverse migration. Upper Cretaceous reservoirs considerably younger than any of the effective source rocks produce oil (Figure 2), proving that appreciable transverse migration of UK-subtype oil occurred, if not of the LK-subtype. Within the Powder River Basin, most of the Cretaceous-type oil is in stratigraphic traps. Around the basin margin, anticlinal and combination traps dominate.

Minor quantities of Pennsylvanian-type oil are present in Pennsylvanian reservoirs in a few pools in the southeastern sector of the basin. At least one occurrence has been identified in a Permian reservoir and one in a Jurassic reservoir. This type has been difficult to identify because of oil degradation and limited sample availability but we now have sufficient control to conclusively establish the presence of this third type.

SOURCE ROCKS

Effective oil and gas source beds within the basin principally are lower Upper Cretaceous and Lower Cretaceous shales. In the southeastern part of the basin, where Pennsylvanian-type oil occurs, local thin

Pennsylvanian shale beds expelled oil. The other systems do not contain thermally mature rocks with significant oil-generating capability. Locally, part of the Jurassic has thermal gas-generating potential and small quantities of recoverable biogenic gas have been generated in the Cretaceous and Tertiary systems.

Several analytical methods were used to evaluate source rocks and to correlate oils to their respective sources. Quantitative methods provided information on the amounts of organic carbon, extractable bitumen, and extractable C_{15+} saturated hydrocarbons in the source rock samples. Oil-rock correlations were established by a variety of analyses on the extracts and oils. Quantities of hydrocarbons already generated and quantities remaining to be generated were obtained from pyrolysis. Qualitative methods were used to determine the convertibility of organic matter to oil and natural gas, and also to determine the extent of organic matter degradation. Thermal carbonization stages of organic matter and the maturity of generated bitumen and hydrocarbons were interpreted from qualitative methods. In addition, several methods indicated whether organic matter was preferentially oil- or gas-generating, or non-generating.

PHOSPHORIA OIL SOURCE

A large quantity, perhaps as much as 2 billion (2×10^9) barrels, of Phosphoria-type oil migrated into northeastern Wyoming before the Powder River Basin formed. The rich

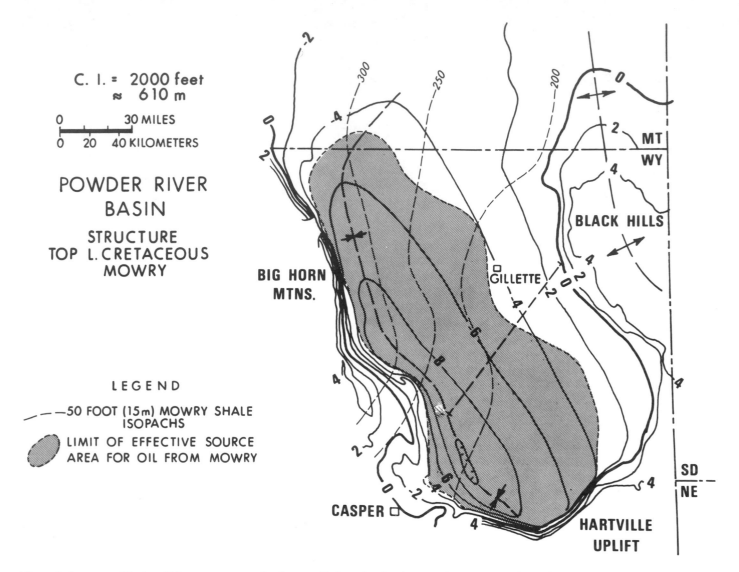

C. I. = 2000 feet
≈ 610 m

0   30 MILES
├───────────────┤
0   20   40 KILOMETERS

POWDER RIVER
BASIN

STRUCTURE
TOP L. CRETACEOUS
MOWRY

LEGEND

— — —50 FOOT (15m) MOWRY SHALE
ISOPACHS

▨ LIMIT OF EFFECTIVE SOURCE
AREA FOR OIL FROM MOWRY

Figure 3. Interpreted limits of Mowry source rocks that expelled crude oil, relative to Mowry isopachs and present-day structure on the top of the Mowry Shale.

Phosphoria source sequence centered in southeastern Idaho expelled oil during Late Jurassic time. An almost ideal, westward-tilted, sealed carrier system comprised mainly of permeable Permian and Pennsylvanian sandstone enabled Phosphoria oil to migrate long distances across Wyoming (Sheldon, 1967) along several migration paths. In western Wyoming, the oil migrated mainly through Pennsylvanian Tensleep sandstone, moving up-section into and updip through Wolfcampian Tensleep-Casper sandstone in eastern Wyoming where that section is preserved.

Phosphoria-type oil has been identified in four separate areas of the basin. It entered northeastern Wyoming via the Casper Arch along one path (Figure 1), and was trapped mainly in Permian Tensleep and Minnelusa reservoirs in three of the areas. Terminal entrapment occurred where permeable sandstone carrier beds change facies into tight evaporites and red beds flanking the Black Hills Uplift. Nevertheless, oil staining and heavy oil residues in outcrops indicate that a large volume of oil escaped and a significant amount was degraded.

A second migration path delivered a considerably smaller amount of oil to Tensleep reservoirs and a Mississippian reservoir in the area now on the northwestern margin of the basin in Montana. This oil migrated through mid-Pennsylvanian-age sandstone and also terminated at an evaporite facies. The truncated subcrop of the Upper Pennsylvanian and Lower Permian is located southeast of this northernmost migration path. Some redistribution of Phosphoria-type oil occurred during basin-subsidence and the Laramide orogeny (Late Cretaceous-early Tertiary).

In the deep, relatively hot southern part of the basin and elsewhere where geothermal gradients are sufficiently high, Phosphoria oil has been cracked to gas. Oil pools have survived in cooler locations as deep as 15,000 ft (4,572 m) in the Reno Field area and more should be discovered along the original migration and remigration paths if shallower, cooler traps are present. Most of the gas accompanying Phosphoria-type oil during its Late Jurassic migration has escaped. Possibly much of the gas resulting from the later thermal cracking is in undiscovered traps.

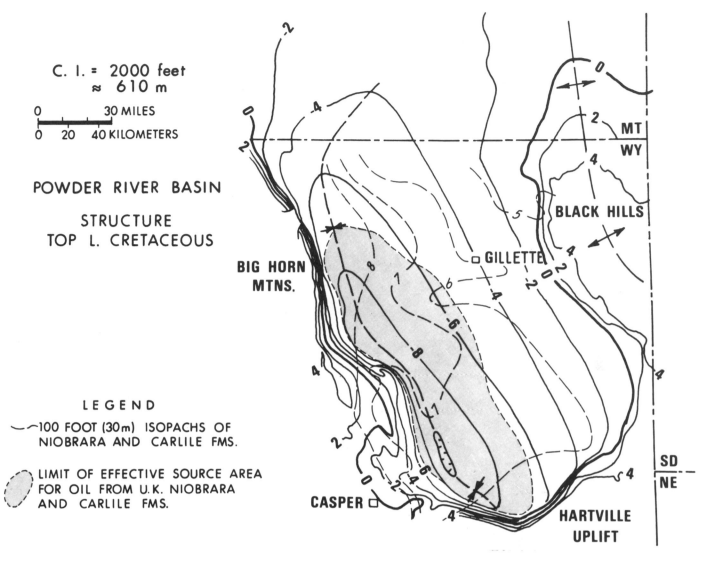

Figure 4. Isopachs of combined Niobrara Formation and Sage Break Shale Member of the Carlile indicate that the interval within the area of effective oil source rocks ranges from about 550 ft (168 m) to more than 800 ft (244 m). However, on the flanks of the basin, the actual thickness of the effective source rocks is less than the total thickness of these units, as shown on Figure 2. This section accounted for most of the UK-subtype oil.

CRETACEOUS SOURCE BEDS

Several effective oil-sources are present in the Cretaceous (Figure 2). The Lower Cretaceous Mowry Shale evidently expelled at least as much oil as all other sources combined. The next most important oil sources were the Upper Cretaceous Niobrara and subjacent Carlile formations. The Frontier, Skull Creek, and Steele shales (and possibly the Fuson Shale) were relatively minor, effective oil source sections.

Mowry Shale

The effective Mowry Shale oil source area is delineated on Figure 3. Superimposed on a structural map of the Lower Cretaceous are 50-ft (15.2-m) isopachs of the Mowry. The Mowry is highly siliceous. It is more than 350 ft (107 m) thick along the western margin of the basin and is beveled by erosion toward the southeast. Geochemical analyses and

subsidence profiles (using exposure time and burial temperatures) both indicate that the Mowry in the shaded area expelled oil. Times of expulsion from within this area of effective Mowry source rocks depended on the ages and thicknesses of overburden, as well as on the local geothermal gradients, which today range from about 13 to 20°F per 1,000 ft (305 m) of burial depth. Available evidence suggests that the same relationship of high and low gradients has persisted since the Cretaceous in those deeper parts of the section unaffected by cooling phenomena.

Oil expulsion began in late Paleocene or early Eocene time from the basin deep in the southwest. Elsewhere, expulsion was delayed, persisting through much of Miocene time as the oil-generation front moved updip and up-section. Generation eventually ceased as the shallower section in the basin cooled, owing to deep erosion, uplifting, late Cenozoic climatic cooling, and encroachment of relatively cold recharge water. The Mowry and other

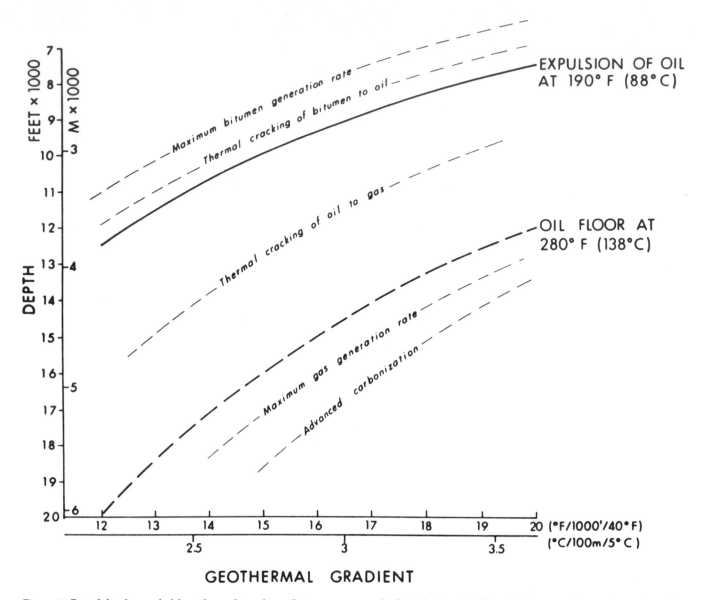

Figure 5. Burial depths needed for oil expulsion from Cretaceous source beds in the Powder River Basin vary with geothermal gradients and exposure times. Expulsion curve shown is for maximum exposure time. Shorter exposure periods required higher temperatures for expulsion. Similarly, the thermal destruction of commercial oil accumulations (oil floor) occurs approximately at the depths indicated for the present range of geothermal gradients and maximum exposure times. In this province, the deepest Cretaceous is less than 15,000 ft (4,572 m) so that the oil floor is encountered only in the highest geothermal gradient settings (that is, generally above 17°F/1,000 ft).

effective source beds remain overpressured today in the deeper part of the basin, although not sufficiently for oil expulsion to occur. The build-up of pressures in these source rock units resulted primarily from fluid-generation during carbonization of organic matter. The shallower parts of the source rocks that had generated substantial quantities of fluids are under-pressured as a result of the cooling trends.

We calculated the Mowry Shale expelled about 11.9 billion barrels of oil (Table 1) from its 10,500 sq mi (27,195 sq km) effective source area, using averages of 3.0 weight percent organic carbon, 240 ft (73.2 m) of shale, and an oil-generating capability of 105 barrels per acre foot (>5,800 ppm).

The expulsion efficiency of 7 percent was obtained by comparing the calculated amount of generated oil with the

Table 1. Total quantity of oil expelled from effective source rocks within Lower Cretaceous Mowry Shale in Powder River Basin.

Organic carbon content	weight percent	3.0
Oil generated	barrels per acre foot	105
Oil expelled at 7% efficiency	barrels per acre foot	7.4
Thickness, average	feet	240
Oil expelled	barrels per acre	1,765
Oil expelled	10^3 barrels per square mile	1,130
Effective source area	square miles	10,500
Total Mowry oil expelled	10^9 barrels	11.9

total of (1) discovered in-place oil, plus (2) an estimate of undiscovered oil, and (3) estimates of the amounts of oil lost by degradation, dispersal, escape, and thermal cracking. Of

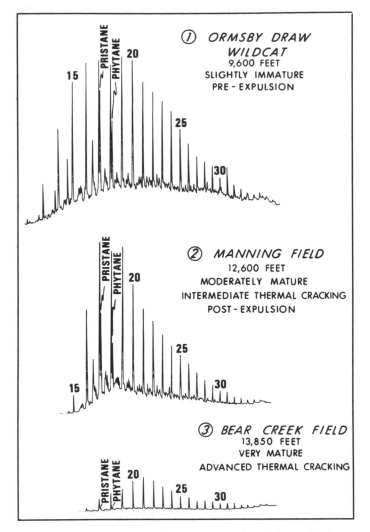

Figure 6. Whole-extract gas chromatograms from Mowry Shale samples at different stages of thermal maturity resulting from progressive thermal cracking with deeper burial (higher temperatures) and long exposure times. Well locations are shown on Figure 7.

Table 2. Total volume of oil expelled from effective source rocks within the combined Niobrara Formation and Sage Breaks Shale Member of Carlile Formation, both Late Cretaceous, in the Powder River Basin.

Organic carbon content	weight percent	2.5
Oil generated	barrels per acre foot	75
Oil expelled at 7% efficiency	barrels per acre foot	5.3
Thickness, average	feet	550
Oil expelled	barrels per acre	2,890
Oil expelled	10^3 barrels per square mile	1,850
Effective source area	square miles	4,100
Total volume oil expelled	10^9 barrels	7.6

these estimates, the least accurate are the values assigned to oil dispersed and escaped. Various combinations of values were tried and usually resulted in efficiences in the range of 6 to 8 percent.

Niobrara-Carlile Section

The Niobrara and Carlile formations are much thicker than the Mowry. However, they are generally 800 to 1,000 ft (244 to 305 m) shallower than the Mowry and therefore, the areal extent of effective Niobrara-Carlile source rocks is much more limited (Figure 4; Table 2). The Niobrara is dominantly marl and calcareous shale and the Sage Breaks Shale Member of the Carlile is calcareous. In the northwestern sector, where the Niobrara-Carlile section is 900 to 1,000 ft (274 to 305 m) thick, it is still immature. These units accounted for the bulk of the expelled UK-subtype oil.

Undegraded organic matter in the Niobrara-Carlile section seems to be slightly poorer in quality overall than undegraded Mowry organic matter. Also, the effective source volume of the Niobrara-Carlile is more than 10 percent smaller. Consequently, we calculated that about 7.6 billion barrels of Niobrara-Carlile oil were expelled in the Powder River Basin (Table 2), or about one-third less than the yield from the Mowry.

Minor Cretaceous Sources

The continental, somewhat coaly Lower Cretaceous Fuson Shale (between Fall River and Lakota in Inyan Kara Group; Figure 2) expelled gas and speculatively, 0.1 to 0.2 billion barrels of waxy oil. Control is too sparse to definitely list the Fuson as an effective oil source. Gas-expulsion from the Fuson, other than gas that may have been expelled with oil, was restricted to relatively small areas where the geothermal gradient is high and peak gas-generating conditions have been attained. The marine Lower Cretaceous Skull Creek Shale contains Type II organic matter and contributed about 1 billion barrels of the total Cretaceous oil expelled in the basin. Much of this oil is commingled with oil from the Mowry in Muddy Sandstone reservoirs.

The thickness of the net effective oil source rocks within the mixed marine-continental Frontier Formation is relatively small compared to the total thickness of the Frontier. The Frontier is coaly, in part, and much of the shale contains Type III organic matter. Thus, gas would be the main product of thermal generation but peak gas generation has not been attained in the unit. The amount of expelled oil attributed to the Frontier is 0.8 billion barrels but the calculation is based on less sample data than is needed for an accurate assessment.

A few samples from the marine Upper Cretaceous Steele Shale (above the Niobrara) suggest that it may have expelled some oil in the west-central area of the basin. Most of the samples from the Steele contained organic matter that is immature or has little oil-generating capability. Inadequate data permit only a gross estimate of no more than 100 million barrels of oil expelled from the Steele.

Cretaceous Source Rock Performance

The calculated total volume of oil generated in all Cretaceous source beds is more than 300 billion barrels. The volume expelled is 21.5 billion barrels, roughly one-third of which appears to have been trapped in-place. The remainder was dispersed, degraded, escaped, and cracked to gas. Numerous active oil seeps around the basin—more than a score of seeps were mapped on Salt Creek Anticline alone—attest to the considerable loss of Cretaceous-type oil.

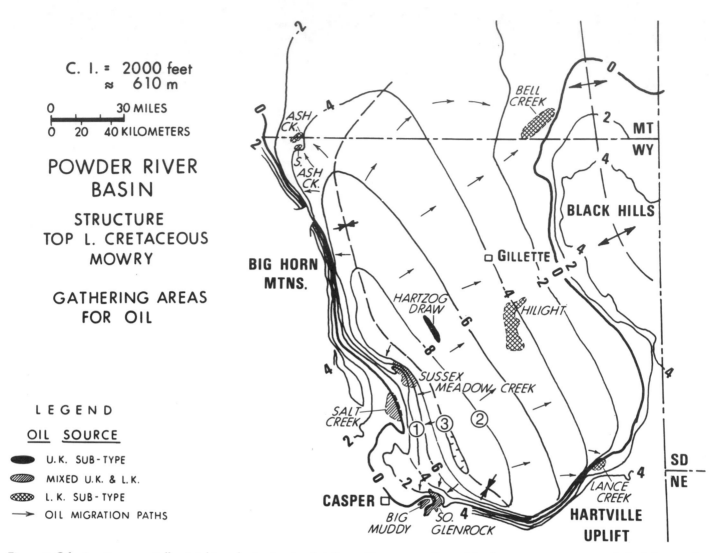

C. I. = 2000 feet
≈ 610 m

0   30 MILES
0   20   40 KILOMETERS

POWDER RIVER BASIN

STRUCTURE TOP L. CRETACEOUS MOWRY

GATHERING AREAS FOR OIL

LEGEND

OIL SOURCE

- ⬭ U.K. SUB-TYPE
- ▨ MIXED U.K. & L.K.
- ▨ L.K. SUB-TYPE
- → OIL MIGRATION PATHS

Figure 7. Oil migration essentially at right angles to structural strike enables structurally positive features to concentrate relatively large volumes of oil; four of five giant fields (Bell Creek, Lance Creek, Sussex-Meadow Creek and Salt Creek) are favorably located with respect to structurally controlled gathering areas. A number of important non-giants are similarly situated. Most of the larger stratigraphic accumulations (for example, Hartzog Draw and Hilight) are parallel or subparallel to strike.

Organic Matter Degradation

Many source rock samples from this basin contain kerogen with below-average oil-generating capability. One-third of the samples analyzed by pyrolysis from the Mowry, Niobrara, and Carlile were rated as of non-source or marginal source-rock quality. This is attributed to degradation of the organic matter after expulsion occurred, the degradation being confined to areas of active meteoric water incursion.

OIL FLOOR

Basinwide data indicate that oil expulsion usually occurred above about 190°F (Figure 5). Depending on local geothermal gradients, this minimum temperature is encountered at depths from about 8,000 to 12,000 ft (2,438 to 3,658 m) in Cretaceous source rocks. Oil in reservoirs is being cracked to natural gas liquids and natural gas at temperatures generally above 280°F, or at depths below

about 12,500 ft (3,810 m) in the high-gradient areas. Because of comparable exposure times, the oil floor in Paleozoic reservoirs generally is at only slightly higher temperature levels than the floor in the Cretaceous because the Paleozoic reservoirs that contain Phosphoria-type oil were at shallow depths until rapid subsidence occurred during Late Cretaceous time. Pennsylvanian-type oil was not expelled until early Tertiary time and thus had the same exposure time as the older Cretaceous-type oils.

Proximity to the oil-floor in the deepest Cretaceous is shown by gas chromatograms of extracts from Mowry samples collected from three relatively deep wells (Figure 6). Identical sample sizes and instrument settings were used. Extract from well 1 (at 9,600 ft, or 2,926 m, and 175°F), is interpreted as slightly immature. Mowry extract from well 2 (at 12,600 ft, or 3,840 m, and 250°F), remaining after expulsion, is moderately mature. The small quantity of thermally cracked extract from well 3 (at 13,850 ft, or 4,221 m, and 275°F) is very mature. Well locations are shown on

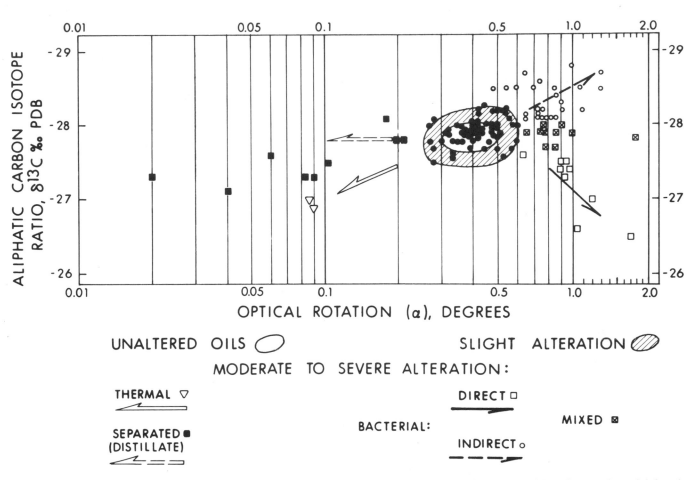

Figure 8. The stable carbon isotope ratios of the aliphatic fraction of crude oils and the optical activities of the oils provide useful data for determining the types and degree of alteration that affected each sample in its reservoir. All data points represent oil samples from Lower Cretaceous, mainly Muddy, reservoirs and indicate that a high percentage of the samples sustained aerobic bacterial degradation.

Figure 7. Wells 2 and 3 are in an area where the geothermal gradient is moderately high; the gradient at location 1 is significantly lower.

FACTORS CONTROLLING MIGRATION

Migration of Cretaceous-type oil was controlled by the Laramide-induced structural configuration which survives essentially unchanged today (Figure 7). Intrastratal migrations of oil are updip, generally at right angles to structural strike, unless diverted laterally along strike by a structural or stratigraphic barrier. Because migration is a dispersal phenomenon, the most favored accumulation sites tend to be at structurally positive gathering areas that locally concentrate oil.

In this basin, three of the five giant fields (Lance Creek, Salt Creek, Sussex-Meadow Creek) are in sizeable anticlines located on even larger, structurally positive salients, as are a number of major fields. Bell Creek and Hilight are stratigraphic traps that were favored by elongated reservoirs oriented parallel or subparallel with structural strike, another preferred setting for large oil accumulations. Most

of the larger stratigraphic accumulations have this orientation. Bell Creek Field also is on a broad anticlinal nose that served as a gathering area.

The oil in Hartzog Draw Shannon reservoir (Figures 2 and 7), a major accumulation, migrated transverse to the bedding. The volume of oil accumulated in Hartzog Draw Field, and in several similar stratigraphic traps located stratigraphically above the shallowest effective source rocks, suggests migration upward along faults or fractures. At Lance Creek, and other fields in its vicinity, cross-fault migration enabled Cretaceous-type oil to be trapped in pre-Cretaceous reservoirs.

ALTERED OILS

Alteration of labile oil begins in most basins soon after the first oil is expelled. The Powder River Basin oils were no exception. One means of recognizing basic oil types and subtypes, and of determining kinds and degrees of alteration, is to plot the stable carbon isotope ratio of the aliphatic fraction of each oil sample versus the optical rotation measurement of the sample. On Figure 8, the large

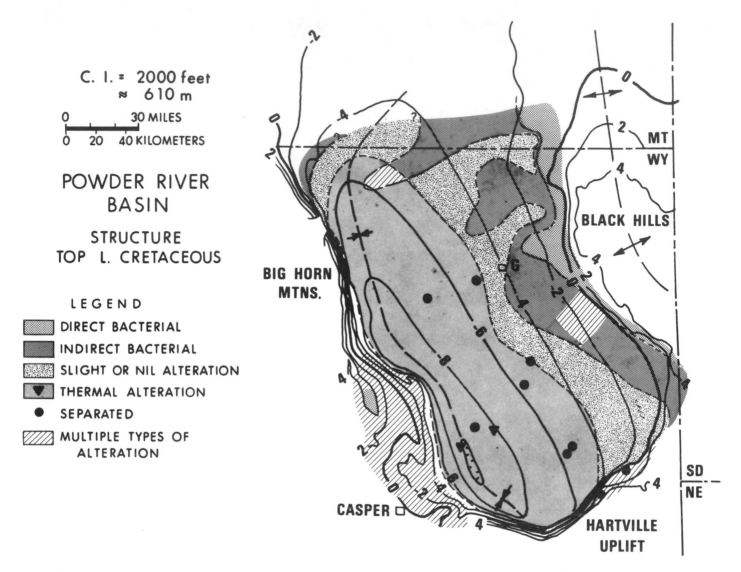

Figure 9. Many of the oil samples from Lower Cretaceous reservoirs show evidence of one or more types of alteration. The areal distribution of unaltered oils and various groups of altered oils are controlled mainly by encroachment of recharge waters and by heat flow.

cluster represents the basic Cretaceous oil type (LK-subtype). The data points in the inner cluster are from oils that evince no alteration. The outer ring of values are from mildly altered oils. To the lower right, the data points represent oils directly degraded by aerobic bacteria in areas and reservoirs invaded by meteoric water. Generally, the farther a point is from the main cluster, the more the oil has been degraded. To the upper right, data points represent oils indirectly altered by the addition of products of biodegradation. They also are from reservoirs in recharge areas, but downdip from directly biodegraded oils. A few of the oils in this group were at reservoir temperatures above 200°F, too high for direct bacterial attack. Intermediate points are for oils that show both types of degradation. The biodegraded products are isotopically light (more negative $\delta^{13}C$), whereas biodegraded oils from which the products were removed are isotopically heavy.

Data points from oils altered by thermal cracking (but not yet condensates) plot to the lower left of the main

cluster. Condensates have little or no optically active components. Points scattered almost horizontally to the left represent oils that have separated, or distilled off, from the basic oil. They have properties similar to condensates but retain heavier components and measurable optical rotations. Distillation does not have much effect on carbon isotope ratios whereas condensates and thermally cracked oils are isotopically heavier than the original oil. Because the distillates are also subjected to cracking, they tend to be shifted slightly toward the heavier isotope ratios. We suspect that this separation occurred in the source beds, mainly the Mowry, during the later phase of oil expulsion, the heavy ends having been retained in the source rocks.

Distributions of unaltered and altered oils in Lower Cretaceous reservoirs are mapped on Figure 9. Basinward from the west flank of the Black Hills, an irregular belt encompasses oils that have been directly altered by bacteria and also are water-washed. Downdip, but still in the recharge area, oils have been indirectly affected by the addition of degradation products transported from updip degraded oil accumulations by migrating meteoric water. These oils have been water-washed and many also were affected directly by aerobic bacteria in the shallower pools. In the next deeper belt, the oils are essentially unaltered; recharge waters have not yet penetrated this belt, part of which is underpressured because of cooling, to a significant extent. Farther downdip, seven oils classed as separated were encountered as were two on the southeast flank. The deeper the oil reservoir, the more these separated oils approach a true thermal distillate. The distillates are similar to, but distinguishable from, condensates caused by thermal cracking.

In the deepest basin, where the Cretaceous is appreciably overpressured, Cretaceous oils have been altered by thermal cracking but retain too much of their heavy components to be classed as condensates. However, thermal cracking and gas generation locally have progressed to a level where condensate is to be anticipated in the hottest deep reservoirs. On the flanks of the basin, particularly the southwest and southeast, a variety of altered types are intermingled (cross-hatched areas).

SUMMARY

The Powder River Basin has a large volume of effective oil source rocks of both Early and Late Cretaceous age that evidently expelled more than 20 billion barrels of oil during the Tertiary. Phosphoria oil migrated eastward across Wyoming from its main source area during Late Jurassic time, entering northeastern Wyoming before the Powder River Basin formed. A relatively small volume of oil was generated and expelled from Pennsylvanian source rocks in the southeastern part of the basin. Much of the trapped oil has undergone one or more types of alteration, and most of the gas expelled with the three types of oil has disappeared.

The present-day Powder River Basin structural configuration controlled the migration of Cretaceous-type and Pennsylvanian-type oils, and the redistribution of Phosphoria-type oils.

Gathering areas around the basin controlled the accumulation of several large oil fields. Geochemical data accurately delineate prospective areas and stratigraphic intervals in the Powder River Basin.

ACKNOWLEDGMENTS

This is an updated version of a summary paper published in the Oil and Gas Journal, December 10, 1979. We thank Amoco Production Company for granting clearance.

REFERENCES CITED

Biggs, P., and R.H. Espach, 1960, Petroleum and natural gas fields in Wyoming: U.S. Bureau of Mines Bulletin 582, 538 p.

Coleman, H.J., et al, 1978, Analyses of 800 crude oils from United States oil fields: BETC/RI-78/14, Technical Information Center, U.S. Department of Energy, November, 447 p.

Curtis, B.F., et al, 1958, Patterns of oil occurrences in the Powder River Basin, *in* Habitat of oil: AAPG Special Symposium Volume, p. 268-292.

Hedberg, H.D., 1968, Significance of high-wax oils with respect to genesis of petroleum: AAPG Bulletin, v. 52, p. 736-750.

Hunt, J.M., 1953, Composition of crude oil and its relation to stratigraphy in Wyoming: AAPG Bulletin, v. 37, p. 1837-1872.

McIver, R.D., 1962, The crude oils of Wyoming—product of depositional environment and alteration: Wyoming Geological Association 17th Annual Field Conference Guidebook, p. 248-251.

Merewether, E.A., and G.E. Claypool, 1980, Organic composition of some Upper Cretaceous shale, Powder River Basin, Wyoming: AAPG Bulletin, v. 64, p. 488-500.

Nixon, R.P., 1973, Oil source beds in Cretaceous Mowry Shale of northwestern interior United States: AAPG Bulletin, v. 57, p. 136-161.

Schrayer, G.J., and W.M. Zarella, 1963, Organic geochemistry of shales, part 1; distribution of organic matter in the siliceous Mowry Shale of Wyoming: Geochimica et Cosmochimica Acta, v. 27, n. 10, p. 1033-1046.

—— and W.M. Zarella, 1966, Organic geochemistry of shales—II; distribution of extractable organic matter in the siliceous Mowry Shale of Wyoming: Geochimica et Cosmochimica Acta, v. 30, p. 415-434.

—— and W.M. Zarella, 1968, Organic carbon in the Mowry Formation and its relation to the occurrence of petroleum in Lower Cretaceous reservoir rocks: Wyoming Geological Association 20th Annual Field Conference Guidebook, p. 35-39.

Sheldon, R.P., 1967, Long-distance migration of oil in Wyoming: The Mountain Geologist, v. 4, n. 2, p. 53-65.

Strickland, J.W., 1958, Habitat of oil in the Powder River Basin: Wyoming Geological Association Guidebook, Powder River Basin, p. 132-147.

Wenger, W.J., and J.S. Ball, 1961, Characteristics of petroleum from the Powder River Basin, Wyoming: U.S. Bureau of Mines, Report of Investigations 5723.

——, and B.W. Reid, 1958, Characteristics of petroleum in the Powder River Basin: Wyoming Geological Association Guidebook, Powder River Basin, p. 148-156.

Reservoir and Source Bed History in the Great Valley, California[1]

D.L. Zieglar
Chevron, U.S.A.
San Francisco, California

J.H. Spotts
Chevron Resources
San Francisco, California

The application of geochemical concepts and relationships of reservoir porosity-permeability-depth help focus exploratory efforts on the favorable parts of geologic trends in partially explored basins. Porosity data from 165 producing reservoirs ranging in age from Late Cretaceous to Pleistocene show that the "best reservoirs" lose porosity at a rate of approximately 1.52 percent per 1,000 ft (0.46 percent per 100 m) of burial. Reservoirs on the large amplitude folds on the west side of the San Joaquin Valley have a more rapid porosity loss with depth. A cross plot of porosity-permeability indicates a "best reservoir" relationship of a tenfold decrease in permeability for each decrease of seven porosity units.

Within the Great Valley, four major depocenters are definable by use of isopach data. Each has had a different source bed history. Continental margin sedimentary rocks of Late Cretaceous age contain organic material that generally is structured and is believed to be the source of gas in the Sacramento Valley. Although a Tertiary depocenter exists in the Delta area, subsidence has failed to place Paleocene and Eocene source beds into the thermal zone thought to be required for oil and gas generation. Gas trapped in Paleocene and Eocene, therefore, must have migrated from more deeply buried Cretaceous source beds.

Tertiary beds in the Buttonwillow and Tejon depocenters in the southern San Joaquin Valley contain large amounts of sapropelic organic material which is believed to be the source of the oil and gas found there. Source beds in the Buttonwillow depocenter have been in the thermal zone for generation for only about 5 million years. In marked contrast, source beds in the Tejon depocenter started subsiding into the thermal zone more than 15 million years ago.

Explorationists who recognize the "best reservoirs" and relate them to source, migration, and trap parameters in undrilled areas will be successful in finding future reserves of oil and gas and might avoid some unprofitable "geologic successes" that are economic failures.

INTRODUCTION

During the past ten years, the literature has grown rapidly concerning the geochemistry of rock units believed capable of being source beds for oil and gas, the effects of temperature and time on the generation of oil and gas from the different types of organic material, and methods and criteria for recognizing paleotemperature effects. Application of these geochemical concepts, together with reservoir relations (porosity-permeability-depth), will help focus exploratory efforts on the more favorable parts of various geologic trends. Four critical parameters are involved in any hydrocarbon accumulation—reservoir, trap, source, and migration (Jones, 1975). All four items must have been effective in order for oil or gas to accumulate. In this paper, we describe and discuss reservoir rock and source bed history of the Great Valley with two objectives in mind: (1) to present some basic data on reservoir quality and some constraints which may be useful in guiding future exploratory efforts; and (2) to test some of the organic geochemical concepts in the "real world" of oil and gas accumulations in the Great Valley. Published reports provided most of the basic data presented. Our company files provide cross checks that show the data to be reasonable.

POROSITY VERSUS DEPTH IN SANDSTONE RESERVOIRS

Prediction of reservoir parameters is an integral part of petroleum exploration geology. Such estimates are

---

[1]This paper was awarded the A.I. Levorsen best paper award at the Pacific Section AAPG 51st Annual Meeting in San Francisco in April, 1976, and was also presented at the 1977 National AAPG Convention in Washington, D.C.

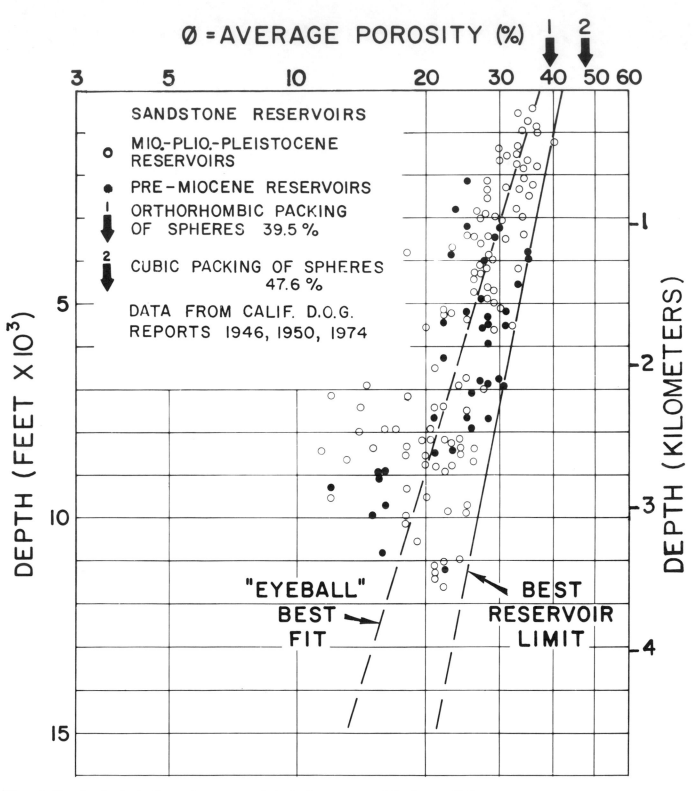

Figure 1. Porosity-depth plot for sandstone reservoirs in the Sacramento and San Joaquin Valley.

especially critical in evaluating prospects and potential objectives where reserves or producibility are critical for economic success. As porosity and permeability information become available during exploration of any basin, rates of porosity loss with depth may be established. Such determinations of rate of loss may be strictly statistical using regression lines, standard deviations, and so forth. The same data also may be interpreted from a geologic point of view with the aim of establishing geologic constraints and limitations. For example, Figure 1 is a plot of average porosity versus depth for 165 producing sandstone reservoirs ranging in age from Late Cretaceous to

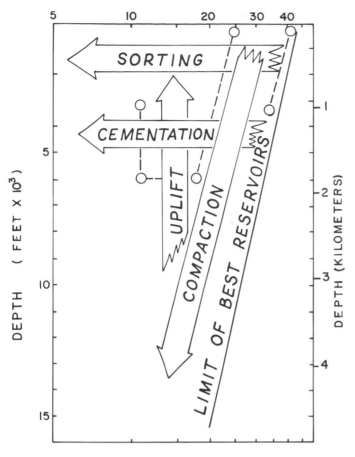

AVERAGE POROSITY (%)

Figure 2. Porosity-depth relationships of sandstone reservoirs indicating geological processes responsible for lower than maximum possible porosity.

Pleistocene from 87 oil and gas fields in the Great Valley (California Division Oil and Gas, 1946, 1950 to 1974). From this set of data, it is apparent that the initial porosity of the rocks is 35 to 40 percent. This is compatible with the theoretical values of 39.5 percent for orthorhombic packing of spheres and 47.6 percent for cubic packing of spheres. It is also apparent that porosity is virtually independent of geologic age. Values from lower Tertiary and Upper Cretaceous reservoirs are comparable to values at the same depth of burial from upper Tertiary reservoirs. This comparison indicates a continuity or a similarity of geologic conditions during the two time periods involved. An "eyeball" best fit of the porosity-depth data indicates an initial porosity of 37 percent with porosity decreasing 1.82 percent per 1,000 ft (0.56 percent per 100 m) of burial. Such a trend line may be useful for basin-wide evaluation where departures above and below the trend line will tend to compensate each other. However, for local areas or prospects, the trend line may cause an overestimation of porosity for areas with "tight" rocks and underestimation of the porosity for areas with porous rocks. A much more significant reservoir parameter can be determined by using the plot to answer the typical question of an optimistic explorationist, "What is the best porosity to be expected at

any given depth?" This limit is apparent from the data, and it indicates an initial porosity of 42 percent with porosity loss with depth of approximately 1.52 percent per 1,000 ft (0.46% per 100 m). In Gulf Coast sand reservoirs, Atwater and Miller (1965) determined a decrease of 1.26 percent per 1,000 ft (0.38 percent per 100 m). The more rapid loss in California reservoirs is probably a function of the poorer sorting of the arkosic sandstones together with generally higher geothermal gradients leading to increased diagenesis.

If one agrees that the limit line is geologically meaningful and indicates the effect of compaction of the "best reservoirs," then a corollary question is "What causes the scattered lower porosities at any given depth?" Figure 2 shows schematically some of the possible geologic processes which can influence the porosity of sandstone reservoirs. When deposited, the units will have a range of values depending on the sorting and shapes of the gains. As they are buried, they lose porosity through compaction, but at any specified depth, cementation may further reduce the porosity below that expected if only gravitational compaction had been the controlling factor. Also, if a sandstone unit is buried to any depth and then uplifted, the porosity will be anomalously low for its present depth of burial. All four of these factors—initial porosity range, compaction, mineralogical cementation, and post-burial uplift—contribute to the scatter of porosity values on Figure 1.

The maximum possible effects of increased temperature on loss of porosity, which Maxwell (1964) believed were significant, have been discussed by Stephenson (1977). Without detailed petrographic studies, such as presented by Hsu (1977) in his study of Ventura Field in California, evaluation of the impacts of the first three factors listed previously is not feasible. However, it is possible to test the significance of postburial uplift in the Great Valley by separating the data into two groups representing (1) fields presently at or near maximum depth of burial, and (2) fields with evidence of significant uplift. Porosity data for 54 reservoirs in 19 fields on the west side of the San Joaquin Valley where some partly eroded folds have amplitudes of 3,000 ft or more than 10,000 ft (about 1 km to more than 3 km) are shown in Figure 3. Additional porosity data from Elk Hills (Maher et al, 1975) are shown to supplement the data from the west side structures. The data suggest a porosity loss of approximately 2.4 percent per 1,000 ft (0.73 percent per 100 m) from an initial porosity of 38 percent. From the trend of the values it appears doubtful that average porosities greater than 15 percent will be present below 12,000 ft (3,670 m).

Figure 4 shows values from fields on the east side of the San Joaquin Valley and from the Sacramento Valley where deformation has been relatively minor and where reservoirs are considered to be at or near maximum depth of burial. Figure 4 also shows that the porosity is independent of the age of the sediment. Lower Tertiary and Upper Cretaceous values are in the same range as the upper Tertiary values. The "eyeball" best fit suggests a porosity loss of about 1.38 percent per 1,000 ft (0.42 percent per 100 m) from an initial values of 37 percent. From these it is clear that average porosities of 15 percent should be expected in the best reservoirs to depths of 20,000 ft (6,100 m) on the east side,

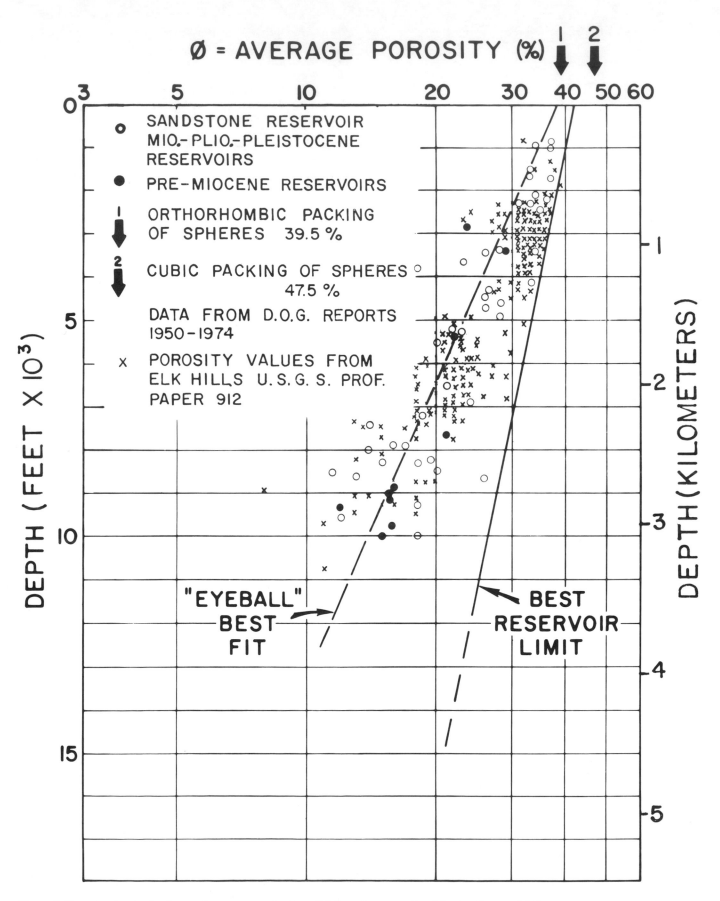

Figure 3. Porosity-depth plot for sandstone reservoirs from fields on the west side of the San Joaquin Valley.

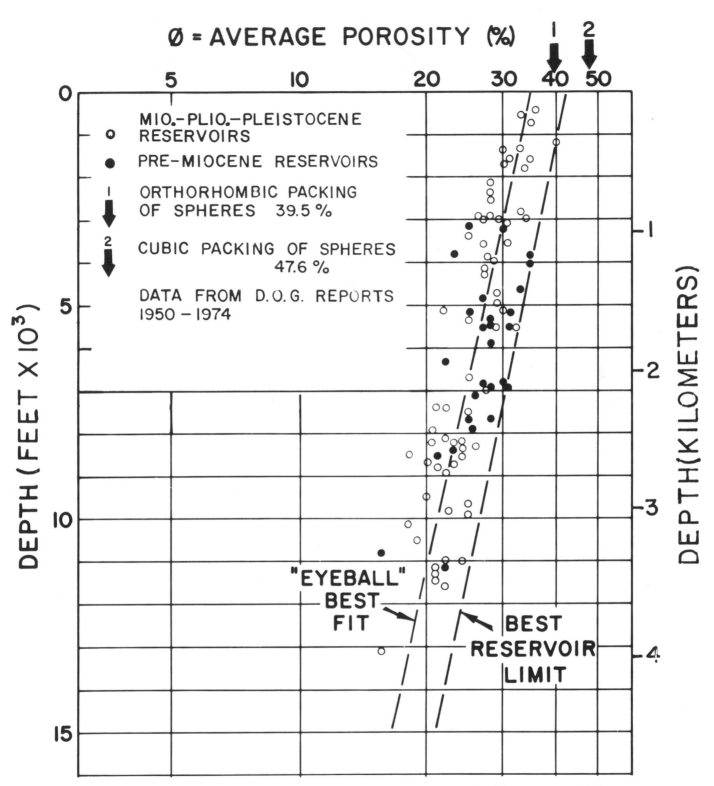

Figure 4. Porosity-depth plot for sandstone reservoirs from the Sacramento Valley and east side of the San Joaquin Valley.

depths significantly greater than on the west side. The more abrupt loss of porosity with depth for reservoirs of the west-side structures is mainly indicative of postburial uplift of 3,000 to 10,000 ft (915 to 3,050 m). The effects of higher geothermal gradients and the effects of deformation during uplift are believed to be only slightly contributory to the more rapid porosity loss.

POROSITY VERSUS PERMEABILITY

Figure 5 is a plot of porosity versus permeability based on several thousand analyses from the Elk Hills field in the San

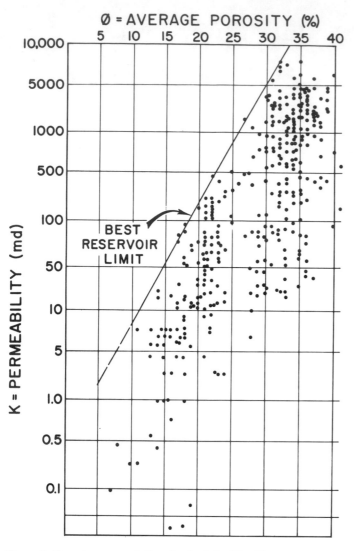

Figure 5. Porosity-permeability plot for Mio-Plio-Pleistocene sandstone reservoirs, Elk Hills Field, Kern County, California.

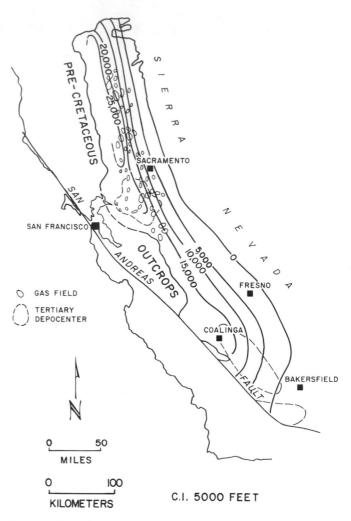

Figure 6. Isopach map of Cretaceous sediments in the Great Valley, California.

Joaquin Valley (Maher et al, 1975). Each datum point shown may represent up to several hundred sample determinations. The best permeability expectable for a given porosity is readily apparent. The indicated limit line suggests that permeability changes by a factor of 10 for each 7 percent change in porosity (permeability from 1 to 10 md, porosity from 4 to 11 percent; permeability from 10 to 100 md, porosity from 11 to 18 percent and so forth). A "best reservoir" averaging 15 percent porosity at a depth of 12,000 ft (3,660 m) on a west-side structure or at a depth of 20,000 ft (6,100 m) on the east-side San Joaquin or in the Sacramento Valley should have permeabilities in the 40 to 50 md range. Such information is basic to estimating fluid production rates using Darcy's equation in which rate of deliverability is proportional to permeability and inversely proportional to viscosity. A reservoir of low permeability containing a low-viscosity fluid may produce at an economic rate, whereas the same reservoir containing a high viscosity fluid may be uneconomic. Prediction of those areas with favorable permeability-viscosity relationships should be possible in partly explored basins where suitable

empirical observations can be made.

These observations concerning reservoir parameters should be useful in evaluating the economics of future oil and gas exploration. Geologic studies of the interrelationships of the four factors influencing porosity versus depth relationships are certainly worthwhile; however, it is surely obvious that if the best predictable conditions won't support acceptable economics, then exploratory investment in the project can hardly be justified.

SOURCE ROCKS AND HYDROCARBON GENERATION

Geochemical studies of sedimentary organic materials suggest that significant hydrocarbon generation in source beds may start with rock temperatures as low as 150°F (66°C) and that at temperatures above 350 to 400°F (175 to 204°C) most liquid hydrocarbons are destroyed. Approximate maximum paleotemperatures can be estimated from the degree of carbonization of palynological organic materials (Staplin, 1969), from vitrinite reflectance and by other methods. The presence of unreworked, black spore and pollen grains indicates formation paleotemperatures in

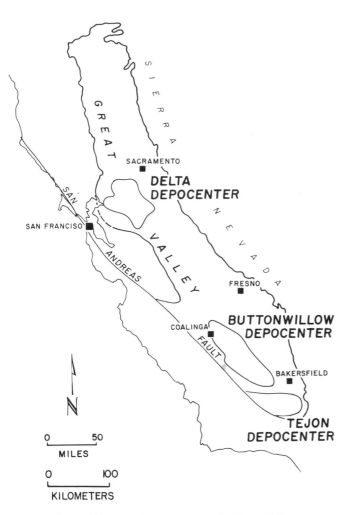

Figure 7. Map of Tertiary depocenters in the Great Valley, California.

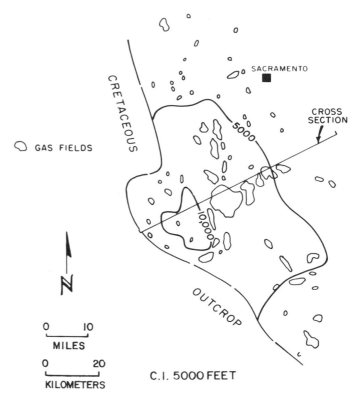

Figure 8. Isopach map of total post-Cretaceous section (Paleocene-Recent) in the Delta depocenter, Sacramento Valley, California.

the 400°F (204°C) range. Such destructive thermal effects have not been noted in the Tertiary rocks of the Great Valley and there is only limited evidence of such elevated temperatures in the Cretaceous beds. From the sparse thermal data available, and assuming a base temperatures of 70°F (21°C), the thermal gradient in the Great Valley ranges from a low of 1.2°F per 100 ft (2.2°C per 100 m) to a high of 2°F per 100 ft (3.6°C per 100 m) with the higher gradient occurring on the west side of the southern San Joaquin Valley. With these gradients, the 400°F (204°C) isotherm will range between depths of 25,000 ft (7,600 m) and 16,500 ft (5,030 m).

Chemical kinetics show that hydrocarbon generation is time and temperature sensitive; older source beds exposed to lower temperatures for longer periods of time may be as effective in generating oil or gas as younger beds subjected to higher temperatures for short periods of time. Without adequate information on the present and past temperature distribution, detailed evaluations or analyses of these relationships are hardly feasible. For general discussion, a depth threshold of 10,000 ft (3,050 m) will be taken as an indicator that the source rocks have attained a critical temperature regardless of the age of the rocks; and those rock units which are now buried or in the past have been

buried deeper than 10,000 ft (3,050 m), will be considered as mature source beds. Beds with favorable organic parameters in less deeply buried intervals will be considered immature. Hydrocarbons present in reservoirs associated with such immature source beds must have migrated into the reservoirs from more mature source beds at greater depths.

GEOLOGIC HISTORY OF SOURCE ROCKS— SACRAMENTO VALLEY

Studies of the geologic history of the Great Valley (Hackel, 1966, and others) show that, during most of the Cretaceous, sediment accumulation was related to a linear continental margin, an "arch-trench gap" in plate tectonic parlance. Thicknesses in excess of 25,000 ft (7,600 m, Figure 6), have been measured along the west side of the Sacramento Valley and thicknesses greater than 15,000 ft (4,575 m) are present in the Coalinga area of the San Joaquin Valley. Therefore, the lower 5,000 to 15,000 ft (about 1,500 to 4,500 m) of this Cretaceous section, a volume of approximately 12,000 cu mi (50,000 cu km) must have passed into or through the temperature regime critical for hydrocarbon generation by the end of Cretaceous time. Our examinations of organic material from Cretaceous beds show that they generally contain less than 1 percent by weight of organic material, that the thermal alteration index (TAI) is favorable, and that most of the organic material is structured (humic or nonsapropelic) and would most likely generate gas. Hydrocarbon production in the Sacramento Valley is almost entirely gas and is therefore compatible

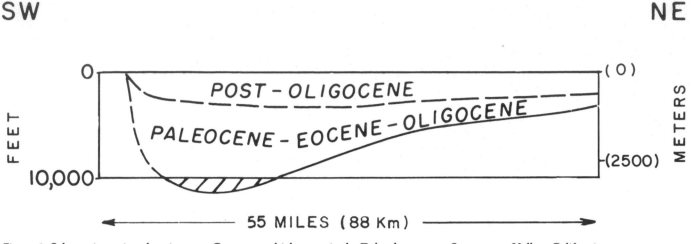

Figure 9. Schematic section showing post-Cretaceous thicknesses in the Delta depocenter, Sacramento Valley, California.

with the geochemical interpretation of source-bed constraints. Assuming that our sampling has been adequate, future discoveries in the Cretaceous objectives of the Sacramento Valley probably will be gas rather than oil.

During latest Cretaceous and early Tertiary time, segments of the Coast Ranges were uplifted along the west side of the Great Valley forming a more or less continuous peninsula or archipelago from the north end of the valley southward to the north part of the Temblor Range. Evidence for these uplifts is recorded in the overlap of upper Eocene and younger units onto Cretaceous and older beds. Within the Great Valley, three significant subbasins formed and became the sites of thick deposition during the Tertiary. These, from north to south, are the Delta depocenter, the Buttonwillow depocenter and the Tejon depocenter (Figure 7). Each of these depocenters has had a different source-bed history as illustrated by gross-isopach differences between the depocenters (Figure 7).

In the Delta depocenter, the lower Tertiary beds reach a maximum thickness of about 8,500 ft (2,600 m). Thus, by the end of Eocene time, potential source-beds in this interval would have been ineffective and immature hydrocarbon sources because they had not attained sufficiently high temperatures. With the added deposition of the Miocene and younger sediments, the maximum gross-thickness in the depocenter (Figures 8 and 9) is slightly more than 11,000 ft (3,350 m). This places a volume of about 20 cu mi (83 cu km) of the Paleocene beds in the thermal zone within which organic material may release oil or gas.

Lopatin Calculations on Source Rocks

Figure 10 displays the stratigraphic thicknesses and zones of thermal oil and gas generation through geologic time. The calculations for oil and gas generation zones are based on the methods developed by Lopatin (1971), which take into account the approximate cumulative effects of both time and temperature on the alteration of organic matter to form hydrocarbons. Involved in the method is the chemical rule of thumb that the reaction rate for a chemical process doubles as the temperature is raised 18°F (10°C). Lopatin has attempted to calibrate such calculations using various other methods of organic alteration indices such as vitrinite

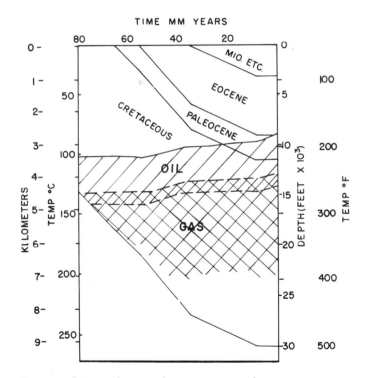

Figure 10. Lopatin diagram showing stratigraphic reconstructions and oil and gas generation windows for the thickest portion of the Delta depocenter, Sacramento Valley, California.

reflectance and coalification stages. Even though Lopatin's scales have been modified several times, we have used his original thresholds on the assumption that the refinements introduce only minor differences in our regional considerations.

At least three major simplifying assumptions underlie our reconstructions. First, it has been necessary to assume a constant temperature gradient with depth (we have used 1.5°F per 100 ft; 2.72°C per 100 m); second, we have assumed that the rate of sedimentation has been fairly uniform within each of the age units shown; and third, we have assumed that the temperature gradient has remained constant through time. As indicated in Figure 10, oil generation in

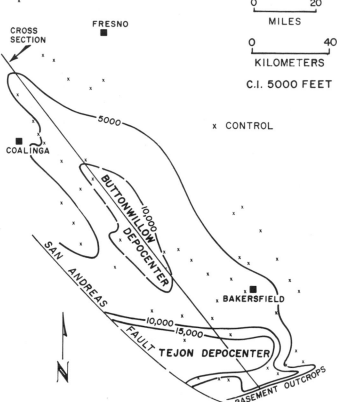

Figure 11. Isopach map from base Pliocene to top Cretaceous or basement in the Buttonwillow and Tejon depocenters, San Joaquin Valley, California.

Cretaceous source beds would not have begun until about 10,000 ft (3,050 m) of sediments had accumulated, or about 90 million years ago. Because of the type of organic material present, oil generation in the Delta depocenter and Sacramento Valley was apparently negligible, so we probably can ignore this oil-generation zone. Important generation of

thermogenic gas would have started between 70 and 80 million years ago and probably continues to the present time where the source bed intervals are buried at least 13,000 to 15,000 ft (about 4,00 to 4,600 m). Hydrocarbon generation in the Delta depocenter as well as in the more northerly parts of the Sacramento Valley evidently has been largely restricted to Cretaceous strata, with perhaps the possibility that some beds in the lower part of the Paleocene may have been in the generating zone for the past few million years. The conclusion that Cretaceous source beds have been the primary source for most of the released gas means that the accumulations trapped in shallower reservoir beds of late Paleocene and Eocene age in the Delta depocenter probably have resulted from migration of the gas from the older and more deeply buried source beds. How this migration took place is beyond the scope of this paper.

GEOLOGIC HISTORY OF SOURCE ROCKS— SAN JOAQUIN VALLEY

Similar analyses of the Tertiary depositional history in the Buttonwillow and Tejon depocenters clearly highlight some basic differences in the geologic history of the depocenters with resultant differences in the source bed histories. Examinations of the organic materials contained in the Tertiary deposits in the San Joaquin Valley indicate that none contain blackened spores and pollen. The absence of blackened spores and pollen is interpreted to indicate that temperatures high enough to have destroyed oil or gas accumulations have not affected the Tertiary strata or have been effective for too short a time. Upper Eocene, Miocene, and Pliocene shaly beds generally contain more than 1 percent by weight of organic material with some of the richer zones containing more than 5 percent. The organic material is largely sapropelic and is interpreted to be oil prone. With 100 percent conversion, each weight percent of organic material would yield approximately 1.3 billion barrels of liquids per 1 cu mi (4.2 cu km) of sediments. It is, obviously, not possible to equate percent organic material in source

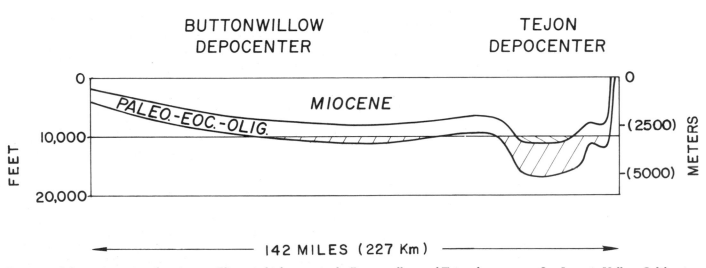

Figure 12. Schematic section showing pre-Pliocene thicknesses in the Buttonwillow and Tejon depocenters, San Joaquin Valley, California.

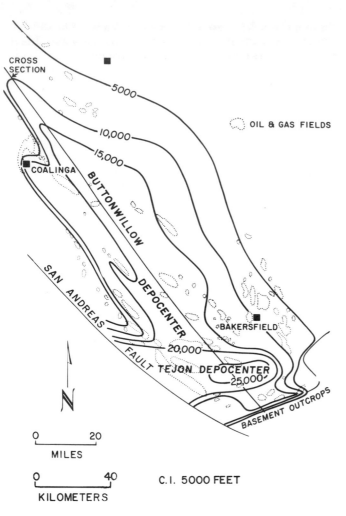

Figure 13. Isopach map of post-Cretaceous units in the Buttonwillow and Tejon depocenters, San Joaquin Valley, California.

rocks with barrels of oil contained in traps because the efficiency of conversion of organic material is unknown, as is the factor of migration from the source beds and also the amount of hydrocarbons which have escaped altogether through geologic time.

Available isopach data for the pre-Miocene lower Tertiary strata show that the Buttonwillow and Tejon depocenters were separate entities. In the Buttonwillow depocenter, the lower Tertiary is up to 3,000 ft (915 m) thick and in the Tejon depocenter it is slightly more than 5,000 ft (1,525 m). Potential source beds in both depocenters would have been immature at the beginning of Miocene time. With the addition of Miocene sediments (Figures 11 and 12) the maximum thickness of Tertiary strata in the Buttonwillow depocenter increased to approximately 10,000 ft (3,050 m) while in the Tejon depocenter, thicknesses of nearly 20,000 ft (6,100 m) occur. Dibblee (1973) reports 19,000 ft (5,790 m) of Miocene alone in the southern Temblor Range which has been uplifted across the western margin of the Tejon depocenter. In the Buttonwillow depocenter, as shown in Figure 12, only the lower portion of the lower Tertiary, a volume of about 100 cu mi (420 cu km), should have subsided into the thermal zone within which source beds would have been generating significant volumes of oil and gas. In marked contrast, in the Tejon depocenter the entire lower Tertiary plus the lower part of the Miocene, a volume of about 900 cu mi (3,750 cu km), appears to have been deeper than the 10,000 ft (3,050 m) threshold and should have been generating hydrocarbons.

With continued basin filling through the Pliocene, Pleistocene, and up to the present, the Buttonwillow depocenter received an additional 6,000 ft (1,830 m) of sediments, resulting in a cumulative thickness of about 17,000 ft (5,180 m). The Tejon depocenter received more than 10,000 additional ft (3,050 m) with a resultant cumulative thickness over 25,000 ft (7,620 m; Figures 13 and 14). In the Buttonwillow depocenter, only the gross volume of about 2,700

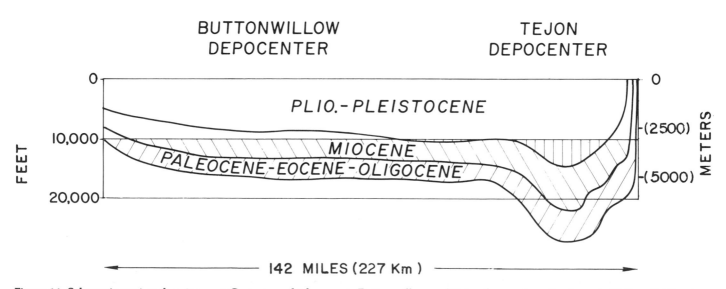

Figure 14. Schematic section showing post-Cretaceous thicknesses in Buttonwillow and Tejon depocenters, San Joaquin Valley, California. For location see Figure 13.

cu mi (11,260 cu km) of lower Tertiary and Miocene beds has been buried deeply enough to experience temperatures sufficienty high to cause hydrocarbon generation during the last 5 million years. In the Tejon depocenter, the lower part of the Pliocene and all the Miocene and older beds have subsided below 10,000 ft (3,050 m) and this gross volume of about 1,800 cu mi (7,500 cu km) may be considered to have entered or passed through the temperature zone for hydrocarbon generation.

Lopatin Calculations on Source Rocks

Figures 15 and 16 show the zones that generate oil and gas in accordance with Lopatin's (1971) methodology. For the Buttonwillow depocenter, oil generation would have begun in Cretaceous beds about 75 million years ago when about 11,000 ft (3,350 m) of sediment load existed in the depocenter. Significant thermogenic gas generation would have started about 55 million years ago when 13,000 ft (3,960 m) of sediments were present. Generation in these older source beds would have continued to the present time. In the Tertiary beds, however, the diagram suggests that hydrocarbon generation has been limited to the last few, perhaps 5, million years. In this brief time, hydrocarbons were generated and were able to migrate into the 34 known accumulations which flank the depocenter. These fields have produced in excess of 2.5 billion barrels of oil.

For the Tejon depocenter (Figure 16), the relationships are considerably different. Eocene-Oligocene beds reached generation temperature 15 million years ago and Miocene beds reached it 5 to 6 million years ago. The three Lopatin reconstructions suggest that Tertiary beds in the Tejon depocenter have been exposed to temperatures appropriate for hydrocarbon generation longer than those in either of

the other two depocenters in the Great Valley. The 41 producing areas flanking the Tejon depocenter have already produced more than 4.3 billion barrels of oil.

SUMMARY

This discussion has focused on the reservoir and source parameters of the critical quartet—reservoir, trap, source, and migration. From the data considered, we conclude that oil and gas in the Great Valley have been generated in four principal, large depocenters. Our interpretations of the data suggest that accumulations in young reservoirs, particularly Pliocene, were generated in older and more deeply buried beds and migrated into these reservoirs by mechanisms involving considerable cross-formational flow. Variations in source rock volumes and lengths of time for generation from source rocks of different ages has been demonstrated. These relations clearly focus attention on the need for explorationists to consider possible times and paths of migration from the depocenters in which oil and gas have been generated. Geologic history and paleostructural studies need to be designed so that they will reflect trapping conditions at the time, or times of migration. As shown by the gross contrasts in timing for generation in the different major depocenters of the Great Valley, generalizations for the entire valley are invalid and separate reconstructions are necessary for at least each depocenter.

The upper limits for average porosity and permeability, as a function of depth, for Great Valley sandstones are principally a function of maximum depths of burial and of postburial tectonism or uplift. In conjunction with fluid mechanics correlations and predictions, these limits aid in setting most favorable limits for exploration ventures in deeper zones and new trends in the valley. If the best prospects will not yield favorable economic results, then the wisdom of proceeding should be questioned. The diverse

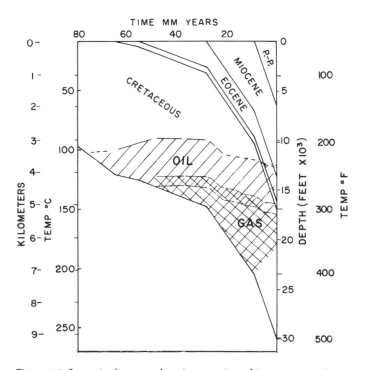

Figure 15. Lopatin diagram showing stratigraphic reconstructions and oil and gas generation windows for the thickest portion of the Buttonwillow depocenter, San Joaquin Valley, California.

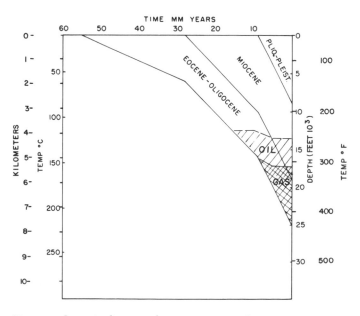

Figure 16. Lopatin diagram showing stratigraphic reconstructions and oil and gas generation windows for the thickest portion of the Tejon depocenter, San Joaquin Valley, California.

causes of porosity loss with depth deserve consideration and study. Those who can recognize rock types in the Great Valley which constitute the "best reservoirs" and can relate those to source, migration, and trap relations in undrilled areas will be successful in finding new oil and gas accumulations.

ACKNOWLEDGMENTS

The authors appreciate the input of many colleagues and the support of Chevron management in the preparation and presentation of this paper. Published with permission of Chevron U.S.A. Inc.

REFERENCES CITED

Atwater, G.I., and E.E. Miller, 1965, The effect of decrease in porosity with depth in future developments of oil and gas reserves in South Louisiana (abst.): AAPG Bulletin, v. 49, p. 334.

California Division of Oil and Gas, 1946, Estimate of the natural gas reserves of the state of California as of January 1, 1946. Case No. 4591, Special Study No. S-525.

—— , 1950 to 1974, Summary of operations: Technical Papers, v. 36 to v. 60.

Callaway, D.C., 1971, Petroleum potential of San Joaquin Basin, California: AAPG Memoir 15, p. 239-253.

Connan, J., 1974, Time-temperature relation in oil genesis: AAPG Bulletin, v. 58, p. 2516-2525.

Dibblee, T.W., Jr., 1973, Stratigraphy of the southern coast ranges near the San Andreas Fault from Cholame to Maricopa, California: U.S. Geological Survey Professional Paper 764, p. 45.

Hackel, O., 1966, Summary of the geology of the Great Valley: California Division of Mines and Geology Bulletin, v. 190, p. 217-238.

Hsü, K.J., 1977, Studies of Ventura Field II: Lithology compaction and permeability of sands: AAPG Bulletin, v. 61, p. 169-191.

Jones, R.W., 1975, Quantitative geologic approach to reserve prediction (abst.): AAPG Annual Meeting Abstracts, v. 2, p. 40.

Lopatin, N.V., 1971, Temperatura i geologicheskoe vremya kan faktory uglefikatsii: Akademiya Nauk SSSR Izvestiya, Seriya Geologicheskaia, no. 3, p. 95-106.

Maher, J.C., R.D. Carter, and R.J. Lantz, 1975, Petroleum geology of naval petroleum reserve no. 1, Elk Hills, Kern County, California: U.S. Geological Survey Professional Paper 912, p. 104.

Maxwell, J.C., 1964, Influence of depth, temperature and geologic age on porosity of quartzose sandstone: AAPG Bulletin, v. 48, p. 497-709.

Morrison, R.R., et al, 1971, Potential of Sacramento valley gas province, California: AAPG Memoir 15, p. 329-338.

Staplin, F.L., 1969, Sedimentary organic metamorphism and oil and gas occurrences: Bulletin of Canadian Petroleum Geology, v. 17, no. 1, p. 47-66.

Stephenson, L.P. 1977, Porosity dependence on temperature: Limits on maximum possible effects: AAPG Bulletin, v. 61, p. 407-415.

Petroleum Source Rocks, Grand Banks Area

J.H. Swift
Amoco Canada Petroleum Company Ltd.
Calgary, Alberta

J.A. Williams
Amoco Production Company
Tulsa, Oklahoma

Thirty-nine exploratory wells drilled on the Grand Banks of Newfoundland failed to encounter commercial petroleum accumulations. The absence of major discoveries is attributed to problems of source rock quality and maturity throughout the sedimentary sequence.

The area contains potential Paleozoic, Mesozoic, and Tertiary reservoir rocks in both sandstone and carbonate facies within a marine and continental stratigraphic sequence. These formations were tested on a variety of structural and stratigraphic configurations that had the capability of holding major hydrocarbon reserves. Several oil and gas shows were reported.

The Avalon Uplift, an Early Cretaceous tectonic feature, had a profound effect on the central Grand Banks area. Potential source rock intervals were deeply eroded and reservoirs that may have received hydrocarbon accumulations during the Jurassic were breached.

Source rock quality and maturity trends show a reasonable correspondence to isopach and facies trends. Source rock characteristics in undrilled areas were predicted from extrapolations of contours. The deeper parts of the South Whale and Jeanne d'Arc Basins may contain oil- and gas-prone source rocks with the quality and maturity necessary to have generated commercial accumulations of petroleum.

INTRODUCTION

The drilling of offshore petroleum exploratory wells in eastern Canada began in 1966 with Amoco-Imperial Tors Cove well on the Grand Banks. The piercement salt dome that was drilled yielded a small recovery of gas from the cap rock and so provided early proof of hydrocarbon occurrence in the area. In the years 1966 through 1975 a total of 39 deep tests were drilled on the Grand Banks. Although additional shows were recorded, no commercial discoveries resulted. The lack of success cannot be attributed to a lack of potential reservoir rocks. Porous units, both sandstones and carbonates, occur within a sedimentary sequence that contains facies variations from marine through continental. Tight units are present to form seals. Nor is there a lack of significant structural closures. Structures tested included salt domes, salt ridges, basement-involved block fault structures and unconformity bounded wedges.

Detailed examination of source rock quality and maturity and estimates of the timing of petroleum generation and expulsion have shown that source rock problems may account for the lack of success. Some of the Grand Banks source rock data have been generalized within the boundaries of major mappable rock units and are presented as contoured maps thus allowing the interpreters to project beyond data control points and to make predictions for areas of little or no well control.

REGIONAL GEOLOGY

The generalized total sediment isopach map, Figure 1, shows two major sedimentary thicks, the Scotian Basin and the East Newfoundland Basin. The extremities of these deep basins extend onto the Grand Banks where they have been called the South Whale and Jeanne d'Arc sub-basins (Amoco and Imperial Oil, 1973). The intervening thin trend extending southeastward from Newfoundland is named the South Bank High and is the result of an early Cretaceous tectonic event known as the Avalon Uplift (Jansa and Wade, 1975a, 1975b). The outlined area is a 14 million hectare (35 million acre) block that has been explored by Amoco, Imperial Oil, Skelly, and Chevron. Geophysical data within this area have been used in the compilation of the maps and cross-sections.

The Grand Banks are underlain by a wedge of Tertiary

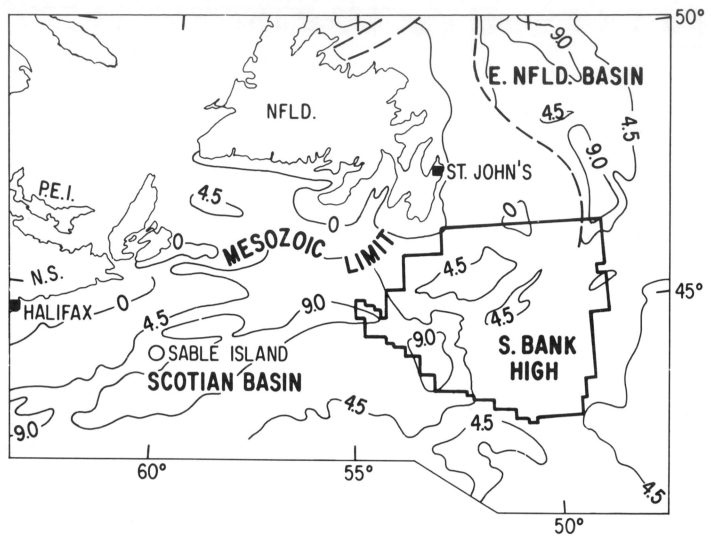

Figure 1. Total sediment isopachs. Contours in kilometers.

and Cretaceous formations separated by an angular unconformity from underlying thick sequences of Jurassic, Triassic, and Paleozoic formations that occur within small structural basins (Upshaw et al, 1974). Figure 2 shows the major stratigraphic units and the correlation to adjacent areas. Figure 3 shows the structural configuration of the base of the Tertiary and Cretaceous wedge. Figure 4 shows the distribution of Jurassic and older formations within structural basins. The Avalon unconformity separates the units displaced by these two maps. Angularity is pronounced along the axis of the Avalon Uplift and diminishes on both flanks, as illustrated by the cross-sections of Figures 5 and 6. Two salt dome structures in the South Whale sub-basin are shown in Figures 7 and 8 as constructed from drilling and seismic data. Angular unconformities are seen at the base of the Cretaceous Logan Canyon Formation as well as the base of the Missisauga Formation.

METHODS FOR EVALUATING SOURCE ROCK PERFORMANCE

The amount of organic carbon in a rock sample is

commonly used as a direct indicator of source rock quality, i.e. its capability for generating petroleum. Under normal conditions, organic carbon values greater than one weight percent are considered to indicate section with good hydrocarbon generating capability. Values of 0.6 to 1.0 represent fair quality source rocks, while lesser values indicate poor or non-source intervals. To generate the large amounts of petroleum required to form commercial accumulations in an offshore area, source rocks with sufficient organic matter must be available also in considerable quantity. It is therefore useful to combine the isopachs and organic content data by contouring values of weight percent organic carbon times meters thickness. On this kind of map, the 300 contour, for example, indicates the equivalent of a 300 m thick section that averages one weight percent organic carbon. Map areas with values 300 and greater are considered to contain potentially significant source rocks, depending on maturity (see below).

A word of caution is in order regarding this type of display in that a high value of organic carbon times thickness is not significant in cases where a very thick section is entirely non-source quality. A non-source interval is considered too deficient in organic matter to generate

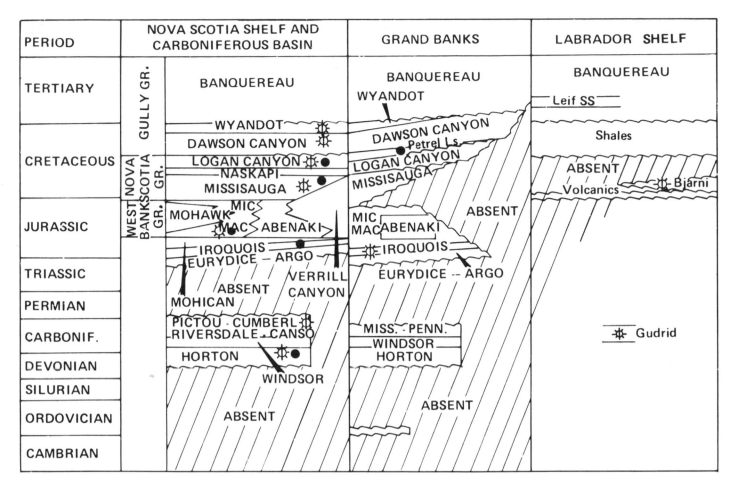

PERIOD	NOVA SCOTIA SHELF AND CARBONIFEROUS BASIN		GRAND BANKS	LABRADOR SHELF
TERTIARY	GULLY GR.	BANQUEREAU	BANQUEREAU WYANDOT	BANQUEREAU Leif SS
CRETACEOUS	WEST NOVA SCOTIA GR. / NOVA SCOTIA GR.	WYANDOT DAWSON CANYON LOGAN CANYON NASKAPI MISSISAUGA	DAWSON CANYON Petrel Ls. LOGAN CANYON MISSISAUGA	Shales ABSENT Volcanics — Bjarni
JURASSIC		MIC MOHAWK MAC ABENAKI IROQUOIS EURYDICE – ARGO	ABSENT MIC MAC ABENAKI IROQUOIS EURYDICE – ARGO	
TRIASSIC		VERRILL ABSENT CANYON MOHICAN		
PERMIAN				
CARBONIF.		PICTOU - CUMBERL. RIVERSDALE - CANSO	MISS. - PENN. WINDSOR	Gudrid
DEVONIAN		HORTON WINDSOR	HORTON	
SILURIAN				
ORDOVICIAN		ABSENT	ABSENT	
CAMBRIAN				

Figure 2. Correlation chart. Formations with significant oil and gas shows are indicated by conventional oil and gas symbols.

enough petroleum for expulsion to occur. Also, this technique does not take into account those portions of the section in which the generating capability of the organic matter may have been adversely affected by some naturally-occurring mild oxidation process.

Maturity

Several methods are available for estimating the maturity of source rocks. Elemental analysis of the organic matter (kerogen) is one method which provides information about maturity. It involves measurement of carbon, hydrogen, oxygen, and nitrogen, the four principal elements in kerogen (Momper, 1971; LaPlante, 1974). Degree of maturity is indicated by weight percent carbon (normalized) in the kerogen. This measurement should not be confused with weight percent organic carbon in the whole sample, which gives petroleum generation capability as described above. Figure 9 shows elemental percent carbon values relative to various generation stages for oil and gas. A minimum values of 82 percent carbon by elemental analysis of a sample is considered necessary for significant oil expulsion to have occurred in the section represented by that sample. Gas may begin to form at slightly lower values, but the most significant stage of gas generation occurs at higher values (85 to 89 percent).

Maps presented in the following section show contours representing weight percent carbon by elemental analysis of

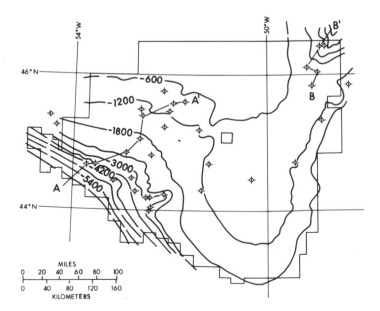

Figure 3. Structure contour map, base of Tertiary-Cretaceous wedge. Contours in meters.

Grand Banks samples. The 82 percent carbon contour is designated as the limit for areas where significant generation could have occurred.

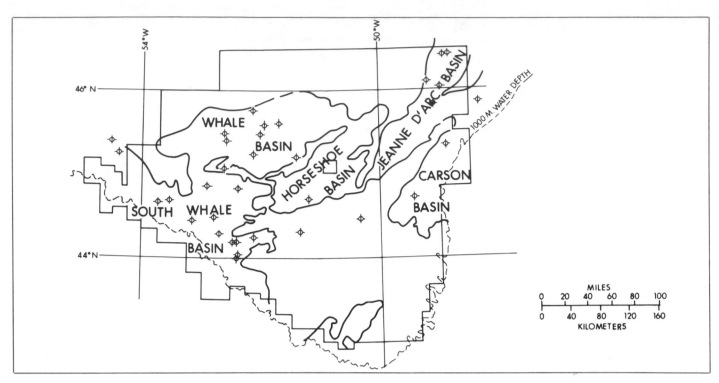

Figure 4. Pre-Cretaceous structural basins.

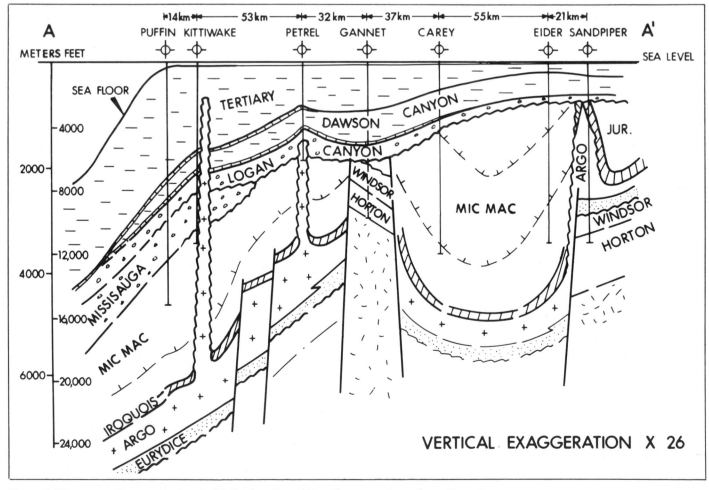

Figure 5. Diagrammatic cross-section A-A′. For location of line of cross-section refer to Figure 3.

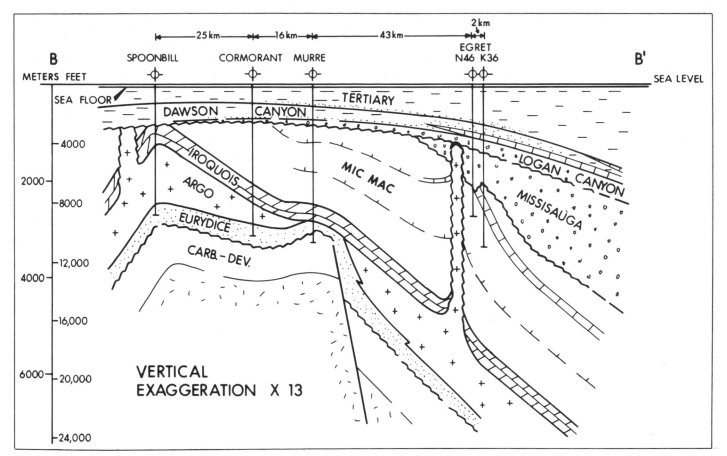

Figure 6. Diagrammatic cross-section B-B'. For location of line of cross-section refer to Figure 3.

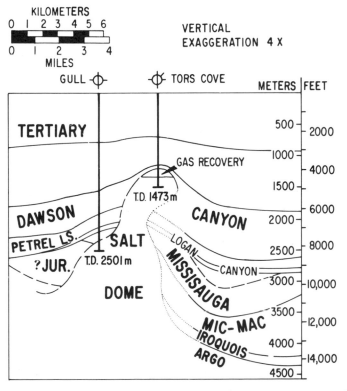

Figure 7. Gull salt dome cross-section. Location shown on Figure 16 (gas show arrow).

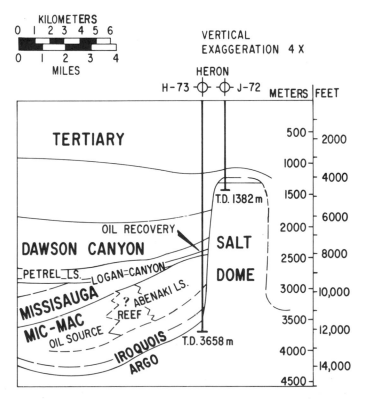

Figure 8. Heron salt dome cross-section. Location shown on Figure 25.

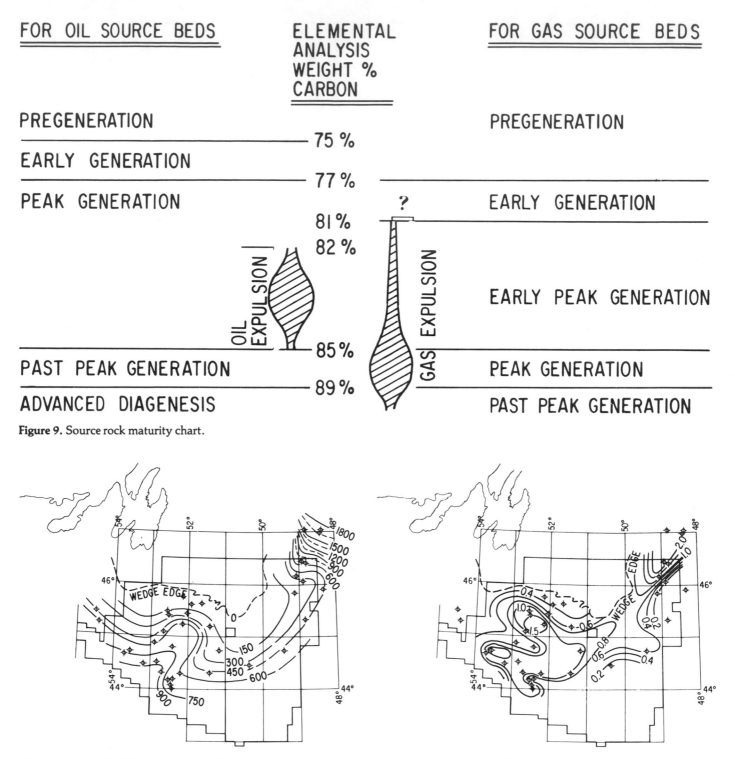

FOR OIL SOURCE BEDS   ELEMENTAL ANALYSIS WEIGHT % CARBON   FOR GAS SOURCE BEDS

PREGENERATION

— 75 %

EARLY GENERATION

— 77 %

PEAK GENERATION

81 %

82 %

OIL EXPULSION

— 85 %

PAST PEAK GENERATION

— 89 %

ADVANCED DIAGENESIS

PREGENERATION

EARLY GENERATION

?

GAS EXPULSION

EARLY PEAK GENERATION

PEAK GENERATION

PAST PEAK GENERATION

Figure 9. Source rock maturity chart.

Figure 10. Isopach of Paleogene part of Banquereau Formation plus Upper Cretaceous Wyandot Formation. Contours in meters.

Figure 11. Paleogene plus Wyandot Formation, organic carbon weight percent in rock cuttings samples.

DISCUSSION OF SOURCE ROCK PERFORMANCE

Lower Tertiary and Uppermost Cretaceous

Isopachs of the Paleogene portion of the Tertiary Banquereau Formation plus the Upper Cretaceous Wyandot Formation are presented as Figure 10; for the same control points, Figure 11 shows weighted average values of organic carbon weight percent. Contours showing the values of

weight percent organic carbon times meters thickness are displayed in Figure 12. This map indicates two areas, in the northeast and southwest, where values greater than 300 are observed. These areas would be considered to contain potentially significant source rocks, provided that adequate maturity levels have been attained. However, this Paleogene-Upper Cretaceous section is immature at all

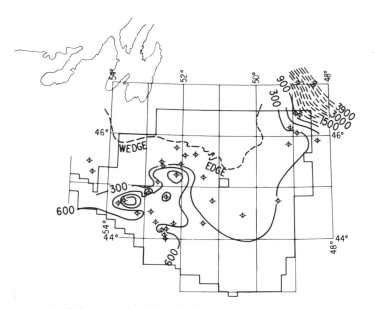

Figure 12. Paleogene plus Wyandot Formation organic carbon weight percent times meters (quality × thickness).

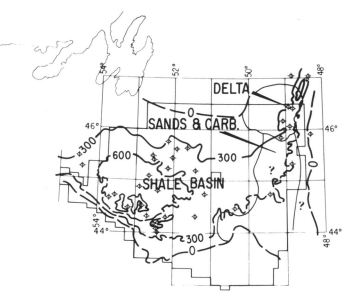

Figure 14. Dawson Canyon Formation isopach and facies map. Contours in meters.

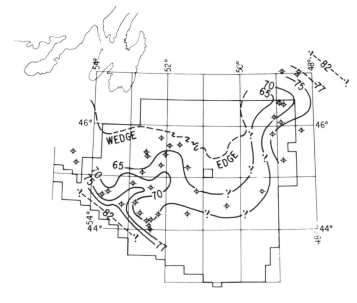

Figure 13. Paleogene plus Wyandot Formation source rock maturity as indicated by weight percent carbon in the kerogen.

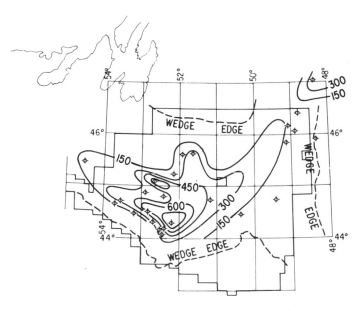

Figure 15. Dawson Canyon Formation organic carbon weight percent times meters (quality × thickness).

control points, as shown on Figure 13. No elemental percent carbon values over 82 percent were observed. Based on the observed increase in carbonization, one may speculate that mature, good quality source rocks are present in this map interval to the southwest and to the northeast of the map area.

A comparison of the source rock maturity map with the isopachs illustrates that the maturity values increase in the areas where the isopachs increase. This situation is to be expected since maturity is partly a function of temperature, which is in turn related to depth of burial.

Dawson Canyon Formation

The isopach map of the Upper Cretaceous Dawson

Canyon Formation (Figure 14) has been compiled from seismic as well as drilling information. The formation is predominantly dark shale with a limestone member, the Petrel limestone, at the base (Swift et al, 1975). Sandstones and carbonates occur along the eastern margin of the Grand Banks and deltaic facies are present in the northeast.

Values of organic carbon weight percent times meters, Figure 15, indicate a combination of source rock quality and formation thickness that results in an area of potentially good source rock in the central basin area. However, source rock maturity as indicated by weight percent carbon in the kerogen, Figure 16, indicates only a small area of possibly effective source within the basin.

The minor gas recovery from the Tors Cove well on the

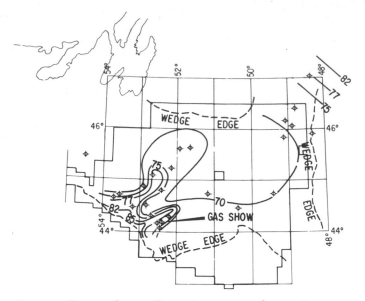

Figure 16. Dawson Canyon Formation source rock maturity as indicated by weight percent carbon in the kerogen.

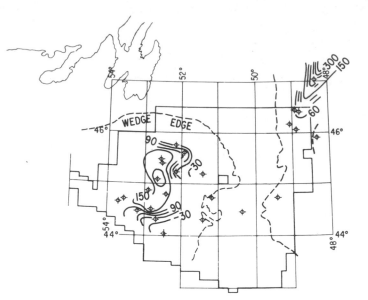

Figure 18. Logan Canyon Formation organic carbon weight percent times meters (quality × thickness).

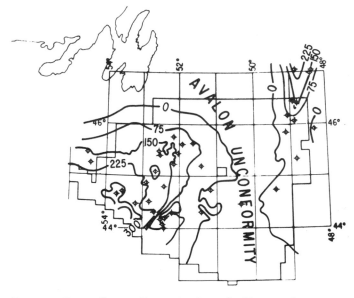

Figure 17. Logan Canyon Formation isopach. Contours in meters.

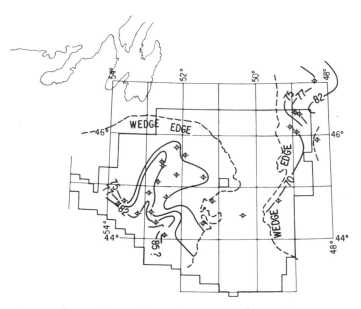

Figure 19. Logan Canyon Formation source rock maturity as indicated by weight percent carbon in the kerogen.

Gull structure (Figure 7) may have been derived from Dawson Canyon source rocks, although deeper sources must be considered as well. The Dawson Canyon shale has reached an early generation stage in the vicinity of this salt dome but not the peak stage required for significant generation from gas source rocks.

The Lower to lower Upper Cretaceous Logan Canyon Formation is a continental to marine wedge of clastic rocks deposited during the transgression over the Avalon unconformity. The isopach map, Figure 17, was constructed from seismic and well data. The Logan Canyon Formation contains good porous sandstone units that should be excellent hydrocarbon reservoirs. The absence of shows in the sands may be related to inadequate hydrocarbon generation from the associated shale as a result of a lack of

organic matter (Figure 18) as well as to a low maturity level (Figure 19). Maturity levels as indicated by the weight percent carbon in the kerogen are sufficient to have generated hydrocarbons in the southwestern Grand Banks but there is insufficient organic matter in the section there.

Missisauga Formation

The Lower Cretaceous Missisauga Formation, Figure 20, was deposited contemporaneously with an adjacent actively rising landmass formed by the Avalon Uplift. The uplift provided various and localized provenance areas. Rock units on the flanks of the uplift include a deltaic sequence, an arkose fanglomerate, volcanic flows and pyroclastics, and marine shoreline facies including clean sandstone and dark shale.

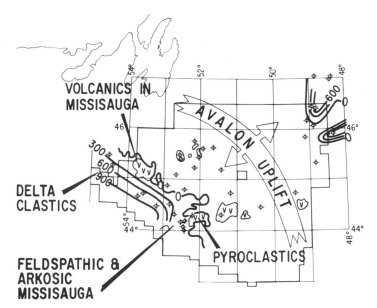

Figure 20. Missisauga Formation isopach and facies map. Contours in meters.

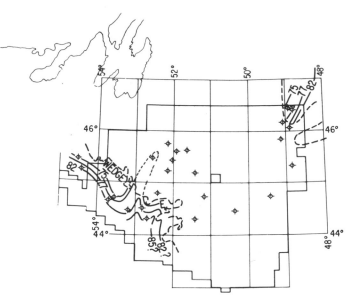

Figure 22. Missisauga Formation source rock maturity as indicated by weight percent carbon in the kerogen.

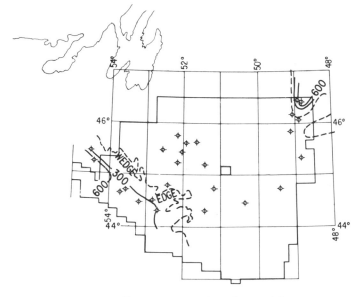

Figure 21. Missisauga Formation organic carbon weight percent times meters (quality × thickness).

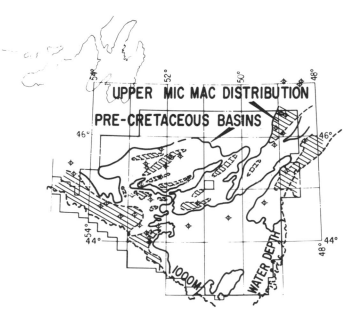

Figure 23. Upper Mic Mac Formation distribution map.

Figure 21 shows that organic matter is present in significant amounts where the formation is most marine. Figure 22 indicates that mature source rocks should be present in the southwestern Grand Banks area.

Upper Mic Mac Formation (Callovian and younger)

Jurassic formations occur in structural basins beneath the Avalon unconformity as shown in Figures 4 through 8. The upper part of the Mic Mac Formation, defined as Callovian and younger Jurassic, is deeply eroded. The remnants, which include marine shales and carbonates, occur primarily in the areas shown in Figure 23.

Figure 24 displays the source rock properties of an interval within the upper Mic Mac Formation identified at the Egret K-36 well. Weight percent organic carbon values

shown on the left indicate source beds of fair and good quality over a 775 m gross interval which includes a 150 m interval with excellent source potential. The graph on the right illustrates an interpretation of the type of hydrocarbons most likely to be generated from the source beds. The indicator used is percent hydrogen in the kerogen, determined by elemental analysis (LaPlante, 1974). Microscopic examination of kerogen in the samples using methods similar to those reported by Staplin (1969) produced a similar interpretation. The maturity map, Figure 25, demonstrates that areas of mature source rocks can be predicted in the northeast and southwest areas. Elsewhere, the small remnants of Callovian and younger Upper Jurassic rocks that remained intact after the deep erosion of the Avalon unconformity were not sufficient to generate

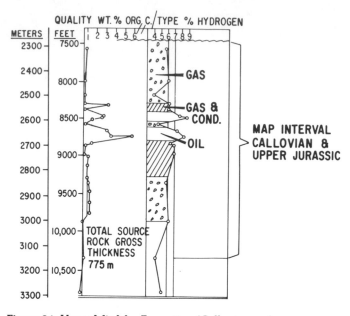

Figure 24. Upper Mic Mac Formation (Callovian and younger Jurassic), Egret K-36 well. Source rock quality by weight percent organic carbon and source rock types by percent hydrogen in the kerogen. For well location see Figure 25.

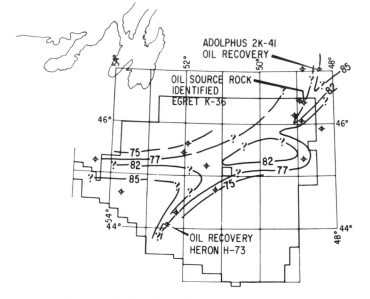

Figure 25. Upper Mic Mac Formation source rock maturity as indicated by weight percent carbon in the kerogen.

significant quantities of petroleum.

Evidence of local effective oil source rocks is provided by the recoveries of oil at two wells, the Mobil-Gulf Adolphus 2K-41 well in the northeast and the Amoco-Imperial Heron well (Swift et al, 1975) in the southwest (Figure 25). Both oil recoveries were from salt dome structures and both were from Cretaceous reservoirs. However, analysis of oil samples from both wells, by techniques such as those described by Williams (1974), have indicated a Jurassic source bed as the most likely origin of these oils, and that source could be the Mic Mac interval identified at the Egret well. Figure 8 shows the Heron salt dome and the

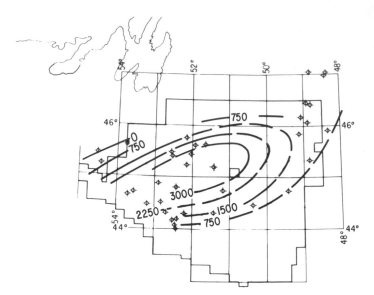

Figure 26. Restored isopach of the lower Mic Mac Formation (pre-Callovian). Contours in meters. Present distribution of Jurassic formations is as remnants within structural basins shown in Figure 4.

relationship of the oil bearing zone in the Petrel limestone to the interpreted Mic Mac source rock. The oil from the Heron well is heavy, with a density of 1,021 to 989 kg/cu m (7° to 11.5° API). It consists of a mixture of bacterially altered oil and a small amount of unaltered oil. Alteration of the oil may have occurred at the end of the Cretaceous, a time when a regional unconformity developed. The caprock of the salt dome may have been sub-aerially exposed at that time. Oil from the Adolphus well is an unaltered oil with a density of 865 km/cu m (32° API).

Lower Mic Mac Formation (pre-Callovian)

The lower part of the Mic Mac Formation, defined as pre-Callovian rocks occurring above the Lower Jurassic Iroquois Formation, includes marine shales and carbonates and correlates in part with the Abenaki carbonate bank and reefal facies (Given, 1977). The isopach map, Figure 26, is an attempt at reconstruction of the lower Mic Mac and Abenaki Basin prior to break-up of the area by the Avalon Uplift.

The organic carbon-times thickness map, Figure 27, suggests significant volumes of source rock are present in this stratigraphic interval. However, this is largely due to the great thickness of formation involved, up to 3,000 m in the center of the basin. Actual organic carbon weight percent values are generally in the range of 0.4 to 0.75, or what are considered to have poor to fair generating capability. Nevertheless, this formation should have generated considerable volumes of hydrocarbon in the area shown on Figure 28 to have reached a mature carbonization stage. Analysis of the type of organic matter present has indicated gas is the most likely generation product.

The Avalon Uplift destroyed the generating capability of these source rocks over the central Grand Banks areas. Block faulting, uplift, and erosion resulted in removal of great volumes of the Mic Mac Formation. Uplift and erosion also would have had the effect of decreasing depth of burial

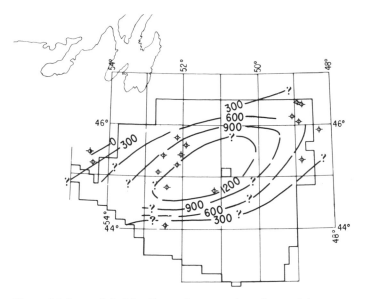

Figure 27. Lower Mic Mac Formation organic carbon weight percent times meters (quality × thickness).

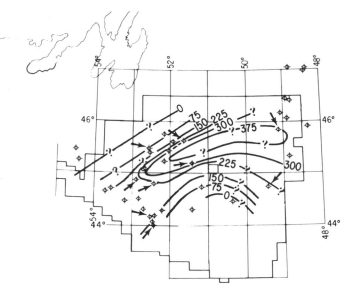

Figure 29. Restored isopach of the Iroquois Formation. Contours in meters. Wells indicated by arrows in South Whale Basin indicate four locations where the Iroquois Formation occurs as a cap rock above piercement salt domes. Other arrows indicate drilled locations at which the Iroquois Formation is breached at the Avalon unconformity.

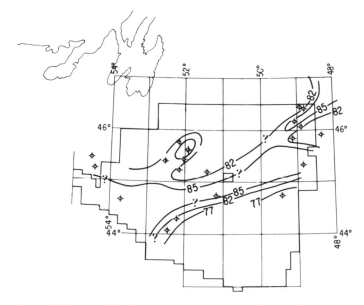

Figure 28. Lower Mic Mac Formation source rock maturity as indicated by weight percent carbon on the kerogen.

with resulting lowering of temperature and suspension of the thermo-chemical generating processes. Significant generating capability may have been detained in the southwest and northeast regions where effects of the Avalon Uplift have been less severe.

Iroquois Formation

Restored isopachs of the Lower Jurassic Iroquois Formation are presented in Figure 29. Carbonate bank facies are common, including porous oolite bank deposits that could be hydrocarbon reservoirs. Both live and dead oil were observed at a number of drill locations, the dead oil occurrences indicating oil that was trapped early and breached at the time of the Avalon Uplift. The source for these oils may have been in Lower Jurassic or Triassic rocks

or perhaps even a Paleozoic source could be considered.

Good organic carbon percentages have been measured in the shaly units of the Iroquois Formation but these shales occur as thin beds, perhaps not thick enough to have generated major reserves (Figure 30). Adequately mature source beds occur over a large area (Figure 31) but, as with the Mic Mac Formation, the Avalon Uplift resulted in breached reservoirs and disruption of generating processes.

Pre-Iroquois Formations

Significant source rocks have not been identified below the Iroquois Formation. The Triassic to early Jurassic Eurydice and Argo Formations consist of continental red beds and salt. Oil source intervals have been identified as thin shaly and dolomitic beds within the salt sequence but these are thought to form too small a volume of source material to have generated the major reserves required for economic oil fields in an offshore area.

The Carboniferous to Devonian section consists of continental red beds with an evaporite in the Windsor Group. All analyses indicate very low organic carbon content and were interpreted as non-source for the Paleozoic section as it is presently known from the sparse well control in that part of the section.

CONCLUSIONS

Problems of source rock quality, maturity, and timing have contributed to the poor results of drilling operations on the Grand Banks of Newfoundland.

Oil shows at the Heron and Adolphus wells and a gas show at the Tors Cove well indicate the presence of some effective source rocks in the area. Geochemical analyses confirm effective oil source rock areas in the northeast and suggest by extrapolation of data, that effective oil source

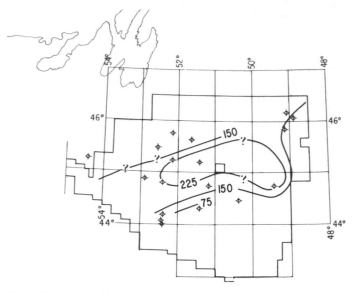

Figure 30. Iroquois Formation organic carbon weight percent times meters (quality × thickness).

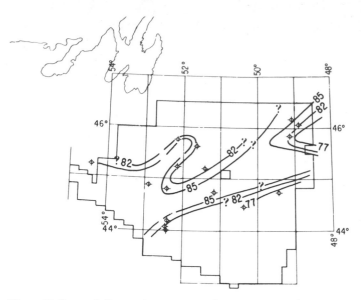

Figure 31. Iroquois Formation source rock maturity as indicated by weight percent carbon in the kerogen.

rocks are present in the southwestern area, i.e. in the areas where the East Newfoundland and Scotian Basins extend onto the Grand Banks.

The potential for discovery of significant oil and gas fields in the central Grand Banks area must be considered low because of the problems of generally poor source material, thin oil source beds, breached reservoirs, and deep erosion of Jurassic formations.

The preferred area for future exploration is off the flanks of the Avalon Uplift. Gas source rocks perhaps occur locally within the Cretaceous Dawson Canyon and Missisauga Formations and in the Jurassic Mic Mac Formation. Effective oil source rocks may be expected within the upper part of the Jurassic Mic Mac Formation and in the Iroquois Formation.

ACKNOWLEDGMENTS

Published with the permission of Amoco Production Company Ltd., Amoco Canada Petroleum Company Ltd., Imperial Oil Limited, Skelly Oil Canada Limited, and Chevron Standard Limited. The authors wish to acknowledge Mr. H.M. Cotten who compiled and reported the overwhelming mass of geochemical data acquired in this area and Mr. G.N. Wright who compiled the regional geology displays. We thank Mr. J.A. Momper for critically reading the manuscript and for his support for the preparation of this paper.

REFERENCES

Amoco Canada Petroleum Company Ltd., and Imperial Oil Limited, 1973, Regional geology of the Grand Banks: Bulletin of Canadian Petroleum Geology, v. 21, p. 479-503. (Also AAPG Bulletin, v. 58, p. 1109-1123.)

Given, M.M., 1977, Mesozoic and early Cenozoic geology of offshore Nova Scotia: Bulletin of Canadian Petroleum Geology, v. 25, p. 63-91.

Jansa, L.F., and J.A. Wade, 1975a, Paleogeography and sedimentation in the Mesozoic and Cenozoic, Southeastern Canada, *in* C.J. Yorath, E.R. Parker, and D.J. Glass, eds., Canada's continental margins and offshore petroleum exploration: Canadian Society of Petroleum Geologists Memoir 4, p. 79-102.

—— and —— , 1975b, Geology of the continental margin off Nova Scotia and Newfoundland, *in* W.J.M. van der Linden and J.A. Wade, eds., Offshore geology of Eastern Canada, Volume 2—Regional Geology: Geological Survey of Canada, Paper 74-30.

LaPlante, R.E., 1974, Hydrocarbon generation in Gulf Coast Tertiary sediments: AAPG Bulletin, v. 58, p. 1281-1289.

Momper, J.A., 1972, Evaluating source bed for petroleum (abs.): AAPG Bulletin, v. 56, p. 640.

Staplin, F.L. 1969, Sedimentary organic matter, organic metamorphism, and oil and gas occurrence: Bulletin of Canadian Petroleum Geology, v. 17, p. 47-66.

Swift, J.H., R.W. Switzer, and W.F. Turnbull, 1975, The Cretaceous Petrel Limestone of the Grand Banks, Newfoundland, *in* C.J. Yorath, E.R. Parker, and D.J. Glass, eds., Canada's continental margins and offshore petroleum exploration: Canadian Society of Petroleum Geologists Memoir 4, p. 181-194.

Upshaw, C.F., W.E. Armstrong, W.B. Creath, E.J. Kidson and G.A. Sanderson, 1974, Biostratigraphic framework of Grand Banks: AAPG Bulletin, v. 58, p. 1124-1132.

Williams, J.A., 1974, Characterization of oil types in Williston Basin: AAPG Bulletin, v. 58, p. 1243-1252.

Paleoenvironment and Petroleum Potential of Middle Cretaceous Black Shales in Atlantic Basins

B. Tissot
Institute Français du Pétrole
Rueil-Malmaison, France

G. Demaison
Chevron Overseas Petroleum
San Francisco, California

P. Masson
Petrofina S.A.
Brussels, Belgium

J.R. Delteil
Société Nationale Elf-Acquitaine
Paris, France

A. Combaz
Compagnie Française des Pétroles
Paris, France

Cores from the Deep Sea Drilling Project in the Atlantic Ocean show widespread organic-rich black shales in the middle Cretaceous. However, geochemical studies indicate that the origin and petroleum potential of the organic matter are highly variable. Three main types of organic material can be recognized in these sediments from kerogen studies: (a) marine planktonic material, deposited in reducing environments; (b) terrestrial higher plants, moderately degraded; (c) residual organic matter, either oxidized in subaerial environments and/or sediment transit, or recycled from older sediments.

Vertical and horizontal variations of these three types of organic matter determined from geochemical logs in each main basin of deposition indicate the paleogeography and environment of deposition of organic-rich shales. The petroleum potentials of the sediments are therefore consequences of their paleogeographic settings. Thus, the zones favorable for oil and gas (given adequate maturation), or those devoid of any potential, can be suggested. Complementary studies of wells on the continental shelf of the North American continent indicate that the organic facies in the deep basins may extend to nearshore locations.

INTRODUCTION

Offshore exploration of oil progressively extends into areas of deeper water, including the deep shelf and the continental slope. Rising exploration costs require the widest possible range of exploration techniques. Furthermore, only large oil fields can be viewed as commercial targets in deeper waters. Such large fields require prolific and widespread oil source beds, plus good reservoirs and seals. Organic geochemistry and paleogeography provide information as to temporal and geographic distribution of such prolific oil sources.

Experience to date in the Atlantic has shown the validity of this approach and the great importance of the information provided by basin-scale geochemistry. In particular, interpretation of the geochemical data obtained from Deep Sea Drilling Project (DSDP) wells in the deep basins and continental slopes of the Atlantic can be linked to the COST B-2, B-3, and GE-1 stratigraphic and test wells drilled by the oil industry on the United States continental shelf (Scholle, 1977, 1979; Smith, 1979).

TYPES OF ORGANIC MATTER

Cores from the Deep Sea Drilling Project in the Atlantic Ocean (Figure 1) have shown the widespread occurrence of organic-rich black shales in the middle Cretaceous. However, geochemical studies indicate that the origin and petroleum potential of this organic matter are highly variable. The analytic results discussed in this paper have been reported by Deroo et al (1977, 1978, 1979a, 1979b, 1979c).

The main types of organic matter in ancient sediments can be recognized by studying the insoluble fraction, usually called kerogen (Tissot et al, 1974). For example, the elementary composition of kerogen (shown in Figure 2 by means of H/C, O/C atomic ratios) permits definition of three main types of kerogen, plus a residual type: type I is present in certain rich oil shales and bogheads, and commonly deposited in lacustrine sediments; type II is derived from marine planktonic organic matter deposited in a reducing environment; it provides at depth a good source potential for oil and gas; type III is derived from terrestrial higher plants, moderately degraded; it provides a comparatively low potential for oil. In addition to these three types, there is residual organic matter, either oxidized in subaerial environments and/or sediment transit, or recycled from older sediments. It has no potential for hydrocarbons.

Each kerogen type shows a progressive maturation or evolution path, with increasing depth (Figure 2). The types and paths have been defined by Tissot et al (1974).

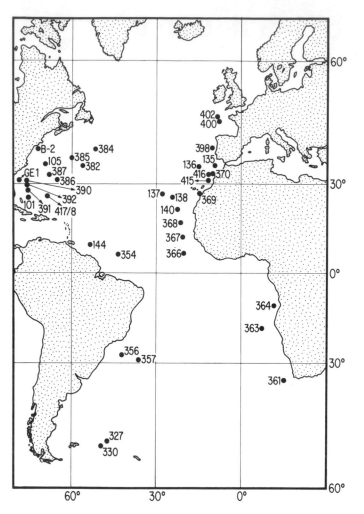

Figure 1. Location map showing DSDP and COST B-2 and GE-1 wells used in this study.

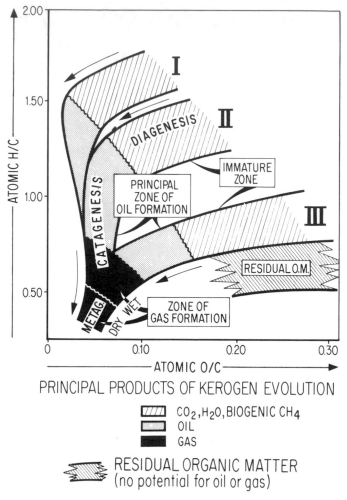

Figure 2. Elementary composition of kerogens showing main types and principal products generated during thermal evolution of kerogen types I, II, and III.

However, the residual type of organic matter shows no evolution path with depth because of its oxidized and frequently recycled character and the erratic nature of its alteration.

Types II, III, residual kerogen, and their associations have been recognized in the Cretaceous sediments of the Atlantic basins (Tissot et al, 1979).

METHODS FOR CHARACTERIZATION OF KEROGEN

The main techniques used to characterize the nature of the organic matter are: (1) kerogen elemental analysis (Deroo et al, 1977, 1978, 1979a, 1979b, 1979c; Castex, 1979); (2) infrared spectroscopy of kerogen (Castex, 1979); (3) pyrolysis, using the Rock-Eval instrument which provides a hydrogen and an oxygen index, directly related to the H/C and O/C ratios respectively, and also a quantitative evaluation of the hydrocarbon source potential (Espitalié et al, 1977); the method is performed directly on crushed rock, without preliminary separation of kerogen, and has been successfully used in oil exploration, both in the laboratory and on well sites (Clementz et al, 1979); (4)

optical examination of selected samples in transmitted light.

The main tools used for characterization (i.e., the elemental analysis of separated kerogen and the direct pyrolysis data) are compared in Figure 3 using samples from DSDP Legs 11, 41, 44, and 48. For example, samples from Leg 41, in the Cape Verde Basin, contain organic matter of planktonic origin (type II) as shown by the high H/C ratios and hydrogen indexes. Under the microscope, kerogen is amorphous, with flakes and pellets, or is finely disseminated. Samples from Leg 11, off the United States, contain terrestrially derived organic matter with lower H/C ratios and hydrogen indexes and higher O/C ratios and oxygen indexes. Samples from Leg 44, in the Blake-Bahama Basin, and from Leg 48, in the Bay of Biscay, contain a high proportion of residual organic matter, shown by abnormally low hydrogen contents associated with relatively high oxygen contents. The hydrogen index obtained by pyrolysis is very low, whereas the oxygen index is higher than that observed from Legs 11 and 41. Optical examination of these samples shows some vegetal tissues plus coaly fragments of oxidized organic matter, or charcoal.

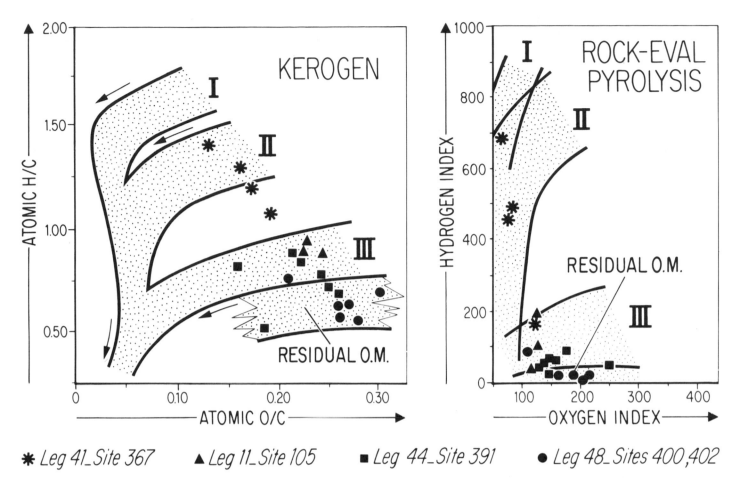

Figure 3. Principal methods for characterization of kerogen. On left, elementary composition is plotted on Van Krevelen diagram for samples from five sites in Atlantic. On right, Rock-Eval pyrolysis data are shown on same set of samples, *I, II, III,* types of kerogen.

PRESENTATION OF DATA

The Rock-Eval pyrolysis is fast and inexpensive and allows the processing of a large number of samples and preparation of well logs (Espitalié et al, 1977; Clementz et al, 1979). Logs of the most significant wells with good core recovery of the middle Cretaceous interval are presented in Figures 4 to 7 and show the organic content, the predominant type of organic matter, and the hydrocarbon source potential for two sites in the South Atlantic and seven sites in the Central and North Atlantic. From each log, samples were selected for separation and analysis of kerogen; these analyses confirmed the interpretations shown in Figures 4 to 7.

In the South Atlantic (Figure 4), the black shales of Aptian to early Albian age contain organic matter of marine planktonic origin (type II) in the Cape and Angola Basins (Sites 361 and 364, respectively). However, in the Cape Basin a high terrestrial content (type III) dilutes the planktonic input, and is locally predominant. Thus, the average oil source potential is lower in the Cape Basin than in the Angola Basin. Conditions were no more favorable for organic matter preservation during middle than during early Albian time. A second occurrence of black shales is recorded in the Angola Basin. There, rich planktonic

material is preserved in a succession of discrete layers of black shales and marls of late Albian to Coniacian or Santonian ages. No comparable occurrence was found in the Cape Basin.

Organic matter of marine planktonic origin is also present off the northwestern coast of Africa (Figure 5) in the Cape Verde Basin (Site 367) where there is a mixture of planktonic and terrestrially derived organic matter in Neocomian to Barremian beds. A later succession of black clay layers, ranging from Aptian to Cenomanian, and possibly to Coniacian in age, contains a large amount of planktonic kerogen (up to 34 percent) with a high source potential. Off Morocco (Site 370), Lower Cretaceous shales and turbidites contain mainly terrestrial organic matter, although several Aptian and Albian beds show a predominance of planktonic material; however, their source potential for hydrocarbons remains lower than in the Cape Verde Basin.

The northeastern Atlantic (off Portugal, Bay of Biscay, Figure 6) shows no significant amount of organic matter of marine planktonic origin, except for a single Cenomanian layer at Site 398 off Portugal. The terrestrial organic material is dominated by residual kerogen (oxidized and/or reworked) except for a few layers in which moderately degraded organic matter of terrestrial origin is abundant.

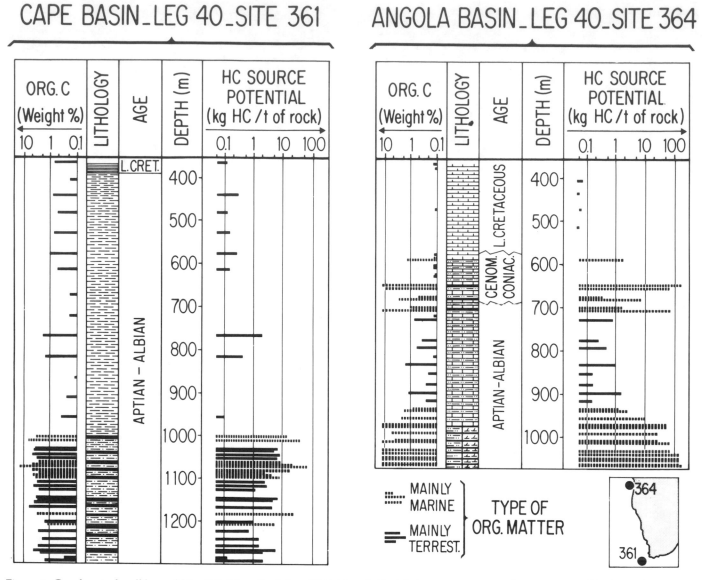

Figure 4. Geochemical well logs of Sites 361 (Cape Basin) and 364 (Angola Basin) showing organic carbon content and hydrocarbon source potential obtained from Rock-Eval pyrolysis. *t*, metric ton; this abbreviation is used throughout text and figures.

The situation is rather comparable in the northwestern Atlantic (Figures 7 and 8) where wells found terrestrial and residual kerogen in the Lower Cretaceous sediments, although a few isolated layers also contain some planktonic material. In the Cenomanian, high concentrations of planktonic kerogen (up to 15%) are present in layers of carbonaceous clays. Somewhat later comparable concentrations are in the Turonian-Coniacian of the COST GE-1 well.

DISTRIBUTION OF ORGANIC MATTER

On the basis of 27 well logs similar to those presented in Figures 4 to 8, the main types of organic matter deposited during the Aptian-Albian and the Cenomanian-Turonian, respectively, were plotted on maps (Figures 9 and 10) to show their geographic distribution. The observed occurrences of organic matter in the North Atlantic basins

can be summarized as follows (Figures 9 and 10).

Aptian-Albian—Only the sites located in the southern and eastern part of the North Atlantic show the presence of Aptian-Albian beds with an important accumulation of planktonic (type II) organic matter: Sites 367, 369, 370, and 417/418. In all other wells, the continents were the main source of organic matter with an admixture of moderately degraded (type III) and residual kerogen; near the northern end of the Atlantic basins at Sites 398, 400, and 402, the predominant organic matter is a residual kerogen.

Cenomanian to Coniacian—Large accumulations of planktonic material occur at Sites 367, 368 (Cape Verde), and 144 (Demerara); in many other places in the Cenomanian, discrete organic-rich beds containing planktonic kerogen (type II) are present; the remainder of this interval, where represented, contains terrestrial organic matter.

Uppermost Cretaceous—The organic matter is terrestrial

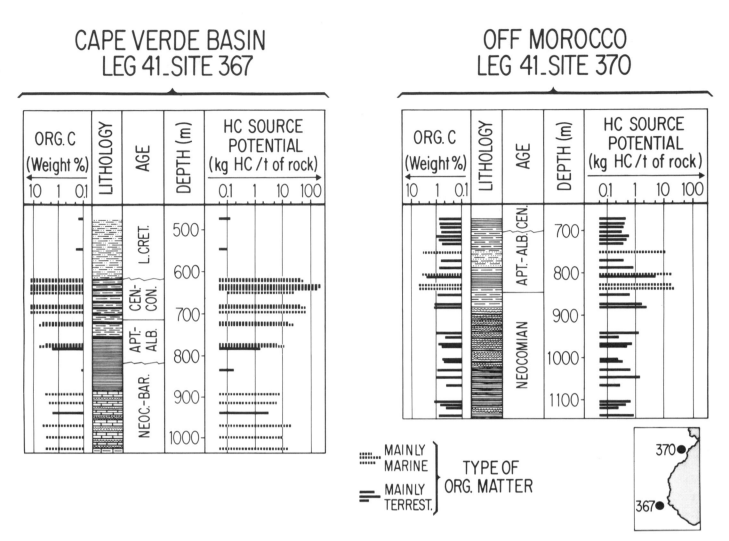

Figure 5. Geochemical well logs of Sites 367 (Cape Verde Basin) and 370 (off Morocco).

at all the investigated sites, and the kerogen content is low.

In the South Atlantic, the geographic and stratigraphic data on organic sedimentation during Cretaceous time are more limited. However, adequate information is available in three widely spaced areas. Planktonic kerogen (type II) is present in Lower Cretaceous rocks of the Angola Basin (Site 364) and also in the Falkland Basin (Sites 327 and 330) and Cape Basin (Site 361) where it is associated with terrestrial organic matter. Some discrete examples of planktonic kerogen are present at the Angola Basin site, in late Albian to Coniacian or Santonian beds. No comparable material has been found in the other wells drilled in the South Atlantic.

DEPOSITIONAL ENVIRONMENTS

The observed distribution of the organic matter in DSDP wells presented in Figures 9 and 10 and discussed previously shows a statistical regularity: although individual beds may not be correlated, accumulation of a certain type of kerogen has occurred in rocks of the same age within relatively large areas. Such regularities of distribution suggest that they reflect paleogeographic conditions. Furthermore,

consideration of the results obtained from other wells drilled by the petroleum industry along the western coast of Africa and the eastern coast of the United States indicates that these regularities may, in some places, extend over the shelf area.

Knowing that the depositional environments have determined kerogen type within a given stratigraphic unit, we can interpret the observed distribution of kerogen types in terms of depositional environments.

Oceanographic factors, more specifically the oxygen content of the water column, directly influence kerogen type. Anoxic conditions, at least periodic, in the water column are critical (Demaison and Moore, 1980) for enhanced preservation of marine planktonic remains that are the precursors of the "oil-prone" type II kerogen.

Widespread and synchronous anoxic conditions, leading to formation of type II kerogen, have occurred in parts of the world ocean several times in the geologic past. The paleo-oceanographic distribution of such "oceanic anoxic events" (Schlanger and Jenkyns, 1976; Arthur and Natland, 1979; Arthur and Schlanger, 1979) in the Atlantic basins, during middle Cretaceous time, resulted from four mutually reinforcing factors: (1) oceans latitudinally rather than meridianally elongated, preventing the deep circulation of

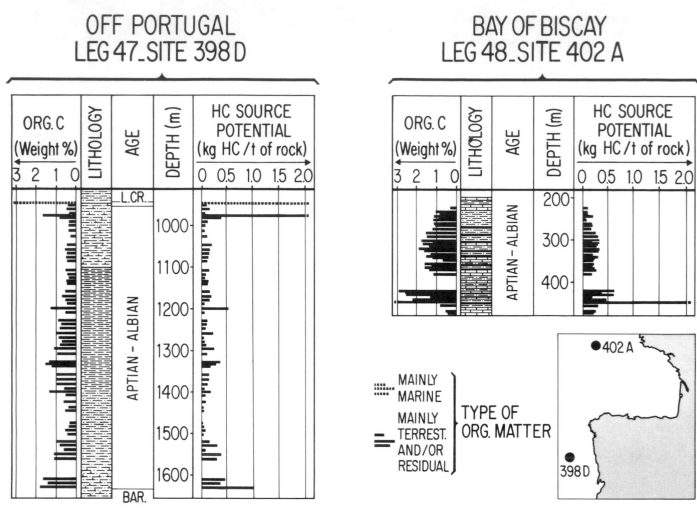

Figure 6. Geochemical well logs of Sites 398D (off Portugal) and 402A (Bay of Biscay).

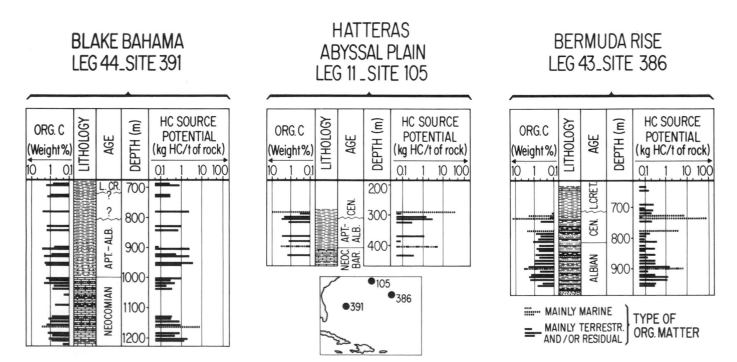

Figure 7. Geochemical well logs of Sites 391 (Blake-Bahama area), 105 (Hatteras Abyssal Plain), and 386 (Bermuda Rise).

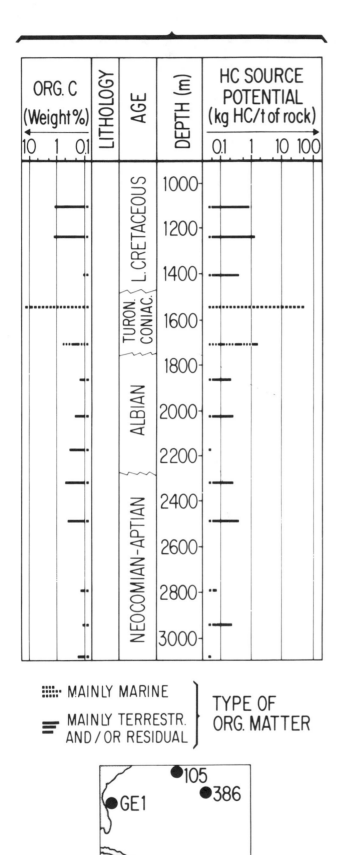

Figure 8. Geochemical log of COST GE-1 well (Georgia Embayment).

polar waters; (2) warm, equable, global climate with warm seawater consequently oxygen depleted; (3) widespread transgression of the world ocean over the continents creating shallow, warm, seas with high biologic productivity; (4) willed- or barred-basin geometry resulting from plate evolution during the early opening of the Atlantic Ocean; this factor favored stagnation.

Conversely, the presence of type III kerogen in marine sediments appears to result largely from transport of terrestrial organic matter into an ocean, where an oxic water column prevents the deposition of planktonic remains. Such conditions lead to comparatively enhanced preservation of the more resistent, hydrogen-poor, terrestrial organic matter, whereas the planktonic input is preferentially biodegraded. In particular, residual oxidized and/or recycled, organic matter cannot be further degraded and thus may be found in any environment. Massive terrestrial organic-matter input into anoxic marine basins, however, may result in the formation of mixed type II-type III kerogen precursors. The paleogeographic distribution of terrestrial organic matter in marine sediments is determined partly by the location of temperate humid climatic belts, where the availability of organic material was highest, and partly by the physiography of the fluvial systems that discharged the organic matter into the sea.

On the basis of these considerations, the observed distribution of organic matter in the Atlantic basins (Figures 9 and 10) may be interpreted in terms of depositional environment (Figure 11).

Aptian and part of Albian—Preservation of large amounts of planktonic material in deep oceanic basins required anoxic conditions, which probably existed continuously in the confined Cape and Angola Basins because of the lack of general oceanic circulation. Anoxic conditions also existed, at least periodically, in the southeastern basin of the North Atlantic east of the Mid-Oceanic Ridge and including the Cape Verde Basin. In other areas, such as the northern embayment (between western Europe and Canada) and the northwestern basin

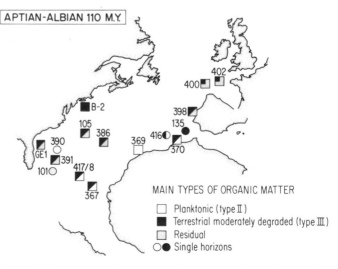

Figure 9. Main types of organic matter observed in Aptian and/or Albian beds of North Atlantic (position of continents is only approximate at 110 m.y.). From Tissot et al (1979), with additions.

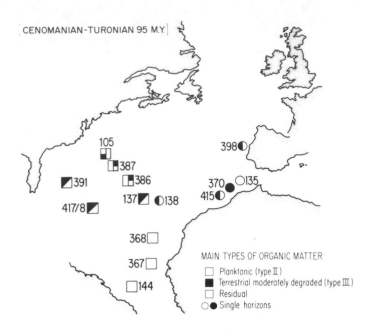

Figure 10. Main types of organic matter observed in Cenomanian and/or Turonian beds of North Atlantic (position of continents is only approximate at 95 m.y.). From Tissot et al (1979) with minor changes.

(west of the Mid-Oceanic Ridge), oxic conditions prevailed, although the oxygen content may have been less than present normal concentration. These conditions prevented the preservation of planktonic material, and the continent was the only source of preserved organic matter, with an important proportion of residual kerogen no longer capable of degradation because of its oxidized and/or recycled character.

Cenomanian—Anoxic conditions prevailed again, at least periodically, in the Cape Verde Basin and extended temporarily over wide areas including part of the northwest Atlantic. To the south, the Cape Basin was open to oceanic circulation with conditions no longer satisfied for preservation of planktonic material. On the contrary, the Angola Basin was protected, at least temporarily, by the Walvis—Rio Grande Ridge and intermittent anoxic conditions were established during the interval.

Turonian and Coniacian—The anoxic basins persisted in the same manner in the South Atlantic, but intermittent anoxia in the North Atlantic was apparently restricted to the Cape Verde—Demerara area, possibly in connection with upwelling conditions (Arthur and Natland, 1979).

Latest Cretaceous—All deep basins were open to general oceanic circulation preventing anoxic conditions and preservation of planktonic material.

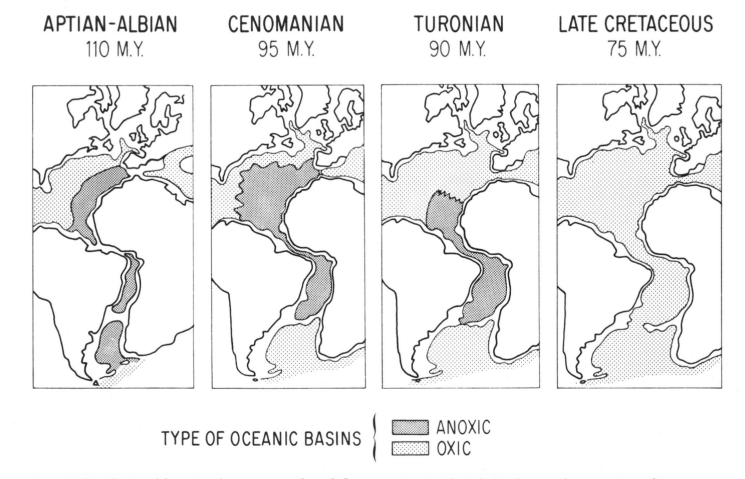

Figure 11. Distribution of depositional environments through Cretaceous time in Atlantic basins. In some places, anoxic conditions may have been intermittent. Oxygen content of oxic water during some intervals was probably less than present concentration in normal oceans.

PETROLEUM POTENTIAL OF MIDDLE CRETACEOUS SEDIMENTS

The distribution of organic matter shows regional regularities, which can be interpreted in terms of paleogeography, and leads us to conclude that an evaluation of the petroleum potential of middle Cretaceous rocks from available wells has some regional significance.

In the following section the hydrocarbon source potential of the sediments will be discussed in terms of the total amount of hydrocarbons that could be generated if the sediments were buried to sufficient depth to insure maturation of the kerogen. This is, of course, not realized at the site of sampling, as the depth of coring is limited by the drilling capability of the vessel. However, where a sufficient thickness of younger sediments has been deposited—as on the upper continental rise, slope, and outer shelf—maturation of the kerogen may be attained. In that respect, the indications of petroleum potential have to be considered as regional (e.g., Cape Basin, Angola Basin) and not local (e.g. Sites 361 and 364).

The amount of hydrocarbons that can be generated is plotted in kilograms per ton of rock in Figures 4 to 8. In addition, if we introduce the respective thicknesses of the successive beds, we can calculate the cumulative hydrocarbon potential—expressed as tons per square meter of sedimentary basin—over the whole Cretaceous interval. Thus, the amount of potential hydrocarbons per square meter of sedimentary basin is shown in Figures 12 to 15 (one metric ton per square meter is approximately 30,000 barrels

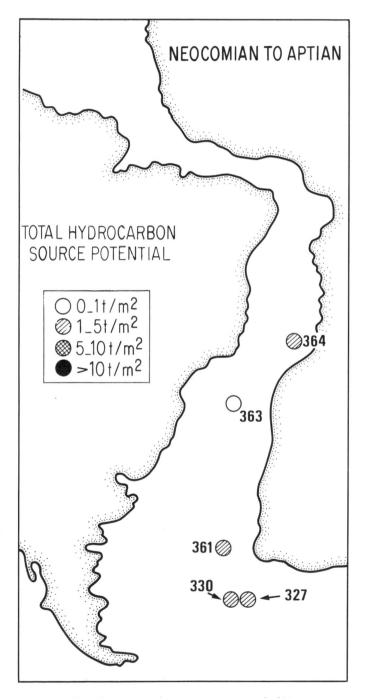

Figure 12. Cumulative petroleum source potential of Neocomian to Aptian rocks in South Atlantic. Cumulative potential is expressed in terms of metric tons of potential oil over a stratigraphic column covering an area of 1 sq m.

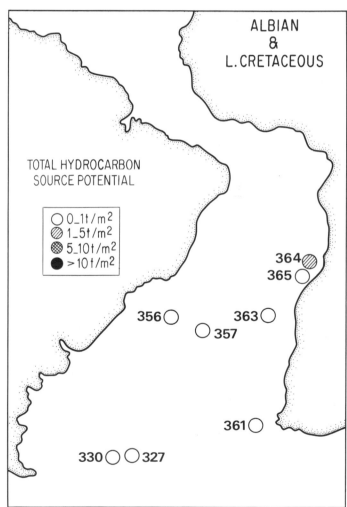

Figure 13. Cumulative petroleum source potential of Albian and Late Cretaceous rocks in south Atlantic (same units as in Figure 12).

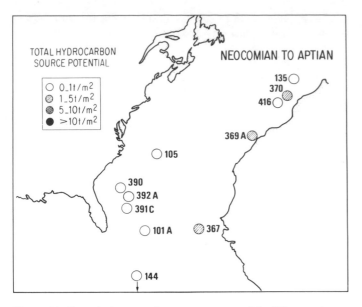

Figure 14. Cumulative petroleum source potential of Neocomian to Aptian rocks in North Atlantic (same units as in Figure 12).

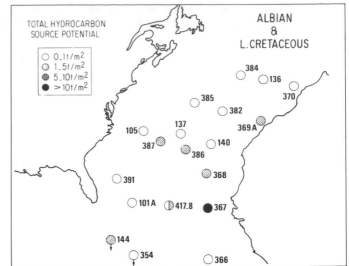

Figure 15. Cumulative petroleum source potential of Albian and Upper Cretaceous rocks in North Atlantic (same units as in Figure 12).

per acre). The following scale of quality for potential hydrocarbons is suggested: low potential, <1 ton/sq m; good potential, 1 to 5 t/sq m; very good potential, 5 to 10 t/sq m; excellent potential, >10 t/sq m.

In the South Atlantic (Figure 12) the source potential corresponding to the lower black shale (Aptian to lower Albian) is good over the Cape and Angola Basins, although Site 364 appears definitely more oil-prone that Site 361. Later, the second black shale provides a good potential for oil in the Angola Basin only (Figure 13).

In the North Atlantic (Figure 14) the oil source potential of the Lower Cretaceous is generally good in the southeastern basin (Cape Verde to Morocco). Elsewhere the potential is low in the western basin and very low at the northern end because of the residual character of the kerogen.

The Upper Cretaceous (Figure 15) shows an excellent oil source potential in the Cape Verde Basin, where it contains abundant planktonic kerogen (Cenomanian to Coniacian). The oil potential is also good in several locations because of the Cenomanian anoxic event. At other places, and in younger beds, low levels of terrestrial kerogen lead only to low or very low potentials.

CONCLUSIONS

Geochemical characterization of the organic matter from the middle Cretaceous black shales of the Atlantic leads to several conclusions which could influence future prospecting in the deep offshore areas:

1. Both marine and terrestrial types of organic matter are present in oceanic basins. The distribution of types is interpreted with plate tectonics, which controls the basin morphology, and with paleo-climatology and paleo-oceanography which control sedimentation.

2. Anoxic conditions in oceanic basins control distribution of the prolific potential oil source rocks. An

understanding of paleogeography through geologic time may lead to prediction of the time and areal distribution of those source beds.

3. Organic richness and oil-source potential are not always related. In particular, some organic-rich beds (containing several percent of organic carbon) deposited in marine basins have proved to contain little or no organic matter of marine origin. Instead, the kerogen is terrestrially derived, either moderately degraded (type III), or deeply oxidized and/or recycled. These types have little or no potential for oil generation, respectively. Under such circumstances the depositional environment is considered to be oxic and a move farther from the continent may not necessarily improve the quality of the organic matter.

4. Analyses of the Cretaceous black shales permit evaluation of their potential for oil generation if the sediments are buried to sufficient depth to insure maturation of the organic matter. Results obtained from wells drilled off the eastern United States seem to confirm the trends observed in DSDP wells.

ACKNOWLEDGMENTS

The writers thank the National Science Foundation for making the core samples available through the Deep Sea Drilling Project, and Chevron Overseas Petroleum Inc., Compagnie Française des Pétroles, Institut Français du Pétrole, Petrofina S.A., and Société Nationale Elf-Aquitaine for their support.

REFERENCES CITED

Arthur, M.A., and J.H. Natland, 1979, Carbonaceous sediments in the North and South Atlantic. The role of salinity in stable stratification of early Cretaceous basins, *in* Deep drilling results in the Atlantic Ocean: continental margins and paleoenvironment: American Geophysical

Union, Maurice Ewing Series, p. 375-401.

——, and S.O. Schlanger, 1979, Cretaceous "oceanic anoxic events" as causal factors in development of reef-reservoired giant oil fields: AAPG Bulletin, v. 63, p. 870-885.

Castex, H., 1979, Spectroscopie infrarouge et analyse élémentaire de quelques kérogènes. Campagne JOIDES-IPOD: Oceanologica Acta, v. 2, p. 33-40.

Clementz, D.M., G.J. Demaison, and A.R. Daly, 1979, Well site geochemistry by programmed pyrolysis: 11th Offshore Technology Conference, OTC 3410, v. 1, p. 465-470.

Demaison, G.J., and G.T. Moore, 1980, Anoxic environments and oil source bed genesis: AAPG Bulletin, v. 64, p. 1179-1209.

Deroo, G., et al, 1977, Organic geochemistry of some Cretaceous black shales from Sites 367 and 368, Leg 41, eastern North Atlantic: Initial Reports of the Deep Sea Drilling Project, v. 41, p. 865-873.

——, ——, 1978, Organic geochemistry of some Cretaceous claystones from Site 391, Leg 44, western North Atlantic: Initial Reports of the Deep Sea Drilling Project, v. 44, p. 593-598.

——, ——, 1979a, Organic geochemistry of Cretaceous mudstones and marly limestones from DSDP Sites 400 and 402, Leg 48, eastern North Atlantic: Initial Reports of the Deep Sea Drilling Project, v. 48, p. 921-930.

——, ——, 1979b, Organic geochemistry of some organic rich shales from DSDP Site 397, Leg 47A, eastern North Atlantic: Initial Reports of the Deep Sea Drilling Project, v. 47, p. 523-529.

——, ——, 1979c, Organic geochemistry of Cretaceous shales from DSDP Site 398, Leg 47B, eastern North Atlantic: Initial Reports of the Deep Sea Drilling Project, v. 47, p. 513-522.

——, ——, 1980, Organic geochemistry of Cretaceous sediments at DSDP holes 417D (Leg 51), 418A (Leg 52), and 418B (Leg 53) in the western North Atlantic: Initial Reports of the Deep Sea Drilling Project, v. 51, 52, 53, p. 737-745.

Espitalié, J., et al, 1977, Source rock characterization method for petroleum exploration: 9th Offshore Technology Conference, OTC 2935, v. 3, p. 439-444.

Schlanger, S.O., and H.C. Jenkyns, 1976, Cretaceous anoxic events: causes and consequences: Geologie en Mijnbouw, v. 55, p. 179-184.

Scholle, P.A., ed., 1977, Geological studies on the COST No. B-2 well, U.S. Mid-Atlantic outer continental shelf area: U.S. Geological Survey Circular 750, 71 p.

——, ed., 1979, Geological studies on the COST No. GE-1 well, U.S. Mid-Atlantic outer continental shelf area: U.S. Geological Survey Circular 800, 114 p.

Smith, M.A., 1979, Geochemical analyses on the COST No. B3 well, U.S. Mid-Atlantic outer continental shelf area: U.S. Geological Survey Open File Report 79-1159.

Thiede, J., and T.H. van Andel, 1977, The paleoenvironment of anaerobic sediments in the late Mesozoic South Atlantic Ocean: Earth and Planetary Science Letters, v. 33, p. 301-309.

Tissot, B., 1979, Effects on prolific petroleum source rocks and major coal deposits caused by sea-level changes: Nature, v. 277, p. 463-465.

——, G. Deroo, and J.P. Herbin, 1979, Organic matter in Cretaceous sediments of the North Atlantic: contribution to sedimentology and paleogeography, *in* Deep drilling results in the Atlantic Ocean: continental margins and paleoenvironment: American Geophysical Union Maurice Ewing Series 3, p. 362-374.

——, et al, 1974, Influence of nature and diagenesis of organic matter in formation of petroleum: AAPG Bulletin, v. 58, p. 499-506.

Van Krevelen, D.W., 1961, Coal: typeology-chemistry-physics-constitution: New York, Elsevier Publishing Company, 514 p.

Source-Rock and Carbonization Study, Maracaibo Basin, Venezuela

Rudolf Blaser
Koninklijke/Shell Exploration and Production Laboratory
Rijswijk, The Netherlands

Christopher White
MARAVEN
Caracas, Venezuela

The Maracaibo Basin of western Venezuela is composed of unmetamorphosed Cretaceous and Tertiary sediments deposited on top of a complex pre-Cretaceous basement consisting of igneous rocks and strongly-to-weakly metamorphosed sediments. All oil accumulations in the Maracaibo Basin originate from oil source rocks intercalated in the Cretaceous to Tertiary sedimentary fill of the basin.

Originally, the organic-rich limestones of the Cenomanian to Coniacian La Luna Formation were considered the most important, if not the only, oil source rocks in this region. They are indeed classical marine oil source rocks developed quite coherently and with considerable thickness throughout the basin. Later, the possible presence of additional oil source rocks (e.g., in the Tertiary rock sequence) was also taken into consideration. To check this possibility, numerous samples of Upper Cretaceous and Tertiary sediments from different parts of the Maracaibo Basin were collected by *MARAVEN*, Caracas, and geochemically analyzed by *KSEPL*, Rijswijk, between 1976 and 1978. The results of the geochemical analyses and carbonization data have been integrated into the geological framework of the Maracaibo Basin.

The results of this geochemical-geological study indicate that source rocks for oil are indeed present in various Tertiary rock formations. Nevertheless, these Tertiary source rocks are characteristically different from those of the La Luna Formation. They are mainly composed of hydrogen-rich land-plant matter. The chemical composition of extracts of these land-plant derived source rocks is markedly different from that of La Luna source-rock extracts. With regard to the regional distribution of the Tertiary source rocks, it has been found that they are restricted to certain areas of the Maracaibo Basin, particularly its western and southern parts.

For an appraisal of the possible contribution of Tertiary source rocks to the oil accumulations in the Maracaibo Basin, it has been necessary to carry out regional carbonization studies and oil/source-rock extract correlations. The degree of maturity was determined by measuring the optical properties of macerals (in particular vitrinite; Patteisky and Teichmuller, 1960) and, additionally, by maturity calculations. The results of this maturity evaluation show that mature Tertiary sediments occur only in two areas: (1) in the northeastern part of the Maracaibo Basin (the Bolivar Coast region), and (2) the Colón District and the North Andean Foredeep. In the first area, Tertiary source rocks for oil have not been encountered. In the second region, land-plant derived Tertiary source rocks are locally well-developed and could have generated oil.

Correlation of oils with source-rock extracts was previously based mainly on the porphyrin content of crude oils and source-rock extracts and on vanadium/nickel ratios. More modern correlations are based on gas-chromatographic and mass-spectrometric methods. Most crude oils investigated correlate well with La Luna source-rock extracts. On the other hand, none of the analyzed oils correlated with extracts of land-plant derived Tertiary source rocks.

INTRODUCTION

The Maracaibo Basin in Western Venezuela is a major area of occurrence of the classical marine La Luna source rocks of Cenomanian to Coniacian age. Until recently, most Shell geologists and geochemists had regarded the La Luna Formation as the only rock formation of the Maracaibo Basin containing significant oil source rocks. Oil-generation studies were therefore concentrated on the facies changes, and particularly the regional maturity variations of this formation.

At the request of *MARAVEN*, a new source-rock study of the Maracaibo Basin was initiated in 1976. This study was aimed at critically reviewing previous concepts and evaluating the significance of previously neglected Tertiary source rocks.

The study embraces the entire Maracaibo Basin. Ample rock-sample material was obtained from 56 wells, strategically distributed over the basin and from several outcrop locations. More than 4,000 rock samples have been analyzed geochemically for source-rock and maturity determination and specific maceral determinations were carried out.

Maturity (VR/E)* calculations were made, using a Lopatin-type calculation method**, for 136 wells and positions along geological or seismic profiles. Furthermore, crude-oil and source-rock analyses for oil/source-rock extract correlation were carried out. A review of the total activities under this project is shown in Figure 1.

RESULTS OF SOURCE ROCK AND MATURITY EVALUATION

The results of source-rock and maturity (VR/E) evaluations are discussed briefly in stratigraphic order, from older to younger rock formations. The main occurrences of oil and gas source rocks and their stratigraphic distribution are shown in Figure 2.

Pre-Cretaceous

Until now there has been no evidence of the presence of pre-Cretaceous oil or gas source rocks in the Maracaibo Basin.

Cretaceous

The Cretaceous carbonate sequence contains source rocks for oil in various stratigraphic intervals, as shown in Figures 2 and 3. By far the most important interval is the La Luna Formation, which is fairly uniformly deposited over the entire Maracaibo Basin. The regional thickness variations of the La Luna Formation are mapped in Figure 4 and are shown to vary between 75 and 150 m. The variation of net source-rock thickness is not known, but may be estimated to range between $1/3$ and $2/3$ of the total thickness of the formation.

La Luna source rocks have been deposited in an extensive, relatively deep-marine area, in which reducing conditions prevailed at the sea bottom. According to Buiskool Toxopeus and Van Lieshout (1979), the organic matter of the source rocks mainly consists of structureless organic matter (SOM).

The extracts of La Luna source rocks are rich in vanadyl porphyrins and organically bound sulfur (Aldershoff, 1953b; and Gransch and Eisma, 1966a). These components are characteristic of source rocks that were deposited in a reducing marine environment with predominant carbonate deposition (Gransch and Posthuma, 1973).

Evaluating the degree of maturity (as expressed in VR/E) of marine source rocks that do not contain vitrinite is difficult. The maturity of such source rocks can be approximately determined by comparison of the calculated VR/E with C_R/C_T values (Gransch and Eisma, 1966b).

The present-day maturity picture for the top of the La Luna Formation resulting from numerous VR/E calculations and VR measurements is shown in Figure 5.

Possible occurrence of source rocks in the Late Senonian Colón and Mito Juan Formations has been investigated in several wells. No evidence of the presence of significant source rocks in these formations could be obtained. It can be

concluded that the depositional environment of these marine intervals was definitely much less favorable than in the La Luna formation.

Tertiary

Paleocene

The Paleocene of the Maracaibo Basin shows a regressive development with different facies belts that overlap each other laterally as well as vertically (Figure 6).

At the beginning of the Paleocene, most of the Maracaibo Basin was covered by a shallow sea. Subsequently, this sea regressed in a northeasterly direction; by the late Paleocene it covered only the northeastern part of the basin. In this sea, mainly shallow-water sediments (shales, marls, calcareous sandstones) were deposited which form the Guasare formation.

During the gradual retreat of the sea, large coastal plains developed with paralic deposits and fluviatile areas in the hinterland. The terrestrial sediments deposited in these areas are represented by the Marcelina and the Orocue formation, which both consist of clastic deposits with locally significant intercalations of coal and carbonaceous shales. Rock-sample sequences (from wells) representing the above-mentioned formations have been analyzed for source rock. The results are summarized in Figure 7, which shows that, in general, the Guasare formation contains only rare organic-rich rocks qualifying as source rocks for gas and some oil.

More promising source rocks have been recorded in the Orocue and Marcelina formations. The source rocks for gas *and* oil (indicated in Figure 7) are relatively hydrogen-rich coals and are comparable with the coaly sediments which have generated plant-derived oil in Borneo and Indonesia.

Because of their large regional persistence, the Paleocene coals could be of considerable importance in areas where the Paleocene is mature for oil generation.

A maturity map for base Paleocene has been constructed on the basis of Lopatin-type VR/E calculations (Figure 8). The comparison of this VR/E map with vitrinite-1 R_o measurements shows a good fit in the north, but a slight misfit in the Colón District and in the southwestern Lake Maracaibo area where the calculated VR/E is too high. It should be mentioned that in the Colón District (situated in the southwest Maracaibo Basin) most Paleocene source rocks occur in the upper part of the Paleocene rock sequence and that, therefore, their degree of maturity probably corresponds better to the base Eocene (Figure 11) than to the base Paleocene (Figure 8).

In the northeast Maracaibo Basin area, the Paleocene had already become mature toward the end of the Eocene. In that region the Paleocene is only represented by the Guasare

---

*VR and VR/E: VR is the light reflectance of the coal maceral vitrinite; it is also written as R_0, absolute, random. VR/E is the equivalent of vitrinite reflectance as established by any other maturity indicator calibrated with vitrinite reflectance.
**According to Lopatin (1972), the reaction rate (in organic metamorphism) doubles for each 10°C increase of temperature. If the temperature history of a sediment is known, Lopatin's "temperature time coefficient" can be calculated. This coefficient can then be calibrated with data of measured VR. Shell's VR/E calculations are based on a worldwide calibration set. Lopatin-type maturity calculations have been discussed in detail by Waples (1980).

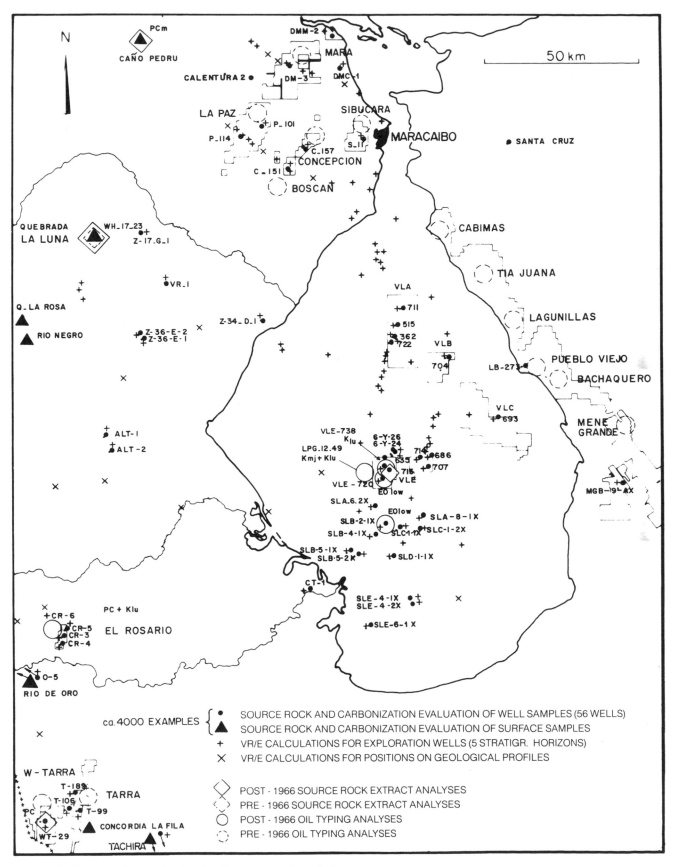

Figure 1. Index map of control points for the source rocks and oils.

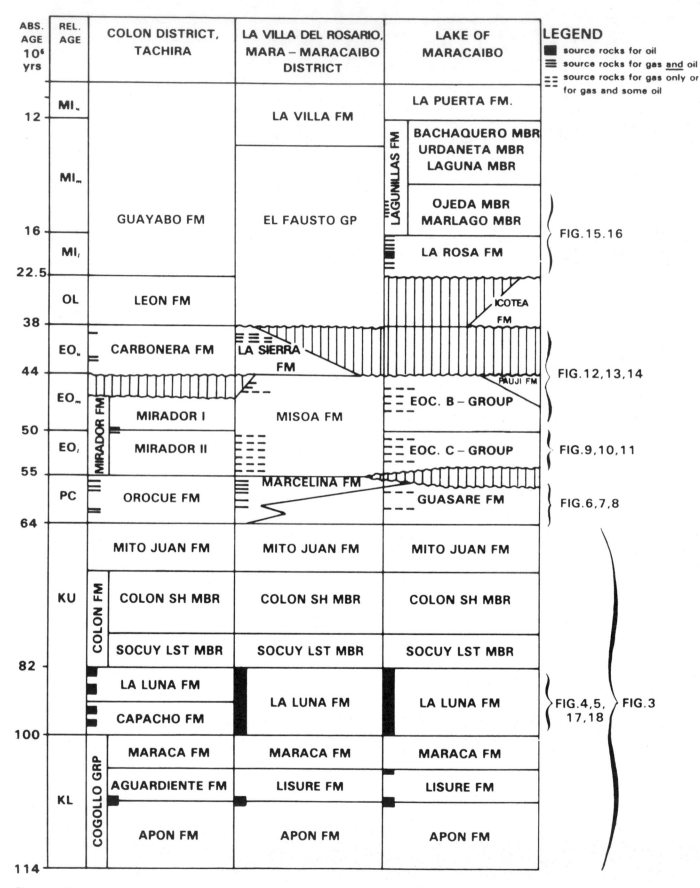

ABS. AGE 10⁶ yrs	REL. AGE	COLON DISTRICT, TACHIRA	LA VILLA DEL ROSARIO, MARA – MARACAIBO DISTRICT	LAKE OF MARACAIBO

LEGEND
- ■ source rocks for oil
- ☰ source rocks for gas **and** oil
- ☲ source rocks for gas only or for gas and some oil

Figure 2. Stratigraphic and regional occurrence of source rocks in the Maracaibo Basin, western Venezuela.

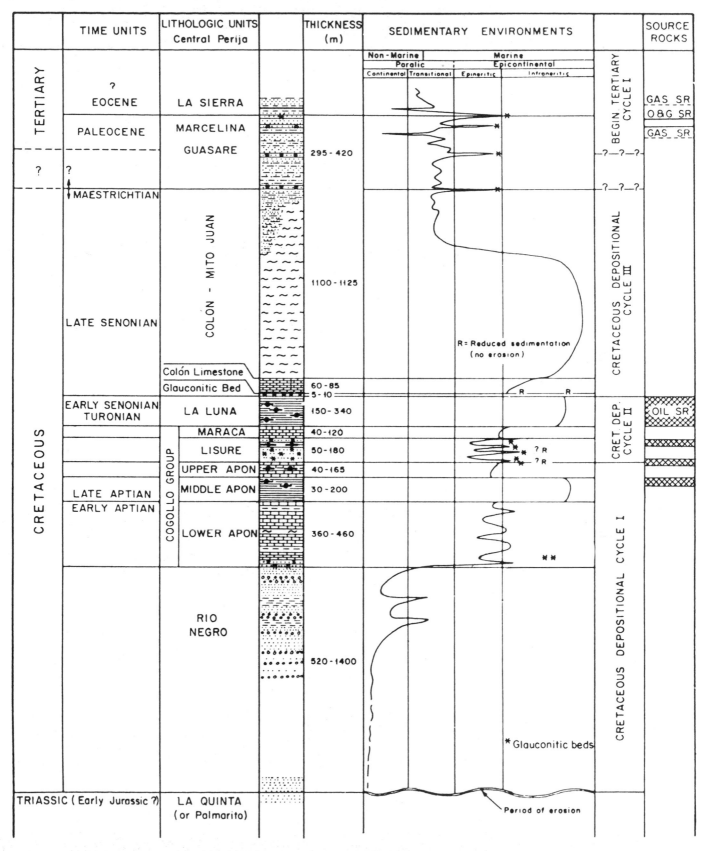

Figure 3. Stratigraphy-lithology-source rock occurrence along the Perija Mountain front (after Rod, 1954).

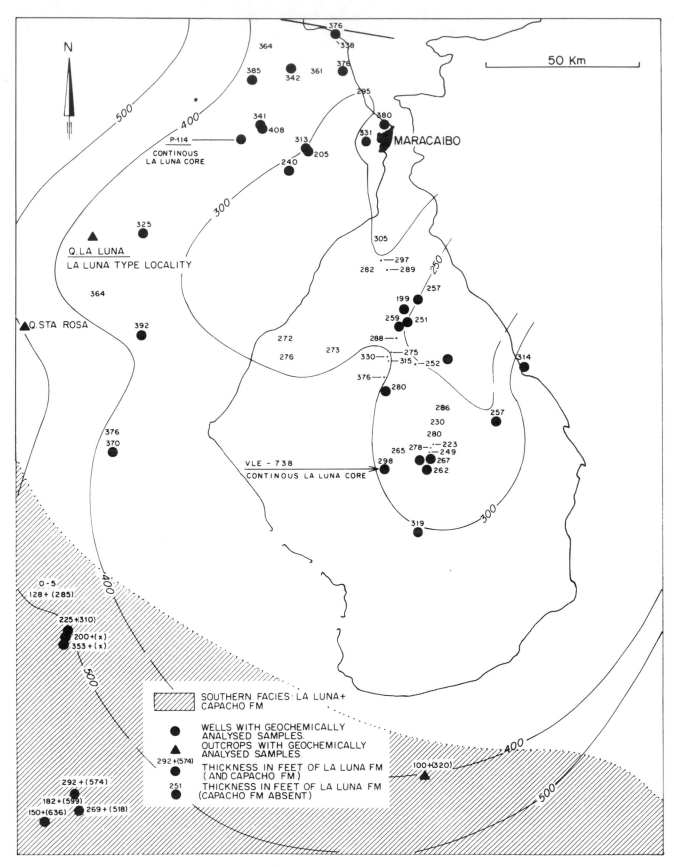

Figure 4. La Luna formation isopachs and geochemical analyses.

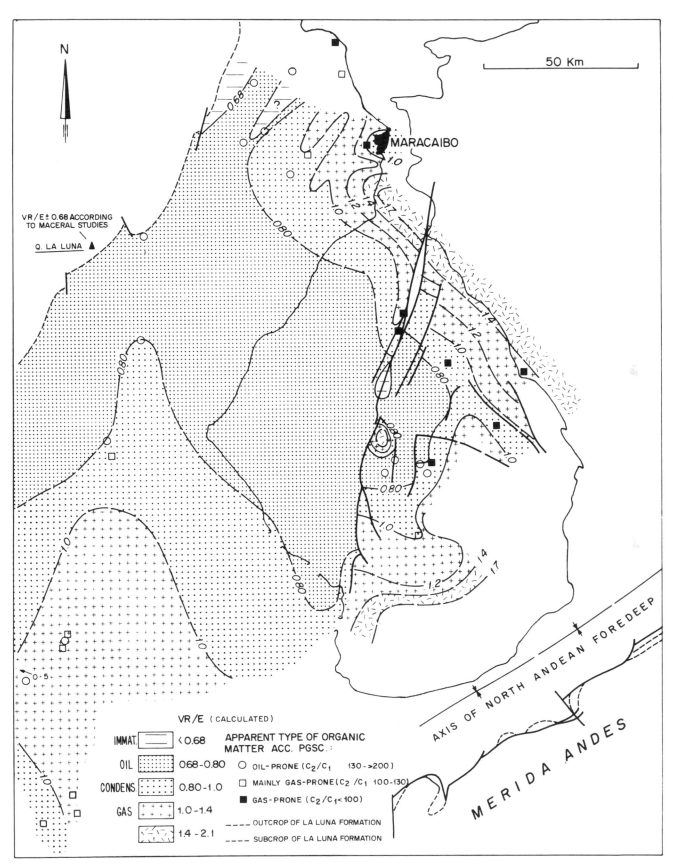

Figure 5. Present-day source rock maturity at top La Luna formation.

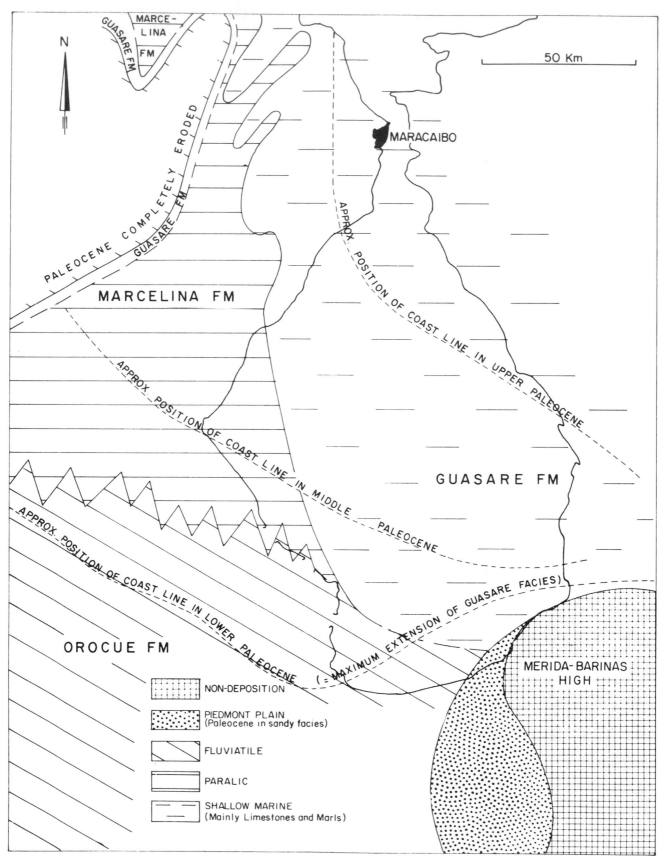

Figure 6. Paleogeographic sketch map of Paleocene sediments.

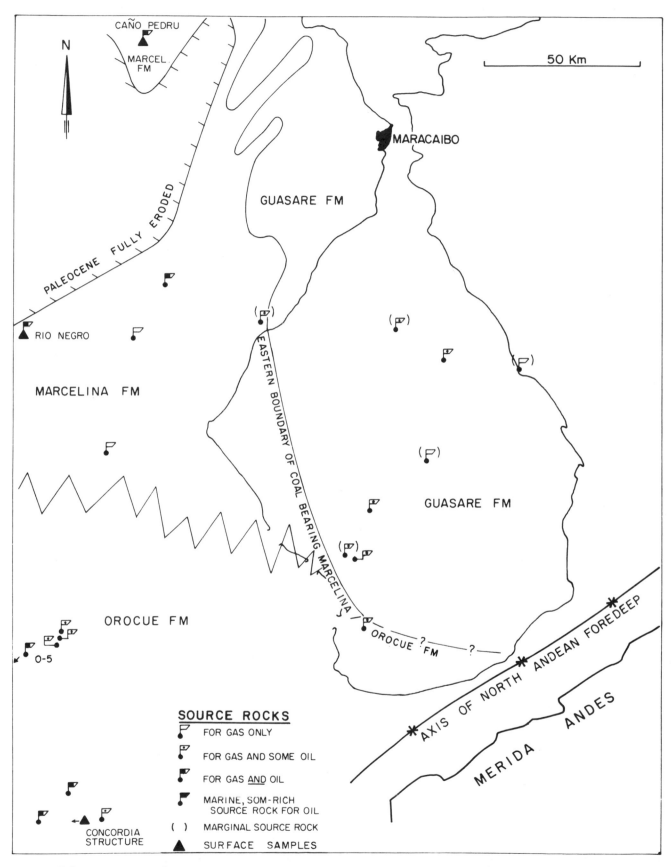

Figure 7. Paleocene source rocks.

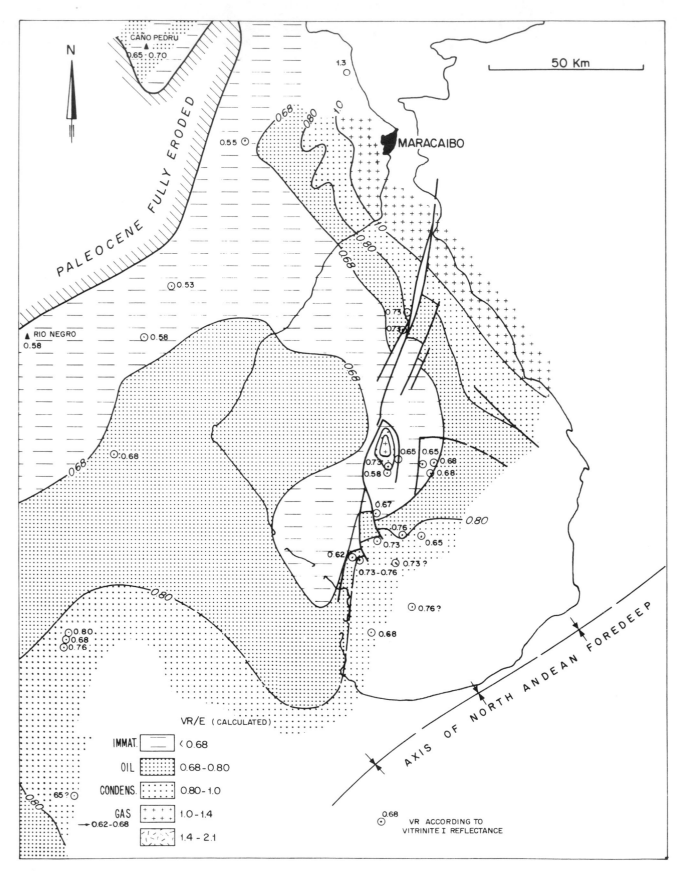

Figure 8. Present-day source rock maturity at base Paleocene.

formation which does not appear to contain significant source rocks.

Based on the above arguments, it appears that mature Paleocene sediments containing significant source rocks may only have reached maturity in the North Andean Foredeep (where source-rock properties are so far unknown) and, perhaps, in parts of the Colón District (in the southwest Maracaibo Basin) and southern Lake Maracaibo.

Early Eocene

The lower Eocene of the Maracaibo Basin can be subdivided into three broad facies belts of marine to terrestrial sediments (Figure 9). Again, sample sequences have been analyzed from the marine and the terrestrial areas (Figure 10).

The marine (deltaic-littoral) formations contain multiple and locally very thick intervals with organic-rich rocks. The total volume of organic matter in early Eocene beds is probably larger than that of the La Luna formation, but consists predominantly of hydrogen-poor land-plant debris. The most favorable organic-rich sediments qualify as source rocks for gas and some oil, whereas many other qualify as source rocks for gas only.

Organic matter of better source-rock quality (land plants with hydrogen-rich components) was recorded in the alluvial plain deposits west of Lake Maracaibo. In these terrestrial deposits source rocks for gas *and* oil are quite common, as is concluded from maceral analysis (Buiskool Toxopeus and Van Lieshout, 1979).

Logically, the maturity picture of the early Eocene source rocks (Figure 11) is even less favorable than that of the Paleocene source rocks. It is almost certain that early Eocene source rocks are mature only in parts of the Colón District (Southwest Maracaibo Basin) and that the maturity/immaturity boundary runs along the northern flank of the North Andean Foredeep (the calculated VR/E values are again somewhat higher as compared with the measured vitrinite-1 reflectance values). The thickness of preserved early Eocene sediments and the presence of good source rocks in the undrilled area of the North Andean Foredeep is speculative. In the northeastern part of the Maracaibo Basin, where early Eocene sediments reached the maximum burial depth in late Eocene times, a large potential oil kitchen and a considerable potential gas kitchen formed. Since the source rocks are mainly composed of humic organic matter in this area, no significant quantities of oil have probably been generated. The gas accumulations in Eocene sandstones of the South Maracaibo area may, however, be derived from the nearby gas kitchen.

Middle and Late Eocene

The facies pattern of the middle Eocene and the extension of late Eocene sediments are shown in Figure 12. The average thickness of the source rocks in the middle Eocene rock formations is smaller than in the early Eocene, but the quality of the organic matter is similar (Figure 13).

In the late Eocene La Sierra formation no source rocks of any importance were found. In contrast, the Carbonera formation contains relatively hydrogen-rich coal which qualifies as source rock for gas *and* oil.

Middle and late Eocene sediments are immature virtually throughout the entire Maracaibo Basin (Figure 14), but may be mature northeast of the Bolivar Coast and in the axial part of the North Andean Foredeep. In this area, however, the thickness and facies of the late Eocene sediments are unknown.

Early Miocene

Major facies belts of the early Miocene sediments in the Maracaibo Basin are shown in Figure 15. In view of the low probability that Miocene sediments are mature anywhere except, perhaps, in the south of the Maracaibo Basin, source-rock analyses of the Miocene La Rosa and Lower Lagunillas sediments have only been carried out for south Lake Maracaibo wells and for well CT-1 (Figure 16).

In four of the eight wells investigated, very promising source rocks were found. Although the sediments investigated are marine, their organic matter content is predominantly land-plant derived, but rich enough in hydrogen that the sediments qualify as good source rocks for gas *and* oil.

Vitrinite-1 reflection measurements from the youngest Eocene rocks clearly indicate that the Miocene source rocks must be immature in the entire Maracaibo Basin, except for the axial part of the North Andean Foredeep (Figure 16).

THE DEVELOPMENT OF OIL AND GAS KITCHENS

Lopatin-type maturity calculations allow calculation not only of the present-day VR/E of a horizon under review, but also the VR/E at different points of time in the course of the burial history. In a similar way, one can also calculate the time at which a certain horizon reached maturity for oil (or gas) generation. Such exercises were carried out. Though the results may not appear to be very accurate, they are good enough to provide a fair picture of the expansion of the oil and gas kitchens in space and time.

Maturity and Depth of Top La Luna at the End of the Eocene

The maturity (VR/E) picture of top La Luna at the end of the Eocene in Figure 17 shows the extension of the oil and gas kitchens of the La Luna source rocks and the estimated depth of the top La Luna at that time. Structural contours and isomaturity lines were still concordant. The figure also shows the spatial relationship between the main Cretaceous oil fields and the oil kitchen. The situation of several oil fields within or at the margin of the La Luna oil kitchen at the end of the Eocene suggests a close relationship between this late Eocene phase of oil generation and these present-day oil accumulations. On the other hand, the oil accumulations in Cretaceous limestones in the Colón District (southwest Maracaibo Basin) are situated very far from the late Eocene La Luna oil kitchen.

Time when the La Luna Source Rocks Reached Maturity

Figure 18 shows the progression of the La Luna oil kitchen during various geological periods. In the northeastern part of the Maracaibo Basin, the La Luna source rocks reached maturity for the first time during the

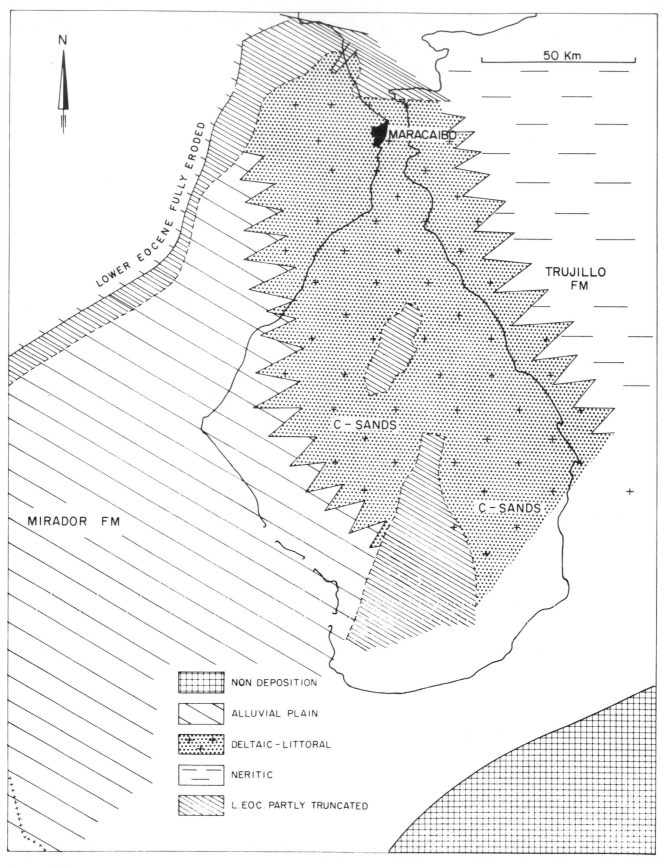

N

50 Km

LOWER EOCENE FULLY ERODED

MARACAIBO

TRUJILLO FM

C – SANDS

C – SANDS

MIRADOR FM

NON DEPOSITION

ALLUVIAL PLAIN

DELTAIC – LITTORAL

NERITIC

L. EOC. PARTLY TRUNCATED

Figure 9. Paleogeographic sketch map of lower Eocene sediments.

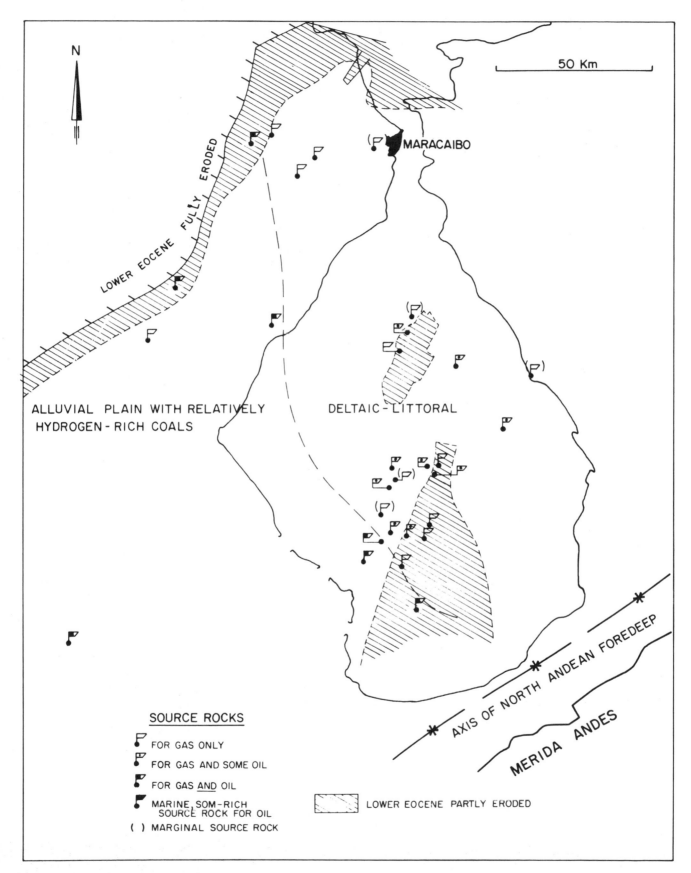

Figure 10. Lower Eocene source rocks.

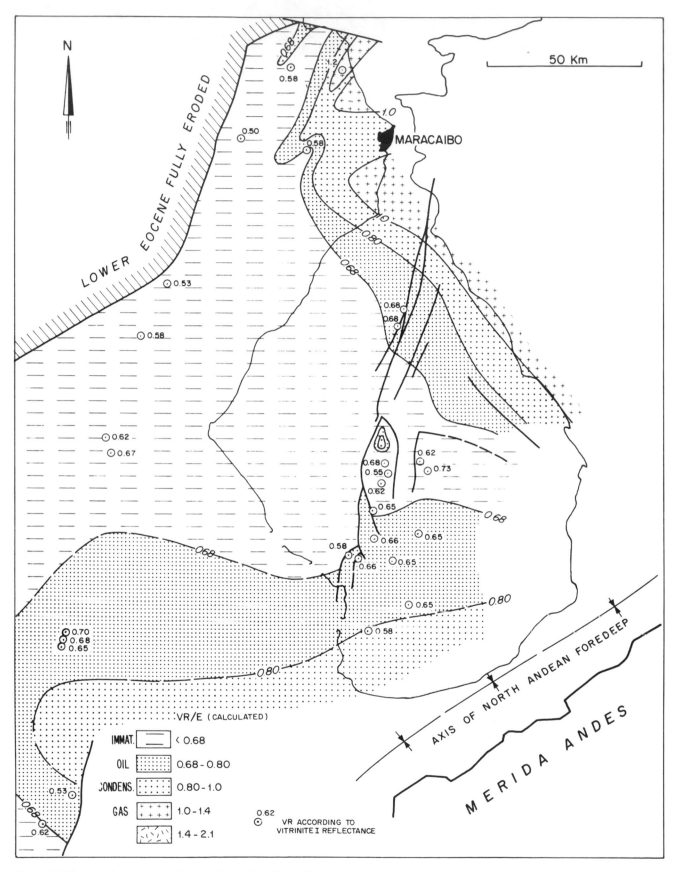

Figure 11. Present-day source rock maturity at base lower Eocene.

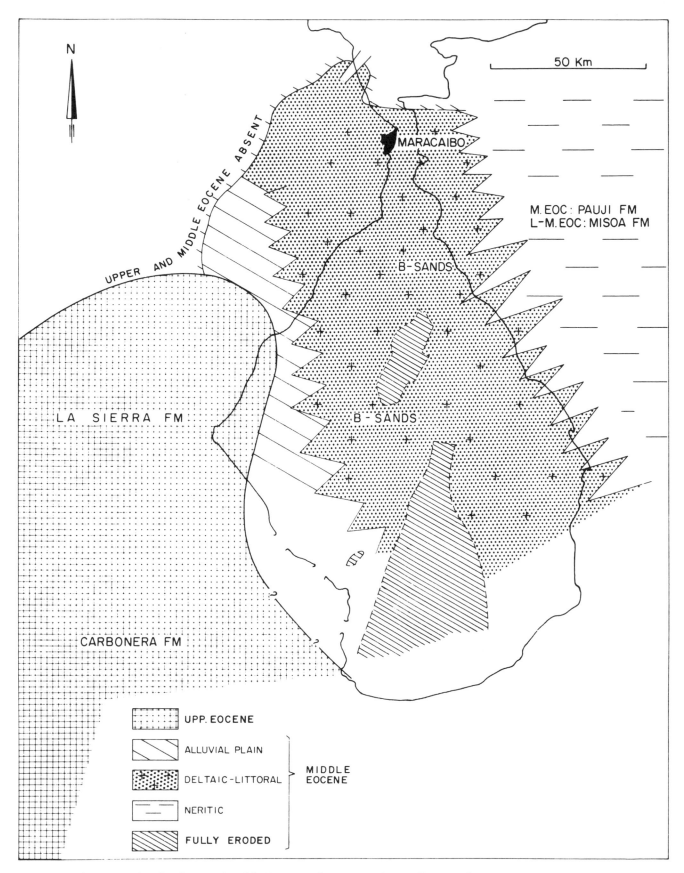

Figure 12. Paleogeographic sketch map of middle Eocene and extension of upper Eocene sediments.

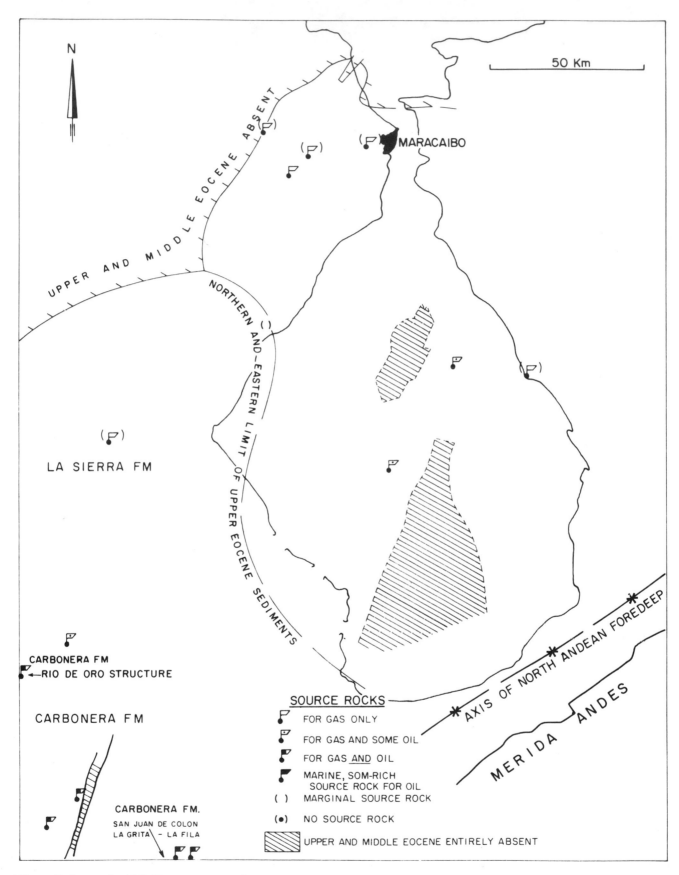

Figure 13. Late and middle Eocene source rocks.

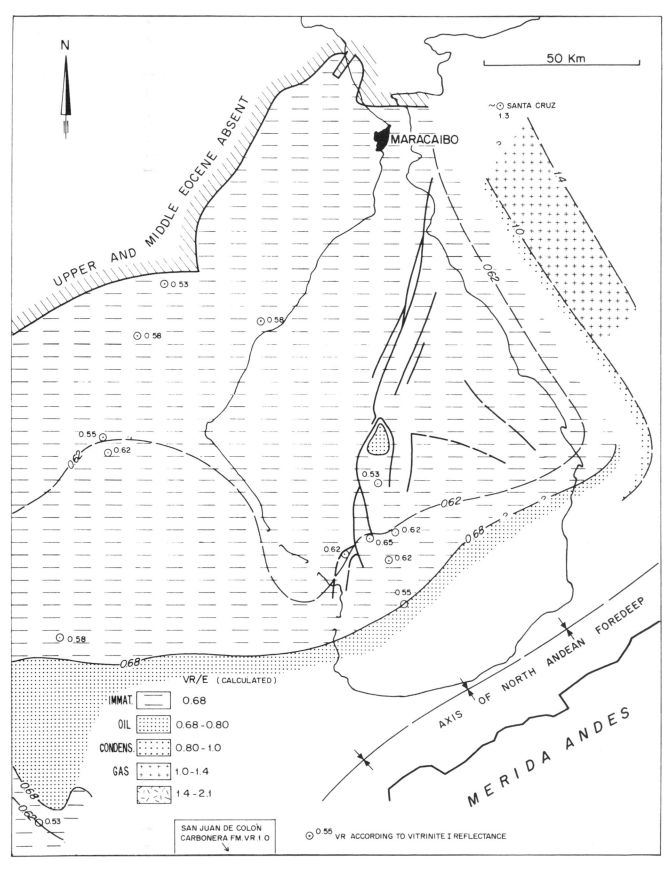

Figure 14. Present-day VR-VR/E at base middle Eocene.

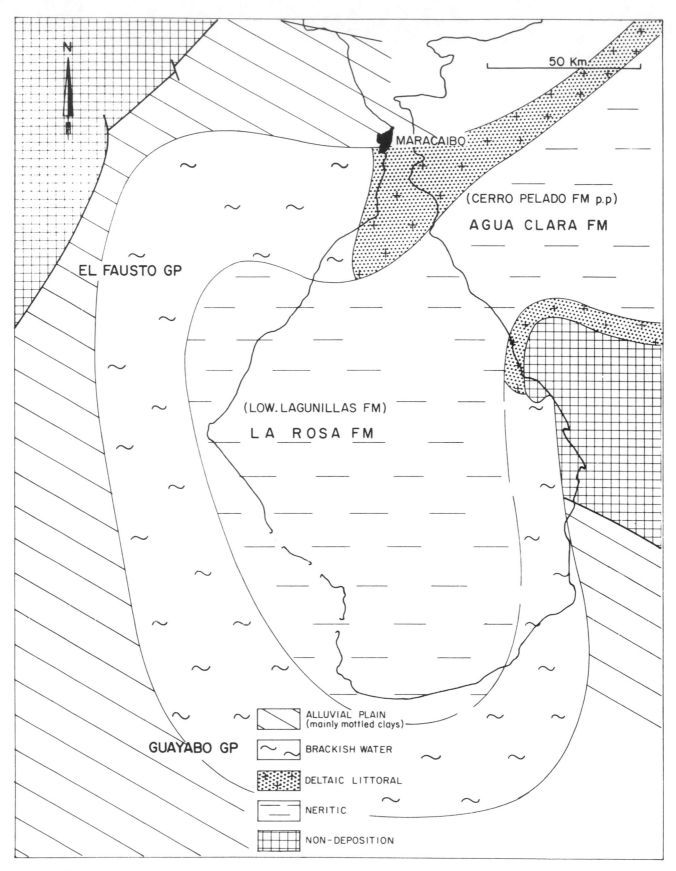

Figure 15. Paleogeographic sketch map of lower Miocene sediments.

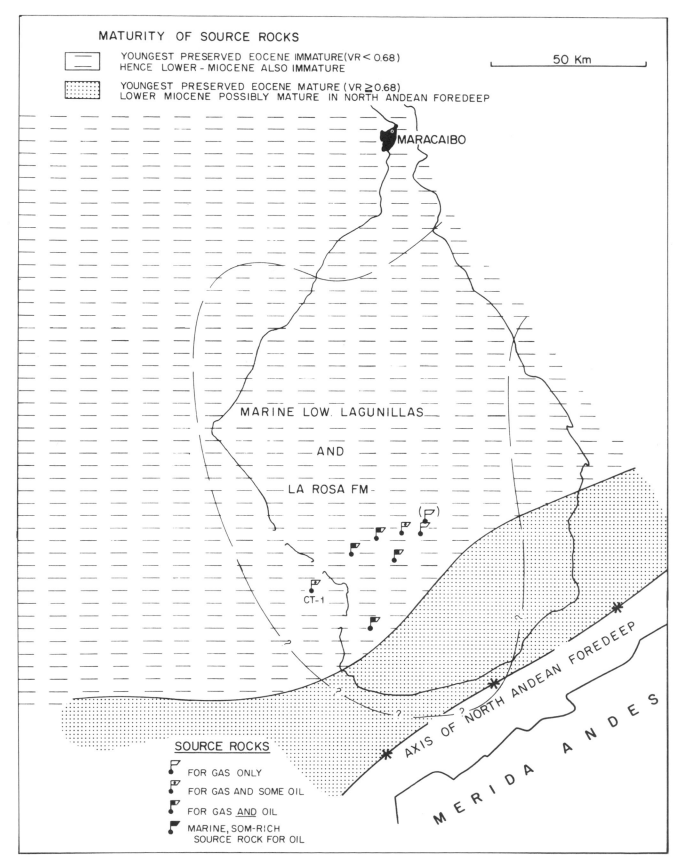

Figure 16. Lower Miocene source rocks.

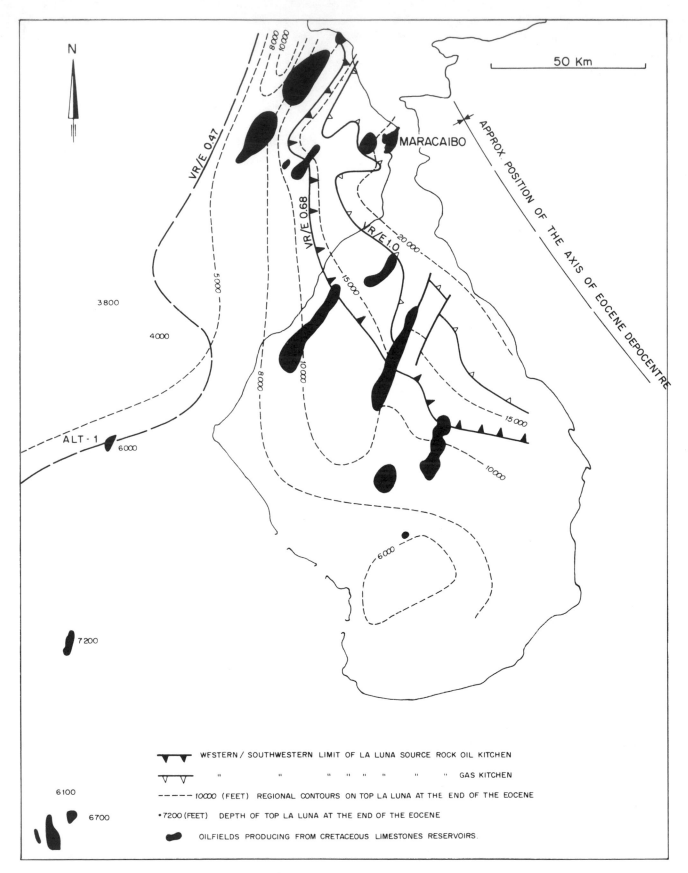

Figure 17. VR/E and depth of top La Luna at the end of the Eocene.

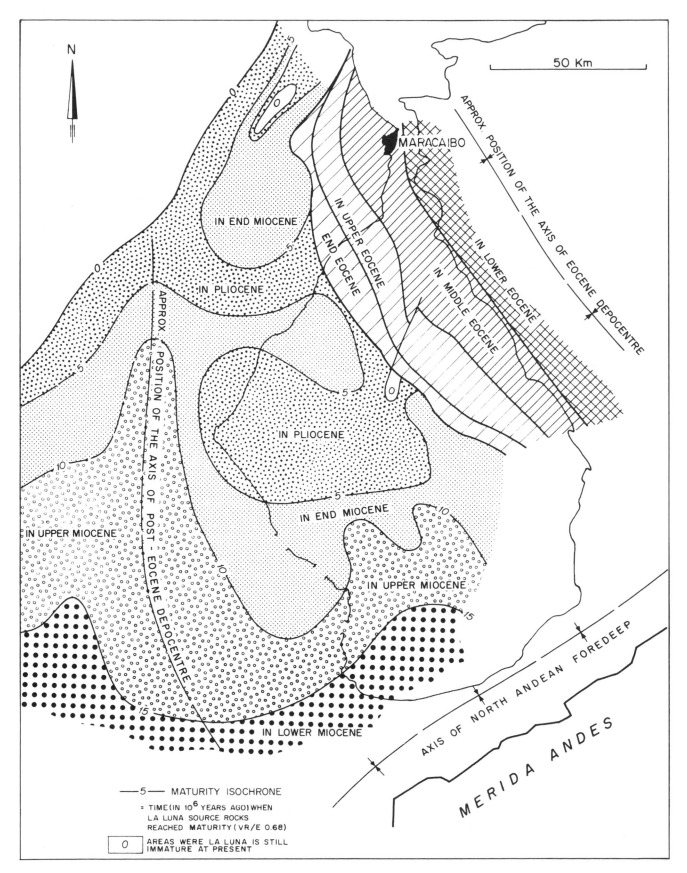

Figure 18. Time when La Luna reached maturity.

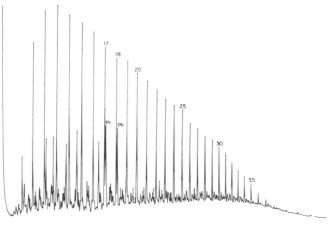

Figure 19. Chromatogram of a crude oil derived from marine, SOM-rich source rocks of the La Luna type.

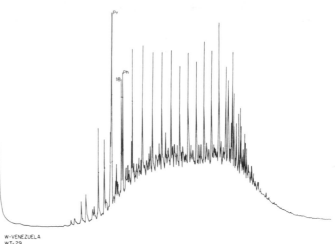

Figure 20. Chromatogram typical of source rocks with land-plant derived organic matter.

early Eocene. Subsequently, the oil kitchen expanded southwestward and reached its widest extension toward the end of the Eocene (Figures 17 and 18). By this time the La Luna source rocks were still immature in the remainder of the Maracaibo Basin.

From late Eocene to early Miocene no significant changes took place in the maturity of the La Luna source rocks. During the Oligocene, large parts of the Maracaibo Basin were affected by uplift and erosion and by reversal of the regional tilt. As a consequence of this tilt reversal, the center of deposition shifted from the northern and northeastern parts of the basin to the southern and southwestern parts. This led to a gradual increase in the burial depth of the La Luna Formation in the southern part of the Maracaibo Basin. During early Miocene times, the La Luna source rocks in the extreme south of the basin became so deeply buried that, according to VR/E calculations, they became mature for oil generation. With continuous post-Eocene sedimentation the critical thickness of the overburden on the La Luna source rock gradually progressed north. In the western Maracaibo Lake area, the La Luna reached maturity only in late Miocene/Pliocene times.

OIL/SOURCE-ROCK EXTRACT CORRELATIONS

Correlation of crude oils with source-rock extracts is an important and powerful method for the recognition of the origin of crude oils. The criteria for such correlation are: similarities between the chemical compositions of crude oils and mature source-rock extracts.

In the Maracaibo Basin, the first crude-oil/source-rock extract correlations were carried out in CSV by Aldershoff (1953a, 1953b) and Gransch and Eisma (1966a). These authors recognized some characteristic properties of the La Luna source-rock extracts, i.e., their high metal-porphyrin content, their V/Ni ratio above unity and their high sulfur content. They analyzed a large number of Maracaibo Basin oils from Cretaceous and Tertiary reservoirs and found that in virtually all cases, the crude oils had the same characteristics as typical extracts from La Luna source rocks. Only in Cretaceous oils from reservoirs situated in

areas where the La Luna source rocks are characterized by a high VR/E (e.g., Sibucara and Concepción) were the contents of sulfur and porphyrins found to be much lower. Based on these results, it was concluded that by far the majority of the oils which had been tested could unambiguously be correlated with La Luna source rocks.

In later years, new analytical methods such as chromatography and mass spectrometry became available and now permit very specific characterizations of the chemical properties of source-rock extracts and crude oils. These methods were applied in some significant cases during the present source-rock project. For example, source rock extracts from typical Tertiary land-plant source rocks (e.g., in well WT-29) were analyzed by chromatographic and mass-spectrometric methods; the results were compared with those of La Luna source-rock extracts and of La Luna-type oils. The chromatograms of Paleocene to Miocene (immature as well as mature) land-plant source rocks are very different from those of the marine La Luna source rock and La Luna derived oils (see Figures 19 and 20). On the other hand, virtually all oils from the Maracaibo Basin produce chromatograms which are comparable with those of La Luna source-rock extracts.

Results of oil/source-rock extract correlation based on porphyrins, sulfur content, and chromatographic and mass-spectrometric characteristics are shown in Figure 21. The figure shows that in our data not a single analyzed crude has appeared to be derived from land-plant source rocks, but that most analyzed oils have a strong affinity with extracts of La Luna source rocks (these results have not been checked by carbon-isotope measurements of crude oils and source-rock extracts). Exceptions are the oils from Concepción and Sibucara which have been expelled at a high VR/E of the source rocks whose chemical composition was already strongly different from the original one (with increasing VR/E, the porphyrin content, V/Ni-ratio and S-content of La Luna extracts decreases significantly).

The combined results of maturity determinations and oil/source rock extract correlations indicate that Tertiary source rocks consisting predominantly of land-plant derived

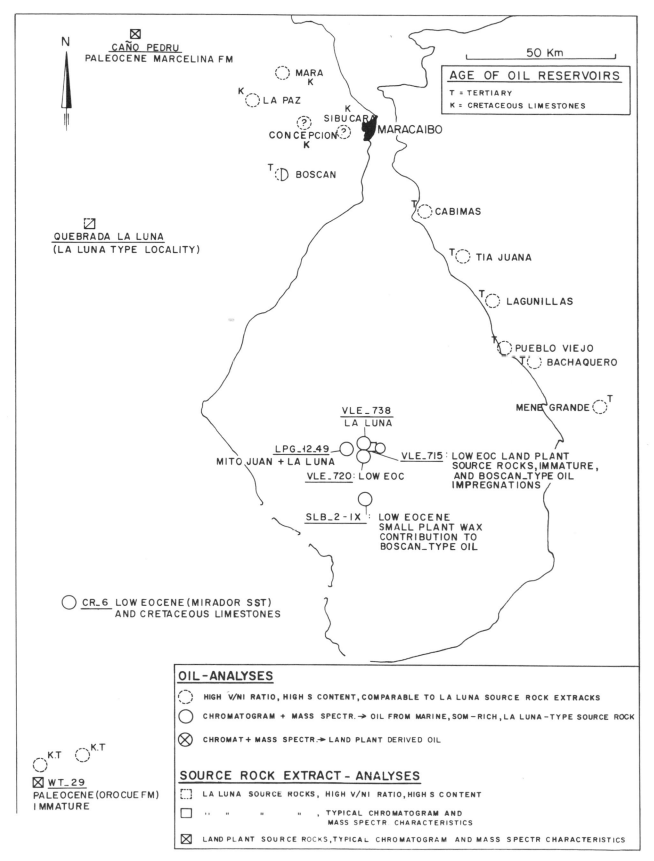

Figure 21. Geochemical crude oil and source rock extract analyses for oil/source rock extract correlation.

organic matter have contributed little, if any, oil to the known oil accumulations in the Maracaibo Basin, and that the bulk of the oil is derived from marine, La Luna-type source rocks. Intercalations with similar but less thick and less consistent marine source rocks have proved to occur in the Lower Cretaceous Cogollo Group (schematically indicated in Figures 2 and 3). Mature Tertiary source rocks occur in the southern part of the Colón District (in the southwest of the Maracaibo Basin) and probably also in the North Andean Foredeep. Nevertheless, until now, no oil accumulations have been found which can be correlated with these potential source rocks.

REFERENCES CITED

Aldershoff, W.G., 1953a, Crude oil studies in Western Venezuela; the establishment of a genetical relation between oil source beds and crude oil by means of tracers (vanadium, nickel, porphyrins): Internally published Shell report.

—— , 1953b, Oil source rock studies in Western Venezuela (Macoa Area, Perigá Trujillo, LB-273 and CQ-1): Internally published Shell report.

Buiskool Toxopeus, J.A.M., and J.B. Van Lieshout, 1979, Maceral analyses of Tertiary and Cretaceous source rocks in the Maracaibo Basin, Venezuela: Internally published Shell report.

Gransch, J.A., and E. Eisma, 1966a, Geochemical aspects of occurrence of porphyrins in West-Venezuelan mineral oils and rocks, in G.D. Hobson and G.C. Speers, eds., Advances in organic geochemistry.

—— , and —— , 1966b, Characterization of insoluble organic matter of sediments by pyrolysis: The C_R/C_T ratio, in G.D. Hobson and G.C. Speers, eds., Advances in organic geochemistry.

—— , and J. Posthuma, 1973, On the origin of sulphur in crudes: Rueil-Malmaison, France, Actes du 6e Congrès Int. Géochim. Organique, p. 727-738.

Lopatin, N.V., 1972, Temperature and geologic time as factors in coalification: Akademiya Nauk SSSR, Seriya Geologicheskaya, Izvestiya, n. 3, p. 95-106 (English translation by N.H. Bostik, Illinois State Geological Survey, February 1972).

Patteisky, K., and M. Teichmüller, 1960, Inkohlungs-Verlauf, Inkohlungs-Masstäbe und Klassifikation der Kohlenauf Grund von Vitrit Analysen: Brennstoff Chemie, v. 41, p. 3-19.

Rod, E., and W. Maync, 1954, Revision of Lower Cretaceous stratigraphy of Venezuela: AAPG Bulletin, v. 38, p. 193-283.

Waples, D.W., 1980, Time and temperature in petroleum formation; application of Lopatin's Method to petroleum exploration: AAPG Bulletin, v. 64, p. 916-926.

The Espirito Santo Basin (Brazil) Source
Rock Characterization and Petroleum Habitat

G. Estrella   K. Tsubone
M. Rocha Mello   E. Rossetti
P.C. Gaglianone   J. Concha
R.L.M. Azevedo   I.M.R.A. Brüning
Petrobras
Centro de Pesquisa e Desenvolvimento Leopoldo A.M. Mello
Rio de Janeiro, Brazil

Two main source rock systems have been identified in the Espirito Santo Basin: the Alagoas shales related to an evaporitic environment of Aptian age, and the Jiquia shales of Upper Neocomian age deposited in a continental to lagoonal environment. The open marine Tertiary to Upper Cretaceous slope sediments, previously believed to have source potential, do not actually constitute a source section.

Almost all the studied oils showed some degree of degradation, probably caused by bacteriological attack. Gas chromatography-mass spectrometry was used in order to provide biomarker information which, together with isotopic studies and gas chromatography, provided reliable oil/oil and oil/rock correlations.

Oil is produced in the basin, mainly from Alagoas sands below the Aptian evaporites and from Tertiary-Upper Cretaceous reservoirs. The proposed oil migration geometry from source rocks to the Alagoas reservoirs is relatively simple, since both source and reservoir rocks have close stratigraphic relationships. Entrapment is provided by structures at the level of the Alagoas anhydrite beds that provide the seal. The generative depression where Alagoas source rocks are presently mature lies a few kilometers east of the discovered accumulations (Rio Itaunas, Sao Mateus), implying relatively short distance migration.

In the case of oil accumulations in Tertiary-Upper Cretaceous turbiditic reservoirs (Lagoa Parda, Fazenda Cedro), the mature Jiquia source rooks are exposed by erosion along the bottom of the submarine canyons along which turbidites were deposited. The complex, coalescent, turbiditic sand bodies acted as a hydrocarbon collecting system and pools were formed where structural or stratigraphic closure exists.

The offshore Cação field is related to a paleogeomorphic closure at the bottom of the submarine canyon. The oil is trapped in Albian reservoirs capped by Upper Cretaceous marine slope shales. It can be explained by a migration geometry including a subcrop of Jiquia mature source rocks below the pre-Alagoas unconformity, and upward migration of the oil through the sandy, permeable Alagoas section, due to the absence of Alagoas anhydrites at this particular location.

INTRODUCTION

The Espirito Santo Basin, located on the eastern Brazilian coast, has an area of about 25,000 sq km (Figure 1). Onshore oil exploration in the basin was initiated during the early 1960s, while offshore operations began in 1968.

More than 100 wildcats have already been drilled in the basin but only a few small-to-medium size oil fields have been discovered. Present-day oil production from the Espirito Santo Basin is about 20,000 barrels per day. The principal production comes from turbiditic reservoirs enclosed in a thick Tertiary and Upper Cretaceous marine slope section, near the bottom of Cretaceous submarine canyons. Thus, both onshore and offshore the main exploration target in this basin has been this type of accumulation, either structurally or stratigraphically closed.

Nevertheless, many apparently low risk prospects, with

regard to oil generation, reservoir, sealing, and closure were drilled and resulted in dry wells.

Reservoir and cap rocks occurred in these wells while seismic data showed excellent structural and/or stratigraphic closures. Previous geochemical characterization of the source rocks in the Espirito Santo Basin, by use of microscopic identification of the type of organic matter, C_1-C_4 light gases for maturation, and C_{15+} gas chromatograms indicated the marine slope shales of Upper Cretaceous to Tertiary age to be very good marine source rocks.

However, with the application of Rock-Eval pyrolysis, vitrinite reflectance, sensitive chromatographic columns and mass spectrometry, these early conclusions had to be radically revised and, furthermore, previously unsuspected source systems became clearly identified. The

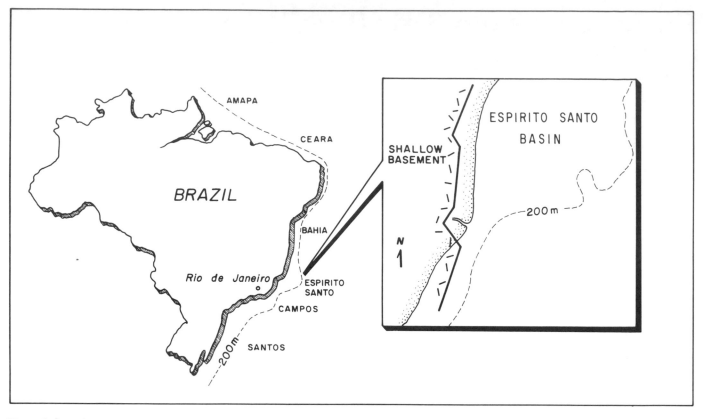

Figure 1. Location map.

explorationist's fundamental question ("why is this structure dry and the other one an oil producer?") can probably, now, begin to be answered. This was the ultimate purpose of this regional geochemical study.

GEOLOGICAL SUMMARY

Tectono-sedimentary history of the Espirito Santo Basin is typical of the Brazilian eastern coastal basins. The stratigraphy and the tectonic style reflect the evolution of a mature, passive, rifted continental margin (Figure 2).

The Espirito Santo Basin is part of a great rift system that resulted from the spreading process between the African and South American continental plates during Early Cretaceous (Neocomian) time.

The lower part of Neocomian series is made up of sandy, conglomeratic continentally-derived sediments, with some volcanic activity. Interpretations suggest that, during the latter part of Neocomian time, shales, sandstones, and limestones were deposited in a marine-influenced (lagoonal) sedimentary environment. These rocks constitute the Jiquia stage, and mark the initial sea incursion that was the precursor of the wide marine transgression that followed in mid-Cretaceous time.

Alagoas (Aptian) time was characterized by tectonic rearrangements resulting in subsidence of the central parts of the basin and the occurrence of relatively widespread unconformity ("pre-Algoas unconformity") along the flanks of the basin, separating the rocks of the Jiquia and Alagoas stages. Evaporites and shales underlain by a sandy, conglomeratic section constitute the Alagoas stage in the basin.

During Albian-Cenomanian time, the already formed proto-oceanic gulf evolved into a narrow seaway along whose margins shallow carbonate platforms developed. This time is characterized by overall tectonic quiescence.

Later, with the continued oceanic crust spreading, strong seaward tilting took place. This tilting in Santonian-Turonian time uplifted the western sediment source areas, and initiated a change from shallow to deeper marine environment, with the consequent widespread deposition of a thick shelf-slope sedimentary wedge (Asmus, 1976; Dauzacker, 1981).

Sedimentation of the shelf-slope marine environment has remained unchanged from Turonian-Santonian time to the present day. Two large fault-controlled submarine canyons (Regencia and Fazenda Cedro) were built during this period, with thick turbiditic sections deposited in them, mainly during the Middle Eocene. The whole of the slope sediments constitute what is called the Urucutuca Slope System. Oceanic volcanism in the form of sea-bottom lavas began near the end of Late Cretaceous time and peaked during the early Eocene, leading to the formation of a large volcanic offshore complex called the Abrolhos High.

Tectonism in the form of eastward gravity sliding of the Albian-Cenomanian shelf sediments over the Aptian evaporites has been, since Late Cretaceous time, an important process for the whole basin. This phenomenon was part of larger halokinetic processes that led to the formation of salt domes in the eastern offshore part of the basin (Araripe et al, 1981) (Figure 3).

Important tectonic reactivation of the main Neocomian faults occurred during Eocene time, regenerating the

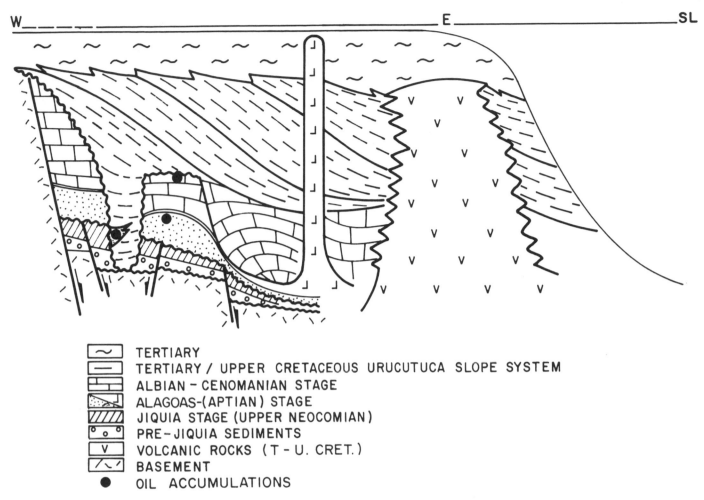

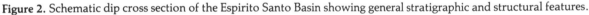

~	TERTIARY
—	TERTIARY / UPPER CRETACEOUS URUCUTUCA SLOPE SYSTEM
▦	ALBIAN - CENOMANIAN STAGE
▨	ALAGOAS-(APTIAN) STAGE
▥	JIQUIA STAGE (UPPER NEOCOMIAN)
∘∘∘	PRE-JIQUIA SEDIMENTS
V	VOLCANIC ROCKS (T - U. CRET.)
⟋⌣⟍	BASEMENT
●	OIL ACCUMULATIONS

Figure 2. Schematic dip cross section of the Espirito Santo Basin showing general stratigraphic and structural features.

sediment source areas. This reactivation resulted in the deposition of a typical prograding sedimentary section.

SOURCE ROCK IDENTIFICATION

Geochemical analyses were carried out by Rock-Eval pyrolysis, total organic carbon, vitrinite reflectance for rock samples as well as gas chromatography, gas chromatography-mass spectrometry, and carbon isotopic analysis for oils and rock extracts. Results of these analyses led to the following conclusions:

Pre-Jiquia Section (Lower Neocomian)

Thin layers of shales from this section have, generally, high organic carbon contents, predominantly Type I kerogen, and therefore good hydrocarbon source potential. Vitrinite reflectance values range from mature to over-mature levels. The carbon-13/carbon-12 isotope ratios and the pattern of C_{15+} gas chromatograms confirm a continental (lacustrine) origin of the organic matter.

Jiquia Stage (Upper Neocomian)

The Jiquia shales are very rich hydrocarbon source rocks (Figure 4), being mature along a large area of the basin (Figure 5). Hydrogen Index and hydrocarbon source potential show very high values (Figures 6 and 7). Type II

kerogen predominates in these rocks (Figure 8). Carbon isotope ratios and gas chromatograms obtained from rock samples suggest a strong marine influence in the depositional environment of the Jiquia shales (Figure 9). The Jiquia shales are now considered the principal source rocks of the Espirito Santo Basin.

Alagoas Stage (Aptian)

Thin layers of black shales associated with evaporites are local hydrocarbon source rocks in the Alagoas Stage (Aptian) of the Espirito Santo Basin (Figure 10).

These rocks are in the right range of maturity only in the northern offshore part of the basin (Figure 11) and probably occur below the basalt flows of the Abrolhos Volcanic High. Hydrogen Index and hydrocarbon source potential values are moderate to high (Figures 12 and 13). Alagoas shales contain Type II kerogen (Figure 14). The whole Alagoas section was deposited in an evaporitic environment and is made of clastic-evaporitic sedimentation cycles (Ojeda y Ojeda, 1982). The black shales were probably deposited in high-salinity, density-stratified, oxygen-poor lagoonal environments.

Albian-Cenomanian Stage

The sediments of this stage are essentially non-source rocks in the basin. This conclusion is based upon very low

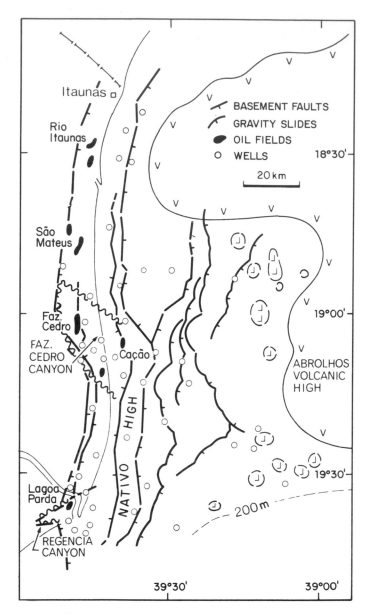

Figure 3. Espirito Santo Basin; structural framework.

values of Hydrogen Index and hydrocarbon source potential, reflecting a shallow, very oxygenated, high energy, marine, shelf environment.

Upper Cretaceous/Lower Tertiary Series

The Urucutuca marine slope system was considered, in the past, to contain effective source rocks. This assumption was reinforced by the presence of the largest oil accumulations (Lagoa Parda, Fazenda Cedro) in Campanian/Maastrichtian and Paleocene turbiditic reservoirs. These Urucutuca shales have moderate to high organic carbon contents but the potential of the rocks to generate hydrocarbons, as demonstrated by Rock-Eval pyrolysis, is extremely poor (Figure 15). These rocks contain predominantly Type III, gas-prone, kerogen (Figure 16).

As shown by vitrinite reflectance, the Upper Cretaceous sediments vary from immature, near the western border, to overmature in the deepest, offshore, parts of the basin. A

paleothermal unconformity, defined within the lower Tertiary section, is discussed later in this report.

The highly oxygenated conditions, prevalent during Late Cretaceous and Early Tertiary, are confirmed by benthonic foraminifera studies. The high number of benthic genera in the studied samples and the absence of other genera characteristic of oxygen-poor environments, show that these rocks were deposited in normal, well-ventilated, marine conditions (Figure 17).

These facts make clear the observation that the Cenomanian-Turonian and Coniacian-Santonian "oceanic anoxic events" (Jenkins, 1980) have had no significant influence on organic deposition in the Espirito Santo Basin. This reinforces the point made by Demaison et al (1983) that "oceanic anoxic events" should not be interpreted as universally leading to ubiquitous oil source bed deposition. "Oceanic anoxic events" need reinforcement by other factors (for instance restricted circulation or upwellings) to become locally or regionally effective.

PALEOTHERMAL UNCONFORMITY

A paleothermal unconformity was defined in the Espirito Santo Basin. It is well depicted in vitrinite reflectance profiles obtained in some wells drilled in the basin (Figure 18). This paleothermal unconformity is well marked along the central offshore part and gradually disappears toward the western margin of the basin. Its occurrence is limited to the lower part of the Paleocene-Lower Eocene section. In some cases there is evidence of an association with a physical unconformity but, generally, this is not clear (Figure 19).

This paleothermal unconformity is interpreted in this report as a consequence of a major tectonic event whose paroxysm occurred during early Eocene time, when the main basement faults were reactivated. As an associated phenomenon, strong magmatic activity took place with an anomalously high geothermal flow. The locally observed stratigraphic unconformity is also interpreted as a result of this basinwide tectonic process.

GEOCHEMICAL CORRELATIONS

In the Espirito Santo Basin, oil has flowed commercially and non-commercially from reservoirs ranging in age from Aptian to Tertiary.

Paleocene-Lower Eocene and Campanian-Maastrichtian turbidites, as well as Albian-Cenomanian limestones-sandstones, both related to submarine canyons, are the main producers in the middle and southern parts of the basin (Lagoa Parda, Fazenda Cedro, and Cacao oil fields). Alagoas conglomerates produce in the northern part of the basin (Rio Itaunas, Sao Mateus, and Rio Preto oil fields).

To establish which source rock generated the produced oils, geochemical correlations through gas chromatography, isotopic analyses, and gas chromatography/mass spectometry of oils and rocks were performed. Due to the moderate-to-severe degradation of the oils (Figure 20), the chromatograms were at a disadvantage when used for oil/rock correlation purposes. Concerning correlation among less degraded oils, saturated hydrocarbon gas

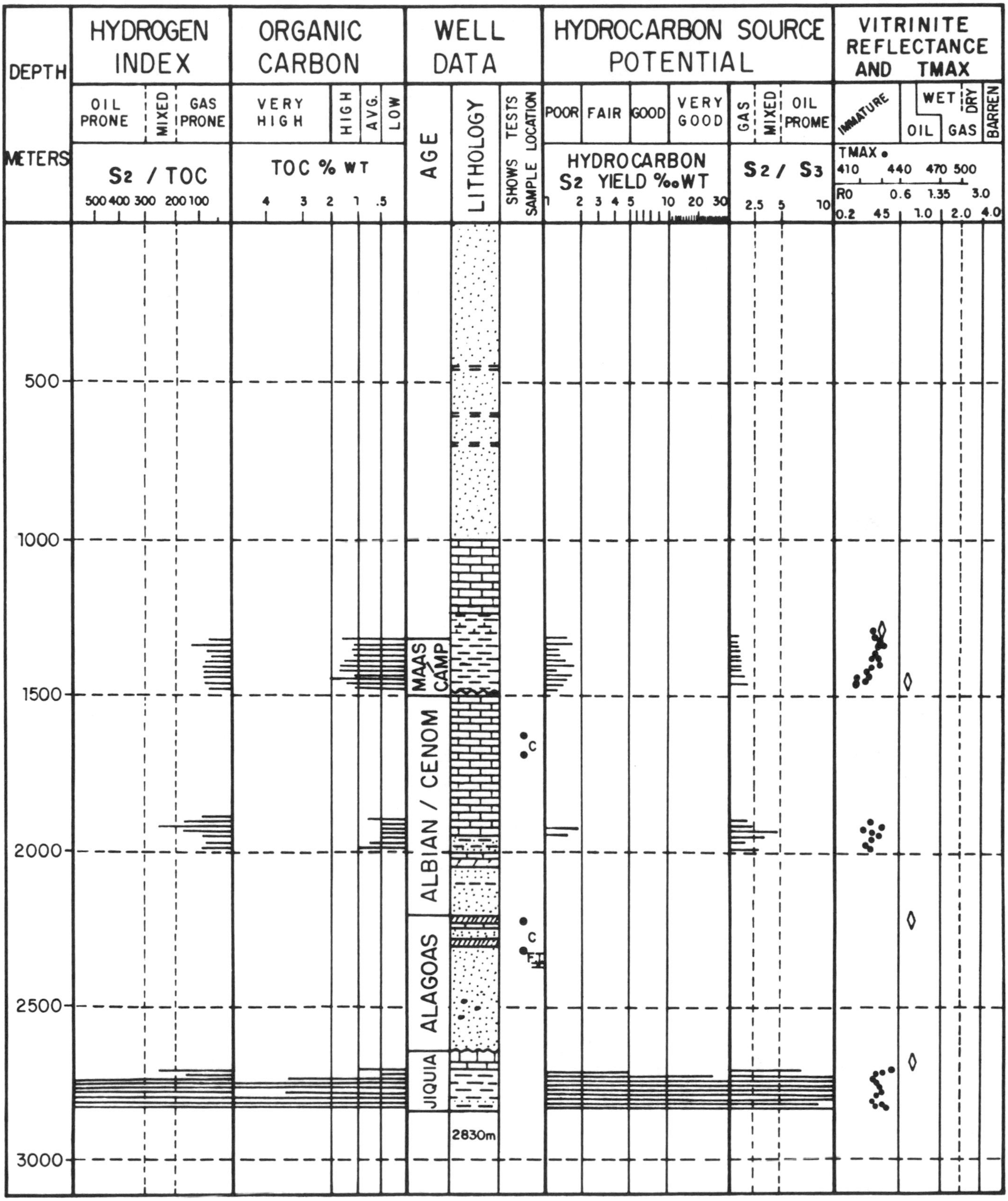

Figure 4. Geochemical log for well 1-PA-1D showing very good source characteristics in Jiquia shales.

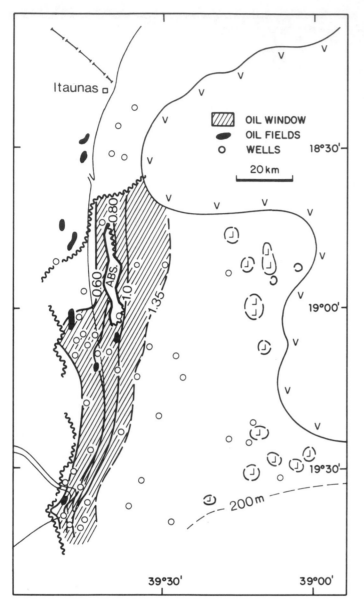

Figure 5. Vitrinite reflectance (Ro); top of Jiquia stage (Neocomian).

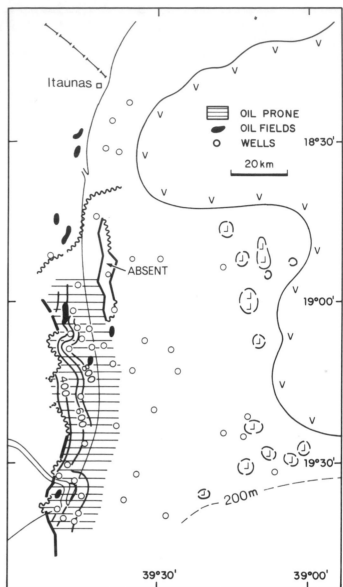

Figure 6. Hydrogen index; Jiquia stage (Neocomian).

chromatograms could be used for typical crude oils from Cacao, Fazenda Cedro, and Lagoa Parda oil fields (Figure 21). These three saturate fractions display nearly replicate gas chromatographic traces implying a great deal of similarity among the crudes. These oils are mature and appear to have derived from the same organic facies, regardless of the age of the reservoir or the location of the oil field in the basin. All these oils have the general appearance of oils generated by marine rather than continental source rocks.

Stable carbon isotope ratios for the analyzed oil samples indicate a single family of oils in Espirito Santo Basin. The marine character is shown by the clustering of the points (Figure 22). An area for typical continental lacustrine Reconcavo type crudes is referred in this same figure (Rodrigues, 1978).

To correlate the various oils of Espirito Santo Basin with

their potential source rocks, the saturate fraction, obtained by liquid chromatography, of oils and extracts was analyzed by gas chromatography/mass spectometry (GC/MS) for biomarkers study purposes. For each sample, mass fragmentograms were obtained for terpanes (Seifert, 1977) and demethylated hopanes. Due to a low concentration of steranes in the samples, mass fragmentograms for these hydrocarbons were not obtained.

Demethylated hopanes present in the majority of the oils, whatever the age of the producing reservoirs, suggest that these oils have possibly undergone biological degradation. However, the analyzed oils contain straight chain paraffins and isoprenoids. In this situation, the presence of hopanes is not expected (Seifert, 1979). The described observations lead to the conclusion that there is a coexistence of unaltered and biologically degraded oils in the same reservoirs. This could be the result of the interruption, or

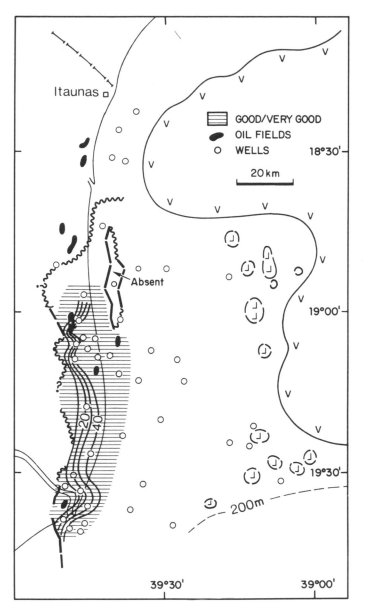

Figure 7. Hydrocarbon source potential; Jiquia stage (Neocomian) (kg HC/ton rock).

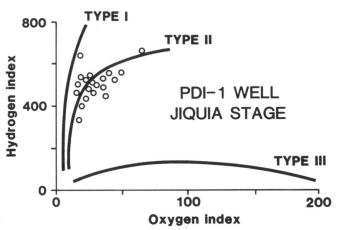

Figure 8. Van Krevelen-type diagram for the Jiquia black shales.

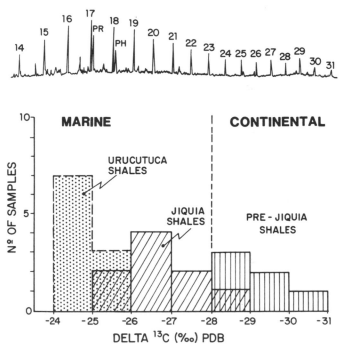

Figure 9. Marine character of Jiquia shales suggested by gas chromatogram, and carbon isotope ratios. Note the comparison of Jiquia with Urucutuca and pre-Jiquia shales.

diminution, of the biodegradation process during the oil charging of the reservoir due, perhaps, to a complex generation-migration history.

A common characteristic of the oils is the presence of gammacerane terpane. This compound is absent in extracts of the marine Cretaceous and Tertiary Urucutuca slope system shales. Large quantities of gammacerane, however, are found in extracts of Jiquia and Alagoas shales, which are documented source rocks (Figure 23). This biomarker evidence provides additional proof that Jiquia and Alagoas shales are the source rocks of the known oils in the Espirito Santo Basin.

HYDROCARBON GENERATIVE AREAS

Two hydrocarbon generative areas ("hydrocarbon kitchens") were outlined in the Espirito Santo Basin through

correlation of geological, geophysical and geochemical data. The Jiquia source rocks are interpreted as mainly responsible for the generation of oils pooled in the Fazenda Cedro, Cacao, and Lagoa Parda fields. The Jiquia effective source rock fairway occurs in the eastern part of the basin only (Figure 24). Considering that there is no definite geological information about the Jiquia black shale depositional basin, no speculation can be made about the eastward extension of the Jiquia prospective area. However, projection of the data into the regional structure characterizes the eastern part of the Nativo High, in the far offshore, as an overmature area. This places it, at Jiquia level, in the oil destruction window (Figure 24).

For the Alagoas section, the mature oil source area is

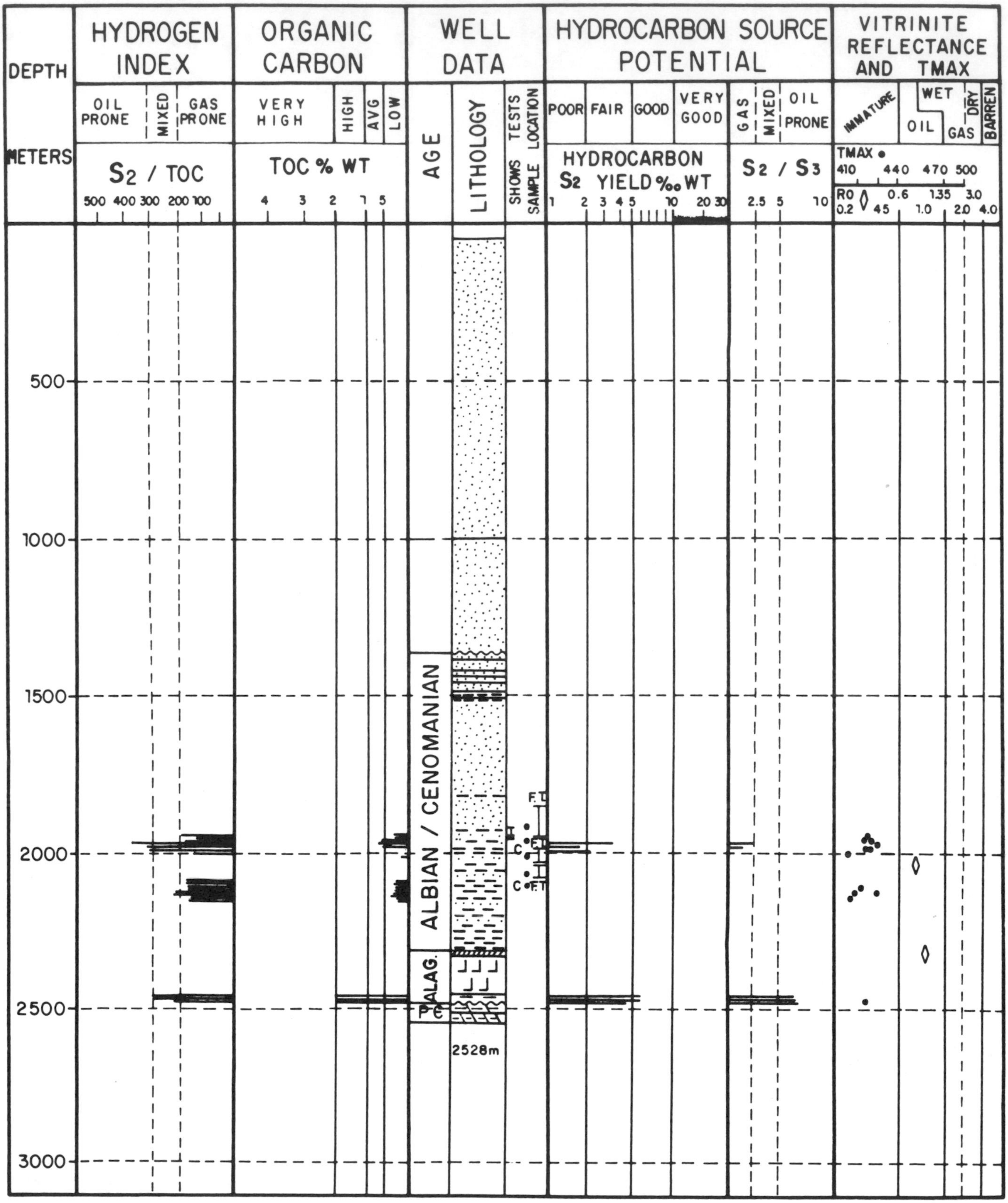

Figure 10. Geochemical log for well 1-ESS-33 showing good source potential for the Alagoas shales.

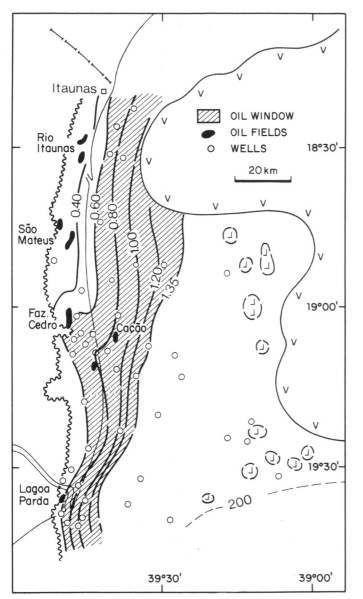

Figure 11. Vitrinite reflectance (Ro); top of Alagoas stage (Aptian) (mg HC/g org. carbon).

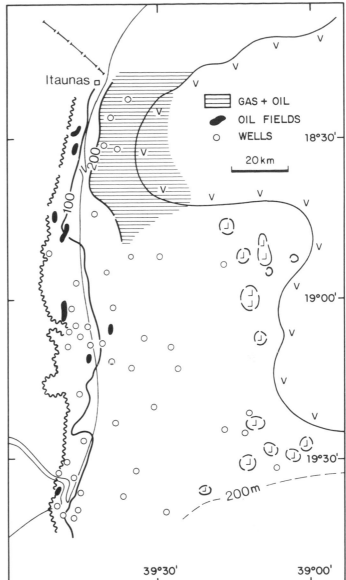

Figure 12. Hydrogen index; Alagoas stage (Aptian).

interpreted as the "hydrocarbon kitchen" for the onshore Rio Itaunas and Sao Mateus fields as well as for some offshore non-commercial flows. The Alagoas hydrocarbon generative area is believed to extend eastward, below the Lower Eocene basalts of the Abrolhos complex (Figure 25).

HYDROCARBON PATHWAY MODELS

Geological interpretations have been made to explain how hydrocarbons migrated from the Jiquia and Alagoas source rocks to their present-day, younger and shallower reservoirs. Accordingly, some speculative hydrocarbon pathway models were constructed.

Hydrocarbons Generated in Jiquia Source Rocks

As stated before, Jiquia shales are interpreted from geochemical data, including biomarker studies, as the source rocks of the submarine canyon related oil fields. This can be geologically explained, in the case of turbiditic reservoirs in Fazenda Cedro and Lagoa Parda oil fields, by the subcrop of mature Jiquia source beds along the bottom of the canyon. These source beds are in direct contact through the walls of the canyons with the turbiditic reservoirs deposited during the Late Cretaceous-Early Tertiary. The widespread, coalescent turbiditic bodies acted as "hydrocarbon collectors" in the bottom of the canyons, allowing updip secondary migration of the oil. Entrapment occurred by structural or stratigraphic closures in the turbiditic lenses. Sealing is provided by marine slope Urucutuca shales (Figure 26A).

In the case of the Cacao oil field, a different geologic interpretation is proposed. Oil is produced from Albian-Cenomanian reservoirs, paleogeomorphically enclosed and sealed by Urucutuca marine slope shales. The

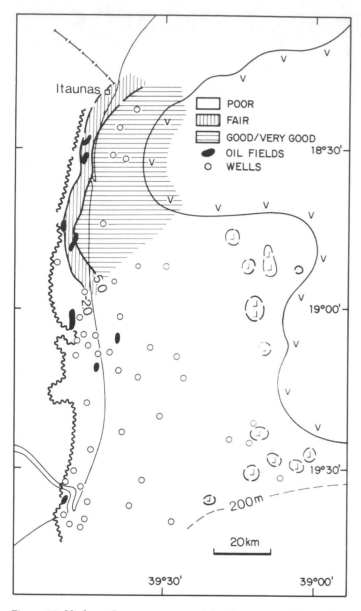

Figure 13. Hydrocarbon source potential; Alagoas stage (Aptian) (kg HC/ton rock).

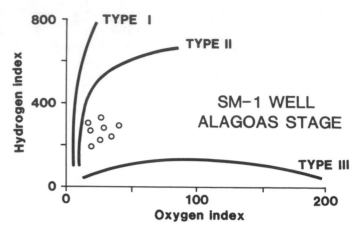

Figure 14. Van Krevelen-type diagram for Alagoas black shales.

Hydrocarbons Generated in Alagoas Source Rocks

The accumulations are easily explained because of a direct stratigraphic relationship between source rocks and reservoirs. In the onshore Rio Itaunas, Sao Mateus, and other producing fields at the north of the basin, the mature Alagoas shales lie deeper, eastward in the offshore. Thus, we can document a few kilometers lateral updip secondary migration of the oil. Sealing is provided by evaporitic, anhydrite beds and the location of accumulations is determined by structural closures occurring at the same levels (Figure 27A).

Prospectivity for Albian-Cenomanian or even younger reservoirs mainly depends upon the occurrence of opened "windows" in Aptian evaporitic seals and of impermeable layers in the section above Alagoas source rocks.

Down to the basin, gravity sliding can provide both, the stretching out of Alagoas evaporites, allowing the upward secondary oil migration, and the structural closure necessary for the entrapment. Overlaying sealing shales occur in the Albian-Cenomanian or higher up in the Urucutuca slope system (Figure 27B).

CONCLUSIONS

In the Espirito Santo Basin, we have demonstrated that deep marine slope Upper Cretaceous/Tertiary shales are not effective petroleum source rocks. The preserved organic matter in the Urucutuca marine slope shales is almost totally composed of terrestrially-derived Type III kerogen. Planktonic preservation was poor due to a persistent open-marine, well-oxygenated environment of deposition.

Jiquia (Late Neocomian) shales are very good and reasonably widespread source rocks. We can document that the oil accumulated in the Regencia and Fazenda Cedro (Cação included) canyons was generated in these rocks. The interpreted migration pathway for these pools consists in a direct contact between source and reservoir rocks along the bottom of the canyons and, in the case of Cacao field, in the absence of impermeable Alagoas anhydrite beds, permitting upward migration of the oil which is sealed by the Urucutuca slope shales.

Alagoas stage shales are the documented source beds for the oil discovered in the northern part of the basin (Rio

hydrocarbon pathway consists of upward secondary migration of the oil from Jiquia source rocks to the Albian-Cenomanian section. Two geological circumstances determine this pathway: a) the subcrop of Jiquia source rocks under the Alagoas section through an angular unconformity, and b) the absence of sealing evaporites at the top of the Alagoas sandy, permeable section. This allowed the upward secondary migration and the oil entrapment under the Urucutuca marine shales, in a paleogeomorphic feature sculptured over the Albian-Cenomanian carbonate platform (Figure 26B). The angular subcrop of Jiquia source rocks seems to be related to truncation on the eastern flank of the Nativo High. The absence of Alagoas sealing evaporites is probably due to the gravity sliding process which created a "migration window" in this otherwise widespread anhydrite seal.

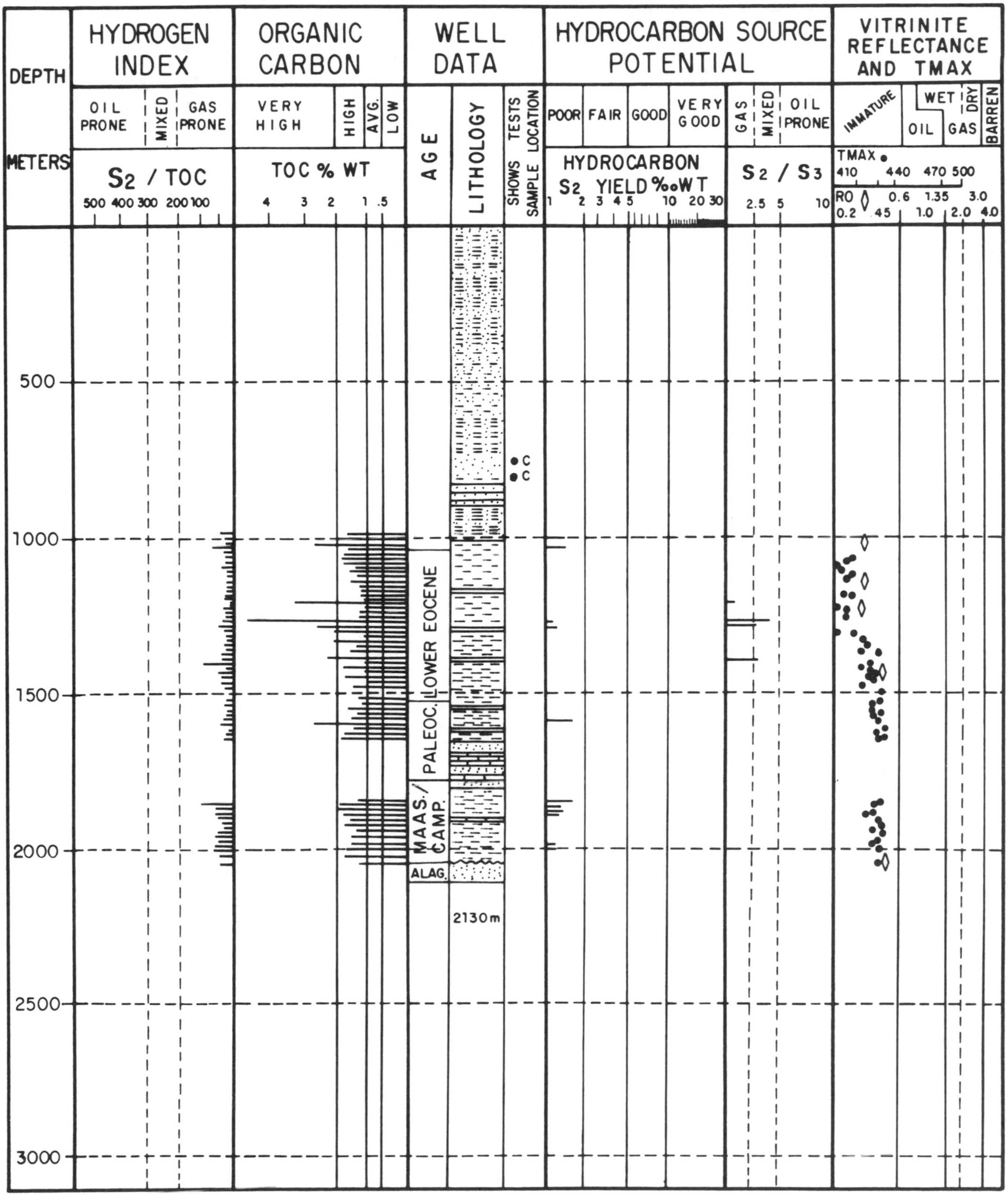

Figure 15. Geochemical log for well 3-LP-22, typical of the Urucutuca slope system deposited in Late Cretaceous to Tertiary. It is void of hydrocarbon source potential.

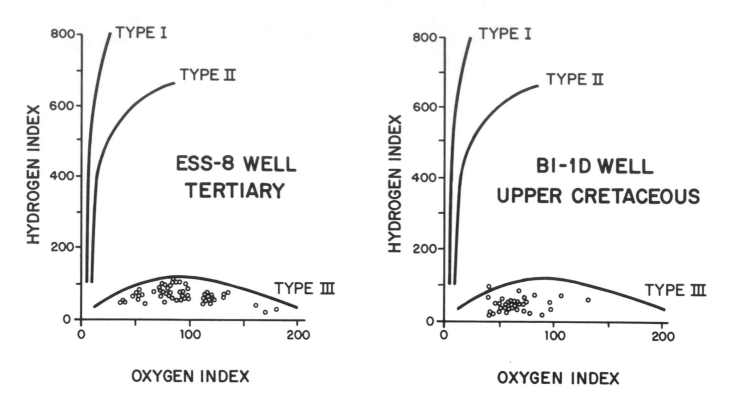

Figure 16. Van Krevelen-type diagrams for the Urucutuca marine slope shales.

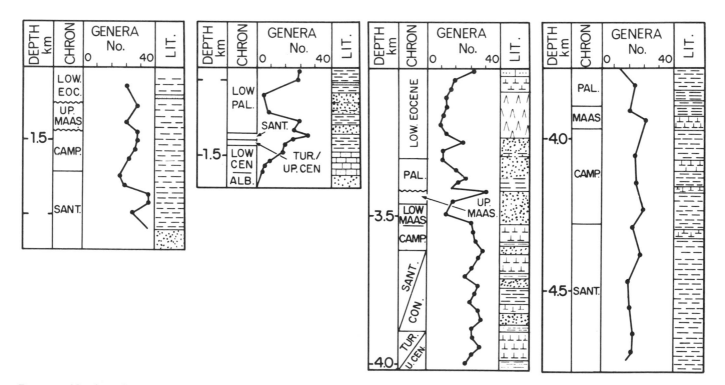

Figure 17. Number of genera of benthic foraminifera in some wells of the Espirito Santo Basin, showing high values along the marine slope shales of the Urucutuca System, reflecting normal oxygenated conditions.

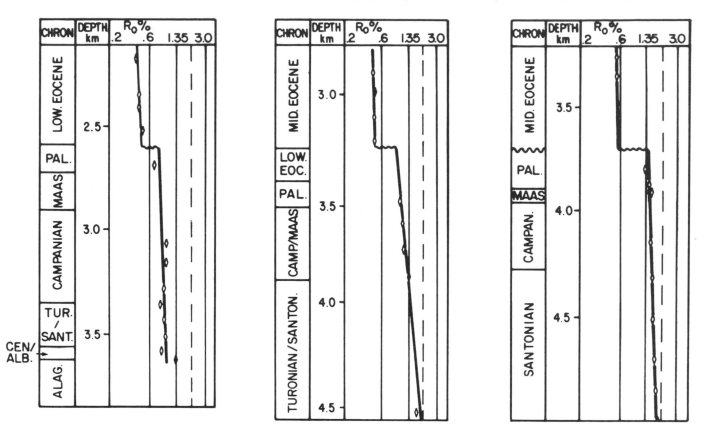

Figure 18. Maturation profiles of some offshore wells showing the occurrence of a paleothermal unconformity in the Paleocene-Lower Eocene section.

Itaunas, Sao Mateus) and the proposed pathways for these accumulations suggest a few kilometers of updip, lateral migration from the east. Jiquia and Alagoas shales contain mainly Type II kerogen.

Regarding maturation, a major paleothermal unconformity was detected by vitrinite reflectance studies in the central, offshore part of the basin, inside the Paleocene-Lower Eocene marine section. It is probably related to an anomalous past heat flow which was a product of the intense volcanic activity associated with the rise of the Abrolhos High. The presence of this paleothermal event has profound implications on the prospectivity of Cretaceous sediments in the far offshore. It could not have been detected by maturation modeling based on present-day geothermal gradients.

There are distinct signs of biodegradation, sometimes very severe, in almost all the studied oil samples. Therefore, biomarker interpretations, together with gas chromatography were simply indispensable to clarify and solve geochemical oil/oil and oil/rock extract correlations. Additionally, analytical procedures involving Rock-Eval pyrolysis and vitrinite reflectance were mandatory to define and characterize with complete clarity the hydrocarbon source rocks in the Espirito Santo Basin.

As a result of this study, we are confident that the same methods can be successfully applied to other Brazilian

offshore basins, and that other generation-migration settings similar to those of the Espirito Santo Basin will be found.

ACKNOWLEDGMENTS

Many other Petrobras earth scientists participated in this project, particularly: L.P. Quadros (vitrinite reflectance), T. Takaki and A.L. Soldan (geochemical analyses), R.L. Antunes (paleontological analyses), and L. Sousa Lins and A.C. Fraga in technical support.

We are specially grateful to Chevron Scientists Gerard Demaison and Wolfgang Seifert whose high level geochemical consulting helped make this study possible. We also thank the management of Petrobras for permission to publish this report.

REFERENCES

Araripe, P., et al, 1981, Projeto domos de sal de Barra Nova: PETROBRAS/DEPEX, Internal Report No. 1030305. (Presented at the 1982 AAPG Annual Meeting, Calgary, Alberta.)

Asmus, A., 1976, Conhecimento atual da margem continental brasileira: PETROBRAS/DEPEX, Internal Report No. 035044. (Presented at the 4th Latin-American

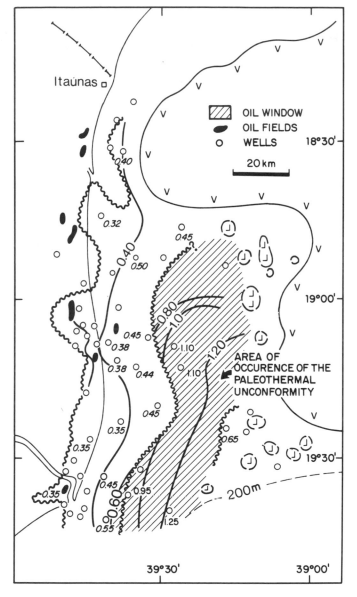

Figure 19. Vitrinite reflectance (Ro); top of Paleocene-Lower Eocene.

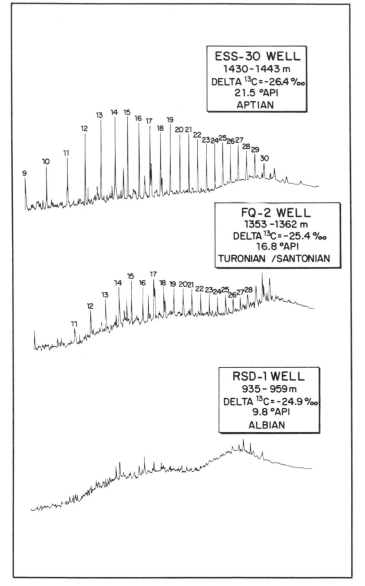

Figure 20. Gas chromatograms of the saturate fraction of crude oils showing different degrees of degradation in Espirito Santo Basin.

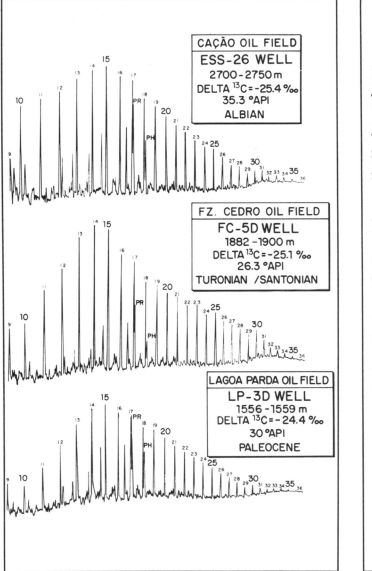

Figure 21. Gas chromatograms of the saturate fraction showing correlation among oils in Espirito Santo Basin.

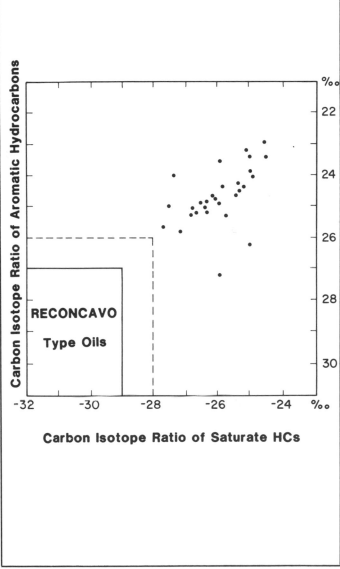

Figure 22. Stable carbon isotope ratios correlating oils from Espirito Santo Basin (black dots), compared with the area of occurrence of continental Reconcavo-type oils.

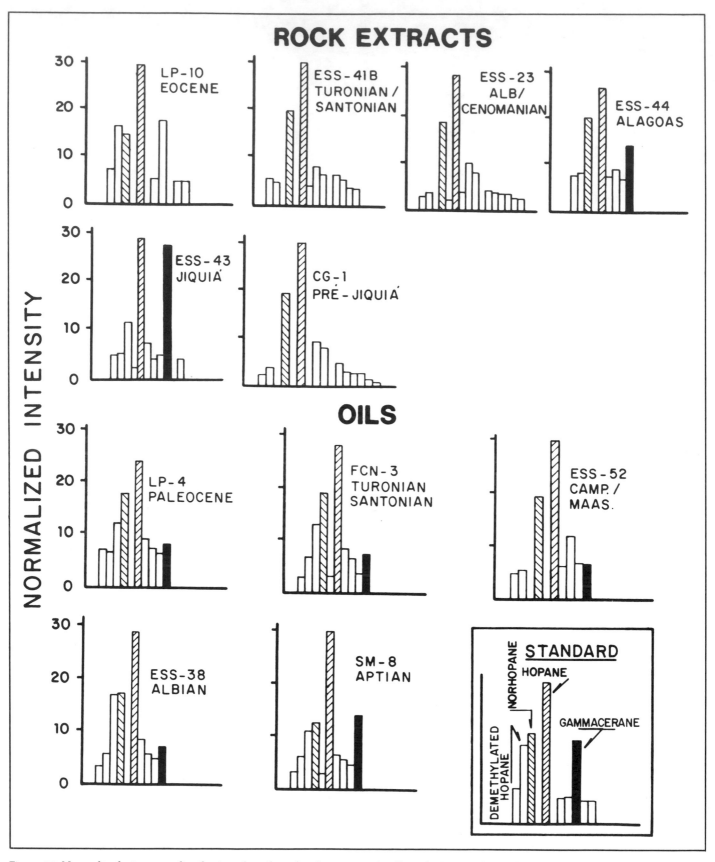

Figure 23. Normalized triterpane distributions for oils and rock extracts of wells in the Espirito Santo Basin. Note that gammacerane, which is present in all the oils, only exists in Alagoas and Jiquiá source rocks.

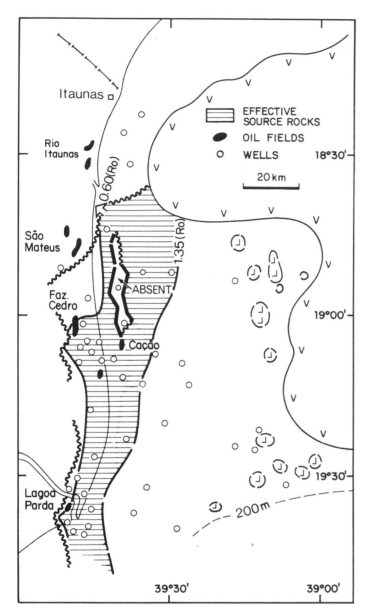

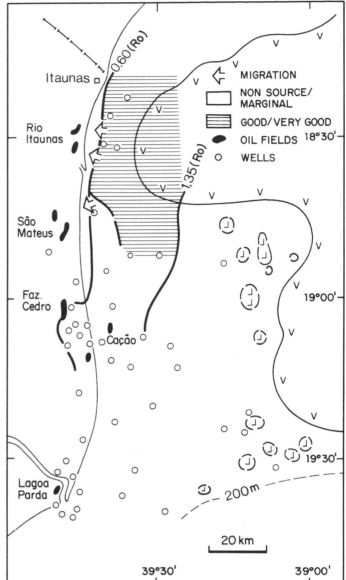

Figure 24. Jiquia stage (Neocomian); hydrocarbon generative areas.

Figure 25. Alagoas stage (Aptian); hydrocarbon generative areas.

Geological Congress, Port of Spain, Trinidad and Tobago, 1979.)

Dauzacker, M., 1981, Basin analysis of evaporitic and post-evaporitic depositional systems, Espirito Santo Basin, Brazil, South America: Austin, University of Texas, Ph.D. dissertation.

Demaison, G., et al, 1983, Source bed stratigraphy; a guide to regional petroleum occurrence: London, 11th World Petroleum Congress.

Jenkyns, H.C., 1980, Cretaceous anoxic events; from continents to oceans: Journal of the Geological Society of London, v. 137, p. 171-188.

Ojeda y Ojeda, H., 1982, Structural framework stratigraphy and evolution of Brazilian marginal basins: AAPG Bulletin, v. 77, p. 732-749.

Rodriques, R., 1978, Aplicacao de isotopos estaveis de carbono e oxigenio no exploracao de petroleo: Rio de Janeiro, 10th Brazilian Petroleum Congress.

Seifert, W.K., 1977, Source rock/oil correlations by C_{27}-C_{30} biological marker hydrocarbons, *in* R. Campos and Gomi, eds., Advances in organic geochemistry: Madrid, p. 21-24.

———, and J.M. Moldowan, 1979, The effect of biodegradation on steranes and terpanes in crude oils: Geochimica et Cosmochimica Acta, v. 43, p. 111-126.

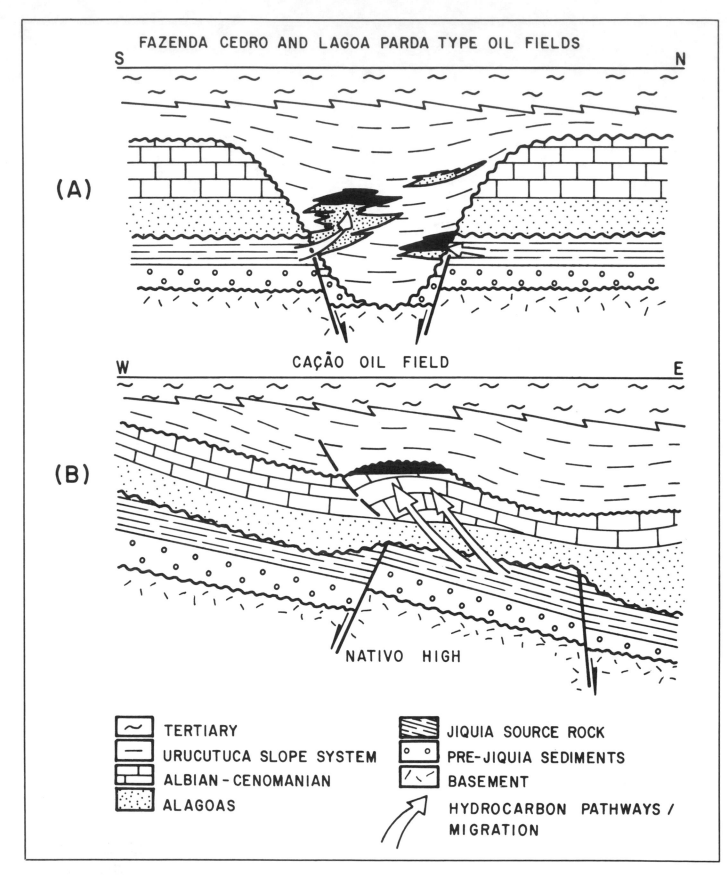

Figure 26. Hydrocarbon pathway models proposed for Jiquia generative areas: (A) oil accumulations associated with submarine canyons, and (B) oil field related to paleogeomorphic closures at the base of Urucutuca slope shales.

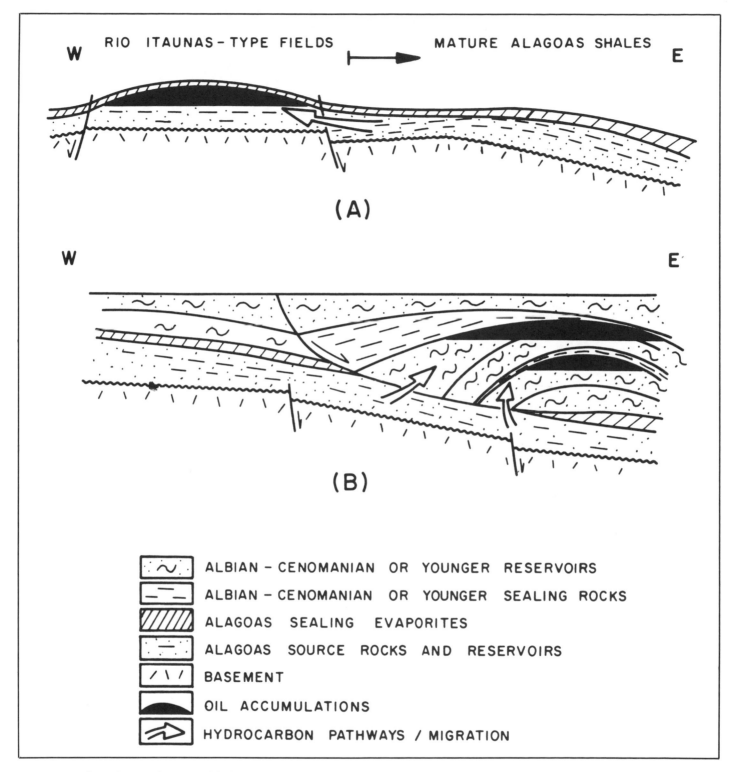

Figure 27. Hydrocarbon pathway models for Alagoas generative areas: (A) oil field in Alagoas reservoirs sealed by anhydrite beds, and (B) oil accumulations in the overlying section, associated to "open windows" in Alagoas anhydrite beds and to gravity sliding structural closure.

Hydrocarbon Generation and Migration from Jurassic Source Rocks in the East Shetland Basin and Viking Graben of the Northern North Sea

J.C. Goff
Department of Energy
London, England

In the East Shetland Basin oil generation began 65 Ma ago; peak oil generation maturity occurs today at 3,250 m (0.7 percent R_o) and was first reached 40 to 50 Ma ago; the oil generation threshold is at 2,500 m. Highest oil saturations in the Kimmeridge Clay occur at 0.8 percent R_o; oil expulsion efficiencies are > 20 to 30 percent. Oil phase migration has probably occurred through oil wet kerogen laminae, and through interconnected large pores aided by low oil/water interfacial tensions. Oil migrated along strong lateral fluid pressure gradients, from overpressured source rocks in half grabens to Jurassic reservoirs in tilted fault blocks.

In the Viking Graben the Kimmeridge Clay is at oil floor maturity below 4,500 m; oil and peak oil generation began 70 to 80 and 55 to 65 Ma ago respectively; 40 Ma ago the Kimmeridge Clay passed through peak generation, and gas generation by cracking of oil had begun. Peak dry gas generation from Brent coals occurs today below 5,000 m, and began 40 Ma ago. The Frigg Field gas, probably generated from late Jurassic source rocks, migrated through microfractures in overpressured mudstones below 3,500 m; above 3,500 m methane probably migrated in aqueous solution and was exsolved in the early Tertiary aquifer.

The East Shetland Basin and the Viking Graben (Figure 1) are located in the northern North Sea Basin between the Shetland Islands and Norway (Figure 2). The Jurassic sandstones in the East Shetland Basin contain 10 billion barrels of recoverable light oil; 3 fields each have recoverable reserves greater than 1 billion barrels (Statfjord, Brent and Ninian). The Viking Graben contains major gas reserves: dry gas and associated heavy oil are trapped in early Tertiary sandstones; the Frigg Field has in place reserves of 270 billion cu m of gas and 790 million barrels of heavy oil (Heritier et al, 1979). Gas condensate has been discovered in deep, high pressure, Jurassic sandstone.

The aims of this study were to determine when these hydrocarbons were generated from their source rocks, and how they migrated from these source rocks to the traps. Both these objectives required integration of geochemical data with knowledge of the stratigraphy, geological structure and history of the study area. The geological framework of the study area, and its development, are reviewed below.

The northern North Sea Basin formed during Permo-Triassic rifting; thick Triassic red beds consisting of alluvial fan, fluvial and lacustrine clastics were deposited unconformably on Caledonian basement. Up to 1 km of early to middle Jurassic shallow-water sediments were then deposited. These comprise Hettangian-Sinemurian fluvial and marginal marine sandstones (Statfjord Formation), Sinemurian to Toarcian shallow marine shelf mudstones, siltstones and thin sandstones (Dunlin Formation), and Bajocian to early Bathonian deltaic/shallow marine sandstones (Brent Formation). Growth faulting occurred during deposition of these rocks along some fault trends, probably at least partly due to differential compaction of Triassic rocks.

The Brent Formation sandstones form a fluvial/wave dominated delta complex, up to 300 m thick, which prograded across a relatively stable shelf at least 12,500 sq km in extent. Major (down to the east) faulting, which occurred during Bathonian to Oxfordian times, created the Viking Graben and a series of westerly dipping fault blocks and half grabens in the eastern East Shetland Basin (Figure 3). Thick Bathonian to early Oxfordian marine mudstones (Heather Formation) conformably overlie the Brent Formation in half grabens; on the crests of the fault blocks, thin Heather Formation is up to 1 km thick.

Further periods of rifting occurred in Oxfordian to early Cretaceous time: Kimmeridgian faulting occurred in the western East Shetland Basin. Late Oxfordian to Portlandian, bituminous, sapropelic, mudstones (Kimmeridge Clay Formation) were deposited throughout

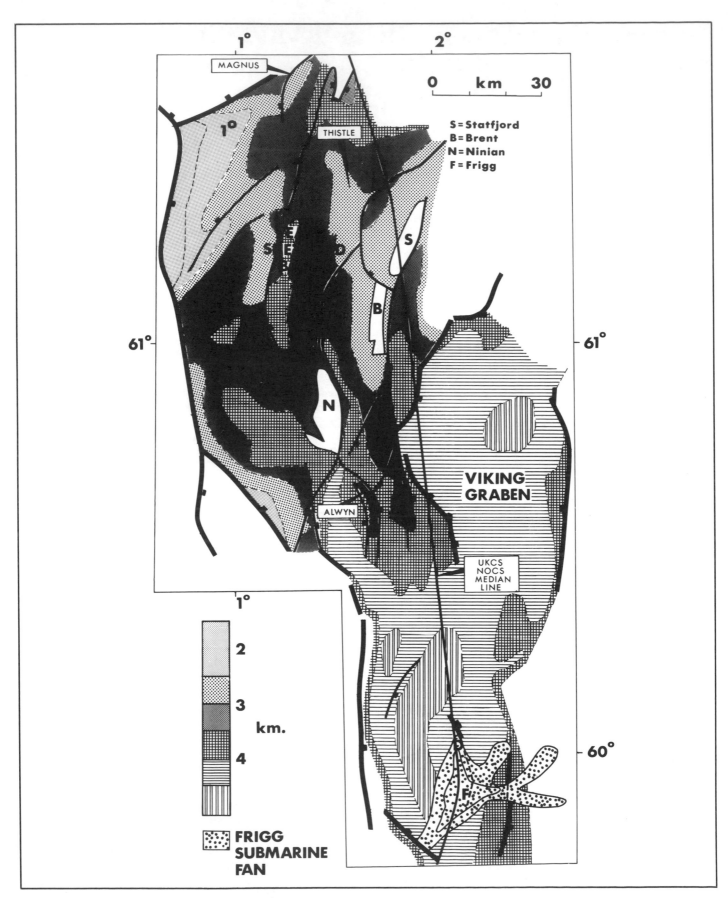

Figure 1. Depth to Base Cretaceous in the study area.

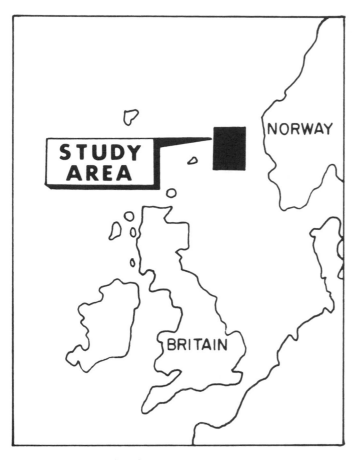

Figure 2. Location of study area.

the area, reaching thicknesses of up to 500 m in the East Shetland Basin. Submarine fan sandstones were deposited in the north-western East Shetland Basin during the Kimmeridgian. Major early Cretaceous faulting occurred in the northwest flank of the Viking Graben. Thick early Cretaceous mudstones (Cromer Knoll Group) are confined to the Viking Graben and some half grabens of the East Shetland Basin. Faulting ceased by mid-Cretaceous times except along the western fault controlled edge of the East Shetland Basin; the geometry of the Jurassic fault blocks was thus established by 100 Ma ago.

Regional subsidence occurred across the East Shetland Basin and Viking Graben during late Cretaceous time, and up to 2,500 m of deep water mudstones and thin limestones (Shetland Group) were deposited. Maximum Tertiary subsidence occurred over the Viking Graben where 2,000 to 2,500 m of sediment accumulated. Palaeocene to early Eocene submarine fan sandstones and mudstones, mid-Eocene to Oligocene mudstones and marine sandstones, the Miocene to Recent lignitic sandstones and mudstones were deposited in the Tertiary basin.

The approach used in this study was first to define the hydrocarbon source rocks and their present day maturity. The present day thermal regime was then determined by calculating heat flow in 4 wells located in contrasting structural positions (Figure 3). Well A is located on a shallow fault block, and Well B in a half graben in the East Shetland Basin. Well C was drilled on a downfaulted block

on the western flank of the Viking Graben; Well D tested a deep fault block in the axial Viking Graben. A range for heat flow history at these 4 locations was deduced from their subsidence histories.

The maturation history of the source rocks in the 4 wells was calculated from their burial history, and the thermal properties of their overburden, for both constant and variable heat flow models. Timing of hydrocarbon generation was deduced from the maturation history using a correlation between calculated maturity and vitrinite reflectance. Timing of generation within the study area as a whole was then estimated from its overall thermal and subsidence history.

The efficiency of oil migration and entrapment in the eastern East Shetland Basin has been estimated by comparing the volume of oil generated with the volume trapped. The mechanism of oil migration in the East Shetland Basin has been deduced by comparing the volumes of oil and compaction water which moved through the source rock during migration, and by considering the physical conditions within the source rock.

Finally the mechanism of oil and gas migration into the Frigg Field, which lies 2 km stratigraphically above Jurassic source rocks in the Viking Graben (Figure 4), is discussed. The variation of pore and fracture pressure, and of methane solubility, with depth in the Viking Graben has been studied to determine the relative importance of microfractures and water movement during migration.

SOURCE ROCKS

The source rock potential of the Cretaceous and Jurassic mudstones in the study area has been reviewed using Total Organic Carbon (TOC) analyses and determinations of Organic Matter Type. Shetland Group mudstones are lean, containing vitrinite and inertinite. Cromer Knoll Group mudstones contain 1 to 2 wt percent TOC which is predominantly inertinite (Barnard and Cooper, 1981).

"Jurassic source rocks" in the northern North Sea have weighted average TOC contents of 5.6 percent wt from 2,600 to 3,200 m, and 4.9 percent from 3,250 to 3,650 m; their non-soluble organic matter contains 80 percent sapropel and 20 percent humic/coaly material (Brooks and Thusu, 1977). Jurassic source rocks with these characteristics have only been reported from the Kimmeridge Clay Formation (Barnard and Cooper, 1981; Fuller, 1980). The Jurassic source rocks analyzed by Brooks and Thusu are thus probably from the Kimmeridge Clay. Immature Kimmeridge Clay organic matter consists predominantly of Type II kerogen (Williams and Douglas, 1980). The Kimmeridge Clay is rated as an excellent oil source rock, generating gas at high maturity levels.

The Heather Formation mudstones contain 1 to 2 percent TOC which consists dominantly of vitrinite and inertinite (Barnard and Cooper, 1981); they are rated as lean dry gas source rocks. The Brent Formation coals and vitrinite rich mudstones are excellent dry gas source rocks; the delta plain facies contains up to 10 m net of coal. The Dunlin Formation is organically lean; mudstones of its lower and middle members contain only 1 percent TOC which is dominantly inertinite (Barnard and Cooper, 1981). The

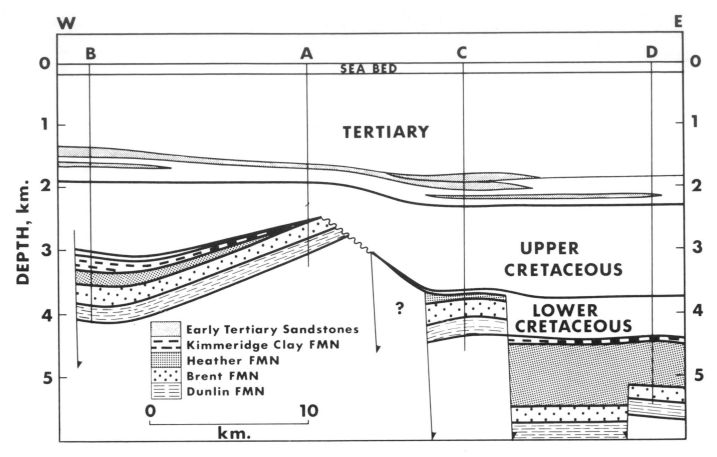

Figure 3. Cross section from the eastern East Shetland Basin to the axial Viking Graben.

upper Drake member, of Toarcian age, contains 2 percent TOC but less than 30 percent of the organic matter consists of sapropel; it thus has only limited oil potential.

The richest source rocks in the study area are thus the oil prone Kimmeridge Clay and the gas prone Brent Formation coals and coaly mudstones. Shetland Group and Heather Formation mudstones, although organically lean, reach thicknesses of 1 and 2 km respectively. They are thus capable of generating large volumes of gas.

The lithology and TOC content of the Kimmeridge Clay are very variable. In its type section in southern England it consists of carbonaceous illitic clays (≤ 10 percent TOC), bituminous shales (≤ 30 percent TOC), oil shales (≤ 70 percent TOC) and coccolithic limestones; the bituminous shale units are up to 2 m thick (Tyson et al, 1979). The limestones are also bituminous and usually interlaminated with the oil shales, which are generally less than 10 cm thick (Gallois, 1976). In the East Shetland Basin the Kimmeridge Clay also contains thin beds of siltstone, fine-grained sandstone and dolomite.

Fuller (1980) reported an average Organic Matter Content for the Kimmeridge Clay in the North Sea Basin of 3.25 percent (equivalent to a TOC of 2.7 percent); the East Shetland Basin Kimmeridge Clay is twice as rich as this. Organic matter type varies laterally and vertically within the Kimmeridge Clay. At shallow burial depths (1,500 to 2,400 m) close to clastic sediment sources, and over stable platform areas, the Kimmeridge Clay contains

predominantly inertinite and vitrinite (Barnard and Cooper, 1981).

The high TOC content and sapropel contents of the Kimmeridge Clay in the East Shetland Basin are probably due partly to deposition in restricted fault bounded half grabens. Oil shale horizons are best developed in the middle part of the Kimmeridge Clay onshore in the United Kingdom (Gallois, 1976). A highly radioactive unit occurs at this stratigraphic level in the East Shetland Basin and Viking Graben (Figure 5); it frequently has a higher than normal electrical resistivity, suggesting it is the organically richest unit in the Kimmeridge Clay. On the crests of the fault blocks in the East Shetland Basin this unit is the only part of the Kimmeridge Clay present, indicating it was deposited during the peak Kimmeridgian transgression.

MATURITY

The present day maturity depth gradient in the East Shetland Basin and Viking Graben has been determined from vitrinite reflectance measurements on Brent Formation coals and early Jurassic to mid-Cretaceous mudstones in 12 wells. The wells were drilled in the Norwegian sector of the eastern East Shetland Basin, the axial Viking Graben, and on shallow fault blocks on the eastern flank of the Viking Graben east of the study area. The data (Figure 6) indicate a uniform present day maturity gradient across the eastern half of the study area. For a type II kerogen (Tissot and

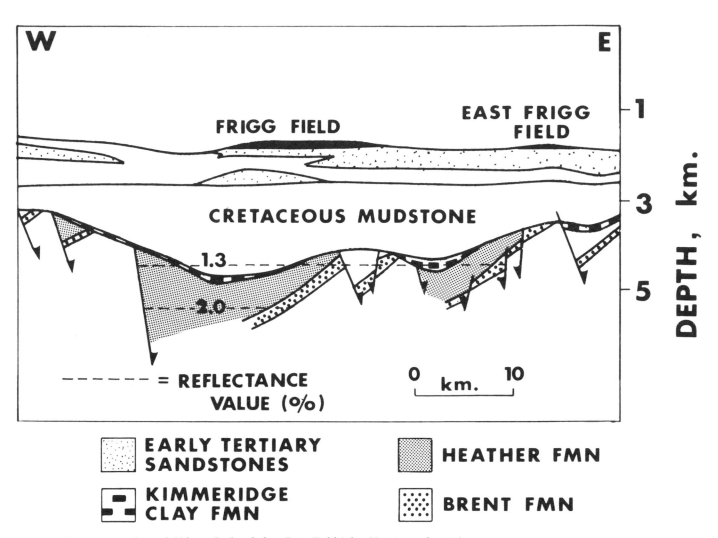

Figure 4. Cross section through Viking Graben below Frigg Field (after Heritier et al, 1979).

Welte, 1978) this gradient indicates the oil window extends from 2,550 to 4,500 m, the wet gas zone from 4,500 and 5,400 m, and that the dry gas zone occurs at depths greater than 5,400 m. Peak gas generation from coal corresponds to the boundary of the medium and lower volatile bituminous coal ranks (Hunt, 1979; Figure 5.7), corresponding to a vitrinite reflectance of 1.5 percent. This is equivalent to a burial depth of 5,000 m in the Viking Graben.

Hydrocarbon/TOC ratio data for the Kimmeridge Clay have been correlated with the vitrinite reflectance gradient to determine the vitrinite reflectance level corresponding to peak hydrocarbon generation (Figure 7). The maturity corresponding to peak hydrocarbon generation is here defined as the maturity when 50 percent of the potential hydrocarbon yield has been generated. For a Type II kerogen, 70 percent of the organic matter is capable of conversion to hydrocarbons (Tissot and Welte, 1978). When peak hydrocarbon generation is reached, 35 percent of the organic matter will thus have been converted to hydrocarbons. This corresponds to a hydrocarbon/TOC ratio in the source rock at peak generation of 0.42, if no expulsion has occurred (assuming 1 gm of organic matter is equivalent to 0.83 gm of TOC).

However, because some of the generated hydrocarbons are expelled from the source rock, the hydrocarbon/TOC ratio corresponding to peak generation will be less than 0.42. The expulsion efficiency can be estimated from the observed decrease in TOC with depth for the Kimmeridge Clay of 0.7 percent from about 2,900 to 3,400 m (assuming the TOC decrease is due to expulsion of generated hydrocarbons rather than early diagenetic processes). If ≤ 50 percent of the potential hydrocarbon yield is generated over this depth interval, which corresponds to a narrow vitrinite reflectance range of 0.6 to 0.8 percent, an expulsion efficiency ≥ 0.25 is indicated. The hydrocarbon/TOC ratio in the Kimmeridge Clay at peak generation would thus be ≥ 0.35 which corresponds to a vitrinite reflectance of 0.7 percent by extrapolation of the graph on Figure 7.

Organic matter colouration (Staplin, 1969) and source rock electrical resistivity (Meissner, 1978) can also be used to estimate source rock maturity. From 2,600 to 3,200 m plant material in the Kimmeridge Clay is light to medium brown, indicating it is moderately mature; between 3,200 and 3,650 m it is dark brown indicating it has achieved peak generation (Brooks and Thusu, 1977). Electrical resistivity increases from 2 to 3 Ωm at 2,500 to 2,600 m to a maximum

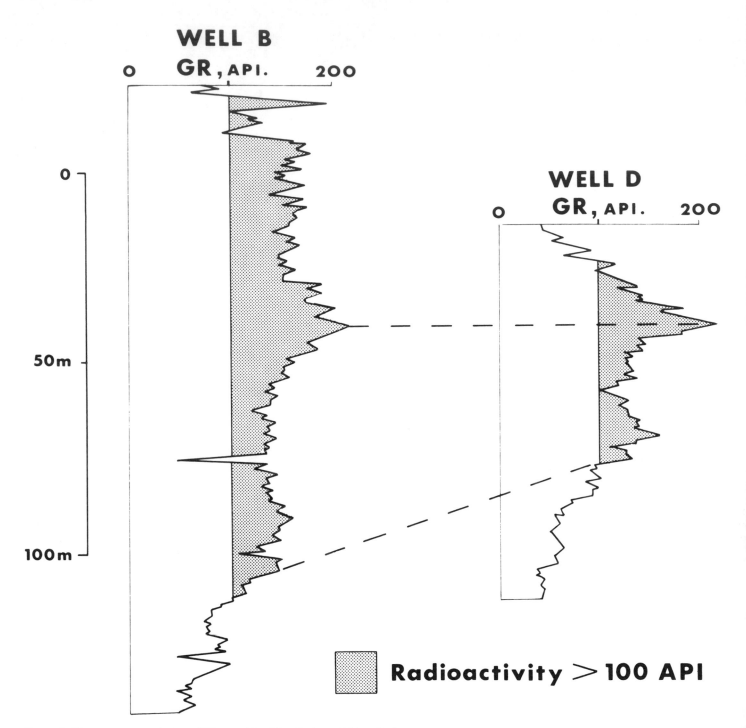

Figure 5. Gamma ray correlation of Kimmeridge Clay Basin to Viking Graben.

of 10 to 25 Ωm at 3,500 to 3,600 m (Figure 8). These data indicate peak generation has occurred between 3,200 and 3,500 m at a reflectance level of 0.7 to 0.8 percent. At a reflectance level of 1.2 to 1.3 percent (base oil window maturity) in Well D (Figure 8), the Kimmeridge Clay is highly overpressured, and has a very high transit time (115 to 135 μsec/ft) and low resistivity (< 3 μm). These log responses suggest it has a high water content, and that oil, formerly present in the rock, has cracked to gas, most of which has been expelled. Low resistivities thus do not necessarily indicate immaturity.

PRESENT DAY THERMAL GRADIENTS AND HEAT FLOW

Bottom hole temperatures have been calculated in Wells B, C, and D from electric logging run temperature measurements, drilling mud circulation history, and borehole diameter, using the method of Oxburgh et al (1972). The temperature for Well A was measured during a drill stem test. Heat flows were derived from the bottom hole temperatures and thermal conductivity estimates for the rocks penetrated by each well. The thermal conductivity

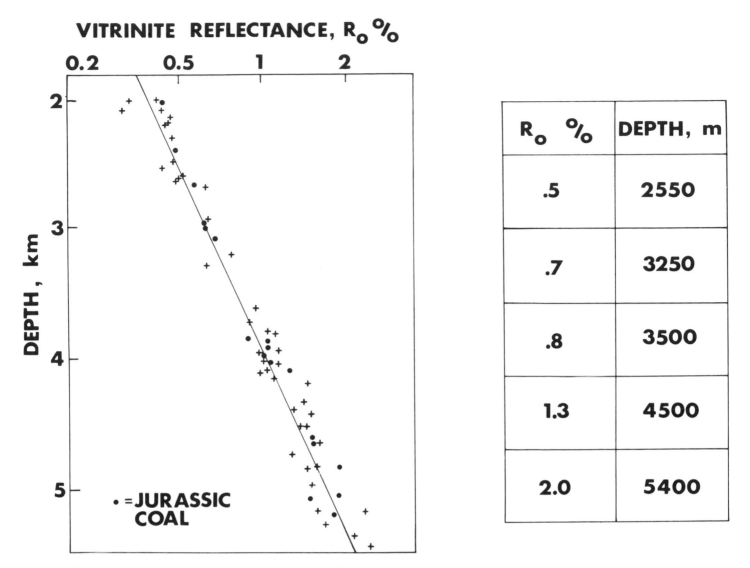

R_O %	DEPTH, m
.5	2550
.7	3250
.8	3500
1.3	4500
2.0	5400

Figure 6. Vitrinite reflectance versus depth for Jurassic to early Cretaceous coals and mudstones.

of a rock (K) is a function of lithology and porosity; it varies with porosity according to the following equation (Oxburgh and Andrews-Speed, 1981);

$$K = K_w^\phi \cdot K_m^{1-\phi}$$

where K_w is the conductivity of water, and K_m is the conductivity of the rock matrix. Rock matrix conductivities have been determined by BP's Geophysical Research Division. The standard equation for steady state conductive heat flow (q) across a uniform rock layer of thickness, L, which has no heat production within it is given by:

$$q = \frac{k \, dT}{dL}$$

where dT/dL is the thermal gradient across the layer in the direction of heat flow.

Because of the variation of conductivity with lithology and porosity it is convenient to calculate the heat flow using

the following equation (Oxburgh and Andrews-Speed, 1981):

$$q_z = \frac{T_z - T_o}{R_z}$$

where q_z, T_z, and R_z are the heat flow, temperature and thermal resistance at the depth of temperature measurement, z. T_o is the surface temperature.

The thermal resistance at depth z is defined as:

$$R_z = \frac{L_1}{K_1} + \frac{L_2}{K_2} + \frac{L_3}{K_3} + \ldots$$

where $L_{1...}$ are the thicknesses of the rock layers in the overburden, and $K_{1...}$ are their thermal conductivities (Oxburgh and Andrews-Speed, 1981).

Thermal resistance was calculated as a function of depth in each well using the observed variation of lithology with

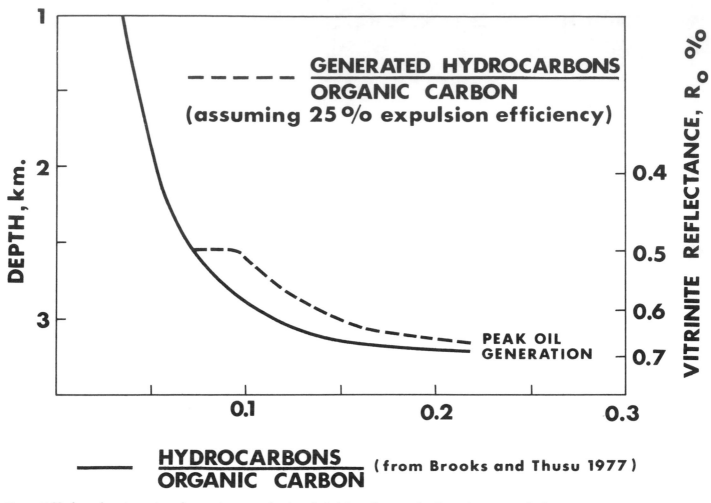

Figure 7. Hydrocarbon/organic carbon ratio versus depth and vitrinite reflectance for 'Jurassic source rocks.'

depth, and porosity/depth gradients derived from density logs and shale density measurements.

Porosity (ϕ) was calculated from the density measurements using the following equation:

$$\phi = \frac{P_m - P_r}{P_m - P_w}$$

where P_m, P_r, and P_w are the rock matrix, whole-rock and water densities respectively, using a matrix density of 2.7 gm/cc for mudstone and 2.65 gm/cc for sandstone.

Porosity/depth relationships have been modelled for each lithology assuming an exponential decrease in porosity with depth defined by:

$$\phi = \phi_o e^{-Z/m}$$

where ϕ_o is the surface porosity and m is the compaction gradient. Surface porosities of 65 percent for mud and 43 percent for sand were selected (Sclater and Christie, 1980). Calculated porosity gradients are shown in Table 2; the variation in porosity with depth for mudstones in wells C and D is shown in Figure 9.

Calculated heat flows in the 4 wells range from 0.051 to 0.061 Wm^{-2}. Heat flow varies with depth because of heat generation from decay of radioactive isotopes of potassium, uranium, and thorium in the sedimentary rocks. Surface heat flow (q_o) has been calculated for each well, using average values of heat production for mudstone and sandstone, from the following equation:

$$q_o = q_z + AZ$$

where A is the average heat generation/unit volume of rock above depth Z. The mean surface heat flow in the 4 wells is 0.060 Wm^{-2} (Table 1). This is the same as the mainland UK average heat flow (Richardson and Oxburgh, 1979).

Interval thermal gradients for Wells A-D were calculated from the thermal resistance profile and calculated bottom hole temperatures in each well. The interval gradients decrease with depth from 37 to 40°C/km for the Tertiary, 31 to 33°C/km for the Cretaceous, to 26 to 27°C/km for the Jurassic. This decrease in interval gradient with depth is attributed to the decrease in porosity, and thus increase in thermal conductivity, with depth, and to the distribution of heat sources.

Carstens and Finstad (1981) also observed a decrease in interval gradients with depth down to the base Cretaceous in the East Shetland Basin and Viking Graben. They also

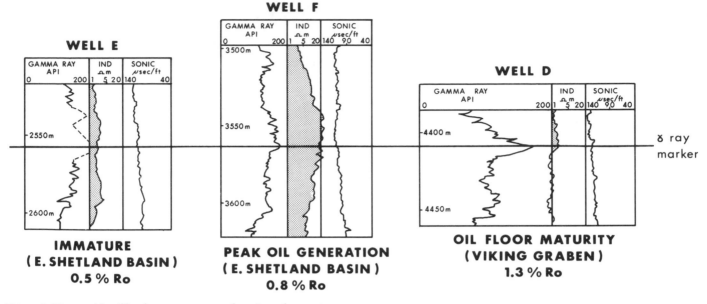

Figure 8. Kimmeridge Clay log response as a function of maturity.

calculated very high Jurassic interval gradients > 35°C/km. These high apparent Jurassic gradients probably result from the lack of correction for borehole diameter in the Horner Plot method, used by Carstens and Finstad, to calculate true bottom hole temperatures. The Horner Plot method can underestimate true formation temperature by as much as 20 percent in a 12¼ in diameter borehole, and by about 5 percent in a 8 in hole (S.W. Richardson, personal communication).

Consider a 500 m Jurassic section at 3,000 to 3,500 m depth with a true interval gradient of 26°C/km, and true bottom hole temperatures of 113°C at 3,000 m, and 126°C at 3,500 m. A 20 percent error in the Horner Plot derived temperature in 12¼ in hole at 3,000 m, and a 5 percent error in 8 in hole at 3,500 m, will give an apparent Jurassic interval gradient of 59°C/km (which is more than 100 percent too high!).

Average thermal gradients in Wells A, B, and C in the eastern East Shetland Basin are 35 to 36.5°C/km. Well D in the Viking Graben has an interval gradient of 35°C/km. Carstens and Finstad (1981) reported average gradients of 30 to 35°C/km in the central and western East Shetland Basin. These data indicate a fairly uniform present day thermal

Table 2. Compaction data for gross lithologies.

Well	Lithology	Surface Porosity	Compaction Gradient (meters)
A, B, C, D	Sandstone	0.43	5,000
A, B	Mudstone	0.65	2,000
C	Mudstone	0.65	2,350
D	Mudstone	0.65	2,500

regime in the study area.

SUBSIDENCE AND HEAT FLOW HISTORY

McKenzie (1978) proposed that the North Sea Basin formed as a result of stretching, and consequent thinning, of the continental lithosphere. He predicted that the thinning of the crust was caused by listric normal faulting. In his model an initial *fault controlled* subsidence occurred which maintained isostatic compensation; it increased with the amount of stretching, β, where

$$\beta = \frac{\text{initial lithospheric thickness}}{\text{lithospheric thickness immediately after stretching}}$$

Table 1. Thermal data for wells.

Well	Depth (mBRT)	Temperature (°C)	Time Since Circulation (hrs)	Calculated Bottom Hole Temperature (°C)	Thermal Resistance ($Wm^{-2}T^{-1}$)	Measured Heat Flow (Wm^{-2})	Surface Heat Flow (Wm^{-2})
D	5,180	150	15				
		153.3	21	161.5	3,055	0.051	0.057
C	4,775	154.4	7.75				
		157.2	12.5	161.7	2,554	0.061	0.065
B	3,940	115.6	5				
		121.1	9	128	2,214	0.055	0.058
A	2,725	94.2	Drill stem test temperature		1,554	0.056	0.058

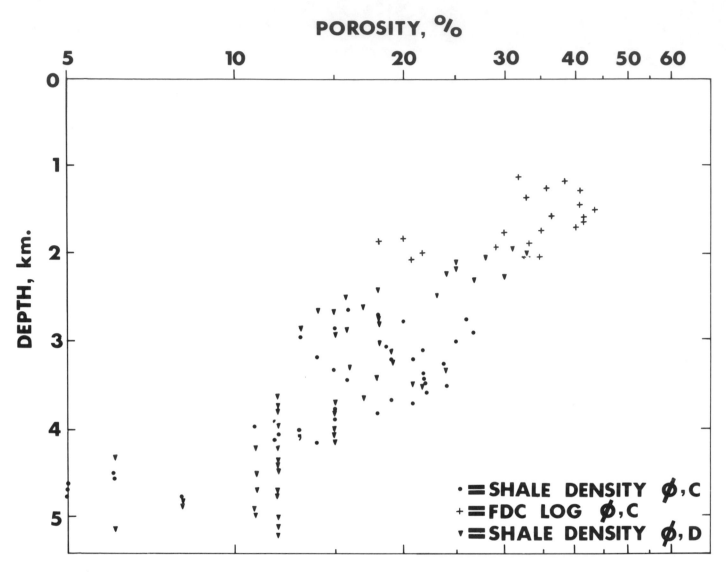

Figure 9. Mudstone porosity versus depth in Wells C and D.

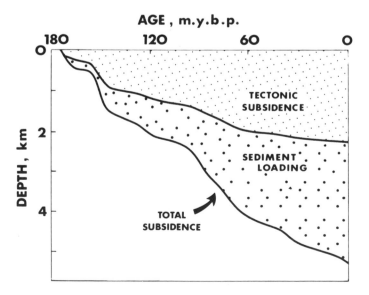

Figure 10. Subsidence history at Well D since late early Jurassic.

The initial fault controlled subsidence varies across rifted basins because of local uplift of tilted blocks. The fault controlled subsidence in the axial grabens is most likely to be representative of the amount of stretching.

Cooling of buoyant hot asthenosphere below the thinned lithosphere creates a thermal anomaly reflected by a high surface heat flow. The heat flow declines with time as the asthenosphere cools. This decline in heat flow with time can be calculated if the age and amount of stretching is known (McKenzie, 1978; Figure 2). The *age* of stretching is defined by periods of normal faulting; the *amount* of stretching, β, can be calculated from the amount of tectonic subsidence that has occurred since faulting ceased.

Deposition of sediments in a basin is an additional important subsidence mechanism (Turcotte, 1980). The subsidence of sediment loaded crust is two to three times greater than the water loaded tectonic subsidence, depending on the sediment density. The total subsidence is related to the water loaded tectonic subsidence by the following equation:

$$d_{Total} = \frac{(pm - pw)}{(pm - ps)} d_w$$

where d_{Total} is the total subsidence, d_w is the water loaded tectonic subsidence, pw is the water density, ps is the sediment density, pm is the density of the mantle.

Geophysical evidence indicates thinned crust below the northern North Sea as predicted by McKenzie's model. Seismic refraction studies across the Viking Graben indicate a value of β of about 2 (Sclater and Christie, 1980); gravity profiles across the East Shetland Basin and Viking Graben suggest the crust has been thinned from 30 km to 20 km, i.e. $\beta = 1.5$ (Donato and Tully, 1981). The initial fault controlled subsidence predicted by McKenzie's model for $\beta = 1.5$ is 800 m (Sclater and Christie, 1980). The thickness of Bathonian to early Cretaceous mudstones in the axial Viking Graben is about 2 km (Figure 3). This thickness is consistent with an initial fault controlled subsidence of about 800 m (after removing the effects of sediment loading).

There are two major problems in applying McKenzie's model to determine the heat flow history of the East Shetland Basin and Viking Graben. Firstly, several phases of faulting occurred between late Bathonian to early Cretaceous time; secondly, water depth during deposition in late Jurassic to early Tertiary time is not accurately known.

Major phases of rifting occurred in Bathonian to early Oxfordian time along the western flank of Viking Graben (Figure 3) and in late Oxfordian to Kimmeridgian time along the western edge of the East Shetland Basin; both phases of faulting occurred within the East Shetland Basin. Early Cretaceous rifting occurred in the northern part of the East Shetland Basin and possibly also along the western flank of the Viking Graben (Figure 3). Fault controlled subsidence ceased at the end of the early Cretaceous (100 Ma ago) in the East Shetland Basin and Viking Graben except along the eastern bounding fault of the East Shetland Platform. Similar diachronous rifting occurred in the East Greenland Jurassic rift (Surlyk, 1968), with main phases in the Bathonian, late Oxfordian, early Kimmeridgian and at the Jurassic/Cretaceous boundary. The faulting history suggests that several phases of crustal thinning occurred between 165 and 100 Ma ago in the study area.

The water loaded tectonic subsidence at Wells A-D has been calculated from the subsidence history (corrected for compaction) by removing the loading effects of the sediments, assuming local isostatic compensation. The tectonic subsidence and the subsidence due to sediment loading for Well D are shown in Figure 10.

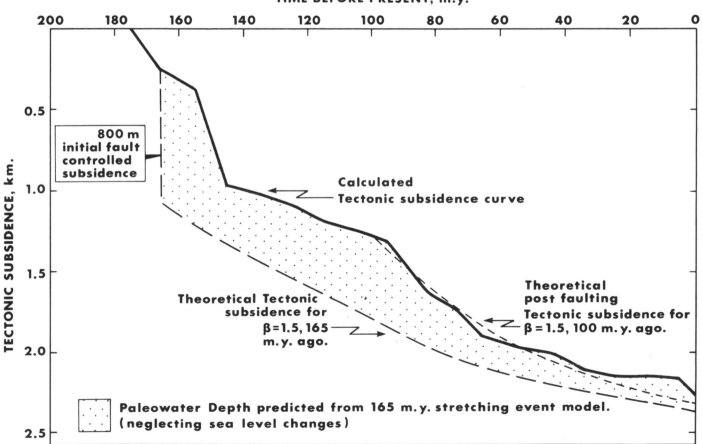

Figure 11. Interpretation of tectonic subsidence at Well D.

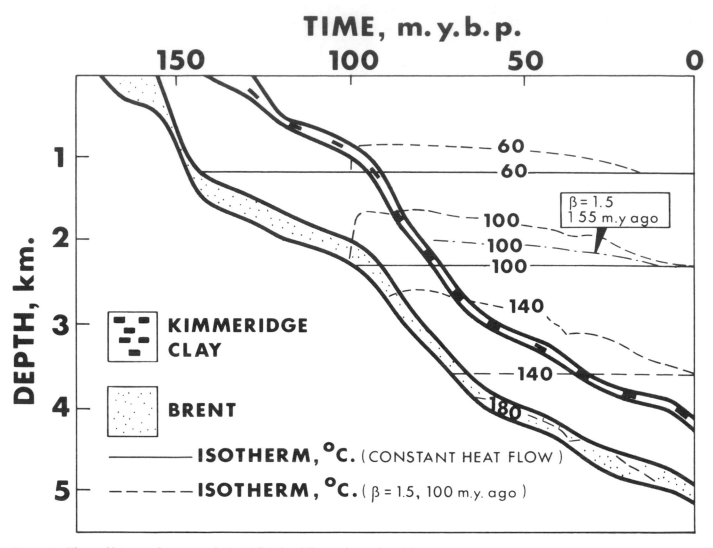

Figure 12. Thermal history of source rocks in Well D for different thermal models.

Modelling of the post-faulting tectonic subsidence in Well D gives a good fit to a stretching event 100 Ma ago with $\beta = 1.5$ (Figure 11) assuming constant water depth in the late Cretaceous and Tertiary. However, sedimentation rates were slow during the Kimmeridgian to early Cretaceous, and water depths were probably much greater than at the present day. A more realistic model for the tectonic subsidence at Well D is obtained with a stretching event 165 Ma ago (corresponding to the main Viking Graben rifting), with a stretching factor, (β), of 1.5 (Figure 11). This model predicts Kimmeridgian to early Cretaceous water depths of 500 to 700 m in the Viking Graben, and progressive shallowing during the late Cretaceous and Tertiary to the present day value of 150 m.

The variation in the age of crustal stretching (and consequent increase in heat flow) within the study area indicates that heat flow was probably very variable between 165 and 100 Ma ago. However, Jurassic source rocks were not deeply buried enough for hydrocarbon generation to occur prior to 100 Ma ago. Consequently hydrocarbon generation in the study area is insensitive to early variation in heat flow.

Two extreme heat flow models can be used to determine the timing of hydrocarbon generation in the study area. A *constant heat flow model* will underestimate maturity at a given time because it fails to account for higher heat flows during the Cretaceous and Tertiary than at the present day. The model with a *heating event 100 Ma ago* will overestimate maturity because it fails to account for early heat loss from the crust during late Jurassic and early Cretaceous rifting. The true maturity of the source rocks will lie *between* the values predicted by these two models.

THERMAL AND MATURATION HISTORY OF SOURCE ROCKS

The thermal history of the source rocks is related to the variation in heat flow, and the increase in the thermal resistance of their overburden, with time. The thermal resistance in Wells A-D was calculated as a function of time using the depositional history, and the compaction gradients for mudstone and sandstone. The maximum range for the thermal history of the Jurassic source rocks is bounded by the constant heat flow and 100 Ma heating event models.

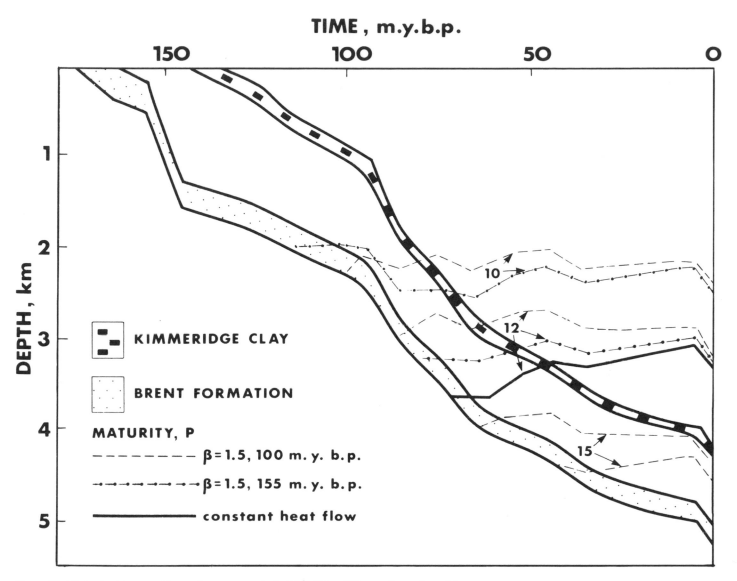

Figure 13. Maturity history of Jurassic source rocks in Well D for different thermal models.

The thermal history of the source rocks in Well D in the Viking Graben is shown in Figure 12. Source rock temperatures rose rapidly during late Cretaceous burial. During the Tertiary the constant heat flow model predicts a moderate increase in source rock temperature; the 100 Ma heating event model predicts that source rock temperatures remained relatively constant, because of a decrease in heat flow with time during the Tertiary. A model with a heating event in the late Jurassic 155 Ma ago predicts source rock temperatures intermediate between the constant heat flow and 100 Ma heating event models (see the 100°C isotherm on Figure 12).

The maturation history of the source rocks in Wells A-D was calculated by integrating time and temperature using the following equation (Royden et al, 1980):

$$P = \mathrm{Ln} \int_o^t 2^{T/10} dt$$

where P is the maturity parameter, t = time (Ma) and T = temperature (°C). This equation allows for the

approximate doubling of the reaction rate of hydrocarbon-forming reactions for every 10°C increase in temperature, in a similar way to 'Lopatin's Time Temperature Index' used by Waples (1980) to determine maturity.

The maturity parameter, P, was calculated as a function of time for each source rock, and as a function of depth at the present day in each well, using the constant heat flow and 100 Ma heating event models. The maturation history of the source rocks in Well D as predicted by the different thermal models is shown in Figure 13. The 100 Ma heating event model predicts earlier maturation of the source rocks than the constant heat flow model. The 155 Ma heating event model predicts intermediate maturation levels, except for the Brent Formation maturity prior to 90 Ma ago. The maturities predicted by the different models converge with time so that *present day* maturities predicted by the different models are similar.

The maturity/depth gradient calculated using the 100 Ma heating event model is shown in Figure 14. Note that the

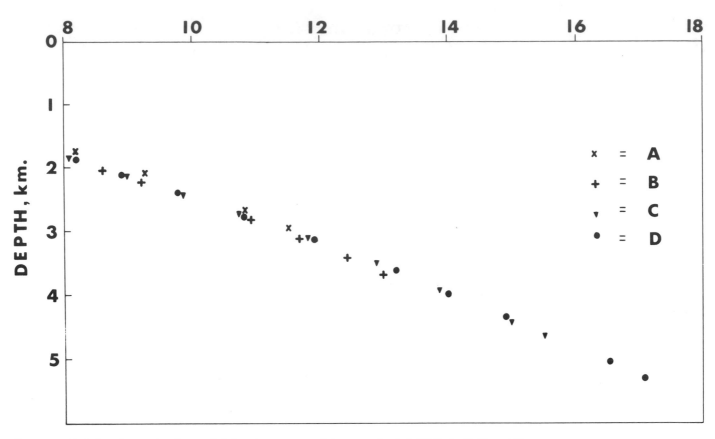

Figure 14. Calculated maturity, for 100 Ma heating event model, versus depth in Wells A, B, C, and D.

gradient is almost the same for each well, despite their different structural locations. This is consistent with the uniform maturity gradient indicated by the vitrinite reflectance data (Figure 6).

In order to determine the timing of generation from the calculated maturity, vitrinite reflectance has been correlated with P. Vitrinite reflectance measurements in Wells C and D have been correlated with P value calculated at the same depths. Vitrinite reflectance measurements from other wells have been correlated with P values obtained from the calculated regional maturity/depth gradient. In this way P has been correlated with vitrinite reflectance, and thus generation stage, for both *the constant heat flow* and *100 Ma heating event models* (Table 3). The correlation of vitrinite reflectance with P, calculated using the 100 Ma heating event model, is shown in Figure 15.

TIMING OF GENERATION

The timing of oil generation from the Kimmeridge Clay in Wells A-D has been determined from its maturation history using the correlations of vitrinite reflectance with P (Table 4). The constant heat flow model predicts *later* generation than the 100 Ma heating event model. The onset of oil generation (0.5 R_o) predicted by the constant heat flow model is up to 10 Ma later than predicted by the 100 Ma heating event model. For Well D, the top of the Kimmeridge Clay began to generate oil between 69 and 77 Ma ago and reached peak generation between 53 and 65 Ma ago.

The maturity distribution of the Kimmeridge Clay at the

end of the Cretaceous (65 Ma ago) has been determined by correlating its maturity 65 Ma ago in Wells B, C, and D, with the present day Cretaceous thickness in these wells (Figure 16). This correlation shows that the Kimmeridge Clay had become mature by the end of the Cretaceous, where the present Cretaceous thickness is > 1,700 m. The area of mature Kimmeridgian at the end of the Cretaceous was then deduced from a Cretaceous isopach map of the study area. The maturity distribution of the Kimmeridge Clay 40 and 20 Ma ago was estimated using the present day base Cretaceous structure contour map and correlations of maturities at these times in Wells A-D with present day depth to the Kimmeridge Clay (Figure 17, Table 5).

Table 3. Correlation of vitrinite reflectance with calculated maturity (P), and oil and gas generation from source rocks.

Vitrinite Reflectance R_o%	P for 100 Ma Heating Event Model	P for Constant Heat Flow Model	Generation Stage
0.5	10.1	9.7	Oil threshold
0.7	11.9	11.5	Peak oil generation for Kimmeridge Clay and gas generation threshold for coal
1.3	15.0	14.4	Oil floor
1.5	15.8	15.0	Peak gas generation for coal
2.0	17.2	16.3	Wet gas floor

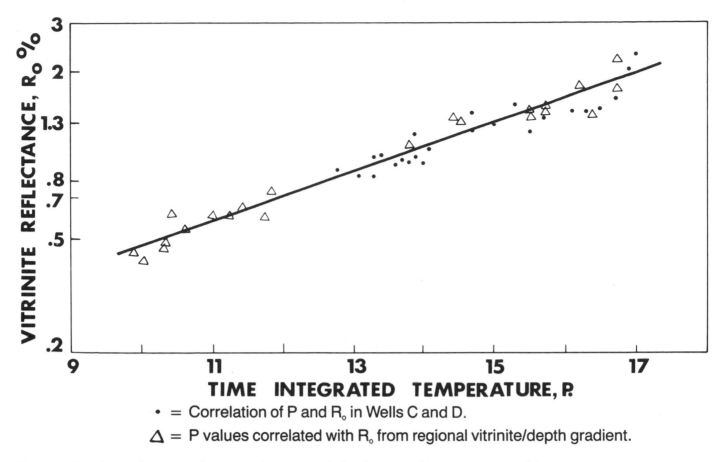

• = Correlation of P and R_o in Wells C and D.

△ = P values correlated with R_o from regional vitrinite/depth gradient.

Figure 15. Correlation of vitrinite reflectance with maturity calculated using 100 Ma heating event model.

The generation history of the Kimmeridge Clay during the Tertiary is summarized in Figure 18. At the end of the Cretaceous the Kimmeridge Clay was mature throughout the Viking Graben, and in the deepest troughs and half grabens of the East Shetland Basin. In the Viking Graben peak generation was reached during the Palaeocene. By the end of the Eocene, peak generation was occurring in the trough west of the Ninian Field in the East Shetland Basin. During the Oligocene, peak generation was reached in the half grabens of the East Shetland Basin, and oil floor maturity was attained in the deepest parts of the Viking Graben. By the present day, peak generation had been established throughout the axial region of the East Shetland

Basin, and a maturity just above or greater than the oil floor was attained in the Viking Graben. A generation front has thus moved progressively updip with time within the Kimmeridge Clay from the deepest parts of the Viking Graben during Campanian time, to the structurally highest fault blocks in the East Shetland Basin at the present day. The zone of intense oil generation (0.65 to 0.9 percent R_o) has moved from the Viking Graben to the axial East Shetland Basin since the end of the Eocene.

The generation history deduced above is consistent with diagenetic studies in the Brent Formation sandstone reservoirs in the East Shetland Basin (see Figure 1 for the locations of fields discussed below). Hancock and Taylor

Table 4. Timing of maturation for Kimmeridge Clay in Wells A-D for different thermal models. a, Heating event 100 Ma ago. b, Constant heat flow.

a Maturity, P	Time Reached, Ma Before Present				b Maturity, P	Time Reached, Ma Before Present				
	A	B	C	D		A	B	C	D	
10.1	–	47	63	77	9.7	–	37	53	69	To Kimmeridge Clay
	3	57		79		3	47		71	Base Kimmeridge Clay
11.9	–	–	34	65	11.5	–	–	27	53	Top Kimmeridge Clay
	–	–		67		–	–		56	Base Kimmeridge Clay
12.5	–	–	25	59	12.1	–	–	15	44	Top Kimmeridge Clay
	–	–		62		–	–		47	Base Kimmeridge Clay
15.0	–	–	–	–	14.4	–	–	–	–	Top Kimmeridge Clay
	–	–	–	–		–	–	–	–	Base Kimmeridge Clay

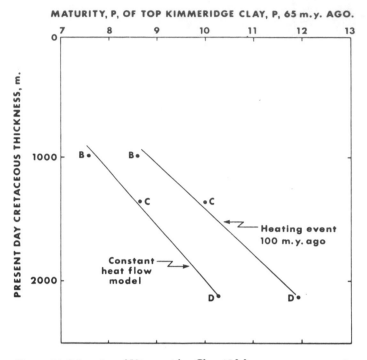

Figure 16. Maturity of Kimmeridge Clay 65 Ma ago versus present day burial depth.

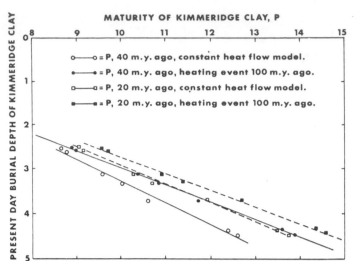

Figure 17. Maturity of Kimmeridge Clay 40 and 20 Ma ago versus present day Cretaceous thickness.

(1978) observed an upward transition from authigenic illite to authigenic kaolinite within a thick oil column in an East Shetland Basin oil field. They concluded that oil migration occurred synchronously with illite diagenesis. Sommer (1978) observed a more abrupt change from illitic cement in the oil/water transition zone to predominantly kaolinite in the oil zone in an East Shetland Basin oil well drilled by Total (presumably in the Alwyn area). The youngest radiometric dates obtained for pure illite cements in this well were 45 to 55 Ma, indicating that this field had filled up by mid-Eocene times. This age of migration is consistent with the calculated age of generation in the trough west of the Alwyn field (Figure 18).

In the Thistle Field (Hay, 1977), porosity/depth gradients within individual oil saturated sandstone units range from 1.5 to 5 percent porosity loss/100 m. These very rapid reductions in porosity with depth probably indicate progressive cessation of diagenesis in the sandstones as they filled with oil, over a long period of time. This suggests that oil migration occurred over an extended period of time in the Thistle area, which is consistent with the calculated progressive updip movement of the oil generation front of this part of the northern East Shetland Basin (Figure 18). A probable palaeo-OWC in the Magnus Field (De'ath et al, 1981) suggests that this field had filled up prior to late Tertiary tilting, which is consistent with the early onset of generation in the half graben downdip to the west (Figure 18).

The timing of gas generation from Brent Formation coals in the Viking Graben can be inferred from the maturation history of this formation in Wells C and D. Maturation analysis in Well D, using the constant heat flow and 100 Ma heating event models and the appropriate correlations of P

and vitrinite reflectance (Table 3), indicates gas generation began 70 to 85 Ma ago. Peak dry gas generation from the coals in Well D began 30 to 40 Ma ago. In Well C, on the flank of the Viking Graben, gas generation began 27 to 34 Ma ago. In the deepest part of the Viking Graben, where the Brent Formation is buried to 6 km, gas generation probably began 100 Ma ago.

SOURCE AND TIMING OF GENERATION OF FRIGG FIELD HYDROCARBONS

The Frigg Field oil is a relatively heavy (24 API), napthenic oil consisting dominantly of hydrocarbons with a carbon number $> C_{17}$; condensate dissolved in the overlying dry gas column comprises 87 percent $C_{11}+$ (Heritier et al, 1979). Heritier et al (1979) suggested, using pristane/phytane ratio data for the East Frigg Field oil, that the Frigg Field oil was sourced from Lower-Middle Jurassic rocks. However, good oil source rocks have not been described from the Dunlin and Brent formations in the Viking Graben. An alternative hypothesis is that the Frigg

Table 5. a. Correlation of palaeomaturity of Kimmeridge Clay from 100 Ma heating event model with present burial depth.

Maturity 40 Ma Ago	Present Depth	Maturity 20 Ma Ago	Present Depth, m
10.1	3,050	10.1	2,800
11.9	3,700	11.9	3,500
12.5	4,000	12.5	3,700
15.0	5,000	15.0	4,600

b. Correlation of palaeomaturity of Kimmeridge Clay from constant heat flow model with present burial depth.

Maturity 40 Ma Ago	Present Depth	Maturity 20 Ma Ago	Present Depth, m
9.7	3,100	9.7	2,800
11.5	4,000	11.5	3,600
12.1	4,300	12.1	3,800
14.4	5,400	14.4	4,800

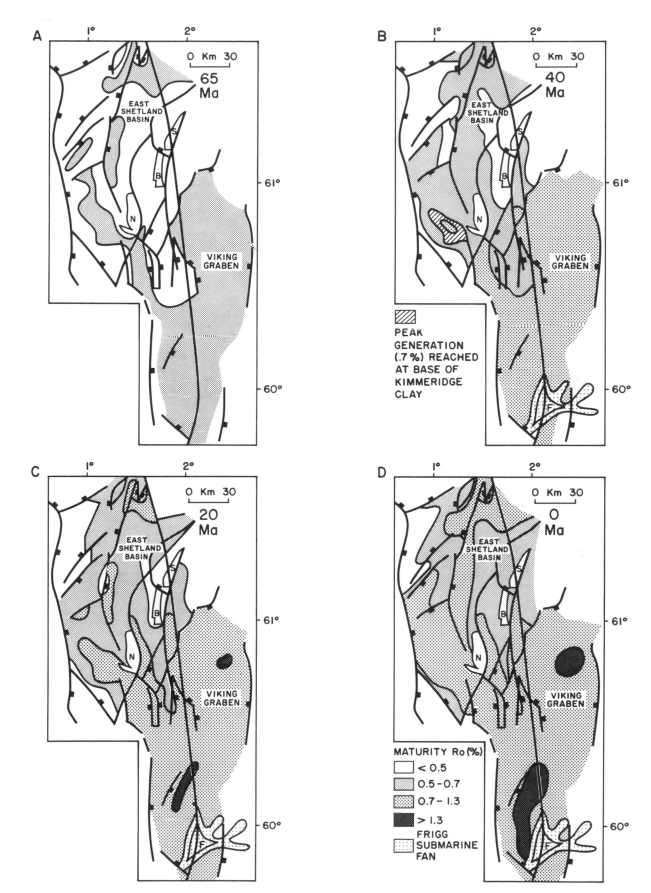

Figure 18. Maturity of Top Kimmeridge Clay (A) 65 Ma ago, (B) 40 Ma ago, (C) 20 Ma ago, vnd (D) at present day.

oil was generated from the Kimmeridge Clay.

The drainage area of the Frigg Field is estimated to be at least 1,500 sq km from the contour map of the top of the Frigg Sand published by Heritier et al (1979). The Kimmeridge Clay within this drainage area is buried below 4 km (Figure 4). The maturation history of the Kimmeridge Clay within the drainage area of the Frigg Field can be inferred from the maturation history of this formation in Well D (Table 4), since they have had similar burial histories. The Cretaceous to early Eocene thickness is 2,500 m in Well D, compared to 2,500 to 2,700 m in the main drainage area of the Frigg Field. The Cretaceous thickness in Well D is 2,100 m, which corresponds to the Cretaceous thickness in the deepest part of the Frigg drainage area.

The Kimmeridge Clay within the drainage area of the Frigg Field thus began to generate oil 70 to 80 Ma ago. Peak generation was reached 55 to 65 Ma ago. At the time the trap was sealed, 45 Ma ago, it had reached a maturity of 0.8 to 0.9 R_o percent equivalent, corresponding to a transformation ratio of 70 to 80 percent. The Kimmeridgian had thus been expelling oil at peak generation 10 to 20 Ma ago before the trap was sealed.

The heavy, napthenic, nature of the oil may thus result from migration of a normal light oil into a leaky trap in which sea water was circulating, with consequent water washing and biodegradation removing the light hydrocarbons and n-alkanes.

The Frigg Field methane has a δC^{13} value of 43.3 percent (Heritier et al, 1979), which is characteristic of the carbon isotope composition of methane derived from sapropelic organic matter in the maturity range 0.5 to 1.3 percent (Hunt, 1979, p. 376). This suggests the gas was sourced from the Kimmeridge Clay and not from coaly material in the Brent Formation. Unfortunately there is little analytical data on the formation of methane from sapropel at maturities less than 1.0 percent R_o (Tissot and Welte, 1978, p. 218). However, much of the methane derived generation from kerogen must be formed by cracking of previously formed oil remaining in the rock in the lower part of the oil window (0.9 to 1.3 percent R_o). Hunt (1979, p. 1964) stated that the gas yield from sapropelic kerogen is 1.5 to 2 times that of humic kerogen because of this cracking of the previously formed oil. The Kimmeridge Clay passed through the 0.9 to 1.3 percent R_o maturation interval between 50 Ma ago and the present day, largely after the trap was sealed.

VOLUMETRIC MODELLING OF OIL GENERATION AND ENTRAPMENT

The ratio of oil generated from the Kimmeridge Clay to oil trapped in Jurassic sandstones has been studied in the central and eastern area of the East Shetland Basin (Figure 19). The amount of oil generated within this area was calculated using a Geochemical Mass Balance Equation. The Geochemical Mass Balance method of reserves calculation has been discussed by White and Gehman (1979). Waples (1979) pointed out that a major problem in quantitative evaluation of oil source rocks is lack of knowledge of oil expulsion efficiencies. In this section I shall use the Geochemical Mass Balance method to determine expulsion efficiency.

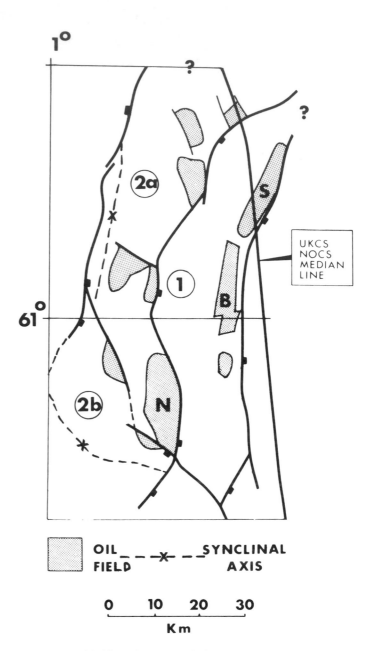

Figure 19. Oil fields and source rock drainage areas in the eastern and central East Shetland Basin.

The volume of oil generated from a unit volume of source rock is related to the amount, type, and maturity of its kerogen. The volume of generated oil that is ultimately trapped is related to the expulsion efficiency of oil from the source rock, the migration efficiency, and the sealing efficiency of the traps.

The average TOC content of the Kimmeridge Clay is 5.6 percent wt at 2,600 to 3,200 m at pre-peak generation maturity. The average TOC content at the onset of oil generation (2,550 m) was thus probably about 6 percent wt, which is equivalent to an organic matter content of 7 percent wt. The Kimmeridge Clay source rock is estimated to have a porosity of 15 percent at the onset of oil generation, using the compaction gradient for mudstone in Well B. The volume percentage of organic matter in the

Kimmeridge Clay calculated using a wt percent organic matter content of 7 percent, a porosity of 15 percent, and rock matrix and water densities of 2.7 and 1 gm/cc respectively, is 15 percent.

The amount of organic matter capable of conversion to hydrocarbons is given by its 'genetic potential.' The theoretical genetic potential of a Type II kerogen is 70 percent (Tissot and Welte, 1978). The relative proportions of sapropel and humic/coaly material in the Kimmeridge Clay kerogen indicate that the oil content of the hydrocarbons generated from the kerogen during maturation is about 80 percent.

The maturity of a kerogen can be expressed by its 'transformation ratio,' which is defined as the ratio of the amount of hydrocarbons generated to the total amount of hydrocarbons that the kerogen is capable of generating. The transformation ratio corresponding to peak generation is thus 0.5. The transformation ratio for the Kimmeridge Clay kerogen as a function of vitrinite reflectance is given in Figure 20 based on the correlation of peak oil generation with a reflectance of 0.7 percent.

Generation of a typical East Shetland Basin light oil (specific gravity = 0.84) from kerogen involves a volume increase of about 20 percent. The volume of oil generated from a unit volume of source rock is given by the following equation:

$$\text{Oil generated} = \frac{\text{bulk rock volume}}{\text{of source rock}} \times \frac{\text{organic matter}}{\text{content by volume}}$$

$$\times \frac{\text{genetic}}{\text{potential}} \times \frac{\text{fraction of oil in}}{\text{hydrocarbon yield}}$$

$$\times \frac{\text{transformation}}{\text{ratio}} \times \frac{\text{volume increase on}}{\text{oil generation}}$$

For the Kimmeridge Clay, the volume of oil generated (V_o) becomes:

$$V_o = \frac{\text{bulk rock}}{\text{volume}} \times 0.15 \times 0.7 \times 0.8$$

$$\times \frac{\text{transformation}}{\text{ratio}} \times 1.2 = 0.1 \frac{(\text{bulk rock volume} \times}{\text{transformation ratio})}$$

The eastern East Shetland Basin has been divided into two drainage areas (Figure 19). The eastern boundary of drainage area 1 is defined by the Brent/Statfjord fault trend, the boundary of drainage areas 1 and 2 by the Ninian/Hutton fault trend. These fault trends converge to the north. The northern boundary of drainage area 2 is not well defined, but is probably defined by a series of faults and dip reversals. The overpressure in the Brent Formation in the eastern East Shetland Basin is 1,700 to 1,900 psi. However, the author has not studied enough pressure data to determine whether there is fluid communication within the Brent across the Ninian/Hutton fault trend.

The bulk rock volume of Kimmeridge Clay in each drainage area has been calculated from an isopach map based on well control and geological structure maps. Weighted average thicknesses for the Kimmeridge Clay are 70 m in Areas 1 and 2a, and 200 m in Area 2b.

The oil generated and trapped within each drainage area is shown in Table 6. The average entrapment/generation ratio is 25 percent, which indicates a very high oil expulsion efficiency from the Kimmeridge Clay (>25 percent) and extremely high migration and sealing efficiencies.

The apparent entrapment/generation ratio for drainage area 1 is 54 percent. However, the sonic log response of mudstones in the upper 150 m of the Heather Formation in Well B (Figure 21) in drainage area 1 suggests that this interval is a source rock. It might be placed in the Kimmeridge Clay on the basis of its sonic velocity. Recalculating the amount of oil generated in drainage area 1, including this extra interval, gives an entrapment/generation ratio of 25 percent; the average entrapment/generation ratio becomes 20 percent.

The average entrapment/generation ratio has also been calculated using average values of organic richness and hydrocarbon yield determined for immature Kimmeridge Clay samples from onshore UK and the North Sea by the BP's Geochemical Research Division. Average organic matter content is 8 percent by wt; average hydrocarbon yield during pyrolysis is 30 kg/ton rock, equivalent to a genetic potential of 37 percent. The average entrapment/generation ratio obtained using these data, and the additional source rock interval in drainage area 1, is 30 percent.

The oil entrapment/generation ratio in the eastern East Shetland Basin is 20 to 30 percent. Oil expulsion efficiency is ≥ 20 to 30 percent, since some oil must be lost as residual oil along migration paths, and oil generated in the upper part of the source rock may have been expelled into the Cretaceous mudstones.

OIL SATURATION IN COMPACTION FLUIDS EXPELLED FROM KIMMERIDGE CLAY

The oil saturation of the compaction fluids expelled from the Kimmeridge Clay in the drainage areas (Figure 19) during migration has been estimated by calculating the volumes of oil and water expelled. By the end of the Cretaceous the Kimmeridge Clay was mature in the deepest parts of the East Shetland Basin; migration is thus considered to have begun 65 Ma ago. The volume of oil generated is (15.4 to 20.2) $\times 10^9$ cu m (Table 6); the volume of oil expelled using an expulsion efficiency of 30 percent is thus (4.6 to 6.1) $\times 10^9$ cu m. The water filled porosity loss from the Kimmeridge Clay has been estimated assuming that the mudstone compaction gradient during the Tertiary was the same as at the present day. The average porosity loss for the Kimmeridge Clay is estimated to be 10 percent since the end of the Cretaceous. This porosity loss is a maximum value, since the Kimmeridge Clay is over-pressured (under-compacted) in the East Shetland Basin. The bulk rock volume of the Kimmeridge Clay is 280 $\times 10^9$ cu m (Table 6); the maximum amount of water expelled from the Kimmeridge Clay during oil migration is thus about 30 $\times 10^9$ cu m. The minimum oil saturation in the compaction fluids expelled during migration is thus about 13 to 17 percent.

However, 50 percent of the oil is generated over a limited

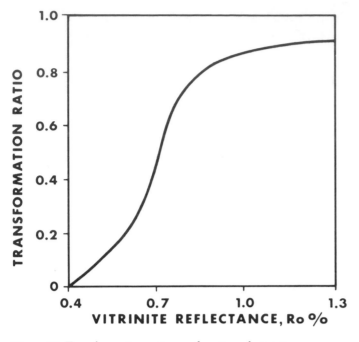

Figure 20. Transformation ratio as a function of vitrinite reflectance for Kimmeridge Clay.

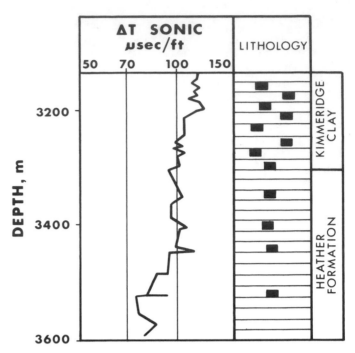

Figure 21. Compaction gradients in Kimmeridge Clay and Heather Formation in Well B.

maturation range (0.65 to 0.8 R_o percent maturity) during peak generation (Figure 20). The maturation history of the Kimmeridge Clay in Well C is representative of deeply buried Kimmeridge Clay in the East Shetland Basin. The Kimmeridge Clay in Well C passed through this maturity range 25 to 5 MA ago; its maximum water filled porosity loss is estimated at 1 to 2 percent during this time. The minimum oil saturation of compaction fluids expelled during peak generation is thus about 30 to 50 percent. These very high oil saturations indicate that oil phase primary migration occurred.

MECHANISM OF OIL MIGRATION

Several mechanisms of oil phase primary migration have been proposed: McAuliffe (1979) suggested that oil migrates through a 3 dimensional, oil wet, kerogen network. The high average organic matter content of the Kimmeridge Clay (15 to 18 vol percent) suggests that much of the rock may be oil wet. Momper (1978) and Meissner (1978) suggested that oil expulsion is a direct consequence of maturation, and that oil migrates through microfractures created by abnormal pore pressures resulting from generation. They pointed out that generated oil and residual

kerogen occupies a greater volume than the immature kerogen. Dickey (1976) suggested that oil is able to flow through source rocks at relatively low oil saturations (1 to 10 percent of the total porosity) because much of the pore water is absorbed on clay mineral surfaces and is thus not part of the *effective* porosity of the rock. Du Rouchet (1981) considered that generated oil is expelled from kerogen by compaction and that it migrates through laminae with relatively large pore throats, or through microfractures. Momper (1978) noted that about 90 to 95 percent of the pore volume of mudstones comprises small subcapillary pores (1 to 3 nm diameter). Abnormally large mudstone pores are associated with lenses of silt, microvugs, and leached zones.

Two other factors that may be important in primary migration are *creation of porosity* by conversion of kerogen to oil, and the reduction in oil/water interfacial tension with increasing temperature. The change in porosity during maturation of the Kimmeridge Clay source rock, with 30 percent oil expulsion, is summarized in Table 7. The total porosity of the source rock at first decreases as water is expelled from the rock, but then increases slightly near peak oil generation. Oil saturation in the effective porosity of the rock (comprising the abnormally large rock matrix pores

Table 6. Volume of oil generated and trapped in the central and eastern East Shetland Basin.

Drainage Area	Bulk Rock Volume ($\times 10^9$ cu m)	Weighted Average Maturity = Transformation Ratio	Oil Generated $\times 10^9$ Barrels	Trapped Oil $\times 10^9$ Barrels	Entrapment/ Generation Ratio
1	98 (180)	0.43 (0.5)	27 (57)	14.5	0.54 (0.25)
2a	102	0.51	33	9.5	0.14
2b	79	0.75	37		
Total	279 (361)	0.55 (0.56)	97 (127)	24	0.25 (0.19)

Figures in parentheses calculated using additional source rock in upper part of Heather Formation in drainage area 1.

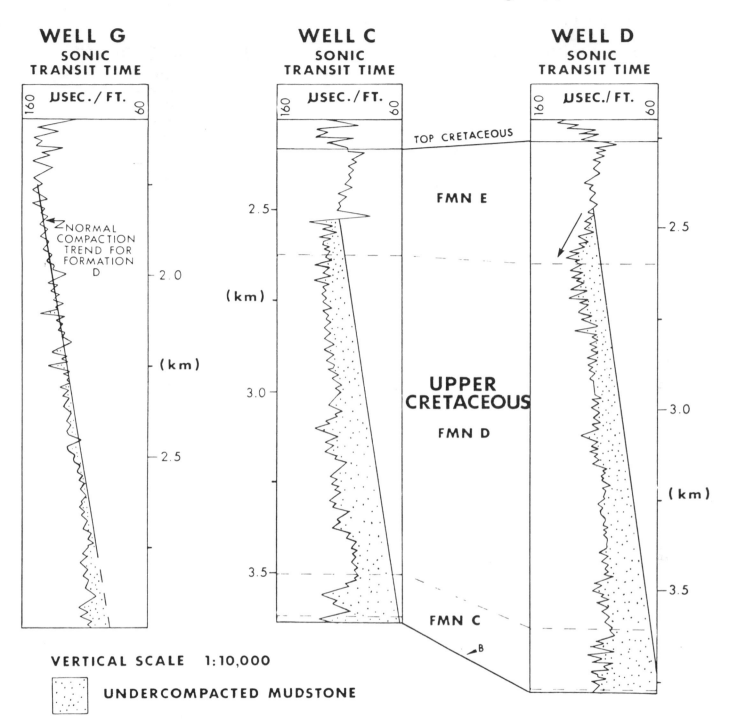

Figure 22. Sonic transit time of Cretaceous mudstones as a function of depth for Well G in the Sogn Graben, and Wells C and D in the Viking Graben.

Table 7. Estimated porosity and oil saturation of mature Kimmeridge Clay source rock.

Transformation Ratio of Kerogen	Water Filled Clay Porosity (%)	Oil Content of Rock (Vol %)	Oil Saturation in Effective Porosity† (Vol %)	Total Porosity
0.1	18	1	35	19
0.25	14	1.7*	55*	15.7
0.5	13	3.5*	70*	16.5
0.75	12	5*	80*	17

*Calculated assuming 30 percent oil expulsion efficiency.
†Effective porosity = oil-filled porosity + 0.1 (water-filled clay porosity).

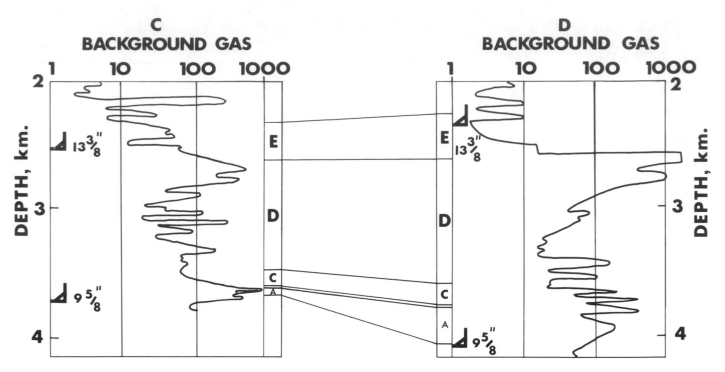

Figure 23. Gas readings in drilling mud in early Tertiary to late Cretaceous mudstones in Wells C and D.

and porosity in the kerogen laminae) increases from 35 percent at a transformation ratio of 0.1, to 55 percent at 'quarter' generation. These high oil saturations suggest that oil may be able to migrate out of the rock at relatively low maturities prior to peak generation.

Oil/water interfacial tension decreases with temperature by 0.2 to 0.4 dyne/cm°C from 25 to 70°C (Schowalter, 1979). At temperatures >70°C the oil/water interfacial tension for light oils is <5 dyne/cm. No data is available for oil/water interfacial tension at temperatures of oil generation from the Kimmeridge Clay (95 to 140°C). Clearly very low oil/water interfacial tensions <5 dyne/cm may exist in source rocks in this temperature range. Capillary pressure (the resistant force to oil phase migration through water wet rock) is directly proportional to interfacial tension:

$$P_c = \frac{2\gamma \cos \theta}{R}$$

where P_c is the capillary pressure, γ is the interfacial tension, θ is the contact angle between oil and water, R is the radius of the pore throat.

The pressure required to inject oil through a pore throat is termed the displacement pressure. Mercury/air displacement pressures for mudstones are >5,000 psi and 1,700 to 5,000 psi for siltstone. For an interfacial tension of 5 dyne/cm these correspond to oil/water displacement pressures of about 20 to 50 psi for silty laminae and >50 psi for mudstone (from Schowalter, 1979, Figure 23).

These low displacement pressures estimated for the larger interconnected pores of the source rock suggest that microfractures, created around isolated kerogen laminae, would tend to inject their oil into the more coarsely porous

laminae of the source rock. The main oil migration routes within the Kimmeridge Clay and underlying Heather Formation are thus probably the continuous oil wet kerogen laminae, and water wet, silty and sandy laminae.

The mature Kimmeridge Clay source rocks in the East Shetland Basin are overpressured. Drilling exponent data in Well F indicates approximately normal pore pressures in the late Cretaceous mudstones above 3,100 m. The estimated pore pressure gradient increases with depth to 0.63 psi/ft in the Kimmeridge clay source rock at 3,500 m.

The Brent reservoirs in the East Shetland Basin were probably normally pressured, or slightly overpressured, prior to oil migration and Tertiary loading. Oil has probably migrated along strong *lateral* fluid pressure gradients set up from actively generating, overpressured source rocks in the half grabens, towards the lower pressured Brent reservoirs.

MECHANISM OF HYDROCARBON MIGRATION INTO THE FRIGG FIELD

Primary migration into the early Tertiary aquifer of the Frigg Field requires vertical migration through 2,000 m of Cretaceous and Palaeocene mudstones (Figure 4). Possible migration mechanisms include: (1) buoyant hydrocarbon phase flow through the interconnected largest pores of the mudstone, (2) migration in aqueous solution with exsolution of hydrocarbons in the early Tertiary aquifer, (3) migration through microfractures created by abnormal pore pressures. In order to assess the relative importance of these possible migration mechanisms it is first necessary to determine the physical conditions in the Jurassic and Cretaceous mudstones, and the hydrocarbon distribution within them.

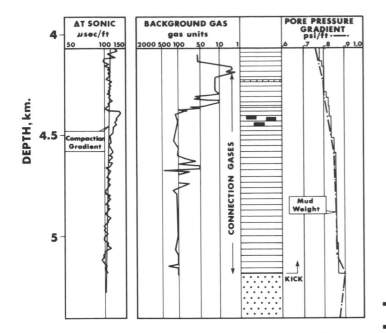

Figure 24. Pore pressure gradient, sonic transit time and drilling mud background gas in the early Cretaceous and Jurassic section of Well D.

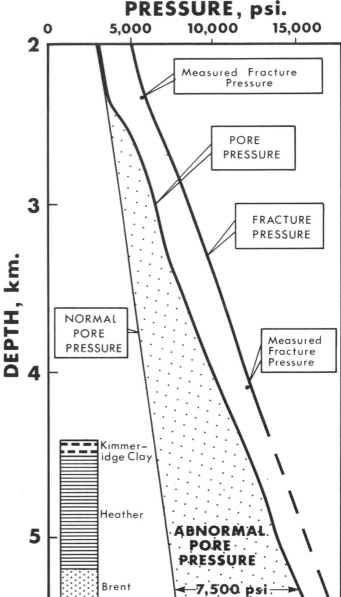

Figure 25. Pore and fracture pressures in Well D.

Pore Pressure

The pore pressure distribution in the Tertiary to Jurassic mudstones below the Frigg Field can be inferred from a study of pore pressures in Well D drilled in the axial Viking Graben. Pore pressures were estimated from gas readings, mud weights required to balance minor flows of formation fluid during drilling, and from mudstone sonic transit times.

Low sonic transit times and gas readings indicate the late Cretaceous mudstones are normally pressured above 2,350 m (Figures 22 and 23). Between 2,350 m and 2,500 m transit times are less than those of normally compacted late Cretaceous mudstones in the Sogn Graben to the NNE of the study area (Figure 22). Between 2,400 and 2,600 m sonic transit time increases from 100 to 120 μsec/ft, which suggests an increase in porosity with depth. Gas readings increase gradually from 2,350 to 2,550 m and dramatically below 2,550 m. These two observations indicate that an abnormal pore pressure transition zone occurs from 2,350 to 2,550 m.

Between 2,550 m and 3,900 m mudstone transit time decreases with depth; the transit time diverges from the normal compaction trend (Figure 22), indicating a progressive increase in overpressure with depth. Between 3,950 to 4,000 m the borehole is badly caved and the sonic readings are thus unreliable. High gas readings at the base of the 12¼ in hole (Figure 23) indicate a pore pressure gradient of 0.76 psi/ft at 4,050 m at the base of the late Cretaceous mudstones.

The increase in mudstone transit time between 3,950 to 4,070 m near the base of the late Cretaceous mudstones reflects an increasing pore pressure gradient but may also be caused by a change in mudstone mineralogy. The transit time of the early Cretaceous and late Jurassic mudstones decreases with depth at a very slow rate (Figure 24), indicating a further progressive downward increase in

overpressure to 0.83 psi/ft at the base of the Cretaceous. The Kimmeridge Clay transit time is abnormally high relative to the overlying and underlying formations because of its high organic matter content. Mud weights required to control the pore pressure in the Jurassic sandstone at 5,200 m indicate a pore pressure gradient of 0.89 psi/ft at the base of the late Jurassic mudstones.

The pore pressure profile in the Cretaceous and Jurassic mudstones of Well D (Figure 25) thus comprises a compacted 'caprock' mudstone (2,250 to 2,350 m), a pore pressure transition zone (2,350 to 2,550 m) and a progressive slow increase in overpressure (2,550 to 5,200 m). The general form of this profile is characteristic of thick laterally extensive mudstones buried below normally pressured permeable overburden (Bishop, 1979). Chiarelli and Duffaud (1980) concluded that overpressures in the

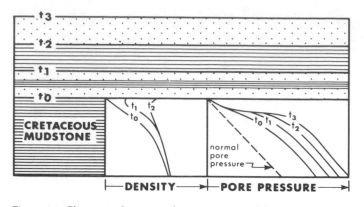

Figure 26. Change in density and pore pressure of Cretaceous mudstones in Viking Graben during successive increments of loading by Tertiary sediments (t_o-t_3) (based on Bishop, 1979).

Frigg area are associated with undercompaction. Bishop numerically modelled the evolution of the pore pressure of abnormally pressured, compacting, shale masses. The qualitative evolution of the pore pressure of the late Cretaceous mudstones as predicted by Bishop's model is shown in Figure 26.

Fracture Pressure

The Fracture Pressure Gradient is given by the sum of the least principal stress gradient and the pore pressure gradient (Eaton, 1969):

$$\frac{Pf}{Z} = \frac{(S - P_p)}{(Z)} \frac{M}{(1 - M)} + \frac{P_p}{Z}$$

where Z = depth
Pf = fracture pressure
P_p = pore pressure
S = overburden pressure
M = Poisson's Ratio

This equation was used to estimate fracture pressure gradient during drilling of Well D. The variation in Poisson's ratio with depth was assumed to be the same as that reported in the Gulf Coast by Eaton (1969). The calculated fracture pressure as a function of depth below 2,000 m is shown in Figure 25.

The fracture pressure gradient was also measured directly during formation leak off tests at 2,345 and 4,065 m. Poisson's ratio was calculated at these depths using the measured fracture gradient and estimated pore and overburden pressure gradients (Table 8).

For comparison Poisson's Ratio at 2,000 m and 4,000 m in the Gulf Coast are 0.42 and 0.47 respectively (from Eaton's plot of Poisson's ratio versus depth).

Microfracture Distribution

The 'fracture pressure' calculated using Eaton's Method, and measured during formation leak off tests, is the pressure required to propagate large scale fractures. Palciauskas and Domenico (1980) reviewed work on experimental microfracturing of sandstones, siltstones, and carbonates. They concluded that these rock types begin to microfracture

at fluid pressure about 20 percent less than the 'fracture pressure.' They suggested that this microfracture criterion probably also applies to mudstones.

The difference in pore and fracture pressure in Well D has been calculated to determine whether microfractures exist in the overpressured Cretaceous and Jurassic mudstones, using this microfracture criterion. The pore pressure (3,400 psi) at the top of abnormal pore pressure transition zone (2,350 m) is 42 percent less than the fracture pressure (5,900 psi). At the base of the transition zone (2,550 m) the pore pressure (5,000 psi) is 26 percent less than the fracture pressure (6,800 psi). These data indicate the transition zone is not microfractured.

Between the base of the transition zone and 3,600 m the pore pressure is 20 to 25 percent less than the fracture pressure. This suggests that this interval may be microfractured. Below 3,600 m the pore pressure is within 20% of the fracture pressure which indicates that active microfracturing is probably occurring.

Within the Kimmeridge Clay and Heather Formation source rocks the pore pressure is within 10 to 13 percent of the fracture pressure. The pore pressure at 5,300 m within the Brent Formation (15,200 psi) is only 10 percent less than the fracture pressure (16,800 psi).

Pore pressure within the high pressure Brent reservoirs of the Viking Graben converges updip with the calculated seal fracture pressures. Figure 27 shows the extrapolated Brent pore pressure updip of Well D, and the corresponding seal fracture pressure. At 4,500 m the extrapolated Brent pore pressure equals the fracture pressure. A Brent pore pressure measurement, reported by Lindberg et al (1980) in the Frigg area is within 5 percent of the calculated seal fracture pressure at that depth (Figure 27). These data suggest that either the microfracture criterion of Palciauskas and Domenico is incorrect, or that the calculated fracture pressures are too low, or that microfracturing does not prevent further increases in pore pressure gradient.

Well C penetrated a high pressure gas condensate column in the Brent Formation; the pore pressure is within 10 percent of the calculated fracture pressure. Retention of this gas column in the reservoir suggests that either its seal is not microfractured, or that the rate of migration into the trap at least balances leakage of gas through microfractures.

Methane Solubility

The aqueous solubility of methane in the early Tertiary to Jurassic section of Well D has been estimated from the calculated variation in pore pressure and temperature with depth using methane solubility data given by Magara (1980), and Jones (1980) (Figure 28a). Methane solubility increases from 3 standard cu m/cu m water at the top of the abnormal pore pressure transition zone to 13 cu m/cu m water at 5,400 m.

Table 8. Calculated values of Poisson's Ratio in Well D.

Depth (m)	Pore Pressure Gradient (psi/ft)	Fracture Pressure Gradient (psi/ft)	Overburden Pressure Gradient (psi/ft)	M
2,345	0.45	0.76	0.89	0.41
4,065	0.76	0.91	0.95	0.44

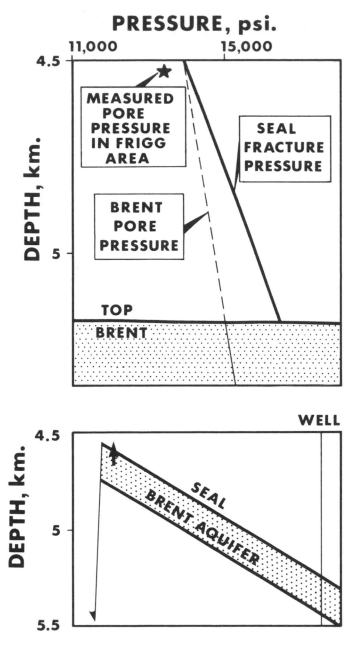

Figure 27. Convergence of reservoir pore pressure and seal fracture pressure in deep high pressure fault blocks.

Cumulative dissolved methane within a 1,500 sq km area around Well D (equivalent to the drainage area of the Frigg Field) is shown in Figure 28b. Cumulative dissolved methane was calculated from the solubility/depth plot and the mudstone compaction gradient, assuming all mudstone pores contained fresh water saturated with methane. The actual solubility of methane in pore water with a salinity of 35,000 ppm total dissolved solids is 10 to 20 percent less than the fresh water solubility (Hunt, 1979).

Volume of Gas Generated in Drainage Area of Frigg Field

The amount of gas generated within the drainage area of the Frigg Field has been estimated assuming that the average thickness and maturity of Kimmeridge Clay and Heather formations within the area are the same as in Well D. The

amount of gas generated from the Heather Formation has been calculated assuming it contains dispersed coaly material, using the methane generation/maturity plot for coal (Hunt, 1979). The amount of gas generated from the Kimmeridge Clay has been estimated assuming it is derived from cracking of oil remaining in the source rock after oil generation ceased. Estimated gas yields from the Heather Formation and Kimmeridge Clay are 100×10^{10} cu m and $(100 \text{ to } 200) \times 10^{10}$ cu m respectively. Ten m of Brent coal within a drainage area equivalent to the Frigg Field would have generated 200×10^{10} cu m of methane assuming the same Brent maturity as in Well D. However the carbon isotope data suggests the Frigg gas was not sourced from Brent coals.

Variation in Light Hydrocarbon Composition and Concentration with Depth (Figure 29)

In the normally pressured early Tertiary and late Cretaceous (2,000 to 2,350 m), hydrocarbon concentrations are low; gas wetness increases from 10 percent at the top of the interval to 30 percent at the base. In the abnormal pore pressure transition zone (2,350 to 2,550 m) total gas, gasoline range hydrocarbon (C_5-C_7) concentration, and gas wetness increase dramatically with depth (Figure 29).

Very high background gas was measured in the 250 m interval below the abnormal pore pressure transition zone (2,550 to 2,800 m). Oil shows were recorded in limestone cuttings in this interval. A similar zone of high background gas occurs below the abnormal pore pressure transition zone in Well C. The much higher background gas values over this interval in Well D are due to underbalanced drilling. Analysis of canned cuttings from this interval indicate the gas is wet (40 to 80 percent C_2-C_4); C_5-C_7 hydrocarbon concentration exceeds 5,000 vol ppm of rock between 2,550 to 2,650 m.

Within the late Cretaceous mudstones below 3,300 m total gas and gas wetness increases with depth. Mud background gas increases markedly over a 50 m interval above the top of the Kimmeridge Clay (Figure 24). Total gas in the late Jurassic mudstones is nearly half an order of magnitude greater than in the Cretaceous mudstones (3,000 to 4,387 m). This reflects higher organic richness and maturity of the late Jurassic mudstones.

Evaluation of Hydrocarbon Migration Mechanisms

Methane/water interfacial tension is about 25 dyne/cm in the temperature range 100 to 300°F and pressure range 5,000 to 15,000 psi (Schowalter, 1979). Gas-water displacement pressures are > 350 psi for mudstones which have mercury/air displacement pressures > 5,000 psi for this interfacial tension. These data indicate that vertical gas columns > 300 m in height are required to inject gas through mudstone pores in the late Cretaceous section of the Viking Graben. There is no evidence for the existence of such columns; gas phase migration through mudstone pores is unlikely.

Comparison of the pore and fracture pressures strongly suggests that gas is migrating through microfractures below 3,600 m. From 3,600 to 5,150 m there is a positive correlation between total gas and gas wetness which suggests migration is occurring in gas phase. The lack of evidence of microfractures above 3,500 m, and especially

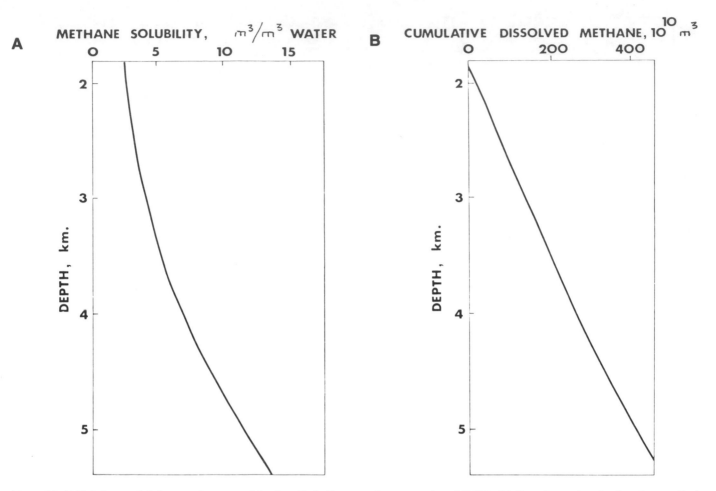

Figure 28. (A) Methane solubility as a function of depth in Early Eocene to Jurassic rocks in Well D. (B) Cumulative dissolved methane in Early Eocene to Jurassic rocks within the Frigg Field drainage area (1,500 sq km).

within the abnormal pressure transition zone and overlying normally pressured mudstones, suggests another migration mechanism occurs in the upper part of the late Cretaceous section.

Any hypothesis for the origin of the Frigg Field gas must account for its very low wet gas content (3.5 percent ethane, 0.05 percent propane + butane). This composition can be explained by migration in aqueous solution within the upper part of the late Cretaceous mudstones, and subsequent exsolution in the early Tertiary aquifer. This hypothesis requires that methane/saturated pore water exsolves sufficient methane, when it is expelled into the early Tertiary aquifer, to account for the in-place reserves of the field.

The amount of gas generated from Jurassic source rocks within the drainage area of the Frigg Field (200 to 300 $\times$ 10^{10} cu m) exceeds that required to saturate the water in the late Cretaceous mudstones between 2,250 m and 3,500 m (150 $\times$ 10^{10} cu m) (Figure 28). This suggests that pore waters in this interval are saturated with respect to methane.

The aqueous solution hypothesis can be tested by calculating the amount of pore water expelled into the aquifer from the Frigg Field drainage area since the trap was sealed, and the volume of methane exsolved (Table 9). The water loss from the Jurassic and Cretaceous mudstones in

the last 40 Ma has been calculated from the mudstone compaction gradient in Well D. The pressure drop across the early Tertiary aquifer has been estimated form Magara's solubility data for the temperature and pressure conditions prevailing in the Frig aquifer in the last 40 Ma. The calculated volumes of exsolved methane is more than enough to account for the in-place reserves of the field (27 $\times$ 10^{10} cu m).

To conclude, the migration mechanism of the Frigg gas probably involved three processes: 1, Migration as gas phase in microfractures in the over-pressured late Jurassic

Table 9. Evaluation of aqueous solubility mechanism for migration of Frigg Field methane.

*Drainage area of field	1,500 sq km
*Decrease in thickness of late Cretaceous mudstones since field was sealed	450 m
*volume of water expelled into early Tertiary aquifer from below within field drainage area	$(1,500 \times 10^6 \times 450$ cu m $= 70 \times 10^{10}$ cu m
*Pressure drop between top and bottom of aquifer	900 psi
*Methane exsolved in aquifer due to this pressure drop	0.5 m^3/m^3 water $= (70 \times 10^{10} \times 0.5)$ cu m $= 35 \times 10^{10}$ cu m.

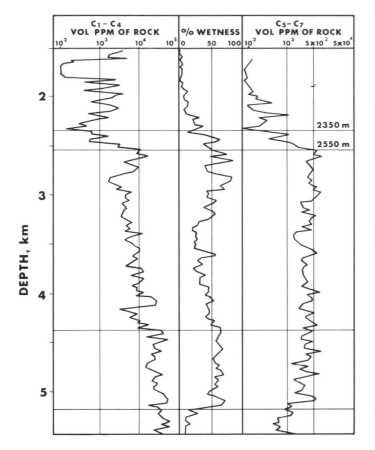

Figure 29. Total gas, gas wetness and C_5-C_7 hydrocarbons in canned cuttings samples as a function of depth in Well D.

mudstones and overlying late Cretaceous mudstones at depths greater than about 3,500 m. 2, Migration in aqueous solution through the late Cretaceous mudstones at depths shallower than about 3,500 m. 3, Exsolution of methane in the Frigg Field aquifer and subsequent buoyant flow of gas to the trap.

An aqueous solubility mechanism for the Frigg Field oil has also been evaluated. About 5×10^9 cu m of oil is estimated to have been expelled from the Kimmeridge Clay source rock within the drainage area of the Frigg Field assuming a source rock thickness of 100 m. The in-place reserves of the field are $\cong 0.1 \times 10^9$ cu m. The water loss from the Cretaceous and Jurassic mudstones within the field's drainage area is 70×10^{10} cu m. About 150 ppm of oil would thus have to be exsolved from these compaction waters in the trap to account for the in-place reserves of the field. This amount of exsolution requires an oil concentration of the order of 1,000 ppm in the compaction water. Whole oil solubilities at 100 to 150°C, equivalent to the temperatures prevailing in the Cretaceous mudstones, are <100 ppm (Prie, 1976). An aqueous solubility mechanism thus cannot account for the in-place reserves of the field. However, the heavy polar nitrogen and sulphur bearing compounds in the oil could have migrated by this mechanism.

The relatively high and uniform gasoline range hydrocarbon concentrations in the overpressured Cretaceous mudstones suggest that large scale vertical

migration of hydrocarbons generated from the Kimmeridge Clay has occurred. The most likely migration mechanisms are oil migration through microfractured overpressured mudstones below 3,500 m, and migration through siltstones and the interconnected larger mudstone pores in response to low oil/water interfacial tensions.

CONCLUSIONS

1. The richest source rocks in the East Shetland Basin and the Viking Graben are the Kimmeridge Clay (oil and gas) and the Brent Formation coals (dry gas). Bathonian and Toarcian mudstones may also have generated oil. Thick Heather Formation mudstones are gas source rocks in the Viking Graben.

2. A uniform present day maturity gradient indicates the oil window (0.5 to 1.3 percent R_o) extends from 2,550 to 4,500 m, and the wet gas zone (1.3 to 2 percent R_o) from 4,500 to 5,400 m. Peak oil generation from the Kimmeridge Clay occurs at a reflectance level of 0.7 percent R_o (3,250 m). Highest oil saturations in the Kimmeridge Clay, inferred from resistivity logs, occur at 0.8 percent R_o (3,500 to 3,600 m). Intense gas generation from oil trapped in the Kimmeridge Clay has probably occurred in the reflectance range 1 to 1.3 percent R_o (4,000 to 4,500 m). Peak dry gas generation from Brent Formation coals occurs at a reflectance level of 1.5 percent R_o (4,000 to 4,500 m). Peak dry gas generation from Brent Formation coals occurs at a reflectance level of 1.5 percent R_o (5,000 m). At the present day the Kimmeridge Clay is mature over most of the East Shetland Basin and has reached peak generation throughout the axial region of the basin. In the Viking Graben the Kimmeridge Clay is at a maturity close to the oil floor (1.3 percent R_o).

3. Oil generation from the Kimmeridge Clay began 70 to 80 Ma ago in the Viking Graben; 65 Ma ago the Kimmeridge Clay was generating oil throughout the Viking Graben and in the deepest troughs of the East Shetland Basin. Peak oil generation was reached 55 to 65 Ma ago in the Viking Graben; 40 Ma ago peak generation had occurred in the deepest troughs of the East Shetland Basin, and throughout the Viking Graben. Twenty to 40 Ma ago the Kimmeridge Clay entered the wet gas zone in the deepest syncline. Generation of gas by cracking of oil in the Kimmeridge Clay of the Viking Graben occurred during the last 50 Ma. Gas generation from Brent Formation coals began 100 Ma ago in the Viking Graben; peak dry gas generation occurred during the last 40 Ma.

4. Oil phase primary migration from the Kimmeridge Clay has occurred in the East Shetland Basin. The oil expulsion efficiency from this source rock is <20 to 30 percent. Migration and trapping efficiencies are very high in the East Shetland Basin: 20 to 30 percent of the oil generated from the Kimmeridge Clay is now reservoired in Jurassic sandstones. Oil saturation in compaction fluids expelled from actively generating Kimmeridge Clay was at least 30 percent. Oil migration has occurred along high fluid potential gradients set up from actively generating overpressured source rocks in the half grabens to normally pressured or slightly overpressured Jurassic reservoirs, stratigraphically below, but structurally updip of, the

source rocks. During primary migration oil probably moved through porosity created in the kerogen laminae of the source rock by oil generation, and through the interconnected larger pores of the Kimmeridge Clay and underlying Heather Formation aided by low oil/water interfacial tensions.

5. Migration of the Frigg Field methane probably involves 3 processes: 1, Migration of gas in gas phase in microfractures in the overpressured late Jurassic mudstones and overlying the late Cretaceous mudstones (at depths >3,500 m). 2, Migration in aqueous solution through late Cretaceous mudstones at depth shallower than about 3,500 m. 3, Exsolution of methane in the Frigg Field aquifer, and subsequent buoyant flow of gas to the trap. Most of the Frigg Field oil probably did not migrate in aqueous solution. Migration of oil probably occurred through microfractures, and through siltstone and the larger mudstone pores in response to low oil/water interfacial tensions.

ACKNOWLEDGMENTS

The author acknowledges many valuable discussions with BP geologists, geochemists, and geophysicists, but would emphasize that the views expressed are his own. In particular he wishes to thank Dr. A.M. Spencer for encouragement to publish this paper, Dr. J.R. Bloomer and Dr. S.W. Richardson for invaluable help with the geothermal modelling, and G.C. Speers and Dr. G. Dungworth for discussion of geochemical principles. Anita McKearney kindly typed the manuscript. The author wishes to thank the management of British Petroleum Development Ltd. for permission to publish this paper.

REFERENCES

Barnard, P.C. and B.S Cooper, 1981, Oils and source rocks of the North Sea area, *in* L.V. Illing and G.D. Hobson, eds., Petroleum geology of the Continental Shelf of North-West Europe: Heyden and Son, p. 169-175.

Bishop, R.S., 1979, Calculated compaction states of thick abnormally pressured shales: AAPG Bulletin, v. 63, p. 918-933.

Brooks, J., and B. Thusu, 1977, Oil-source identification and characterisation of the Jurassic sediments in the northern North Sea: Chemical Geology, v. 20, p. 283-294.

Carstens, H., and K.G. Finstad, 1981, Geothermal gradients of the northern North Sea Basin 59 to 62°N, *in* L.V. Illing and G.D. Hobson, eds., Petroleum geology of the Continental Shelf of North-West Europe: Heyden and Son, p. 152-161.

Chiarelli, A., and F. Duffaud, 1980, Pressure origin and distribution in Jurassic of Viking Basin (United Kingdom-Norway): AAPG Bulletin, v. 64, p. 1245-1266.

De'ath, N.G. and S.F. Schuyleman, 1981, The geology of the Magnus Oil Field, *in* L.V. Illing and H.D. Hobson, eds., Petroleum geology of the Continental Shelf of North-West Europe: Heyden and Son, p. 342-351.

Dickey, P.A., 1975, Possible primary migration of oil from source rock in oil phase: AAPG Bulletin, v. 59, 337-347.

Donato, J.A. and M.C. Tully, 1981, A regional

interpretation of North Sea gravity data, , *in* L.V. Illing and H.D. Hobson, eds., Petroleum geology of the Continental Shelf of North-West Europe: Heyden and Son, p. 65-75.

Du Rouchet, J.D., 1981, Stress fields, a key to oil migration: AAPG Bulletin, v. 65, p. 74-85.

Eaton, B.A., 1969, Fracture gradient prediction and its application in oil field operations: Journal of Petroleum Technology, v. 21, p. 1353-1360.

Fuller, J.G.C.M., 1980, *in* J.M. Jones and P.W. Scott, 1980, Progress report on fossil fuels—exploration and exploitation: Proceedings of the Yorkshire Geological Society, v. 33, p. 581-593.

Gallois, R.W., 1976, Coccolith blooms in the Kimmeridge Clay and origin of North Sea Oil: London, Nature, v. 259, p. 473-475.

Hancock, N.J. and A.M. Taylor, 1978, Clay mineral diagenesis and oil migration in the Middle Jurassic Brent Sand Formation: Journal of the Geological Society of London, v. 135, p. 69-72.

Hay, J.T.C., 1977, The thistle field: Paper given at Bergen North Sea Conference.

Heritier, F.E., P. Lossel, and E. Wathne, 1979, Frigg Field—large submarine-fan trap in Lower Eocene rocks of North Sea: AAPG Bulletin, v. 63, p. 1999-2020.

Hunt, J.M., 1979, Petroleum Geochemistry and Geology: San Francisco, W.H. Freeman and Co., 617 p.

Jones, P.H., 1980, Role of geopressure in the hydrocarbon and water system, *in* Problems of petroleum migration: AAPG Studies in Geology, n. 10, p. 207-216.

Lindberg, P., R. Riise, and W.H. Fertl, 1980, Occurrence and distribution of overpressures in the Northern North Sea area: Society of Petroleum Engineers Journal, preprint paper 9339.

McAuliffe, C.D., 1979, Oil and gas migration: chemical and physical constraints: AAPG Bulletin, v. 63, p. 761-781.

McKenzie, D., 1978, Some remarks on the development of sedimentary basins: Earth and Planetary Science Letters, v. 40, p. 25-32.

Magara, K., 1980, Primary migration of oil and gas, *in* A.D. Miall, eds., Facts and principles of world petroleum occurrence: Canadian Society of Petroleum Geologists Memoir, n. 6, p. 173-191.

Meissner, F.F., 1978, Petroleum geology of the Bakken Formation, Williston Basin, North Dakota and Montana: Williston Basin Symposium, Montana Geological Society, p. 207-227.

Momper, J.A., 1978, Oil migration limitations suggested by geological and geochemical considerations: AAPG Course Note Series, v. 8, B1-B60.

Oxburgh, E.R., and C.P. Andrews-Speed, 1981, Temperature, thermal gradients and heat flow in the southwestern North Sea, *in* L.V. Illing and H.D. Hobson, eds., Petroleum geology of the Continental Shelf of North-West Europe: Heyden and Son, p. 141-151.

——— , et al, 1972, Equilibrium borehole temperatures from observation of thermal transients during drilling: Earth and Planetary Science Letters, v. 14, p. 47-49.

Palciauskas, V.V. and P.A. Domenico, 1980, Microfracture development in compacting sediments: relation to hydrocarbon-maturation kinetics: AAPG Bulletin, v. 64,

p. 927-937.

Price, L.C., 1976, Aqueous solubility of petroleum as applied to its origin and primary migration: AAPG Bulletin, v. 60, p. 213-244.

Richardson, S.W. and E.R. Oxburgh, 1979, The heat flow field in mainland U.K.: London, Nature, v. 282, p. 565-567.

Royden, L., J.R. Sclater, and R.P. Van Herzen, 1980, Continental margin subsidence and heat flow: important parameters in formation of petroleum hydrocarbons: AAPG Bulletin, v. 64, p. 173-187.

Schowalter, T.T., 1979, Mechanics of secondary hydrocarbon migration and entrapment: AAPG Bulletin, v. 63, p. 723-760.

Sclater, J.G., and P.A.F. Christie, 1980, Continental stretching: an explanation of the post- Mid-Cretaceous subsidence of the Central North Sea Basin: Journal of Geophysical Research, v. 85, p. 3711-3739.

Sommer, F., 1978, Diagenesis of Jurassic sandstones in the Viking Graben: Journal of the Geological Society of London, v. 135, p. 63-68.

Staplin, F.L., 1969, Sedimentary organic matter, organic metamorphism and oil and gas occurrence: Bulletin of Canadian Petroleum Geology, v. 17, p. 47-66.

Surlyk, F., 1968, Jurassic basin evolution of East Greenland: London, Nature, v. 273, p. 130-133.

Turcotte, D.L., 1980, Models for the evolution of sedimentary basins, *in* A.W. Bally, ed., Dynamics of Plate Interiors: American Geophysical Union and Geological Society of America.

Tissot, B.P., and D.H. Welte, 1978, Petroleum formation and occurrence: Springer Verlag, 538 p.

Tyson, R.V., R.C.L. Wilson, and C. Downie, 1979, A stratified water column environmental model for the type Kimmeridge Clay: London, Nature, v. 277, p. 377-380.

Waples, D.W., 1979, Simple method for oil source bed evaluation: AAPG Bulletin, v. 63, p. 239-245.

——, 1980, Time and temperature in petroleum formation: application of Lopatin's Method to petroleum exploration: AAPG Bulletin, v. 64, p. 916-926.

White, D.A., and H.M. Gehman, 1979, Methods of estimating oil and gas resources: AAPG Bulletin, v. 63, p. 2183-2203.

Williams, P.F.V., and A.G. Douglas, 1980, A preliminary organic geochemical investigation of the Kimmeridgian Oil Shales, *in* Advances in organic geochemistry, 1979: Oxford, Pergamon Press, p. 631-645.

Source Rocks and Oils of the Central and Northern North Sea

B.S. Cooper and P.C. Barnard
Robertson Research International Limited
Gwynedd, United Kingdom

Regional studies of the petroleum geochemistry of the central and northern North Sea have been based on analyses of field sections and samples from approximately 150 released well sections available to Robertson Research International Limited. They show that hydrocarbon source rocks may occur at many levels from Cambrian to Recent, but that the principal source rock is the Kimmeridge Clay Formation and equivalent formations of Late Jurassic to Early Cretaceous age. Methods of evaluating the quality and quantity of this source rock are described and have been integrated with maturity data to obtain estimates of oil productivity. Potentially oil-generating organic matter preserved in the Kimmeridge Clay Formation is derived from both waxy plant detritus and algal material originating in the marine environment. The two important oil provinces of the East Shetland Basin and the Ekofisk Complex contain examples of oil derived from "waxy" land plant and marine algal type source rocks, respectively. We also discuss the presence of gas in this important oil province and the distribution of low gravity oil.

INTRODUCTION

The search for accumulations of oil and gas in the central and northern North Sea is continuing and large accumulations are still being found in recently licensed areas. There is a high probability that substantial reserves of oil and gas are still to be found, but the techniques used are necessarily becoming more sophisticated as the size of structures still to be explored becomes smaller and their complexity increases. Recent literature amply demonstrates that for optimum success in exploration it is essential to build a detailed model of basin history in all its aspects, including geochemistry. This paper includes the results of detailed studies over a number of years of samples of all ages in, and adjacent to, the North Sea. These studies have shown that there is one major North Sea oil source rock presently identified, the Kimmeridge Clay Formation, and its equivalent in the Norwegian-Danish Basin, the Børglum Member of the Bream Formation.

The principal area of study in this paper is loosely termed the central and northern North Sea and is contained within the area between latitude 56°N and 62°N, and longitude 6°E and 6°W. In preparing this paper, we integrated the data obtained from analysis of many field samples and of samples from up to 150 released offshore wells. As a result, we were able to draw generalized maps showing the relationships between depositional environment and kerogen facies and between kerogen facies, maturity/depth of burial, and hydrocarbon accumulations. The data were obtained from material belonging to Robertson Research International Limited, without drawing on any data from wells still confidential to operators.

We have limited the study area to the region north of the mid-North Sea High since we presently do not have data on released wells from the Southern North Sea.

Details of structure used in the maps have been taken from the seismic structure maps of Day et al (1981) and Hamar et al (1980) for the bulk of the study area, and from Chesher and Bacon (1975) for the inner Moray Firth area.

NORTH SEA SOURCE ROCKS

In, and adjacent to, the North Sea there are clastic and carbonate sediments ranging in age from Cambrian to Recent. Potential source rocks are known at a number of horizons and are summarized in Figure 1. Published papers (Barnard and Cooper, 1981; Ziegler, 1980) show that in the central and northern North Sea, oil source rocks are probably located within the Jurassic-Early Cretaceous sediments with the Kimmeridge Clay Formation representing the major oil source rock in the area. This view is upheld by more recent results and discussion of this source rock forms the majority of this paper. However, there is some evidence that Early Jurassic source rocks are

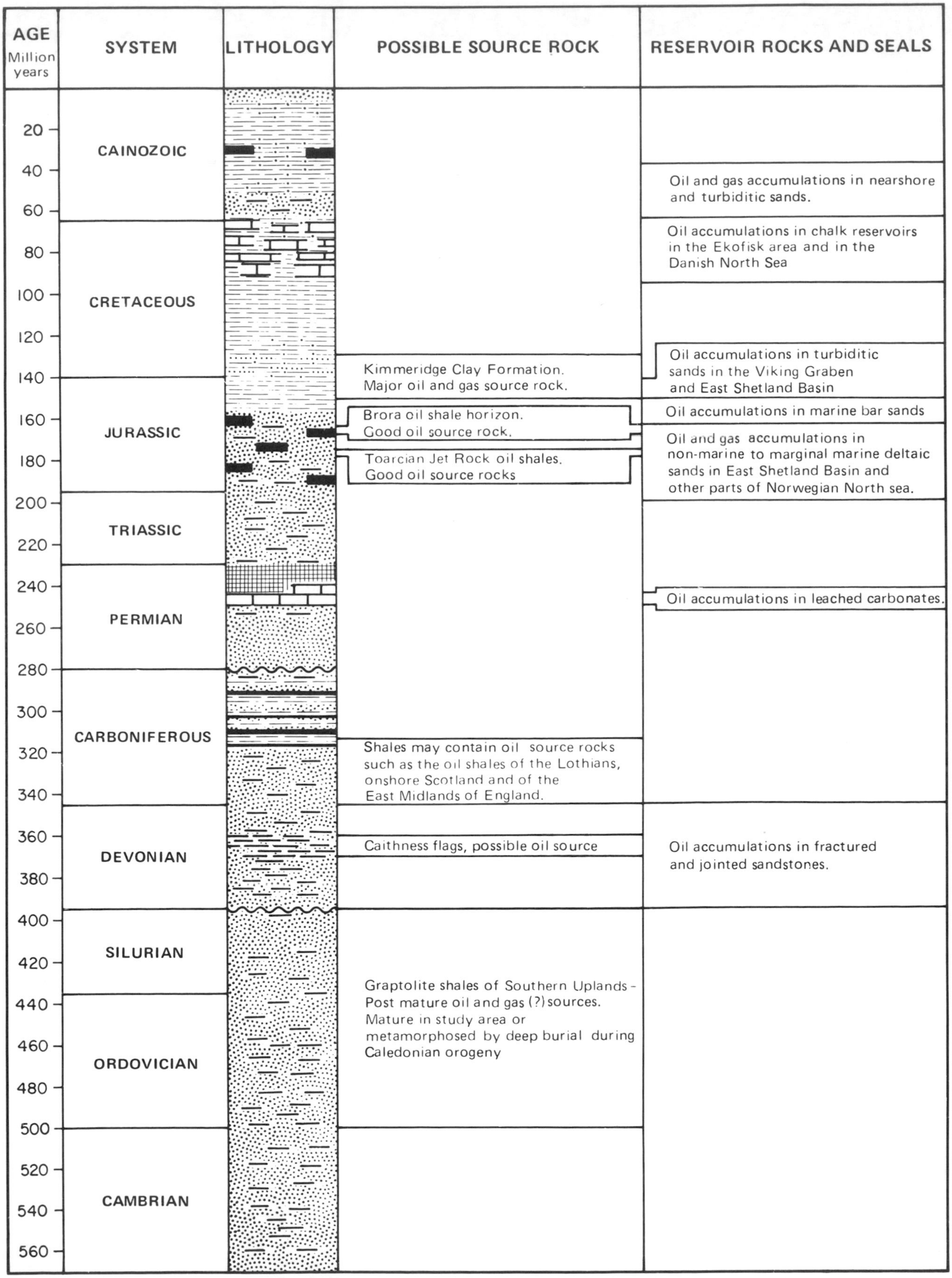

Figure 1. Stratigraphic summary of potential source rocks in the central and northern North Sea.

responsible for some major gas and oil accumulations such as the Frigg field (Heritier, Lossel, and Wathne, 1981). It has been suggested (Ronnevik, Vollset, and Eggen, 1983) that some of the recently discovered accumulations in the Norwegian sector of the North Sea are similarly sourced from Early Jurassic sediments. Although not within the study area, it has been shown (Bodenhausen and Ott, 1981) that the Toarcian age Posidonia Shale is the principal source of oil accumulations onshore the Netherlands. There is also some evidence to show that Carboniferous source rocks may have contributed hydrocarbons to some of the accumulations in the Moray Firth area of the United Kingdom sector.

THE KIMMERIDGE CLAY FORMATION AND ITS EQUIVALENTS

The source rock of Ryazanian to Kimmeridgian age, known as the Kimmeridge Clay Formation over much of the area, and as the Børglum Member of the Bream Formation in the Norwegian-Danish Basin, is the major North Sea oil source rock and the main source of reservoired hydrocarbons. It is always a richly carbonaceous shale with average organic carbon contents reaching 10 percent and thicknesses approaching 2,500 ft (762 m) in basinal areas. The environment of deposition was an extensive but almost land-locked sea, crossed by a system of troughs (where subsidence was faster and water depth greater) and blocks on which only relatively thin veneers of sediment were deposited and sometimes subsequently eroded.

Much of the organic matter enclosed in the sediment is detrital and points to prolific plant growth on the land masses supplying sediment. In this basin the anoxic bottom conditions, which may have extended from sea-bottom high into the water column, were caused by the combination of the high input of organic matter and nutrients and the lack of water circulation leading to stagnancy in deep water areas. In areas of anoxic bottom waters, algal detritus from near surface waters is preserved. The general pattern of distribution of the detrital land-derived kerogen is controlled by the size and density of the component particles being transported across the area of sedimentation. Therefore, there is a sorting process carried out on the organic particles reaching the sediments. Since there are strong contrasts in the chemical composition of the different types of kerogen, their distribution has a marked effect on source-rock quality.

The general pattern of differential trough and platform subsidence continued through the Cretaceous until, by Late Cretaceous times, fairly uniform thicknesses of chalk were deposited over the southern part of the study area and calcareous shale in the northern part. In the Tertiary, a distinct sedimentary depocenter developed at the confluence of the Viking, Moray Firth, and Central grabens. The Kimmeridge Clay Formation presently varies in depth from 3,000 to 18,000 ft (914 to 5,486 m) and shows all stages of maturity from immature to post-mature.

Kerogens and Source Rock Quality

Coal petrologists and palynologists observe kerogen to be composed of various types of fossil plant fragments. In addition, amorphous kerogen, as yellow-brown flocculent masses, is also common and represents degraded plant tissue. It is apparent that a wide variation in chemical composition is represented by these different kinds of kerogen and that different types of amorphous kerogen may be genetically linked to each of the particulate forms. Chemical analysis can be used to distinguish different types of kerogen, and Tissot et al (1974) have used the "Rock-Eval" pyrolysis method to classify kerogens into Types I, II, III and IV. The parameters derived from pyrolysis are the relative amounts of hydrogen and oxygen contained in the kerogen and the maximum rate of thermal degradation, expressed by the terms Hydrogen Index (HI), Oxygen Index (OI), and Tmax, respectively. Each kerogen type follows a different path of compositional change when plotted on an HI-OI diagram. Our own studies show that further distinct kerogen types are present in sediments and these are included in Figure 2. Informal descriptive terms were given to the amorphous kerogens and they have further distinguishing features. In this text, the informal term "waxy" is used to qualify one of the sapropel types frequently found in the Kimmeridge Clay Formation of the North Sea Basin. This sapropel is a Type II kerogen with an important contribution from the waxy membranes of land plant spores and cuticles. The term does not imply, however, that oils generated from this "waxy" sapropel will be high-wax crudes. The "waxy" sapropels are occasionally seen to be composed of very small platelets, and the gas chromatograms of their n-alkanes show an undiminished distribution extending into the $n\text{-}C_{30}$ range suggesting an origin as waxy plant cuticles.

Five of these different kerogen components or types occur in the Kimmeridge Clay Formation of the North Sea Basin, which include one Type I and two Type II kerogens as well as vitrinite and inertinite in their various forms. The sapropels can be found fairly often in an almost pure state as the dominant kerogen component at particular depth horizons and over a wide range of maturity. These, as well as coals and physically separated coal components, vitrinite, exinite, and inertinite, have been analyzed by pyrolysis. From these analyses, maturation paths for the kerogens are drawn on the HI-OI diagram shown in Figure 3 (a and b). Values of Tmax, spore coloration index (for Types I and II), and vitrinite, have been plotted adjacent to the maturation paths.

Kerogen Abundances

Kerogens are most frequently composed of mixtures of components and although estimates of abundance can be made by visual observation, variability in preparation methods, variation in particle thickness, and problems in distinguishing between the different types of amorphous kerogen are sources of significant error.

Pyrolysis data, however, may be used to calculate the abundances of the different constituent kerogen components of a sample using the data displayed in Figure 3. It is assumed that Type Id, anoxic "waxy" sapropel, and IIa, "waxy" sapropel, will not be present in the same environment of deposition and that inertinite will be denoted by zero values for both Hydrogen and Oxygen Index. By having accurate values for spore color index and

Type	Sub-Group	Coal Maceral Equivalent	Amorphous Equivalent	Pyrolysis Parameters			Isoprenoid Abundance	Carbon Isotopes	Usual Environment of Deposition
				HI	OI	TM			
I Sapropel	a	Alginite	Algal Sapropel	800 - 900	5	460	Ph > Pr	?	Freshwater (torbanite)
	b	Alginite	Algal Sapropel	800 - 900	15	435	Ph ~ Pr	- 23	Hypersaline lagoons
	c	Alginite	Algal Sapropel	800 - 900	25	405	Ph > Pr	26	Open marine, anoxic bottom water
	d	Exinite	Anoxic waxy sapropel	800 - 900	10	440	Ph > Pr	- 31	At base of steep shelf, anoxic bottom water
II Sapropel	a	Exinite	Waxy Sapropel	700	60	435	Pr > Ph	- 31	Open marine
	b	Exinite	Waxy Sapropel	700?	60?	435	Pr > Ph	- 22	Shallow marine
	c	Resinite	Resinous Sapropel	250?	20?	445	Pr > Ph	- 31	Deltaic shales
III Vitrinite		Telinite	Collinite	200	50	435	Pr ≫ Ph	- 31	Coal swamp to open marine
IV Inertinite	a	Pseudovitrinite	—	50 - 80	25	460	—	Kerogen - 29?	
	b	Semifusinite	—	20 - 50	10 - 25	460	—	Kerogen - 24	
	c	Fusinite	—	0	0	—	—	Kerogen - 22	

Figure 2. Characteristics and depositional environments of the kerogen types.

vitrinite reflectivity, the Hydrogen Index, Oxygen Index, and Tmax values are used to calculate the relative proportions of each of four kerogen components (Barnard, Collins and Cooper, 1981a).

The Effect of Maturation on Kerogen Composition

During burial, the kerogen enclosed in sediment undergoes alteration caused by increasing temperature. Initially kerogen releases water and carbon dioxide and as temperatures reach values between 70 and 80°C, the products of organic diagenesis, or maturation, are hydrocarbons. The oil-prone kerogens of Type I and Type II generate oil which becomes progressively lighter in composition from API°30 to API°50 as temperature increases. At temperatures of 150 to 175°C, the kerogen is greatly altered and the hydrocarbons being generated are condensate and wet gas. The gas-prone kerogens of Type III generate gas over the same range of temperatures, but usually reach their maximum rate of evolution at temperatures higher than 150°C. The relationships between different maturity indicators and rate of heating are discussed by Cooper (1977) and by Barnard, Collins, and Cooper (1981b).

The loss of hydrocarbons from kerogen during its passage through the "oil window" results in a decrease in mass and an apparent change in the relative proportions of the kerogen components. The kerogens of Types I and II undergo the greatest change, Type III much less, and Type

IV hardly any change. The changes induced by maturation in each of the kerogen types can be calculated using elemental analysis and pyrolysis data and are shown in Figure 4.

Using this method it is possible to calculate the abundances of each kerogen component in a sample at its present level of maturity and, using the data in Figure 4, to calculate abundances at other levels of maturity and prepare kerogen facies maps. Abundances are calculated back to a common level of maturity which is taken to be the beginning of oil generation at SCI 3.5.

Distribution of Kerogen Facies

To obtain the data to prepare kerogen facies maps it is necessary to carry out as many analyses as possible in every available section of potential source rock. This results in a range of values of total organic carbon and abundances of kerogen types. Examination of petrophysical logs, particularly of the gamma ray log, allows the section to be divided into units represented by the analyses. Where close interval sampling of the Kimmeridge Clay Formation has been carried out, rapid changes in source-rock quality are observed which coincide with the cyclic character noted in lithology, fauna and flora, and trace metal chemistry (Aigner, 1980; Cox and Gallois, 1981; Dunn, 1974; Gallois, 1976, 1978; Gallois and Medd, 1979; Tyson, Wilson, and Downie, 1979). The cyclic character seems to be consistent

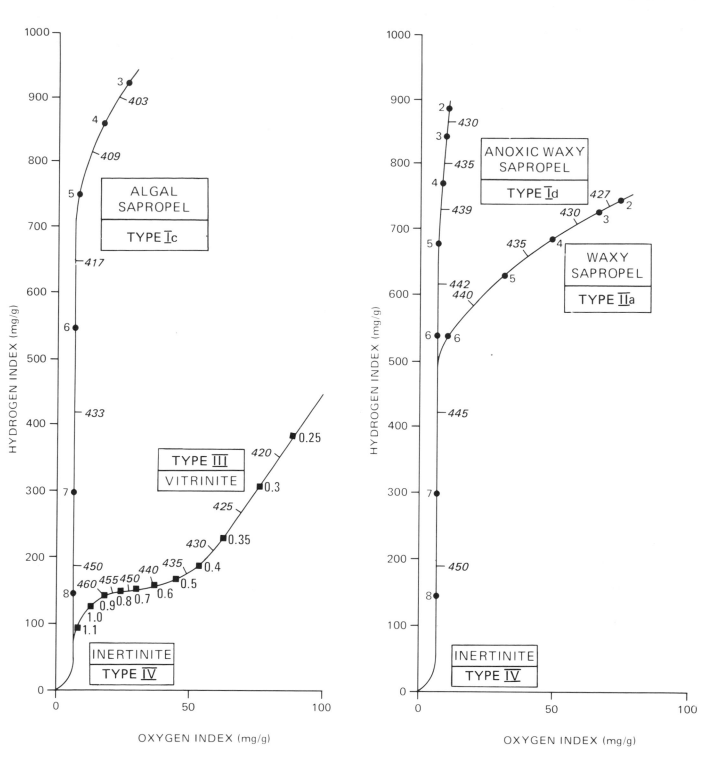

Figure 3. Variation of kerogen pyrolysis properties with maturation.

with episodes of differential trough-block movement and re-equilibration of the erosional base level profile (Aigner, 1980; Barnard, Collins, and Cooper, 1981a).

In each well section containing the Kimmeridge Clay Formation it is possible to identify sub-units, and then to allocate the analytical data to these sub-units after which averaged abundances are made for the kerogen components of the source rock unit. Abundances adjusted to the common base level of SCI 3.5 are then displayed on a map and interpreted values drawn in as contours of equal value. Kerogen facies are then defined by the most abundant kerogen component except for inertinite, Type IV, which, because it is always present in fair quantity, must exceed 40 percent of the kerogen to be considered the dominant kerogen component. Maps of lithofacies and structure are used for guidance when the distribution of data points is sparse. If the source rock unit occupies a well-defined chronostratigraphic zone, then the map of the kerogen facies shows the distribution of autochthonous and allochthonous kerogen. In the case of the Kimmeridge Clay Formation, the top and bottom of the unit transcend time boundaries. But, since in the study area vertical changes in source-rock quality are limited, the map of Kimmeridge Clay Formation kerogen facies gives a representative view of the distribution during deposition of this source rock unit.

The map (Figure 5) shows that inertinite, Type IV kerogen, is prevalent around the margins of the basin close to the ancient shoreline. In this area the lithology is often silty and the organic particles coarse-grained. Inertinite is ubiquitous throughout the basin but becomes fine-grained toward the basin center. Beyond the inertinite zone, vitrinite Type III is predominant in both particulate and amorphous forms but further into the basin gives way to degraded exinite, Type IIa ("waxy" sapropel) or algal sapropel, Type Ic. The greatest depths-of-water occurred in the graben areas and in occasional isolated basins on the platform. In some of these areas, stagnant anoxic water conditions were developed and the preservation of algal sapropel took place and, as a result, forms a predominant part of the organic content. Because the distribution of the clastic components depends on their relative size and density, while the distribution of algal sapropel depends on sea-bottom morphology, the algal sapropel facies can overlap the zones of clastic kerogen facies.

Where structural highs have persisted during deposition, winnowing and reworking by organisms has left only residual inertinite. It is, however, also possible to find thin units of inertinite-rich silty sediment within algal sapropel dominated sequences in the graben areas. It seems probable that these inertinite-rich sediments represent mass gravity flows from the platform (Watkins and Pearson, 1983; Aigner, 1980). Elsewhere, turbiditic sands are seen below and within the Kimmeridge Clay Formation and the shale sequences are usually dominated by anoxic "waxy" sapropel, Type Id. Again, it seems probable that nearshore sediment, in this case containing freshly degraded exinite, has been reworked and rapidly transported down a steeply sloping shelf into anoxic bottom waters where the early stage of organic diagenesis has been completed to produce this type of kerogen.

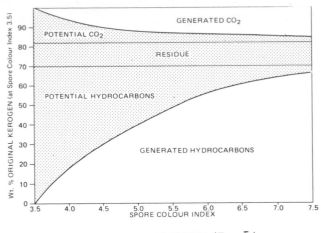

4a – ALGAL SAPROPEL (Type Ic)

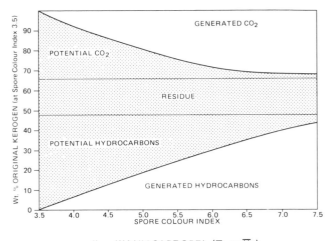

4b – WAXY SAPROPEL (Type IIa)

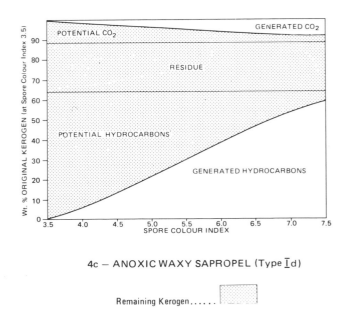

4c – ANOXIC WAXY SAPROPEL (Type Id)

Remaining Kerogen . . . . . .

Figure 4. Change in composition of oil-generating kerogens during oil generation.

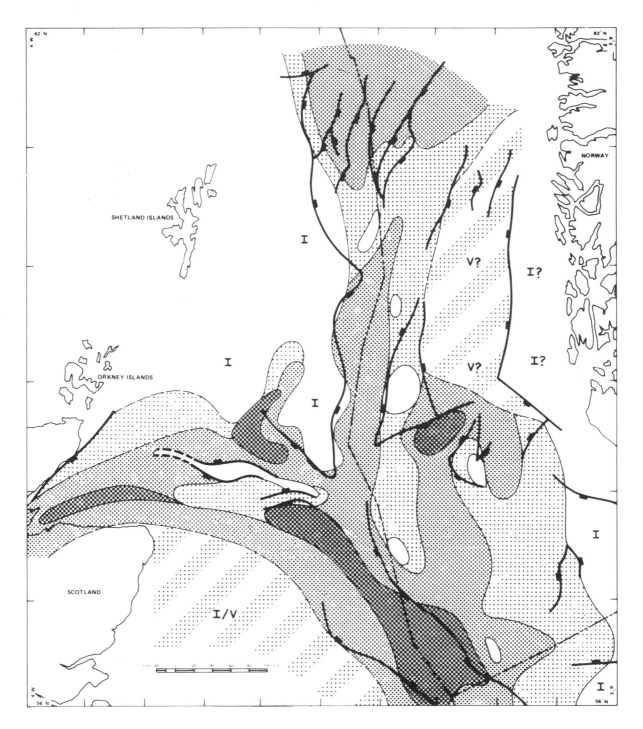

DOMINANT KEROGEN TYPE

ALGAL SAPROPEL	Type Ia	
WAXY SAPROPEL	Type IIa	
VITRINITE (V)	Type III	
INERTINITE (I)	Type IV	

Major faults . . . .

Note: Definitions of kerogen type as per figure 2.

Figure 5. Kerogen facies map of the Kimmeridge Clay Formation and its equivalents in the central cend northern North Sea.

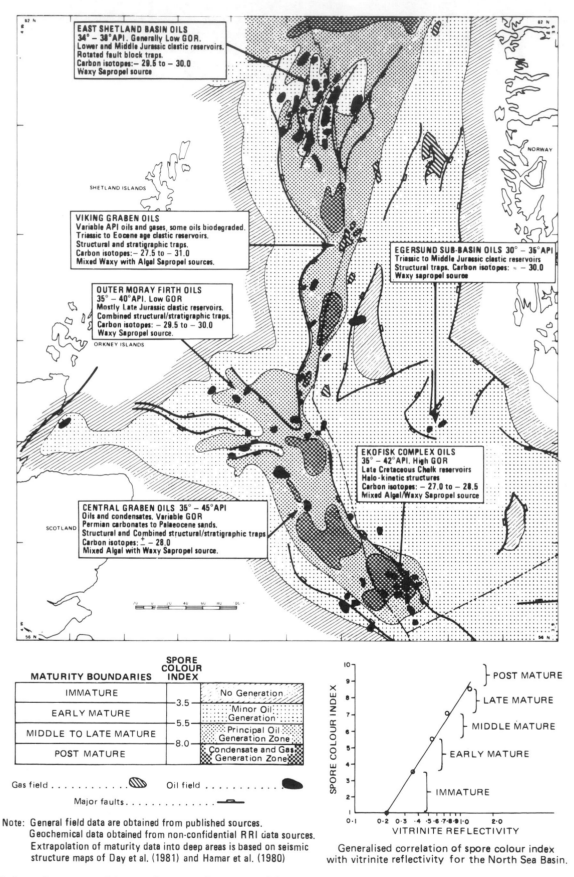

EAST SHETLAND BASIN OILS
34° – 38° API. Generally Low GOR.
Lower and Middle Jurassic clastic reservoirs.
Rotated fault block traps.
Carbon isotopes: – 29.5 to – 30.0
Waxy Sapropel source

VIKING GRABEN OILS
Variable API oils and gases, some oils biodegraded.
Triassic to Eocene age clastic reservoirs.
Structural and stratigraphic traps.
Carbon isotopes: – 27.5 to – 31.0
Mixed Waxy with Algal Sapropel sources.

OUTER MORAY FIRTH OILS
35° – 40° API. Low GOR
Mostly Late Jurassic clastic reservoirs.
Combined structural/stratigraphic traps.
Carbon isotopes: – 29.5 to – 30.0
Waxy Sapropel source.

EGERSUND SUB-BASIN OILS 30° – 36° API
Triassic to Middle Jurassic clastic reservoirs
Structural traps. Carbon isotopes: ≈ – 30.0
Waxy sapropel source

EKOFISK COMPLEX OILS
35° – 42° API. High GOR
Late Cretaceous Chalk reservoirs
Halo-kinetic structures
Carbon isotopes: – 27.0 to – 28.5
Mixed Algal/Waxy Sapropel source

CENTRAL GRABEN OILS 35° – 45° API
Oils and condensates. Variable GOR
Permian carbonates to Palaeocene sands.
Structural and Combined structural/stratigraphic traps
Carbon isotopes: ± – 28.0
Mixed Algal with Waxy Sapropel source.

SHETLAND ISLANDS

ORKNEY ISLANDS

SCOTLAND

NORWAY

MATURITY BOUNDARIES	SPORE COLOUR INDEX	
IMMATURE		No Generation
	3.5	
EARLY MATURE		Minor Oil Generation
	5.5	
MIDDLE TO LATE MATURE		Principal Oil Generation Zone
	8.0	
POST MATURE		Condensate and Gas Generation Zone

Gas field . . . .   Oil field . . . .

Major faults . . . .

Note: General field data are obtained from published sources.
Geochemical data obtained from non-confidential RRI data sources.
Extrapolation of maturity data into deep areas is based on seismic
structure maps of Day et al. (1981) and Hamar et al. (1980)

Generalised correlation of spore colour index
with vitrinite reflectivity for the North Sea Basin.

Figure 6. Hydrocarbon accumulations, oil type, and maturity of the Kimmeridge Clay Formation and its equivalents in the central and northern North Sea.

Maturation

Maturity profiles for well sections give rates of increase of maturity with depth, which have been mapped and lines of equal gradient drawn. The resulting map has been compared with structural maps of the base of the Cretaceous (Day et al, 1981; Hamar et al, 1980) to show the distribution of maturity zones in the source rock unit. The maturity map of the Kimmeridge Clay Formation (Figure 6) shows that the platform areas are immature or just mature; that is, below SCI 3.5 (Ro 0.35 percent). In the graben areas, the source rocks are mature and locally reach late and post maturity (above SCI 8.0; Ro 1.0 percent) so that the whole gravity range of oil through to condensate and gas has been generated.

Timing of Oil Generation

By superimposing the map of maturity gradients on a map of maturation values for the source rock, it is possible to generate a map which reflects the maturity level of a surface above the source rock unit. These isomaturity lines cut across chronostratigraphic surfaces and when compared with structural maps for post-mature source rock units, it is possible to calculate when oil generation began. Such calculations show that in the deepest parts of the northern North Sea, oil generation from the Kimmeridge Clay Formation began in the Late Cretaceous, but only became general in the graben areas in the mid-Tertiary. On the Norwegian Platform, oil generation only began in the Late Tertiary.

In some areas of the Viking Graben and probably elsewhere also, periods of generation and migration were interrupted by periods of uplift and erosion. These periods of erosion often permitted the influx of meteoric waters into reservoirs causing biodegradation, particularly of crude oils, in shallow Tertiary reservoirs. Such tectonic events have probably also led to changes of hydrodynamic gradients resulting in flushing of oils, particularly those in possibly deeper reservoirs and possibly leading to reduction in gas-to-oil ratios. A further effect is changes in migration direction so that some culminations include the product of hydrocarbon generation from more than one drainage area. The Frigg field appears to have suffered such reversal of flow of hydrocarbons.

Limitations of the Kerogen Facies Method

The kerogen facies map is useful in diagnosing and extrapolating kerogen trends. However, it is possible to overlook source rocks. For example, a source rock containing 6 percent TOC may contain 30 percent inertinite, 45 percent vitrinite, and 25 percent sapropel and therefore be classified for mapping as representing the vitrinite facies. However, the sapropelic organic carbon will amount to some 1.25 percent and if the source rock unit is of average thickness, it can be described as a fair oil source rock. A further series of steps, therefore, must be taken to describe the potential and productivity of the source rock which, amongst other things, involves the consideration of maturity and oil kitchens.

Productivity and Amounts of Oil Generated

Maps of abundance of each oil-prone kerogen, used in formulating the kerogen facies map, can be combined with isopach maps of source rock thickness to give a map of source rock potential which may be expressed in tons of sapropel per square kilometer, or similar dimensional unit. By comparing the thickness, quality, and maturity of source rocks, the amounts of hydrocarbon generated can be plotted for each point where an isograd crosses an isopotential line and the yields expressed in tons/sq km or barrels/acre. At this point, however, it is wise to consider how many steps the calculations have passed through; it is estimated that the errors involved are of the order of 2 times so that for a calculated result of 4,000 barrels/acre, the actual value could be between 2,000 and 8,000 barrels/acre. Nevertheless, it is useful to know that an area can yield this figure (rather than say 20 to 80 barrels/acre) in planning and analyzing exploration effort. Minimum values in the North Sea area, which are known to be drainage areas for oil fields, vary between 10,000 and 20,000 barrels/acre, but much lower values are seen in the petroliferous areas of the Middle East where drainage areas are much more extensive.

Relationship of Oil Accumulations to Areas of Mature Source Rocks (Kitchens)

In general, there is a remarkable correlation between maturity of source and oil accumulations (Figure 6). Most oil accumulations are within or adjacent to areas where the oil source rock is within the main oil generating stage (SCI > 5.5; Ro > 0.55 percent). Most of the fields which fall within the early mature zone are clearly adjacent to areas where source rocks are in the major oil generation window. However, there are some notable exceptions in the inner Moray Firth area of the United Kingdom (Beatrice field) and adjacent to the Egersund sub-basin of Norway (Bream, Brisling, and other fields) and these appear to be low-maturity sourced oils. There is also the anomaly of the giant gas field in Norwegian block 31/2 where there is no geochemical control published in that area.

Relationship of Kerogen Facies to Oil Accumulations

Having carried out the analyses and gone through the process of identifying and mapping kerogen facies and predicting areas where oil will be generated, it is obviously necessary to check whether the characteristics predicted for the generated hydrocarbons match those of the accumulated oils. The problem is one of predicting the characteristics associated with, for instance, oil derived from "waxy" sapropel and oil derived from algal sapropel. In the study area it is necessary to compare examples of the hydrocarbons derived from the sapropel Types Ic and IIa of Figure 2 with the characteristics of oils found in areas where these two types are dominant. The kerogen facies map (Figure 5) would suggest significant differences between the oils of the East Shetland Basin (Type IIa source dominant) and the oils of the Ekofisk area (Type Ic source dominant). By reference to Figure 6, it seems that maturity is generally more advanced in the Ekofisk area than in the East Shetland Basin, although some contribution of oil migrating updip from a deep and late mature area to the north of the East Shetland Basin must be considered for some of the oils in this area. In general, however, differences in maturity are not likely to greatly affect bulk properties such as carbon isotopes.

Table 1. Geochemical characteristics of typical sapropelic source rocks.

	SAMPLE A ALGAL SAPROPEL	SAMPLE B WAXY SAPROPEL
Total Organic Carbon (%)	8.18	6.88
Rock-Eval Data		
Hydrogen Index	499	286
Oxygen Index	20	32
T max	409	439
Potential Yield (ppm)	40800	19700
Maturity		
Spore Colour Index	4.5	5.0
Vitrinite Reflectivity (%)	0.8	0.7
Hydrocarbon Content (ppm)	1795	1950
Carbon Isotope Data (‰ PDB)		
Alkanes	− 26.6	− 30.4
Aromatics	− 26.7	− 30.3

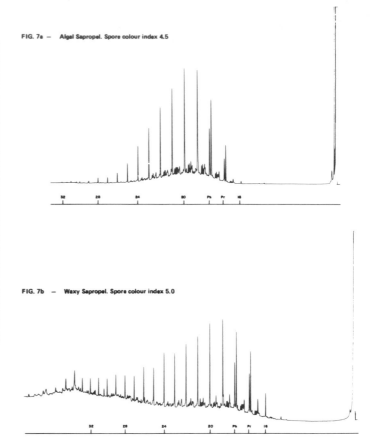

FIG. 7a — Algal Sapropel. Spore colour index 4.5

FIG. 7b — Waxy Sapropel. Spore colour index 5.0

Figure 7. Gas chromatograms of alkanes of different types of sapropelic source rocks.

Using the techniques of oil-to-oil correlation such as alkane gas chromatography, gas chromatography/mass spectrometry (g.c./m.s.), and carbon isotope studies of alkane and aromatic hydrocarbon fractions, it is possible to show that the oils in particular areas show generally similar characteristics. In Figure 6, typical characteristics, including spread of alkane carbon isotope values, are shown for the oils in different areas. In general there is a marked difference in isotope values for East Shetland Basin and Ekofisk oils. If we assume that an oil derived from a sapropel containing equal amounts of algal (Type Ic) and "waxy" (Type IIa) derived material will have an isotope value midway between the values of the two end members (that is, about − 28.5‰), then it follows that the Ekofisk oils are derived from dominantly Type Ic sapropel while the East Shetland Basin oils are derived from dominantly Type IIa sapropel.

While bulk parameters of oils and source rock extracts such as carbon isotopes are extremely useful in defining source type, it has been frequently suggested in the literature that correlation of source rock to oil may be established using the composition of specific chemical markers which migrate from the source rock to the oil pool. Similarities in straight chain (n-) alkane composition and more specifically in the polycyclic steranes and triterpanes identified through use of techniques of gas chromatography (g.c.), and gas chromatography/mass spectrometry (g.c./m.s.), respectively, are the preferred methods. Examples of the geochemical characteristics of source rock containing significant proportions of the two end member sapropel Types Ic and IIa are shown in Table 1. The gas chromatograms and gas chromatography/mass spectrometry traces shown in Figures 7 and 8 are derived from the alkanes of these samples.

The n-alkane composition of a source rock is particularly susceptible to change through maturation but in the immature and early mature stages there are clear differences

between the alkanes and the two most common end-member sapropels of the North Sea (Figure 7). Algal sapropel has a simple n-alkane distribution skewed to the shorter chain length region (around n-C_{20}) while "waxy" sapropels are much richer in the heavier hydrocarbons (up to C_{32+}) and contain relatively larger amounts of branched and polycyclic hydrocarbons shown as the frequent small peaks between the n-alkanes. G.c./m.s. of the end member sapropels similarly shows distinctive differences both in composition and relative abundance of the polycyclic triterpanes and steranes in the C_{27+} region of the alkane trace (Figure 8). Algal sapropel contains, in addition to the hopanes in the C_{30} region, fairly abundant stripped triterpanes in the C_{26-} region. Steranes, by comparison, are present in low concentration and rearranged steranes are hardly present at all. In contrast, "waxy" sapropel having a strong contribution from land plant materials is rich in both triterpanes and steranes, and as a result of bacterial alteration in transport and in early diagenesis is also relatively rich in rearranged steranes. In view of the relative weakness of the chemical signature of algal sapropel, it is often the case that when "waxy" sapropel is present in an algal sapropel facies (even in relatively minor proportions), the g.c./m.s. pattern resembles that of "waxy" sapropels. For this and other theoretical and practical reasons, convincing oil-to-source rock correlation has yet to be demonstrated in the North Sea region using g.c./m.s.

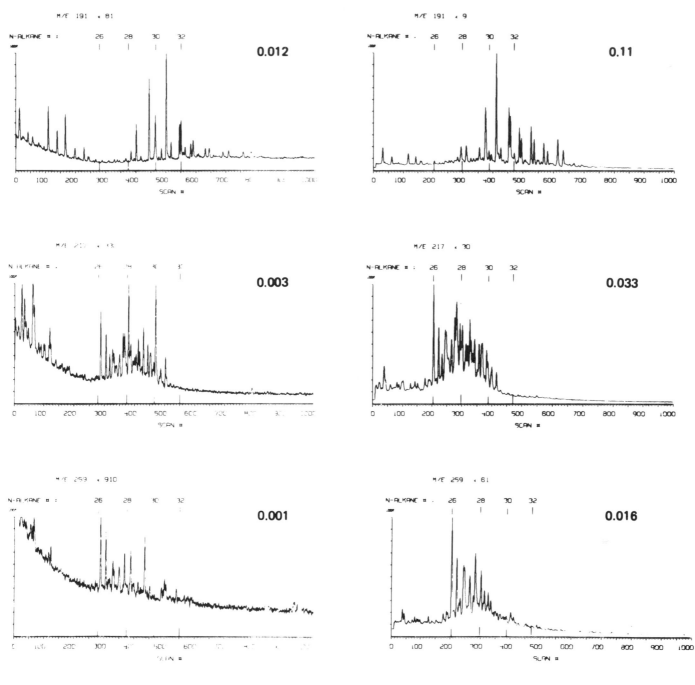

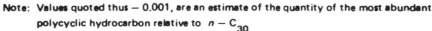

Note: Values quoted thus — 0.001, are an estimate of the quantity of the most abundant
polycyclic hydrocarbon relative to $n - C_{30}$

8a — ALGAL SAPROPEL   **8b — WAXY SAPROPEL**

Figure 8. Gas chromatography/mass spectrometry traces of alkanes of different types of sapropelic source rocks.

CONCLUSIONS

Geochemical analyses show that the principal oil source rock in the central and northern North Sea is the Kimmeridge Clay Formation and its equivalents. Other source rocks are of lesser importance in the area as a whole, but it is possible that several major accumulations in the

Norwegian sector are sourced from Early Jurassic age sediments.

The Kimmeridge Clay Formation varies in its kerogen composition from organically-rich, highly-sapropelic, and oil-prone to organically-lean, inertinitic without hydrocarbon-generating potential. These variations can be explained by a depositional model in which a large part of the organic matter in the sediments is derived from terrestrial plants. The organic residues from these plants, as well as having different original chemical compositions, have suffered variations in preservational environment both before and during transport to the marine environment and during their early diagenesis. Anoxic bottom water conditions persisted through much of the time of deposition of the Kimmeridge Clay Formation, resulting in accumulation of very thick intervals of sapropelic source rocks in some areas.

Using the "Rock-Eval" pyrolysis method, the amounts of the different kerogen components were determined and, by a process involving log interpretation and mathematical calculations, the dominant kerogen facies in well sections through the Kimmeridge Clay Formation were identified. Maps of kerogen facies give a qualitative impression which may be related to the depositional model.

Comparison of kerogen facies with maturity gradients and structure contour maps enables calculation of the amounts of oil generated in the basin and estimation of time of generation of oils. There is generally good correlation between areas of mature source rocks and the occurrence of oil accumulations. The composition of the oils found in different areas generally accords well with predictions based on the distribution of kerogen facies.

ACKNOWLEDGMENTS

We acknowledge the assistance of our colleagues in discussions throughout the long period of time in which the ideas presented in this paper have been formulated. We particularly with to acknowledge the help of A. Collins for the development of the maturation scale and S. Thompson for the drafting of some of the pyrolysis figures and the development of the pyrolysis interpretation method. Finally, we thank the directors of Robertson Research International Limited for permission to publish this paper.

REFERENCES CITED

Aigner, T., 1980, Biofabrics and stratinomy of the Lower Kimmeridge Clay (U. Jurassic, Doreset, England): Neues Jahrbuch fuer Geologie und Palaeontologie, Abhandlungen, v. 159, p. 324-338.

Barnard, P.C., A.G. Collins, and B.S. Cooper, 1981a, Identification and distribution of kerogen facies in a source rock horizon—examples from the North Sea Basin, *in* Brooks, ed., Organic maturation studies and fossil fuel exploration: London, Academic Press, p. 271-282.

———, ———, and ———, 1981b, Generation of hydrocarbons—time, temperature and source-rock quality, *in* Brooks, ed., Organic maturation studies and fossil fuel exploration: London, Academic Press, p. 337-342.

———, B.S. Cooper, 1981, Oils and source rocks of the North Sea area, *in* Illing and Hobson, eds., Petroleum geology of the continental shelf of northwest Europe:
London, Institute of Petroleum, p. 169-175.

Bodenhausen, J.W.A., and W.F. Ott, 1981, Habitat of the Rijswijk oil province, onshore, The Netherlands, *in* Illing and Hobson, eds., Petroleum geology of the continental shelf of northwest Europe: London, Institute of Petroleum, p. 301-309.

Chesher, J.A., and M. Bacon, 1975, A deep seismic survey in the Moray Firth: Institute of Geological Science, Report n. 75/11, 13 p.

Cooper, B.S., 1977, Estimation of the maximum temperatures attained in sedimentary rocks, *in* Hobson, ed., Developments in petroleum geology: London, Applied Science Publishers, p. 127-146.

Cox, B.M., and R.W. Gallois, 1981, The stratigraphy of the Kimmeridge Clay of the Dorset type area and its correlation with some other Kimmeridgian sequences: Institute of Geological Science, Report n. 50/4, 44 p.

Day, G.P., et al, 1981, Regional seismic structure maps of the North Sea, *in* Illing and Hobson, eds., Petroleum geology of the continental shelf of northwest Europe: London, Institute of Petroleum, p. 301-309.

Dunn, C.E., 1974, Identification of sedimentary cycles through Fourier analysis of geochemical data: Chemical Geology, v. 13, p. 217-232.

Gallois, R.W., 1976, Coccolith blooms in the Kimmeridge Clay and origin of North Sea oil: Nature, v. 259, p. 473-475.

———, 1978, A pilot study of oil shale occurrences in the Kimmeridge Clay: Institute of Geological Science, Report n. 78/13.

———, and H.W. Medd, 1979, Coccolith-rich marker bands in the English Kimmeridge Clay: Cambridge, England, Geological Magazine, v. 116, p. 247-334.

Hamar, G.P., et al, 1980, Tectonic development of the North Sea north of the Central High, *in* Norwegian Petroleum Society; the sedimentation of the North Sea reservoir rocks: Geilo Conference Proceedings, 1II, 11 p.

Heritier, F.E., P. Lossel, and E. Wathne, 1981, The Frigg gas field, *in* Illing and Hobson, eds., Petroleum geology of the continental shelf of northwest Europe: London, Institute of Petroleum, p. 380-391.

Ronnevik, H.C., J. Vollset, and S. Eggen, in press, Exploration of the Norwegian Shelf, *in* Brooks, ed., Petroleum geochemistry and exploration of Europe.

Tissot, B., et al, 1974, Influence of nature and diagenesis of organic matter in formation of petroleum: AAPG Bulletin, v. 58, p. 499-506.

Tyson, R.V., R.C.L. Wilson, and C. Downie, 1979, A stratified water column environmental model for the type Kimmeridge Clay: Nature, v. 277, p. 377-380.

Watkins, D., and M.J. Pearson, in press, Organic facies and early maturation effects in Upper Jurassic sediments from the inner Moray Firth Basin, northern North Sea, *in* Brooks, ed., Petroleum geochemistry and exploration of Europe.

Ziegler, P.A., 1980, Northwest European Basin; geology and hydrocarbon provinces, *in* A.D. Miall, ed., Facts and principles of world petroleum occurrence: Canadian Society of Petroleum Geologists, Memoir 6, p. 653-706.

Reprinted from the 6th International Meeting of Organic Geochemistry, 1973, p. 315-334.

Origin and Migration of Hydrocarbons in the Eastern Sahara (Algeria)

B. Tissot
J. Espitalie
G. Deroo
Institut Français du Pétrole
Rueil-Malmaison, France

C. Tempere
D. Jonathan
SNEAP
Pau, France

The source rocks of the oil-bearing series of the Eastern Sahara are very old (up to 440 million years) and have been subjected for long periods to considerable burial. For this reason, the oils exhibit a very advanced character of maturation and their overall composition, as well as that of the saturated hydrocarbons, is fairly uniform. An analysis of the aromatic hydrocarbons, especially of the compounds of the steroid type containing one or several aromatized rings, by mass spectrometry makes it possible to distinguish three groups of oil and to find their source rocks.

A mathematical model of the formation of petroleum and gas which uses the reconstruction of the structural history of the basin and the geothermal data makes it possible to calculate the quantities of hydrocarbons formed in the various sectors of the basin. The results are in agreement with the distribution of the presently known deposits, and these allow us to delimit the regions in which oil deposits may exist.

GEOLOGICAL SETTING

The Eastern Sahara comprises a vast Paleozoic basin, the Illizi basin (Figure 1), folded at the end of the Paleozoic and overlapped unconformably by the Mesozoic series from Triassic to the Cretaceous which extend over a large portion of the Sahara.

Paleozoic basin of Illizi is bounded in its present extent to the south by the outcrops which border on the Precambrian massif of the Hoggar and to the west by a high zone which extends from Amguid to the region of Hassi Messaoud mole of El Biod where the Paleozoic rocks have been largely removed by the Pre-Triassic erosion and where only the horizons of the Cambrian-Ordovician remain.

Beneath the discordance, the Paleozoic series comprise mainly alternations of clay and sandstone (Figure 2). The oil and gas deposits are located in the sandstones of the Paleozoic. The main productive horizons are in the lower Carboniferous (Reservoirs B and D), at the top of the Devonian (Reservoir F2), in the lower Devonian-upper Silurian (Reservoirs F4 and F6), and in the Cambrian-Ordovician.

ANALYTICAL PROCEDURE

The oils from deposits and the organic material extractable from the shale series were investigated by the methods of Oudin (1970) and Durand, Espitalie, and Oudin (1970). The rocks containing organic material are extracted with chloroform. The extracts obtained in this way and the oils are treated by the same operating procedure (Figure 3): After precipitation of the asphaltenes, the saturated aromatic hydrocarbons and the resins are separated by liquid chromatography. The hydrocarbons are analyzed by mass spectrometry and gas phase chromatography. The analysis of the aromatic hydrocarbons by mass spectrometry is preceded by a separation of the mono-, di-, and polyaromatics (Oudin, 1970). This procedure makes it possible, in particular, to reveal the steroid and triterpenoid structures comprising one or two aromatic rings (Tissot, Oudin, and Pelet, 1971; Tissot et al, 1972) which are genetic markers of petroleums.

CHARACTERISTICS OF THE OILS

The crude oils of the Eastern Sahara, although they come from reservoirs whose age varies from the Cambrian-Ordovician to the Carboniferous, show great similarities in overall composition (Figure 4). In fact, all of them have a very mature character, marked by small proportions of heavy products (resins and asphaltenes) and abundance of saturated hydrocarbons. Among the latter, linear, or branched alkanes are predominant, especially the light molecules (less than 15 carbon atoms).

A consequence of this very mature character common to all of the oils is that neither the overall composition nor the

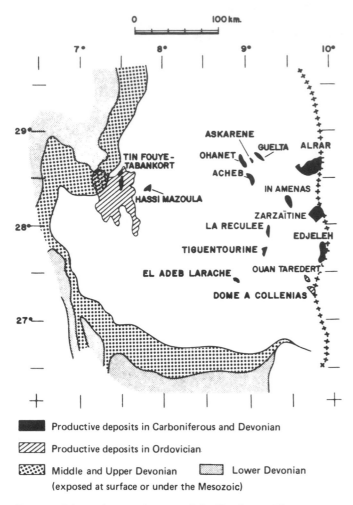

Figure 1. Main oil or gas deposits of the Illizi basin. The present limits of the basin are indicated to the southwest by the outcrops of the Lower Devonian and to the northwest by its limit of erosion under the Mesozoic rocks.

Legend:
- Productive deposits in Carboniferous and Devonian
- Productive deposits in Ordovician
- Middle and Upper Devonian
- Lower Devonian
(exposed at surface or under the Mesozoic)

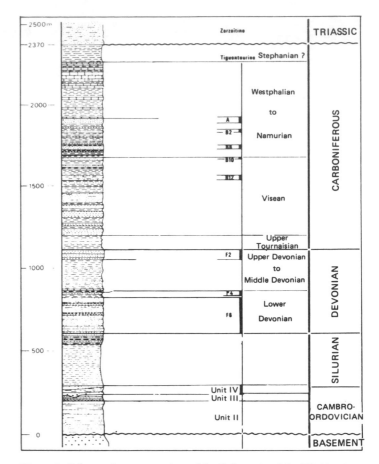

Figure 2. Schematic cross section of the Paleozoic of the Illizi basin showing the main petroleum reservoirs: Carboniferous A and B2 to B12, Devonian F2 to F6, Ordovician IV (according to Beuf et al, 1971).

investigation of the saturated hydrocarbons by mass spectrometry (Figure 4) or by gas phase chromatography (Figure 5) makes it possible to tell whether all of these oils have the same origin or not. It is noted only that, in the northwest part of the basin, the oil from the reservoirs of the lower Devonian contains no gas or only very little gas.

On the other hand, the investigation of the aromatic hydrocarbons by mass spectrometry, conducted up to C_{35}, shows, from the start, a contrast between oils of a very common type, rich in light aromatics and poor in heavy aromatics, and oils showing a very flat distribution, where the heavy aromatics from C_{25} to C_{35} are abundant (Figure 6). It is thus possible to distinguish between the investigated crude oils, from top to bottom of the stratigraphic column:

a) The oils of the Carboniferous and of the Upper Devonian F2, which are very similar to each other, as well as an oil from the Lower Devonian reservoir F6 from the Ohanet deposit; all of them are of the common type.

b) All the other oils of the Lower Devonian reservoirs F4 and F6 (Ozn 1, TFT 101, HMz 1, Edy 1, IKN 1, Tg 40), which are rich in heavy aromatics; the last two, located in the eastern part of the basin and more buried, have a slightly larger amount of light compounds.

c) The oils of the Cambrian-Ordovician reservoir, which again show a common distribution.

A comparison between the superposed deposits in the fields of Tiguentourine and Tin Fouyé-Tabankort confirms these distinctions: at Tiguentourine, the oils of the reservoirs of Carboniferous and Upper Devonian age, F2, are the same and differ from those of the Lower Devonian reservoirs F4 and F6 (Figure 6). At Tin Fouyé-Tabankort, the contrast between the oils of the Lower Devonian reservoir F6 and those of the Cambrian-Ordovician is observed (Figure 7).

At the western edge of the basin, where the sandstone reservoirs of the Triassic are located unconformably on top of the various systems of the Paleozoic, the Gassi Touil oil is similar to that of the Lower Devonian F6 of Tin Fouyé and of the neighboring deposits.

Among the hydrocarbons, the molecules of the steroid and triterpenoid type are of particular interest. They are molecules with 4 and 5 rings inherited from the living substances which have undergone only minor conversions having no effect on the general structure of the molecule. They are compounds of biological origin known at present in plants and animals (sterols, biliary acids, various hormones, etc.) For this reason, they may contain very specific genetic information.

Particularly during the burial process, compounds of this type, of which one ring has been aromatized, seem to be

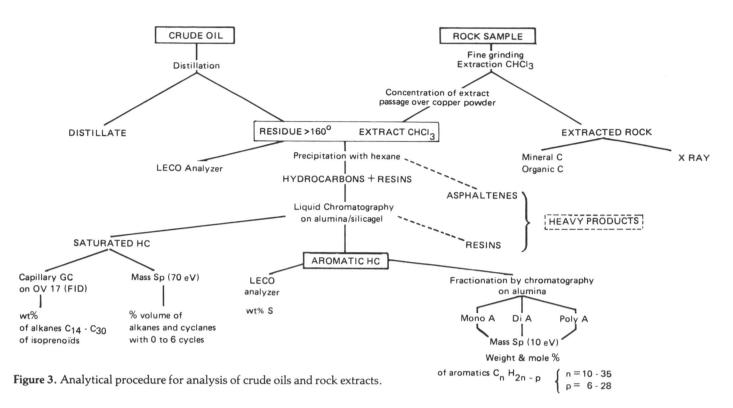

Figure 3. Analytical procedure for analysis of crude oils and rock extracts.

more stable and to be preserved for a longer time in rock extracts and crude oils than the saturated forms (steranes and triterpanes). As an example, Figures 6 and 7 show the distribution of C_nH_{2n-12} steroids in which one ring is aromatic and the other three are saturated.

It is noted that their distribution shows the three groups of oils already mentioned.

a) Those of the Carboniferous, of the Upper Devonian F2, and of the Lower Devonian F6 of Ohanet. The molecules around C_{20} are abundant, while those in the C_{27}-C_{30} range are rare or absent.

b) The other oils of the reservoir F6 show a rather flat distribution where the heavy and light molecules are equally represented.

c) The oils of the Ordovician show mainly molecules around C_{20}, which are extremely abundant at Ouan Taredert (Figure 6).

The oils of the Triassic of Gassi Touil, at the western edge of the basin, are correlated in the same way as previously with most oils of the Lower Devonian F6.

ORIGIN OF THE OILS

In order to search for the origin of the three types of oils just defined, it will naturally be attempted to find geochemical markers in the various possible source rocks included in the sedimentary series. For this purpose, the hydrocarbons of the oils and the hydrocarbons extractable from the organic material of the sediments will be compared. In this search, the saturated hydrocarbons will be of little help, since we have seen that their composition is very similar in all of the investigated oils.

Oil Deposits of the Carboniferous and of the Upper Devonian

We have seen that the oils found in these reservoirs are the same. By studying the aromatic molecules and especially the steroid (Figure 8) and triterpenoid forms, the same distributions are found in these oils and in the extracts of the shales from the Upper Devonian and the underlying Middle Devonian (Frasnian-Givetian): small amount of heavy aromatics, very marked predominance of monoaromatic steroids below C_{22}. By contrast, the shales of the Carboniferous show a predominance of C_{27}-C_{30} steroids.

It may thus be concluded that the oils of the reservoirs of Carboniferous and Upper Devonian age are derived from the shale source rocks of the Middle-Upper Devonian and that the oil has settled in the sandstones of the Carboniferous from the Devonian, due to the faults which are fairly numerous in the deposits of the eastern part of the basin (Tiguentourine, Edjélé, Zarzaïtine). Moreover, the results of the exploration have already shown that the reservoirs of the Carboniferous are only of interest if they are connected to the Upper Devonian reservoir F2 by a fault system. It may also be concluded from the observation of Figure 8 that the oil of the Chanet deposit, although it is located in a reservoir of the Lower Devonian F6, is also derived from the shale rocks of the Upper and Middle Devonian. This is also true for the neighboring small deposits of Guelta and Askarène. We will return to this point later.

Oil Deposits of the Lower Devonian-Upper Silurian

The oils encountered in the reservoir F4 and in the various horizons of the reservoir F6 are the same (with the exception of the Ohanet oil) and show a good correlation with the extracts of the intercalated and underlying clays of Lower Devonian-Upper Silurian age. In these rock extracts, the characteristic distribution of the aromatics rich in heavy molecules is found again. Likewise, the steroid (Figure 8) and the monoaromatic triterpenoids show a more or less

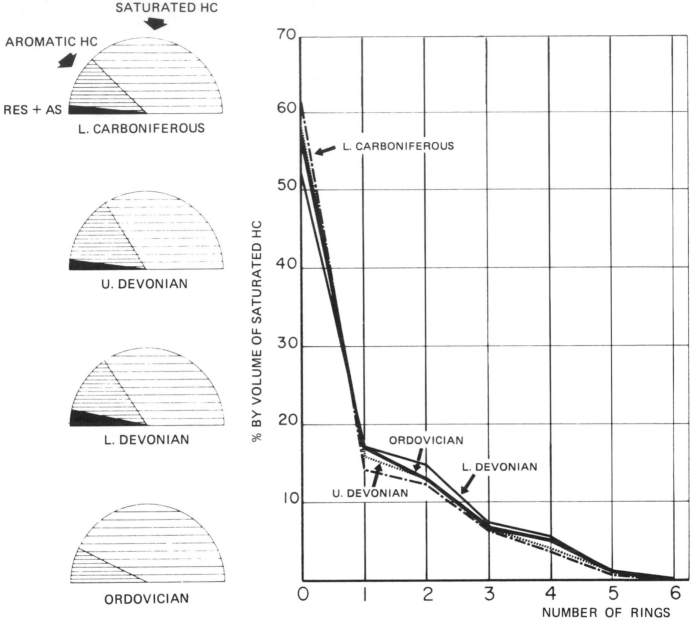

Figure 4. Overall composition of crude oils (left) and distribution of their saturated hydrocarbons by structural type (right).

equivalent richness in the C_{20}-C_{22} range and in the C_{27}-C_{30} range.

On the other hand, it does not seem possible to trace these oils to the clays of the Middle and Upper Devonian or to those of the Lower Silurian, which have very different characteristics (Figures 8 and 10). The only exception consists of the oils of Ohanet and of the small neighboring deposits, of which we have seen that they are derived from the source rocks of the Middle-Upper Devonian. This difference may be compared with an interesting geologic peculiarity (Figure 9): At Ohanet, the reservoir F6 is in contact with the shale series of the Givetian-Frasnian, instead of being separated from it by the clays of the Emsian which are present in the other investigated deposits, where they constitute a shield preventing the clays of the Middle-Upper Devonian from feeding the reservoir.

To the west of the basin, the oil of the Triassic reservoir of Gassi Touil shows the features of most of the oils of the Lower Devonian and is probably derived from the source rocks of the Upper Silurian. It may have found its place with the help of the unconformity, which causes the Triassic to rest on the various Paleozoic rocks, and by the faults, which are considerable in this region.

Oil Deposits of the Ordovician

The aromatic hydrocarbons of the oils of this reservoir may be compared with those extracted from the shales of the Lower Silurian, a level which is very rich in organic material and located immediately above. The steroids (Figure 10), in particular, show a close relationship, while the extracts of the shales and silts of the Cambrian-Ordovician located below the reservoir show a

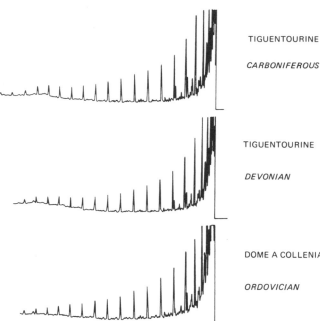

Figure 5. Gas-phase chromatography of total saturated hydrocarbons of the crude oils.

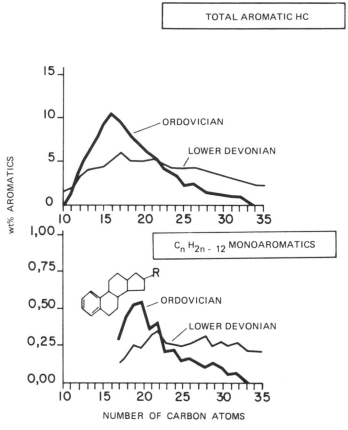

Figure 7. Comparison of the oils of the Lower Devonian and Ordovician from the Tin Fouyé-Tabankort field.

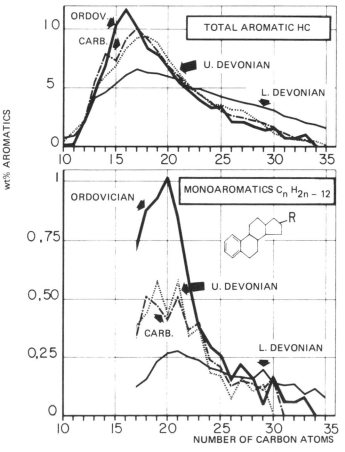

Figure 6. Distribution of total aromatic hydrocarbons (top) and of monoaromatics of steroid type (bottom). The oils of the Carboniferous, the Upper Devonian, and the Lower Devonian come from Tiguentourine field; the oil of the Ordovician comes from Ouan Taredert field.

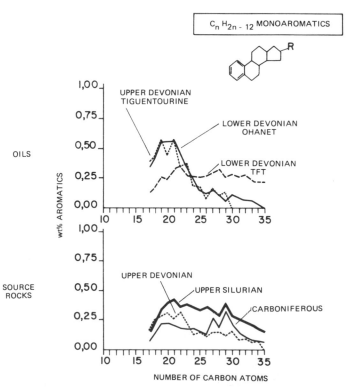

Figure 8. Origin of the oils of the Upper and Lower Devonian. Tentative correlation between oils and extracts from source rocks by means of the monoaromatic hydrocarbons of the steroid type.

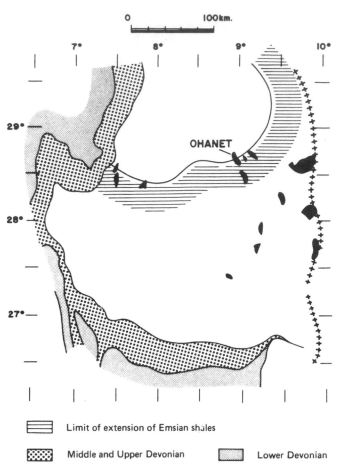

Figure 9. Extension of the Emsian shales and position of the Ohanet deposit.

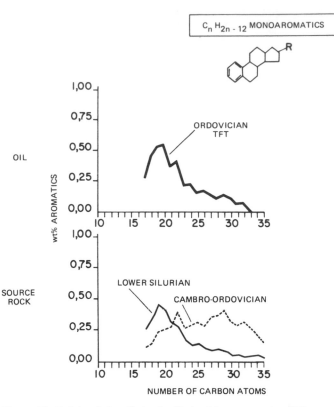

Figure 10. Origin of the oils in the Ordovician reservoir of Tin Fouyé-Tabankort (TFT). Correlation between the oil and the extract of the Lower Silurian source rock.

Table 1. Abundance and composition of extracts of a shale source rock as a function of the distance from the reservoir (Upper Devonian of the Assekaifaff 201 well). (Each value is the average of three or four measurements.)

Distance From Reservoir (m)	Extract/C C. Org. (mg/g)	Hydrocarbons/ Extract (%)	Asphaltenes/ Extract (%)
2	72	54	12.2
4	86	61	11.2
7	90	63	7.5
10.5	112	63	5.7
14	118	64	5.8

different behavior. Thus, it may be assumed that these deposits are derived from the clays of the lower Silurian, which, furthermore, are often very rich in organic material: The organic carbon may reach 30 percent in some horizons.

MATURATION OF THE SOURCE ROCKS DURING BURIAL

The constituents of petroleum, especially the hydrocarbons, are produced by thermal degradation of the organic material of the source rocks (kerogen), which takes place in the course of burial of the sediments. The degree of conversion of the organic material may be measured by the ratios:

$$\frac{\text{Extract}}{\text{Total Organic Carbon}} \quad or \quad \frac{\text{Hydrocarbons}}{\text{Total Organic Carbon}}$$

This ratio is plotted as a function of burial in Figure 11 for the shales of the Middle-Upper Devonian.

In a homogeneous, undrained formation such as the Toarcian of the Paris Basin (Tissot et al, 1971), these ratios depend only on the temperature reached by the source rock, i.e., only on the burial and the geothermal gradient. In a series as complex as the Paleozoic of the Eastern Sahara, the proportion of extract and its composition vary also with the nature of the organic material and with the extent of migration which drains the compounds generated toward the reservoir. Moreover, the geothermal gradient varies considerably across the basin.

For these various reasons, the points representing the organic extracts of the Middle-Upper Devonian constitute a cloud which admits an envelope, traced approximately in Figure 11.

The influence of the migration is particularly noteworthy. The Paleozoic series of this basin consist of repeated alternations of sandstones and clays, from the Cambrian to the Carboniferous. Such a layout facilitates drainage from the source rocks; it may be suspected, therefore, that the proportions of extracts therein will be changed in the vicinity of the reservoirs. Actually, the investigation of a core sample taken from a sandy reservoir (Upper Devonian F2 of Well As 201) and penetrating 15 m

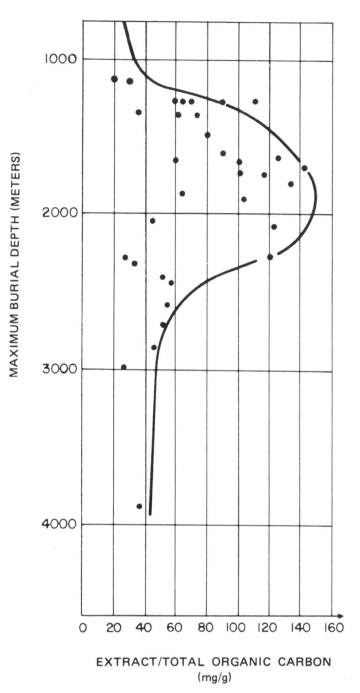

Figure 11. Source rocks of the Middle-Upper Devonian. Ratio of extract to total organic carbon as a function of maximum burial depth.

into the underlying shale source rock shows the nature and the importance of the phenomenon. It is observed (Table 1) in the homogeneous shale that the values of the extract/organic carbon ratio are first stable, far from the sand, and then gradually decrease for the last 10 m, and finally, on contact with the sand, amount to slightly more than half their original value. This distribution shows the withdrawal of the products toward the reservoir. Furthermore, it is observed that the proportions of asphaltenes (heavy compounds strongly absorbed on the mineral phase) are higher in the vicinity of the sand. On the

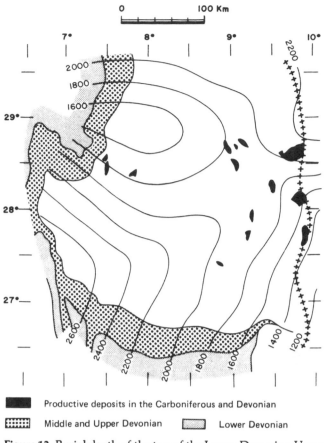

Productive deposits in the Carboniferous and Devonian

Middle and Upper Devonian   Lower Devonian

Figure 12. Burial depth of the top of the Lower Devonian-Upper Silurian source rocks at the end of the Carboniferous (in meters).

other hand, the proportions of hydrocarbons are lower. This difference in behavior is due to the fact that drainage toward the reservoir applies preferentially to the hydrocarbons and that the residual mixture is enriched with asphaltenes.

It is found, in fact, that, among the samples represented in Figure 11, the shale horizons which are intercalated in the reservoir F2 and, consequently, better drained than the other ones, are located on the left of the envelope curve.

With these reservations, the following general development is observed: At a small depth, the ratio

$$\frac{\text{Extract}}{\text{Total Organic Carbon}}$$

is small. It increases rapidly from about 1,200 m on and culminates around 2,000 m. At about 2,500 m, a very marked decrease is observed, corresponding to cracking of the molecules and to the formation of light products (gas). The samples, which are not very numerous, coming from depths of about 3,000 and 4,000 m, show low values: All of the structures capable of supplying hydrocarbons yielded gas by cracking, and the remaining kerogen constitutes a sort of carbonaceous residue.

TIME AND LOCATION OF THE FORMATION OF HYDROCARBONS

The reconstruction of the structural history of the basin,

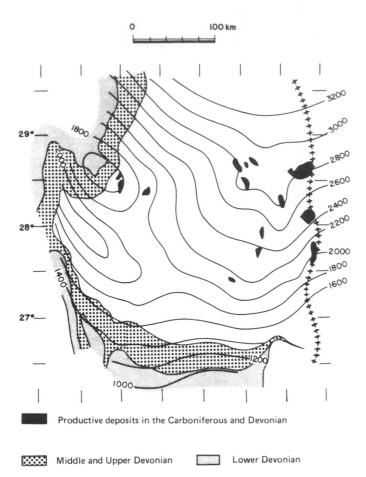

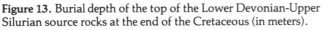

Productive deposits in the Carboniferous and Devonian

▨ Middle and Upper Devonian   □ Lower Devonian

Figure 13. Burial depth of the top of the Lower Devonian-Upper Silurian source rocks at the end of the Cretaceous (in meters).

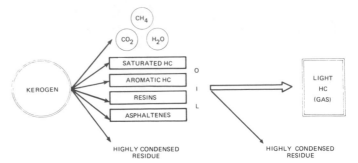

Figure 14. Reaction scheme simulated by the mathematical model of petroleum generation.

together with the use of a mathematical model of generation of petroleum, makes it possible to estimate the time of formation of hydrocarbons and their location. For this purpose, the example of the source rock of the Lower Devonian-Upper Silurian, which are the source of most of the deposits of the Lower Devonian, (the most important reservoir of the basin) will be considered.

The Illizi basin has been the seat of two main cycles of subsidence, with burial of the source rocks. The first cycle lasts almost through the whole Paleozoic and ends in the Upper Carboniferous: The burial of the source rocks of the Lower Devonian exceeds 2,000 m at the north of the basin, but it is particularly important in the southwest, where it reaches 2,600 m (Figure 12). At the end of this cycle, the Hercynian movements are marked by a disconformity. In the central-eastern part of the basin, where several hundred meters of continental shales at the end of the Paleozoic cycle (Tiguentourine shales) are preserved under the Triassic, it may be assumed that erosion has been very slight and, therefore, that the burial has remained almost constant from the Carboniferous to the Triassic. In the other parts of the basin, an erosion steadily increasing toward the west brings the source rocks to a smaller depth. The second cycle of burial starts in the Triassic and ends in the Cretaceous (Figure 13). It is very extensive in the northeast part of the basin, where the depth exceeds 3,000 m for the source rocks of the Lower Devonian. In the central part, the new burial

compensates and more than compensates the Hercynian erosion. In the south and southwest of the basin, however, the burial remains slighter than in the Carboniferous.

In order to make use of the data obtained on the history of the embedment in the various parts of the basin, the mathematical model of petroleum generation was applied, the principles of which were described by Tissot (1969) and Deroo et al (1969). Since then, however, this model has been completed to introduce the formation of gases by cracking: For this last reaction, an activation energy of 50 kcal is assumed (Figure 14).

The diagrams showing the burial of the various layers as a function of time were reconstructed at the vertices of a regular grid formed by the meridians and parallels at a spacing of 0°30'. For the calculation, the Lower Devonian-Upper Silurian limit was taken, considered as the mean point of this assemblage of source rocks. For the temperature-depth relation, the present geothermal data were used (Figure 15). They show a gradual decrease of the geothermal gradient from the south to the north of the basin, with relatively warmer zones above the rises of the basement, such as the mole of El Biod. The adoption of these values for the Mesozoic era seems to be justified. As concerns the Paleozoic, it may be noted that the movements of the basement reflected by these data took place toward the end of the Paleozoic, at the time of maximum burial depth, i.e., at the time when the reaction rates were highest during this cycle of subsidence. It may be assumed, therefore, that the regional geothermal data were more or less comparable to the present data from that time on.

The quantities of oil and gas formed in the source rocks from the Lower Devonian-Upper Silurian until the end of the Paleozoic are calculated by the mathematical model and presented in Figure 16. A certain quantity of oil (100 to 150 mg/g of total organic carbon) was formed in most of the basin. However, in the southwest, where the burial is considerable over a vast synclinal zone, this stage has already been passed, and the cracking phase which permits formation of a large quantity of gas has been reached. In the course of exploration, this region actually only showed signs of gas without any exploitable accumulation. At the time of formation of the gas, there were no structural traps in this region, since they result from the Hercynian folding. This circumstance did not permit the formation of deposits. Moreover, the gas still available certainly moved during the folding from this vast synclinal zone toward the greatly raised zone of the Djebel Essaoui Mellène, which was

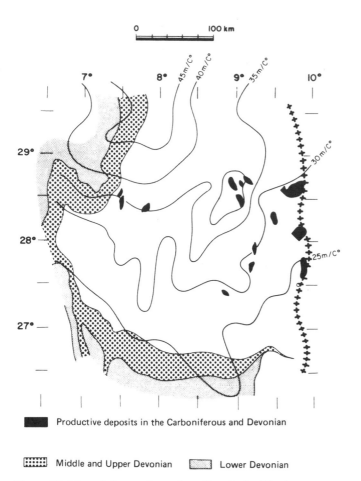

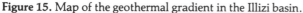

Figure 15. Map of the geothermal gradient in the Illizi basin.

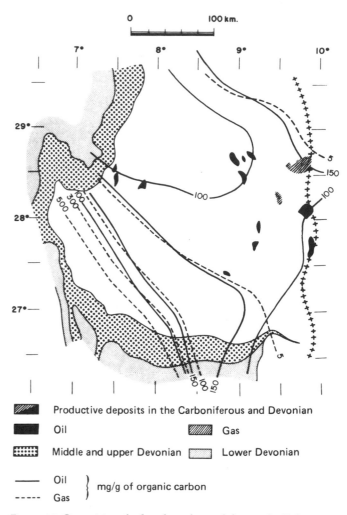

Figure 16. Quantities of oil and gas formed during the Paleozoic in the source rocks of the Lower Devonian-Upper Silurian, calculated by the mathematical model (in mg/g of organic carbon).

subsequently subjected to a very extensive erosion reaching the Ordovician in some places and to a leaching of the reservoirs by fresh water.

In the course of the cycle of Mesozoic subsidence and until the present cycle, the organic material continues to mature except in the western zone, where the depths remain smaller than in the Paleozoic and where, at any rate, the residual kerogen hardly contained any more structures capable of forming hydrocarbons, due to its already very advanced state of degradation. The new hydrocarbons formed can be trapped this time, since the folds have already taken place. The total quantities of hydrocarbons formed up to the present time are shown in Figure 17.

In the northeast part of the investigated area, where the Mesozoic burial is extensive, the organic material reaches the cracking phase at that time and produces large quantities of gas, which formed the gas deposits of Alrar and In Amenas, as well as the gases associated with the oil deposits of Zarzaïtine, Ohanet, etc. The examination of the degree of carbonization of the spores and pollens (Correia, 1967) and the measurement of the reflectance of vitrinite (1.6 to 1.7) (Robert, 1971) confirm the very advanced state of maturation of the organic material, both in the northeast part and in the southwest part of the investigated area.

Between these two regions, a northwest-southeast-oriented area retained a more moderate state of maturation and, consequently, an oil potential. It is in this area that all of the oil deposits of the Lower Devonian reservoir are

located. It may be noted, however, that the eastern deposits which are in the vicinity of the gas zone have a high GOR, as indicated above, while the western deposits (Oued Zenani, Tin Fouyé, Mazoula) contain practically no gas, which is in agreement with the particularly small proportions of gas calculated (less than 5 mg per g of organic carbon). The time of formation of these oils extends over both the Paleozoic and the Mesozoic in the eastern region, while it is mainly Paleozoic in the western deposits. It may be noted that the structural conditions of this last region provided for conservation of at least part of the Paleozoic oil, in contrast to what occurred for the gas in the southwest of the basin. In fact, it is found (Figures 12 and 13) that there exists constantly a high zone in the region of Tin Fouyé-Tabankort, in both the Paleozoic and the Mesozoic. This circumstance prevented leakage of all of the hydrocarbons toward the Djebel Essaoui Mellène, as was the case for the gases.

The source rocks of the Lower Silurian were subjected at all times to temperatures above those reached by the Upper Silurian-lower Devonian. For this reason, the only oil deposits known in the Ordovician are located in the zones where the state of maturation is the least advanced: In the

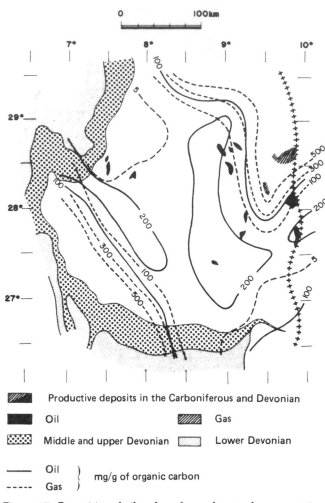

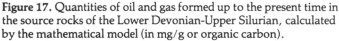

Figure 17. Quantities of oil and gas formed up to the present time in the source rocks of the Lower Devonian-Upper Silurian, calculated by the mathematical model (in mg/g or organic carbon).

southeast part of the investigated area (Dome à Collenias, Ouan Taredert) and in the northwest part (Tin Fouyé-Tabankort). Nevertheless, not far from this last deposit, the organic material reaches the cracking phase: This is why the oil is associated in the deposit with a very large accumulation of gas.

CONCLUSION

The Illizi basin offers an example of very old source rocks (up to 440 million years) subjected for long periods to an extensive burial. As a result, the source rocks have reached a sufficiently advanced state of maturation in some parts of the basin so that oil can no longer be found in the corresponding reservoirs. Moreover, the oils have a generally highly mature character, and it becomes difficult to distinguish the different types of oil derived from different source rocks. The overall composition is fairly uniform. The same applies to the distribution of the saturated hydrocarbons, which shows the abundance of short-chain alkanes in all of the oils, regardless of whether they result from cracking of the kerogen or of the previously

formed oil. On the other hand, the cyclic structures comprising one or several aromatic rings seem more stable and retain their genetic characteristics which make it possible to separate the different types of oil and to relate them to their source rocks.

On the other hand, the use of a mathematical model of formation of petroleum and gas based on the reconstruction of the structural history of the basin made it possible to evaluate the time of formation of the hydrocarbons in the various parts of the basin. The results obtained are in agreement with the observed distribution of the oil and gas deposits. In particular, they allow us to delimit the regions where oil deposits may exist.

REFERENCES CITED

Beuf, S., et al, 1971, Les grès du Paléozoïque inférieur au Sahara: Paris, Editions Technip, 484 p.

Correia, M., 1967, Relations possibles entre l'état de conservation des éléments figurés de la matière organique (microfossiles palyonplanctologiques) et l'existence de gisements d'hydrocarbures: Revue de l'Institut Français du Pétrole, XXII-9, p. 1285-1306.

Deroo, G., et al, 1969, Possibilité d'application des modèles mathématiques de formation du pétrole à la prospection dans les bassins sédimentaires, in P.A. Schenck and I. Havenaar, eds., Advances in organic geochemistry: Oxford, Pergamon Press, p. 345-354.

Durand, B., J. Espitalie, and J.L. Oudin, 1970, Analyse géochimique de la matière organique extraite des roches sédimentaires, III Accroissement de la rapidité du protocole opératoire par l'amélioration de l'appareillage: Revue de l'Institut Français du Pétrole, XXV-11, p. 1268-1279.

Millouet, J., A. Rast, and R. Vincent, 1962, Le champ d'Ohanet, I: Exploration Revue de l'Institut Français du Pétrole, XVII-3, p. 396-403.

Oudin, J.L., 1970, Analyse géochimique de la matière organique extraite des roches sédimentaires, I. Composés extractibles au chloroform: Revue de l'Institut Français du Pétrole, XXV-1, p. 3-15.

Robert, P., 1971, Etude pétrographique des matières organiques insolubles par la mesure de leur pouvoir réflecteur, Contribution à l'exploration pétrolières organiques insolubles par la mesure de leur pouvoir réflecteur, Contribution à l'exploration pétrolière et à la connaissance des bassins sédimentaires: Revue de l'Institut Français du Pétrole, XXVI-2, p. 105-135.

Tissot, B., 1969, Premiéres données sur les mécanismes et la cinétique de la formation du pétrole dans les sédiments. Simulation d'un schéma réactionnel sur ordinateur: Revue de l'Institut Français du Pétrole, XXIV-4, p. 470-501.

—— , J.L. Oudin,and R. Pelet, 1972, Critères d'origine et d'évolution des pétroles; Application à l'étude géochimique des bassins sédimentaires, in H.R. Von Gaertner and H. Wehner, eds., Advances in organic geochemistry: Oxford, Pergamon Press, p. 113-134.

—— , et al, 1971, Origin and evolution of hydrocarbons in early Toarcian shales, Paris basin, France: AAPG Bulletin, v. 55, n. 12, p. 2177-2193.

Hydrocarbon Habitat of Tertiary Niger Delta

B.D. Evamy
J. Haremboure
P. Kamerling
W.A. Knaap
F.A. Molloy
P.H. Rowlands
*Shell-BP Petroleum Development
    Company of Nigeria Ltd.*
Lagos, Nigeria

The Tertiary Niger delta covers an area of about 75,000 sq km and is composed of an overall regressive clastic sequence which reaches a maximum thickness of 30,000 to 40,000 ft (9,000 to 12,000 m). The development of the delta has been dependent on the balance between the rate of sedimentation and the rate of subsidence. This balance and the resulting sedimentary patterns appear to have been influenced by the structural configuration and tectonics of the basement.

Structural anlaysis of the Tertiary overburden shows that individual fault blocks can be grouped into macrostructural and eventually megastructural units. Such megaunits are separate provinces with regard to time-stratigraphy, sedimentation, deformation, generation and migration of hydrocarbons, and hydrocarbon distribution. A recurrent pattern emerges in the distribution both of absolute volumes of hydrocarbons and the ratio of volume of gas-bearing reservoir rocks to the volume of oil-bearing reservoir rocks within megaunits and macrounits.

The maturity of potential source rock in a given fault trend was achieved when sedimentation had almost reached the present surface, and when the active depocenter had been advanced seaward by several trends. Thus, migration started when deposition, together with the intrinsically synsedimentary structural deformation, had almost come to a halt in that particular trend.

The source rocks of the Niger delta yield a light waxy paraffinic oil, which is transformed bacterially to a heavier nonwaxy crude at temperature below 150 to 180°F (65 to 80°C). The coincidence of the boundary between transformed and unaltered oils, within a rather narrow temperature range on a delta-wide basis, implies that little or no subsidence with concomitant increase in geotemperature of the oil-bearing reservoirs has occurred after migration.

The conclusion that migration took place after the structural geometry of a given trend had been determined originates from several independent lines of evidence. The observed uneven distribution of oil and gas in the delta therefore cannot be explained in terms of the passage of the source rocks through the oil-generating zone into the gas-generating zone (oil and gas "kitchens," respectively), with early structures receiving mainly oil and late traps receiving mainly gas. Rather, the hydrocarbon distribution probably is the result of original heterogeneity of the source rock and of segregation during migration and remigration.

INTRODUCTION

The megatectonic setting of the Niger delta has been discussed by Stoneley (1966) and by Burke et al (1972). The basement configuration deduced from geophysical data, was presented by Hospers (1965, 1971), and the synsedimentary tectonics of the Cenozoic delta were described by Merki (1972).

Sedimentologic aspects of the upper Tertiary deltaic deposits, derived from subsurface data, were described by Weber (1971) and Weber and Daukoru (in press). The importance of longshore drift and submarine canyons and fans in the development of the Niger delta also has been emphasized by Burke (1972).

The most comprehensive articles on the Niger delta, from a petroleum geologic point of view, are the works of Short and Stauble (1967) and Frankl and Cordry (1967).

MEGATECTONIC BASEMENT FRAME

Probably the most reliable onshore information on the configuration of the basement has been obtained from gravity data (Figure 1A). Hospers (1965, 1971) described the various basement blocks and pointed out the predominance of northeast-southwest and northwest-southeast trends in the megatectonic framework. The outline of the main depocenter as deduced from gravity and magnetic data (Figure 1A, B) appears to reflect on interaction of these two

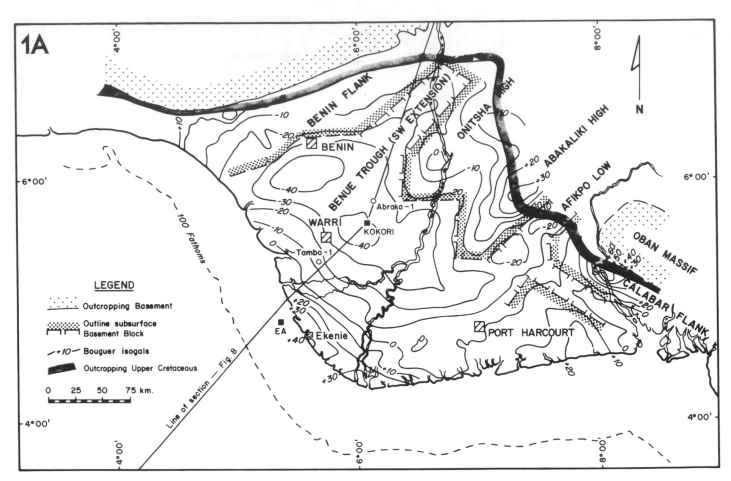

Figure 1. Basement configuration based on **A**, gravity data, and **B**, magnetic data.

main directions.

As has been pointed out by Emery et al (1975), the occurrence of these two trends may be related to the unique position of the Niger delta during the opening of the southern Atlantic, at the boundary between the southern area of crustal divergence and the equatorial zone of crustal translation. The northeast-southwest basement trends appear to indicate possible extensions within the African continent of the Charcot and Chain oceanic (transform) fracture zones. The northwest-southeast trends may be the result of block faulting that occurred along the edge of the African continent during the earlier stages of divergence.

An interesting feature is the regional high (Ekenie gravity high) on the southwest flank of the central low. Gravity readings in this area are considerably higher than those recorded in the northwestern area of outcropping granitic basement. However, evidence for shallow basement has not been found on seismic reflection lines nor from any nearby wells. A refraction seismic line, shot over the Ekenie area, suggests basement is present at a depth of about 25,000 ft (7,500 m).

The Ekenie high thus is interpreted as an area of rising basement having considerably greater density than the granitic basement cropping out in the northwest. A transition from continental to oceanic crust beneath the Niger delta has been suggested by several authors (e.g., Hospers, 1965, 1971) and would explain the anomalously

high gravity values above the Ekenie high.

STRATIGRAPHY OF TERTIARY NIGER DELTA

Biostratigraphic Subdivision

In the Niger delta, hydrocarbons have been found in rocks of Paleocene to Pliocene age. This time span is subdivided biostratigraphically into 29 palynologic zones and subzones, each having an alphanumeric code (e.g., P630, see Figure 2). The zones generally can be recognized in all facies types from continental to marine. The zonal and subzonal boundaries are quite sharply delineated, not only by marker species, but also by changes in the distribution of other species.

Paleogeographic Evolution

The growth of the Tertiary Niger delta is depicted by a series of maps showing the principal depocenters for selected microfloral units between the Paleocene and the Pliocene (Figure 2).

Hydrocarbons are concentrated along the updip or proximal edge of the successive depocenters. This undoubtedly reflects the predominance of suitable traps associated with the major growth faults, which delineate the proximal margin of the depocenters.

Figure 3 shows the vertical subdivision of the delta into broad facies units. An upper series of massive sands and

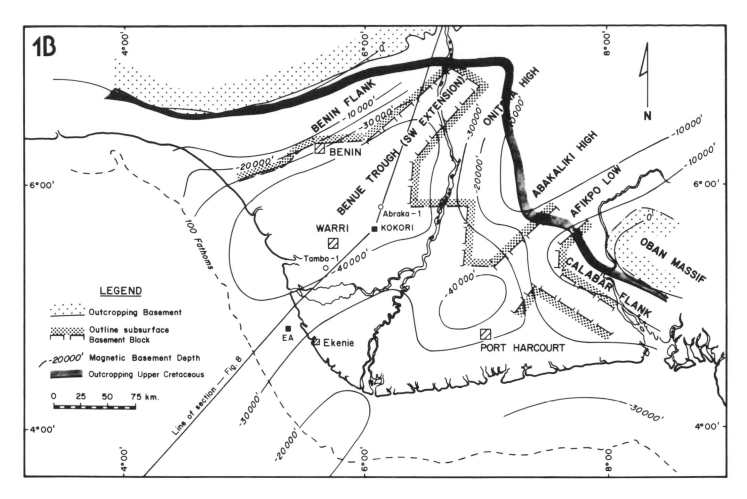

gravels, deposited under continental conditions, grades downward through a transitional series composed mainly of sand but with some shale, into an alternation of sandstone and shale of roughly equal importance, deposited under paralic conditions. Lower in the section, marine shale predominates and the associated sandstone units are very likely turbidites.

Paleocene-Eocene (P200; Figure 2a)—During the Paleocene and earliest Eocene times, marine shales were deposited over much of the southern Nigerian sedimentary basin. Paralic and marine/paralic sediments are present only over a restricted area (i.e., where the present Cross River flows between the Abakaliki anticlinorium and the Oban massif) and, in this area, are considered to represent the deposits of the incipient Cross River delta.

The delta of the Niger River itself first becomes apparent during the time designated as P330 to P430. Figure 2a shows that the delta lay west of the present course of the Niger River. The isopach pattern reflects the deltaic form of the depositional area, and the facies distribution conforms with expected change from continental through transitional and paralic to marine in a downdip direction.

The Niger delta continued to grow in the Eocene, initially in response to the positive epeirogenic movements along the Benin and Calabar flanks (Murat, 1972). Near the end of the late Eocene (P480), a major regression commenced which accelerated the expansion of the Niger delta. This regressive

phase has continued until the present, frequently interrupted by generally minor transgressions.

Oligocene and earliest Miocene (P520-P630; Figures 2b, 2c)—During this period the successive depocenters in the west considerably overlapped, reflecting in general a pronounced subsidence and a relatively slow advance of the delta front toward the west and southwest. This pattern of overlapping depocenters resulted in a rather thick development of paralic sediments over the western part of the delta. By contrast, the successive depocenters in the east generally do not overlap, and they reflect a more rapid advance of the delta front.

The eastern depocenters are clearly separated from those of the west by a paleohigh area, characterized by a period of erosion or nondeposition (P580 to P650). Significantly, this area lies on the southern extension of the Abakaliki anticlinorium, which probably was undergoing some positive readjustment during the Oligocene and early Miocene.

Miocene to present (P650-P900; Figures 2c, 2d)—Later in the early Miocene, the hitherto separated depocenters gradually merged, and the enlarged delta began to prograde along a wide front (P680; Figure 2c). The more rapid subsidence, and corresponding slower rate of advance of the delta front, continued to characterize the western part of the delta.

Depocenters continued to develop during the later

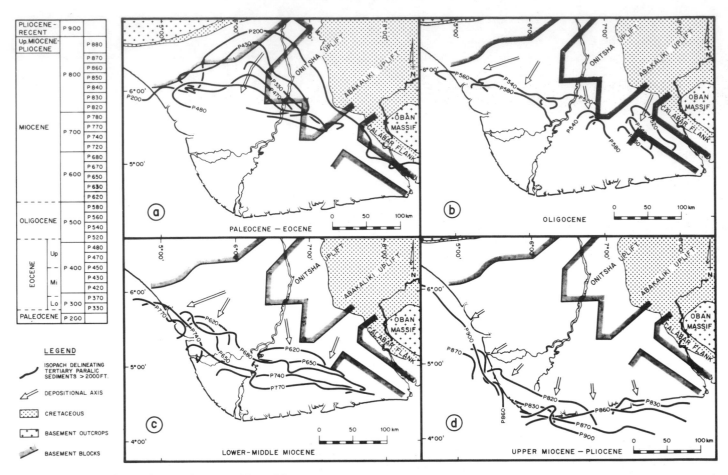

Figure 2. Stratigraphic evolution of Tertiary Niger delta.

Miocene and Pliocene (P830 to P900; Figure 2D). A large depocenter of late Miocene age (P860) is present in the eastern offshore. The youngest depocenter (P900) is in the western offshore.

At several stages during the Miocene, sedimentation in the eastern part of the delta was interrupted by periods of uplift and erosion. Erosional surfaces have been widely recognized and are locally in the form of entrenchments, which subsequently have become filled with clay deposits. An example is the Afam Clay Member.

Sedimentologic Evolution

Conceptual model—The model of delta development described by Curtis (1970) for the United States Gulf Coast has been applied to the Niger delta. Deltaic sedimentation is seen as a function of the rate of deposition (Rd) and the rate of subsidence (Rs). Depending on this function, the delta builds out or progrades (Rd > Rs), remains stationary and builds up (Rd ≈ Rs) or retreats (Rd < Rs). Important variations in the relation of Rd and Rs result in the development of distinct sedimentary megaunits of different shape, extent, and thickness.

In delta building where the delta gradually progrades seaward, such as that of the Niger, regressive phases (Rd > Rs) obviously predominate. Regional transgressions form relatively short-lived interruptions in the general advance of the delta, occurring under conditions of Rd < Rs

along the seaward margin of sedimentary megaunits.

Synsedimentary tectonics—Growth faults are regarded as a product of gravity sliding during the course of deltaic sedimentation. As shown in Figure 4 (modified after Bruce, 1973), down-to-the-basin growth faulting is initiated when relatively heavy, sandy deposits of a regressive cycle (Rd > Rs) prograde over little compacted clays with low shear strength.

Figure 4b illustrates the development of growth faults under conditions of a prograding delta. The amount of space created by individual growth faults is insufficient to accommodate the supply of sediment. New, fault-controlled depocenters therefore are formed progressively in a seaward direction. If no important change in Rd/Rs has occurred, all successive depocenters constitute part of the same sedimentary megaunit.

Where Rd ≈ Rs (Figure 4c), that is, under conditions of neither regression nor transgression, the amount of sediment is not in excess of the space available for accumulation and the depocenter will continue to be active until Rd again surpasses Rs. Under these conditions, each new fault-controlled depocenter is the result of a change in the relation of Rd and Rs, and is considered to form the beginning of a new sedimentary megaunit.

The gradual development of a long regional south flank, under conditions of Rd ≈ Rs, may have been interrupted by a temporary return to a condition of Rd > Rs. In the western part of the offshore "K" Block (Figure 5), for example, steep regional flanks exist at depth, reflecting initial conditions of a relatively high rate of subsidence (Rd ≈ Rs). The flanks, however, are overlain by a new cycle of sediments with low dips and are affected by numerous closely spaced faults. The faulting could be due to gravitational instability when renewed progradation of the delta took place over the relatively steep, preexisting slope of the older cycle of sediments. Farther seaward, the two cycles merge in a steeply dipping flank, indicating the transient nature of the return to conditions of Rd > Rs.

Sedimentary shale ridges—The shale ridges at the distal ends of the long, regional south flanks are thought to be the product of a strongly diachronous southward facies change from the sandstone and shale alternations of the paralic facies to purely marine shale, developed when Rd ≈ Rs. Differential loading is believed to have caused some synsedimentary movement between the heavier sandy and silty sediments and the marine shales farther seaward, thus creating a counter-regional growth-fault contact along the zone of facies change (Figure 4c). Evidence is lacking for a diapiric origin of these shale ridges; no warping up, thinning, or wedging out of the paralic sediments adjacent to these shale ridges has been observed. On the contrary, the section north of a shale ridge seems to dip, almost without disturbance, into the massive shales and shows expansion rather than thinning against it (Figure 6).

Large-scale diapiric upheavals seem to be present in the proximal part of the continental shelf only in the eastern offshore area. They are, however, typical farther seaward, along and beyond the outer margin of the continental shelf. The nature of these diapiric structures is discussed later.

Regional sedimentary patterns—The relation between the various sedimentary megaunits which constitute the Niger delta is presented on Figure 7 which shows areas where the delta prograded without significant interruptions. The long and relatively steep south flanks also are indicated. The dentated symbol represents counter-regional (i.e., north-hading) growth faults. These form the northern boundaries of regionally developed shale ridges, which in turn mark the distal ends of the various megasedimentary units.

An intermediate geologic condition also is depicted, where the older part of the section has been deposited under conditions of Rd ≈ Rs and the younger cycle represents renewed progradation of the delta (where Rd > Rs), giving rise to the heavily faulted structures exemplified by Figure 5.

Conditions of Rd > Rs apparently prevailed in the more updip parts of the delta where the provenance was nearby. Farther seaward, however, in the present coastal and offshore areas, progradation was interrupted frequently, resulting in the development of several clearly defined megaunits.

Basement influence on Tertiary overburden—The disposition of the basement blocks has affected the sites of the successive Tertiary depocenters (see Figure 2). In particular, the Oligocene and early Miocene depocenters seem to correspond to the low areas between basement blocks (e.g., at the extension of the Benue Trough and in the low between the Onitsha and Abakaliki blocks).

The central area of erosion or nondeposition during late

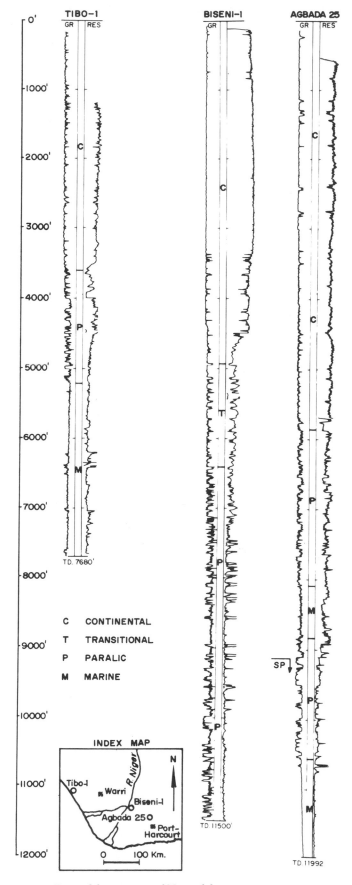

C   CONTINENTAL
T   TRANSITIONAL
P   PARALIC
M   MARINE

Figure 3. Typical facies units of Niger delta.

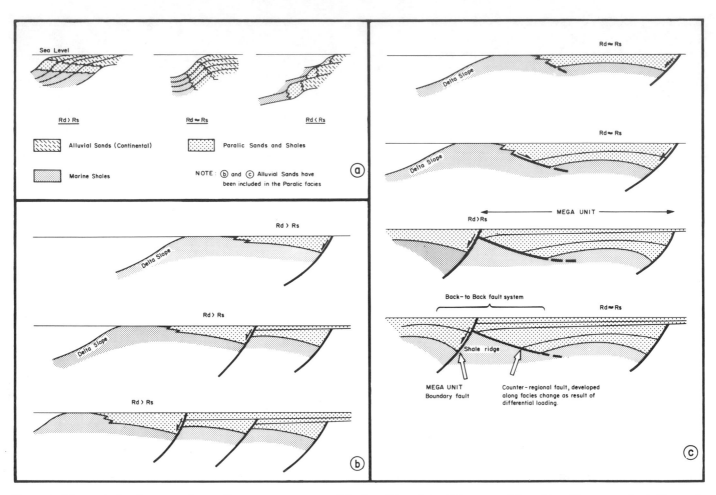

Figure 4. Schematic development of synsedimentary structures of Tertiary Niger delta. Rd = rate of deposition; Rs = rate of subsidence. b, Growth-fault development when rate of deposition exceeds rate of subsidence. c, Growth fault development when rate of deposition is in balance with rate of subsidence.

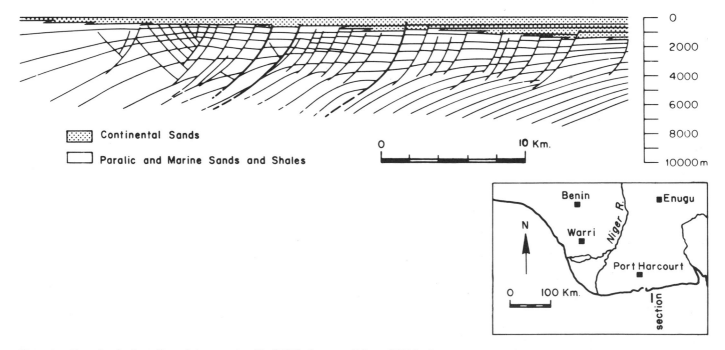

Figure 5. Complex fault pattern characterizing Shell-BP's former offshore "K" block.

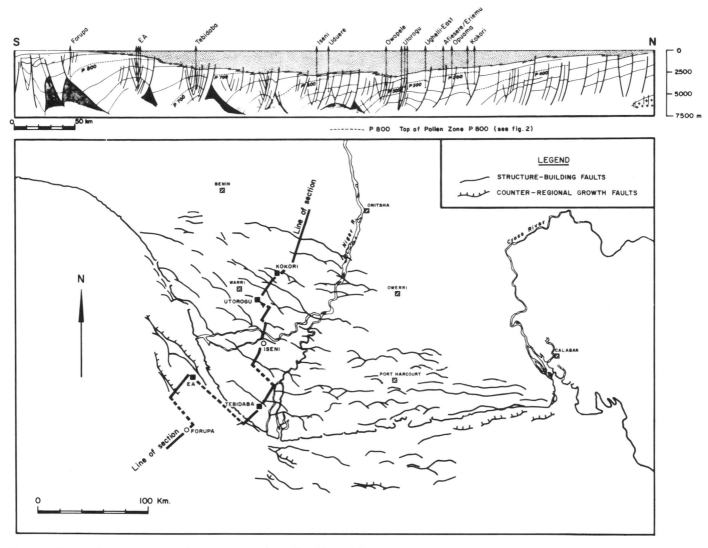

Figure 6. Regional cross section and main structural trends in Niger delta.

Oligocene and early Miocene times (P580 to P650) has been attributed to the positive effects of the underlying Abakaliki high.

Basement control in the eastern parts of the delta is less obvious. Only the P520 depocenter corresponds to a prominent basement feature—the subsidiary northwest-southeast trough cutting in the Calabar flank (Figure 2). This trough probably determined the early course of the Cross River.

Tectonic movement of the continental basement blocks in the northeast seems to have continued to some extent during the deposition of the Tertiary, particularly along the western fault boundary of the Onitsha block and its presumed extension southward (Figure 7). That this movement continued during the Tertiary is apparent from the following facts.

1. The extension of the fault bounding the Onitsha high on the west separates a western area with predominantly east-southeast regional dips along the structural plunge in the Tertiary overburden from an eastern area with west-northwest dips. This fault extends as far south as the main divide of the Niger River (see Figure 7).

2. The north-bounding faults of depocenters have different trends on either side of the deep-seated basement fault. In most places, furthermore, these depocenters do not extend laterally across the basement-fault trend.

STRUCTURE OF TERTIARY NIGER DELTA

Structural Evolution

A conception model for the evolution of the Tertiary Niger delta is presented in a series of diagrams (Figure 8). Basement tectonics related to crustal divergence and translation during the Late Jurassic(?) to Cretaceous continental rifting probably determined the original site of the main river and controlled the early development of the delta. In view of the present isostatic balance (suggested by the gravity data), the Cenozoic development of the delta also could have taken place under approximately isostatic equilibrium (Hospers, 1971). The main depocenter could have been at the junction between the continental and oceanic crust, where the delta reached a main zone of crustal instability.

A more passive involvement of basement tectonics has

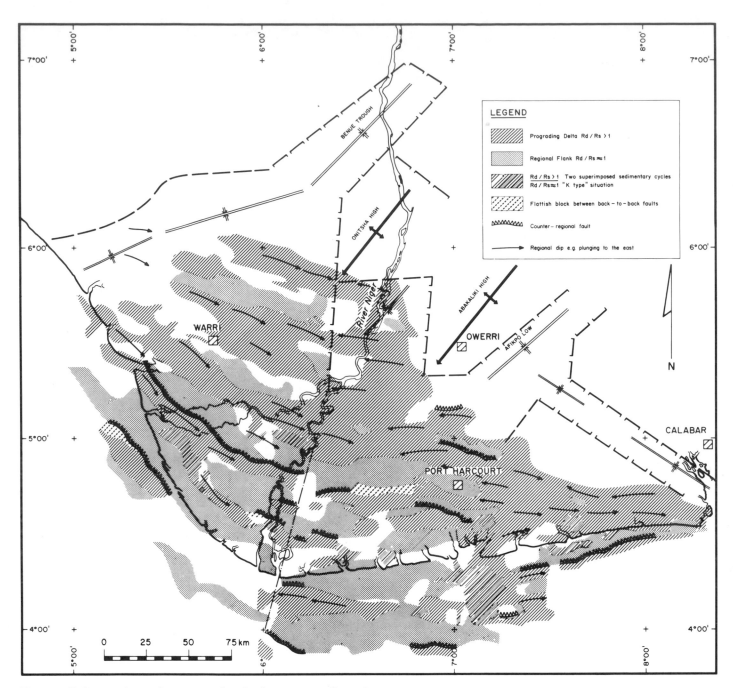

Figure 7. Sedimentation and structure related to basement configuration.

been envisaged for the development of the delta farther south during its progradation over the oceanic crust. In this area, increases in the rate of subsidence are indicated in the Tertiary overburden by the regularly spaced regional flanks associated with shale ridges and counter-regional faults. These increases in the rate of subsidence may be attributed to repeated failure of the oceanic crust under the thick sedimentary pile of the outbuilding Tertiary delta (Figure 8). The concept of normal faulting under a deltaic load forms a refinement of Hosper's model of crustal subsidence (Hospers, 1971).

Vertical subsidence of the basement probably provided only part of the space needed to accommodate the deltaic sediments. Gravitational faults, which developed during the sedimentation of the paralic sequence, must have caused a considerable horizontal extension of the section. The faulting and related extension probably were associated with lateral flowage and extrusion of prodelta clays on the continental slope in front of the developing depocenter of paralic sediments. This would explain the occurrence of diapiric structures which have been noted on the continental slope off the Niger delta (Mascle et al, 1973).

Synsedimentary Faults

Structure-building faults—These are the faults which define the updip limits of major rollover structures (Figure

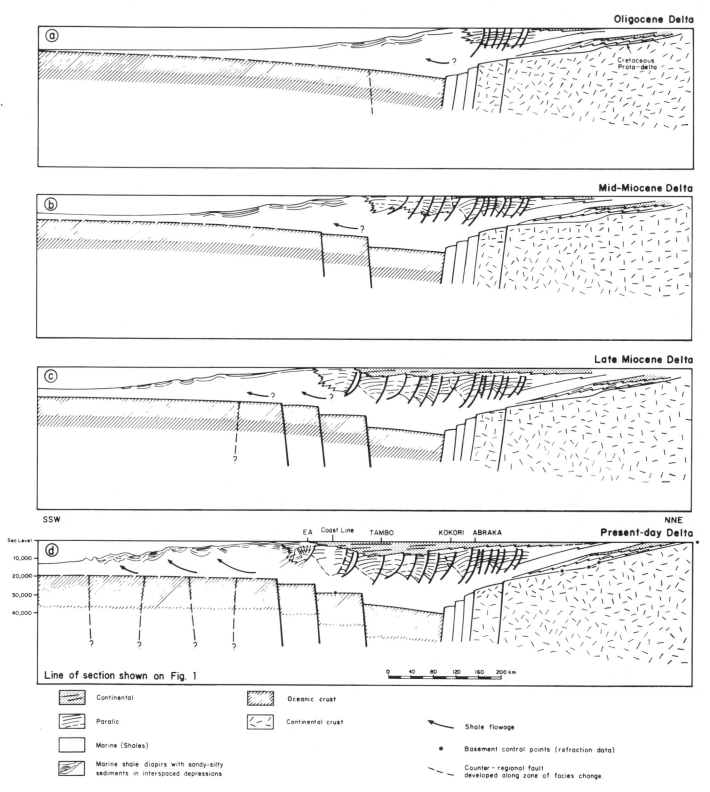

Figure 8. Tertiary evolution of Niger delta.

9). In the horizontal plane, they are essentially concave in a downdip direction. The degree of curvature varies from being rather linear in the east to truly crescent-shaped in the western and southern parts of the delta (Figure 10).

The curvature of structure-building faults at their lateral extremities creates a mapping problem because of the way in which they repeat each other in the strike direction. In some places the structure-building faults repeat each other en echelon (Figure 10). Where this occurs, the structure-building faults die out in the flanks of the adjacent

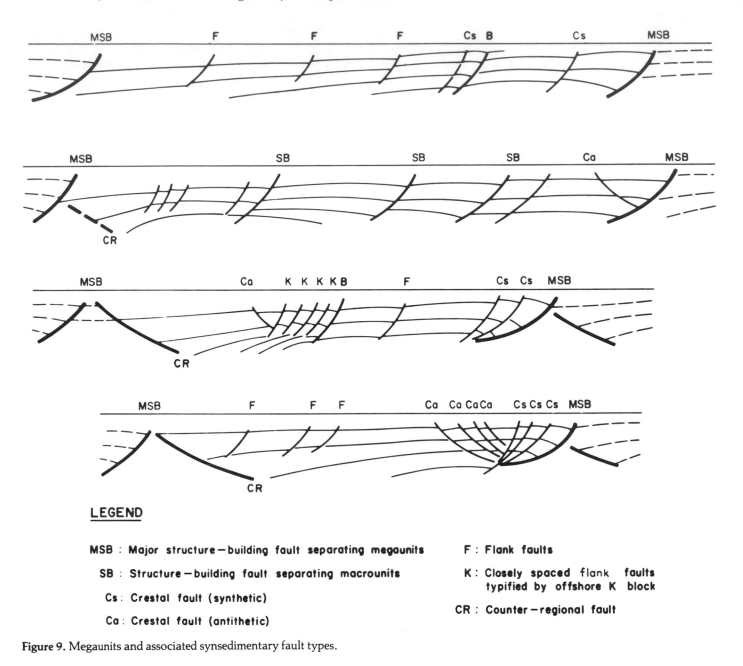

Figure 9. Megaunits and associated synsedimentary fault types.

rollover structure.

Crestal faults—A rollover structure may contain one or more crestal faults (Figure 9). These characteristically parallel the axis of the structure and differ from structure-building faults in that they (a) show less curvature in the horizontal plane; (b) are generally steeper in the vertical plane; and (c) display less growth, which also tends to be less continuous.

In some structures the crestal faults have very large vertical displacements. At depth, they may ring a sandy sedimentary section on the downthrown side into juxtaposition with older marine shales. Some crestal faults even cut the slip plane of the structure-building fault.

Flank faults—These faults, as their name suggests, are located on the southern flanks of major rollover structures. Although they may show some rollover deformation at shallow levels, southerly dips are typical on either side of

the fault at depth (Figure 9).

Other faults—Major counter-regional growth faults, already discussed, are located at the southern end of regional flanks. Antithetic faults also have a counter-regional hade, but they are of secondary structural importance and display no growth (Figure 9), being simple compensations for extension in the overburden. K type faults are essentially flank faults. They are considered as a separate class only because of their extremely close spacing, which gives rise to a multiplicity of narrow fault blocks (see section on regional sedimentary patterns and Figures 5 and 9). They are common (as their name implies) in Shell-BP's original offshore "K" Block.

Macrostructures and Megastructures

A series of fault blocks generally can be grouped together to form a macrostructure (Figure 11), which is essentially

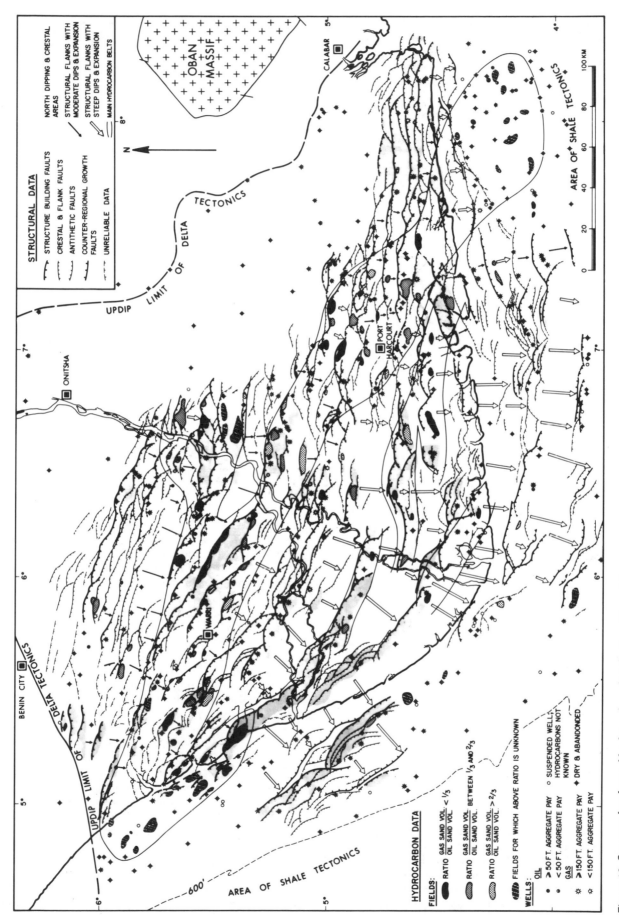

Figure 10. Structural styles and hydrocarbon distribution, Niger delta.

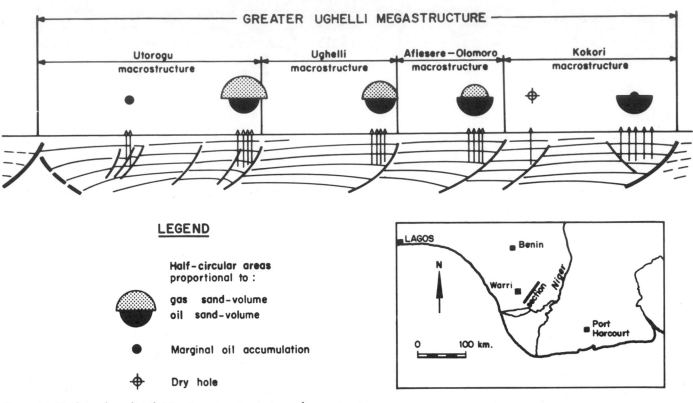

Figure 11. Hydrocarbon distribution in macrostructures and megastructures.

one large rollover deformation characterized by: (a) one or several fault blocks with predominantly north dips; (b) a zone of symmetrical (anticlinal) dips, generally associated with a deep anticlinal axis (at about 15,000 ft or 4,500 m); and (c) a southerly dipping flank of variable extent.

The unit is bounded updip by the structure-building fault, against which most of the sedimentary expansion has taken place. Most macrostructures show several culminations in a strike direction.

The delta is made up of many macrostructures, which vary greatly in areal extent and complexity (Figure 10). Particularly along the axis of the delta, macrostructures can be grouped into sets designated as megastructures (Figures 9 and 11). The boundaries of a megastructure correspond to major breaks in the regional dip of the delta. Their southern limits are defined most commonly by large counter-regional faults, but some are delineated by the next structure-building fault in a seaward direction. It becomes difficult to distinguish these megastructural units near the lateral extremities of the delta, because they tend to narrow and eventually to coalesce.

Classification of Structural Types

Simple Structures

Simple rollover structures with predominant landward dips—These structures are associated typically with a structure-building fault, and therefore are the northernmost (or most landward) fault blocks in a macrostructure. Their southern boundary is in general an important crestal fault. The structures are characterized by a shift of culmination with depth, generally giving rise to both shallow and deep closures. Where these fault blocks are extremely large and show proportionately large shifts of culmination, they tend to be bounded downdip by another structure-building fault rather than a crestal fault. As such, they are considered as macrostructural entities.

Simple rollover structures with anticlinal dips—These are called "crestal fault blocks" because their symmetrical dips emphasize the crest of the macrostructure. They also typically are cut by crestal faults. The structures generally show only moderate shifts of culmination, and furthermore generally are related directly to the deep axis of the macrostructure, common on seismic below 3 sec.

Simple rollover structures with predominant seaward dips—These are the flank structures. They may have a moderate rollover deformation at shallow to intermediate levels, with accompanying landward dips, but at depth the dips are characteristically seaward and closure relies on the sealing capacity of faults alone.

Complex Structures

Complex rollover structures with symmetrical faulting—Collapsed crests are representative of this class of structure. These are characterized by an overall domal aspect, with marked anticlinal dips at depth. Two swarms of crestal faults, one hading seaward and the other hading landward, typically "collapse" the structural crest to compensate for overburden extension.

Complex structures with seaward faulting—These are found in the former Shell-BP "K" Block offshore (see the section on synsedimentary tectonics and the section on

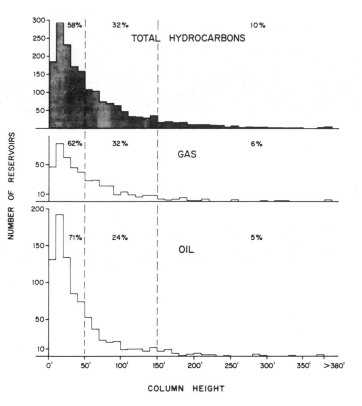

Figure 12. Hydrocarbon column heights, Niger delta.

other faults). Two sedimentary cycles seem to be involved. The upper cycle is intensely faulted with most of the closely spaced faults hading seaward. Low northern dips are present within each fault sliver and growth into each fault is small. The regional dip of the low sedimentary cycle is steep to very steep southward.

Structural Changes in Strike Direction

As structure-building faults are mostly arcuate in plan view, the well-defined, deep anticlinal axis of the macrostructure in the central part tends to die out toward its lateral extremities. This causes a progressive reduction of the landward dips, apparent first in the deeper layers. Eventually, in dip section, a flank structure is all that remains. The more arcuate the structure-building fault, the more abrupt the change.

HYDROCARBON DISTRIBUTION

Hydrocarbon-Column Heights

Most reservoirs in the Niger delta are not filled to their synclinal spill points. On the downthrown side of faults, at least in the hydropressured regime, the spill point of a reservoir tends to lie at its intersection with the fault. The plane of the fault on its upthrown side, however, appears to be capable of sealing and retaining significant columns of hydrocarbons.

Data available from Shell-BP fields at the end of 1971 showed that, out of 1,400 reservoirs, 71 percent have oil-column heights of 50 ft (15 m) or less and only 5 percent are greater than 150 ft or 45 m (Figure 12).

Oil-column height, of course, has a prime influence on

the recoverable reserves of a trap. A relatively few reservoirs, with hydrocarbon columns greater than 300 ft (100 m), contribute the largest part of the recoverable oil and gas in the Niger delta. To attain a significant hydrocarbon-column height, and thus large reserves, a large structure with vertical closure in the strike direction is a critical parameter. The reverse, however, is not necessarily true, and many large structures are known to contain only small reserves.

General Pattern

The following pattern concerning the hydrocarbon distribution throughout the delta has been observed (see Figure 10).

1. A hydrocarbon-rich belt cuts across the depositional and structural trends of the delta from southeast to northwest. North and east of the hydrocarbon-rich belt, the gas-to-oil ratio[1] is higher, and the recoverable oil reserves of the accumulations found to date are smaller. South of the main hydrocarbon-rich belt is a series of narrow oil-rich zones; otherwise, a predominantly gassy province occupies a fairly wide part of the central delta.

2. Known commercial oil accumulations occur predominantly in the structurally highest part of a given macrostructure in the strike sense, despite viable trapping conditions down plunge.

3. Dry holes and marginal oil and gas finds are located predominantly on the south flanks of macrostructures.

4. In a given macrostructure, the gas-to-oil ratio increases down plunge and in a generally seaward direction. Hence northern blocks with pronounced landward dips are regarded as being highly prospective.

5. The more downdip a macrostructure is within a megastructure, the greater is the probability of a higher gas-to-oil ratio (Figure 11).

6. Because of the intensity of faulting and steeply dipping reservoirs of collapsed crest structures, the recoverable reserves can be rather low. Exceptions are reservoirs such as that of the Forcados Estuary field, where a reservoir with a long oil column and concomitantly large areal extent exists at shallow depth. Hydrocarbons in collapsed-crest structures tend to be trapped behind crestal and antithetic faults, whereas the collapsed crest itself is commonly barren.

7. Although the probability is high of finding hydrocarbons behind the numerous faults in structures such as occur in the "K" Block, commercial discoveries have not been found as yet. It seems that the available hydrocarbons have been widely distributed among the many fault blocks present.

Hypothesis of Hydrocarbon Distribution

It has been argued that macrostructures and megastructures can be considered as entities with respect to stratigraphy, structural building, and hydrocarbon distribution. The megaunits generally are bounded on the north by very important structure-building faults, and on

---

[1]Gas-to-oil ratio denotes the ratio of the volume of gas-bearing reservoir rock to the volume of oil-bearing reservoir rock and *not* solution gas volume to recoverable oil (GOR).

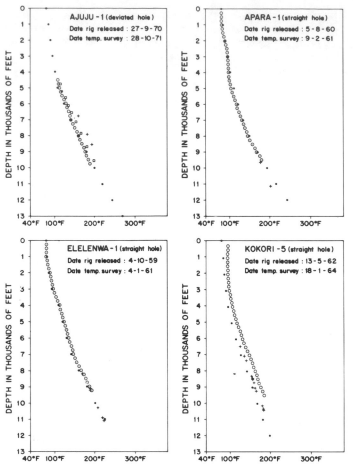
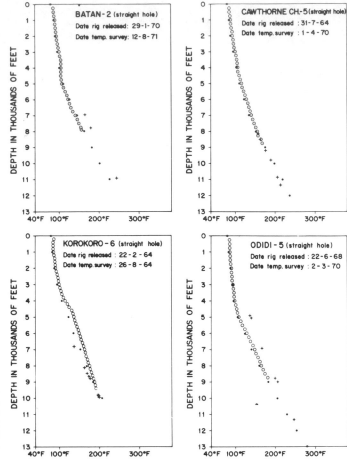

Figure 13. Temperature profiles from selected wells, Niger delta.

the south by prominent counter-regional faults (Figure 9). Therefore, it would be difficult for hydrocarbons generated in one megaunit to migrate updip into the next megaunit. This would explain why each megaunit has its individual hydrocarbon-distribution pattern, being in fact a separate hydrocarbon province.

It was hypothesized initially that the pattern of a southerly increase in the gas-to-oil ratio within a macrounit or megaunit might be explained in terms of the time of trap creation relative to that of oil and gas migration. The older structures, associated with the main boundary faults, might trap the product of early migration—presumed to be predominantly oil—whereas the younger structures, farther south within the macrounit or megaunit might trap progressively more gas when the source rocks subsided into the gas-generating zone. To test this idea, a more precise method of dating the traps was sought.

Time of Trap Creation

The time that the rollover traps became effective can be assessed by determining the period of activity of the associated faults. This is reflected by the expansion in the principal downthrown block.

The good-quality seismic data, which are available for a large part of the delta, provided a means of measuring the amount of this expansion. Dip lines were selected and the

position of the faults, together with several distinctive reflections, were marked on the sections.

The seismic time intervals (converted into thicknesses in feet) between each of the horizons selected were measured, first in a crestal position and then adjacent to the fault. A growth index was derived by dividing the thickness adjacent to the fault by the crestal thickness. The intervals then were tied into wells for stratigraphic control.

In many rollovers, the structure-building fault, as seen on seismic sections, is unique rather than composite, and appears as a discrete slip plane along which movement and resulting growth have occurred throughout the formation of the structure.

More commonly, however, the early growth of a structure associated with the main boundary fault is continued and relayed progressively southward into the younger part of the section by successive crestal faults.

Within the area studied, no obvious relation exists between the time and intensity of growth along a fault and the distribution of oil and gas. Comparable growth patterns are present along faults associated with large fields and those associated with traps containing few or no hydrocarbons.

Expansion toward the faults is, however, common at depths considerably greater than penetrated by wells which are known to have entered thick marine shales. This

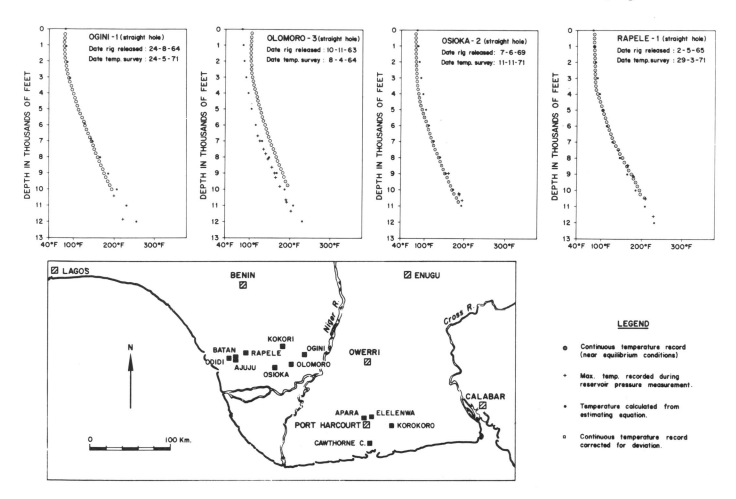

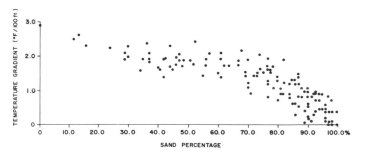

suggests that the basic structural configuration was determined at an extremely early stage—even before deposition of the principal reservoirs.

GENERATION AND MIGRATION

Geotemperatures in Niger Delta

Relation between sand percentage, depth, and temperature gradients—Convection currents, set up in the freely moving groundwaters of the continental sandstones, result in a low temperature gradient in the shallow part of the sedimentary section.

Plots of subsurface temperature against depth and lithology show a distinct relation between temperature gradient and sandstone/shale ratio. Equilibrium temperature data were gathered from continuous temperature surveys conducted at the start of completion operations in wells (with a delta-wide distribution), each well having been closed-in for a period of a few months to 2 years (see Figure 13).

A plot of temperature gradient against sand percentage is shown in Figure 14. The temperature gradient increases with diminishing sand percentage from less than 1.0°F/100 ft (1.84°C/100 m) in the continental sands, to about 1.5°F/100 ft (2.73°C/100 m) in the paralic section, to a maximum of about 3.0°F/100 ft (5.47°C/100 m) in the

continuous shales of the Niger delta.

The equation which best fits the equilibrium data reflects the observed slight increase of the temperature gradient with depth:

$$T = a - bS^2 + cD,$$

where T is the temperature gradient (°F/100 ft), S is the sand percentage of the selected interval, and D is the average depth of the selected interval. The coefficients were determined as follows: $a = 1.811 \pm 0.161$; $b = (1.615 \pm 0.146) \times 10^{-4}$; $c = (7.424 \pm 1.749) \times 10^{-5}$.

Figure 14. Relation between geotemperature gradient and sand percentage, Niger delta.

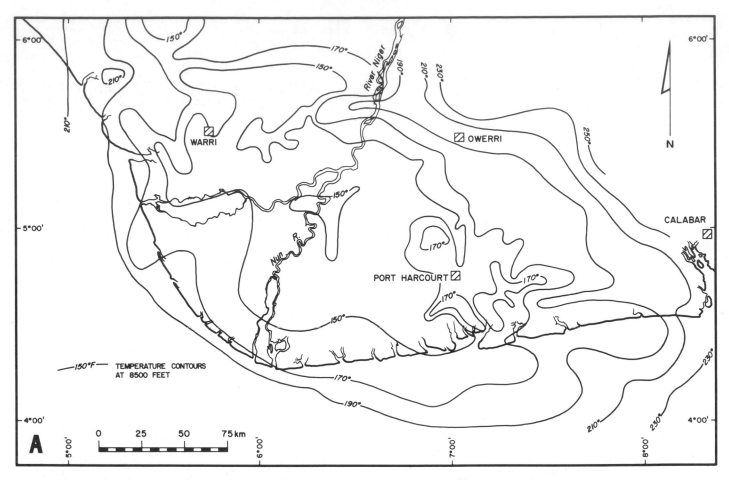

Figure 15. A, Geotemperatures at 8,500 ft (2,591 m). **B**, Subsurface depth to 240°F (116°C) temperature.

The vertical distribution of sand percentage is available on computer tape for most exploration wells and many other wells drilled by Shell-BP and competitors (about 500 wells total). A computer program has been written to calculate and list subsurface temperatures at 100-ft intervals for all wells on file. A surface ambient temperature of 80°F (27°C) has been assumed. Downward extrapolation below a well's total depth (e.g., 9,180 ft) is achieved by employing the average sand percentage over the basal 1,000 ft of known section until the next depth of an integral number of thousands of feet is reached (e.g., 10,000 ft). Thereafter 10 percent is subtracted from the sand percentage, averaged for each 1,000 ft of section, until a figure of 10 percent of less is achieved, below which depth the sand percentage is assumed to remain at 10 percent.

The calculated depth/temperature profiles for the wells used to derive the estimating equation have been plotted beside the original data on Figure 13 to compare input and output. Maximum temperatures recorded during reservoir-pressure measurements also have been plotted where available, and a generally good agreement is evident.

Plotting of temperature data on maps—In addition to listing the geotemperatures, further programs are available to plot geotemperatures, at a desired depth, or depths to a given temperature on maps, according to the coordinates of the wells concerned. The plotted outputs of temperatures at 8,500 ft (2,591 m) and depths to 240°F (115°C) have been contoured—a useful method of screening inconsistencies. The contoured patterns are shown in relation to the features of the Niger delta on Figure 15.

Tracing evolution of oil "kitchen"[2]—The temperatures of 240°F (115°C) and 300°F (150°C) are considered to represent respectively the top of the oil and gas "kitchens" for Tertiary provinces (unpublished Shell research). Maps showing the subsurface distributions of these temperatures have been used to investigate the evolution of the oil "kitchen" of the Niger delta.

The position of the oil and gas "kitchens" may be traced back through geologic time from the present situation by employing the computerized pollen-zone data. The computer program to calculate subsurface temperatures from sand percentages has been extended to allow this calculation to be made with the sedimentary section above a selected pollen subzone stripped away. This procedure is shown schematically on Figure 16. The temperature at the top of a given subzone automatically reverts to the present average ambient temperature of 80°F (27°C) in the program and the temperature profile is recalculated for the sand percentages of the underlying section, extrapolated below total depth, as previously described.

---

[2]The oil- and gas-generating zones are referred to as oil and gas "kitchens."

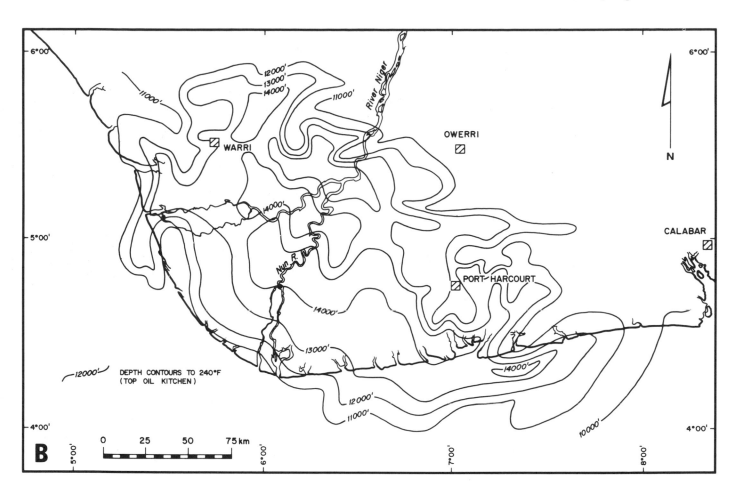

DEPTH CONTOURS TO 240°F
(TOP OIL KITCHEN)

Source-Rocks of Niger Delta

Samples from several wells have been analyzed for their source-rock properties, and were found to be consistently very poor.

Although samples from a wide variety of depositional environments (ranging from fully marine shales, through marine/paralic, to paralic shales) were analyzed, not only was the organic content low, but it was of the humic and mixed types which are purported to be precursors for gas and light oil, respectively. This supports the contention, discussed in the section on hydrocarbon properties, that the primary oils in the delta are indeed light and paraffinic.

It has been assumed that the most effective source rocks are the marine shales and the shales interbedded with the paralic sandstones, particularly in the lower part of the paralic sequence where the shales are at least volumetrically more important. Where the "kitchen" lies well below the top of the continuous shales, any oil generated is considered to have only a remote chance of finding its way into the overlying reservoirs, as the faults at depth within the shales are not considered to provide effective migration paths. It is of importance, therefore, to know the facies at and directly above the "kitchen."

Most of the data on subsurface temperature, and the depth to the top of the continuous shales, are derived from wells on the updip part of each trend, and it is realized that conditions may be different down flank. The top of the

"kitchen" may be expected to rise with respect to the top of the continuous shales as the overlying paralic section becomes more shaly, as this would have the effect of increasing the local temperature gradient. For this reason, the length and the slope of the south flanks of macrostructures or megastructures are important.

Oil and Gas "Kitchens"

The depth to the top of the present-day oil "kitchen" and its position in relation to the continuous shales are shown on Figure 17.

The difference between the eastern and western parts of the delta is striking. Over a large part of the area west of the Niger and Nun Rivers, the top of the oil "kitchen" lies well above the continuous shales, within the paralic and paralic to marine sequences. In the east, however, the "kitchen" generally lies entirely within the continuous marine-shale sequence. This pattern results from the paralic sequence being generally thinner over the eastern part of the delta (see section on Oligocene and earliest Miocene).

Of greater significance than the position of the oil "kitchen" today is its position at various times in the past. A reconstruction therefore was made to show the depth to the oil "kitchen" and its relation to the marine shales at the end of deposition of selected pollen subzones (see Figures 18 to 20).

The contrast between the western and eastern parts of the

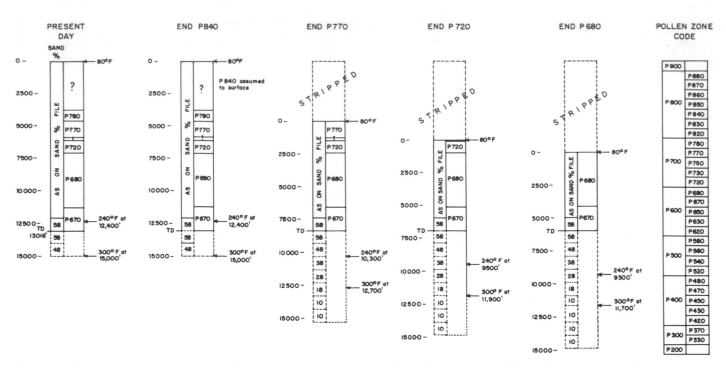

Figure 16. Derivation of paleotemperatures by stripping away successive pollen subzones in the well Bomadi-1. Average sand percentage over basal 1,000 ft of known section is assumed to continue downward until the next depth of an integral number of thousands of feet is reached. Thereafter, reduction of 10 percent is made for each 1,000 ft below total depth.

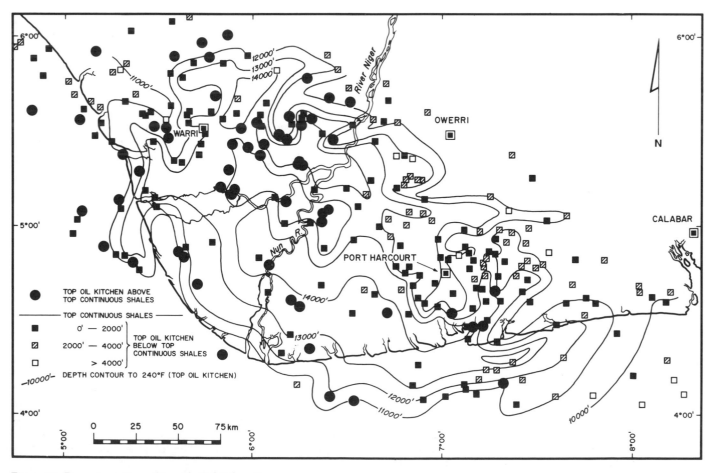

Figure 17. Present position of top of oil "kitchen."

delta is more pronounced where the younger zones are stripped away. The position of the "kitchen" relative to the continuous shales remains more or less constant in the west, whereas in the east the "kitchen" is progressively deeper within the shale section. Thus, in the east, it appears that only in comparatively recent times did the oil "kitchen" reach a level permitting the free expulsion and migration of the oil generated. Over much of the area in the west, however, the top of the continuous shales and even the overlying paralic deposits attained maturity at an earlier stage.

An objection that could be raised against the interpretation of very late migration of oil in the eastern part of the delta is that the true levels of maturity may not be compatible with those interpreted from the temperature data. The geologic setting of the delta suggests that the eastern area might have a higher than average heat flow owing to the proximity of basement and the Cameroon volcanic province. However, the few analyses carried out on samples from the east indicate maturity levels well in line with the delta-wide temperature interpretation.

The major growth faults bounding the depocenters of the selected pollen subzones have been traced on the maps depicting the changing position of the oil "kitchen" (Figures 18 to 20). In the western part of the Niger delta, the top of the oil "kitchen" remains well above the marine shales,

although only several depositional trends (megaunits) north of the active depocenter. During the time when areas were receiving sediments and being deformed, the "kitchen" characteristically lay well below the top of the marine shales.

Thus generation and migration—although a continuous process in the west—must have occurred for any one megaunit at a time when deposition had moved several trends downdip. Indeed, the evidence leads to the conclusion that the present depth of burial had been approached before the onset of migration and accumulation.

Because the structures of the Niger delta are essentially induced by growth faulting, it must be concluded that generation and migration for a particular trend postdated the structural deformation of the entire megaunit concerned.

Hydrocarbon Properties

Niger Delta crudes and their origin—Two basically different crudes are present in the delta, a light crude, which is characteristically paraffinic and waxy with pour points in the range of about 20 to 90°F (-7 to $+32$°C) and a medium crude with specific gravity greater than 0.90 (i.e., less than about 26° API). The latter is dominantly naphthenic, nonwaxy, and has pour points generally lower than -13°F (-25°C).

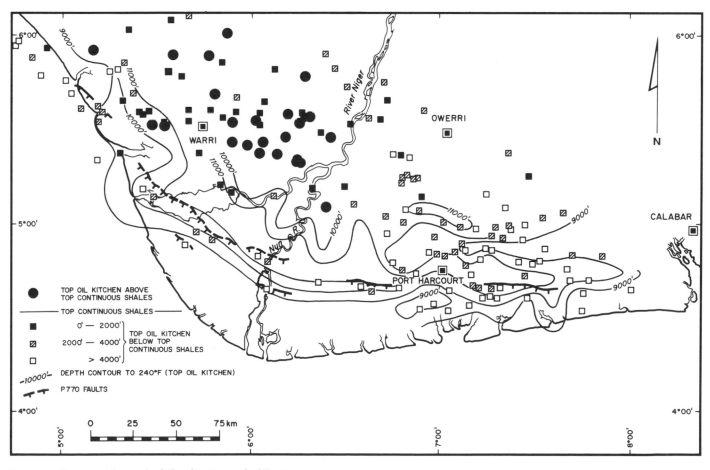

Figure 18. Position of top of oil "kitchen" at end of P770.

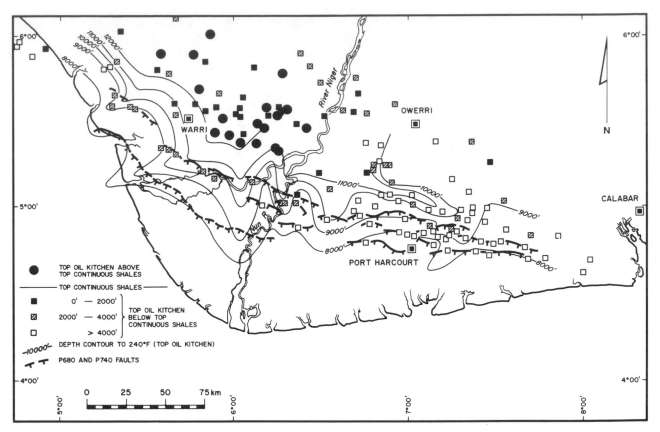

Figure 19. Position of top of oil "kitchen" at end of P720.

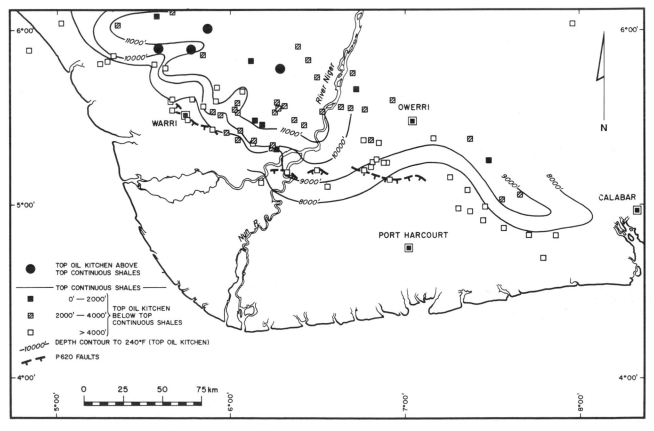

Figure 20. Position of top of oil "kitchen" at end of P620.

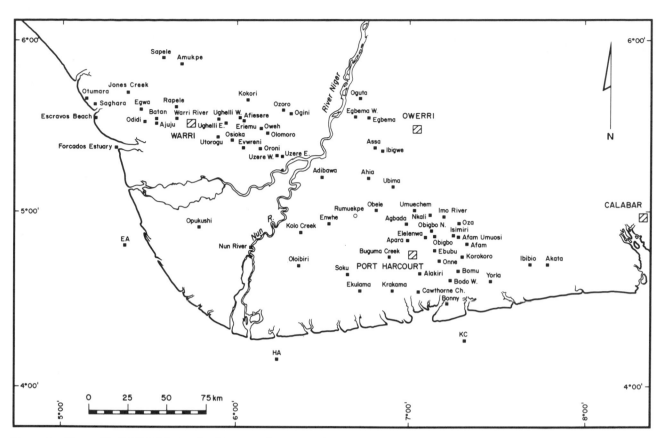

Figure 21. Key to Shell-BP fields involved in maps depicting oil properties.

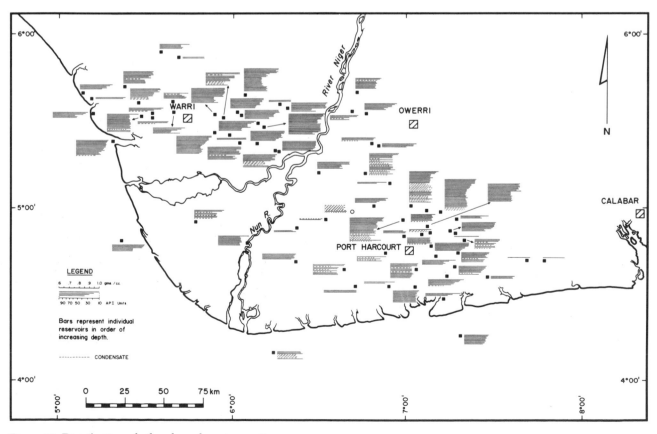

Figure 22. Distribution of oil and condensate gravity.

Hedberg (1968) has proposed, from a worldwide statistical review, that waxy crudes are derived from terrigenous vegetable matter in nonmarine source rocks. C.C.M. Gutjahr and L. Leine (unpublished research) offered a plausible reason for Hedberg's observations. They considered the precursor of paraffin wax in crude oil to be wax in the protective cuticles of leaves of plants, found in tropical to subtropical climates. The waxy cuticles and resins are more resistant to decay than the woody tissues; the latter soon decompose to leave a concentrate of waxy plant material in the then potential source rock. More specifically, K. de Groot, K. Reiman, and J.J. Hartog (unpublished research) have demonstrated that the Nigerian crudes originated from source rocks containing land plant matter.

Figure 21 shows the Shell-BP fields used in depicting the distribution of oil properties. The distribution of crudes according to their tank-oil specific gravity and pour point (Figures 22 and 23) shows the heavy, low pour-point crudes to be consistently above the lighter, high pour-point varieties.

The change from heavy to light oil with depth is particularly abrupt in fields with both types of crude. There is, however, no critical depth for the change applicable to all fields, as can be seen from the plots on Figure 24. The temperature of the transition is more consistent, when considered on a delta-wide basis, and is between 150 and 180°F (60 and 80°C).

The dependence of the quality of crude oil in the Niger delta on temperature rather than depth is demonstrated particularly well by maps (Figures 25 and 26) showing the oil-gravity and pour-point distributions for a particular depth slice (e.g., 8,000 to 9,000 ft or 2,400 to 2,700 m). For such a depth slice, the heavy, low pour-point oils are clustered in the cooler parts of the delta, whereas the lighter oils, with variable pour points above about 20°F (−7°C) are found where subsurface temperatures are relatively higher within the same depth slice.

A simple explanation for the regional distribution of heavy oils, in terms of subsurface temperatures, is that they are the products of bacterial transformation. The bacteria responsible for the transformation, like other forms of life, survive only below a certain temperature. Geochemical investigations of the Niger delta crudes by K. de Groot (unpublished research) confirm that the heavy crudes originate through bacterial transformation. Earlier, Weeks (1958, p. 51 and 52) had noted that certain bacteria rapidly destroy paraffinic compounds. He offered this as a plausible reason for the absence of heavy paraffins in natural seepages.

Although the pattern of oil properties related to temperatures was recognized initially from an investigation of specific gravity and pour point, it not surprisingly applies to other crude oil properties as well. Graphs comparing the

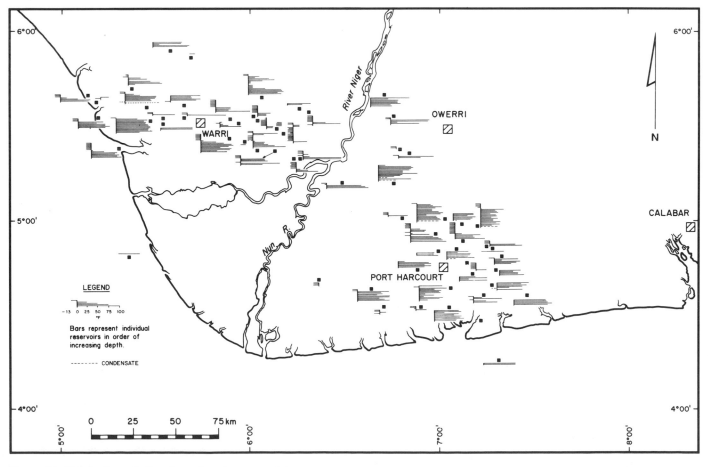

Figure 23. Distribution of oil and condensate pour point.

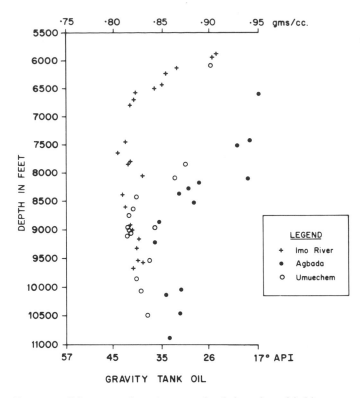

Figure 24. Oil gravity plotted against depth for selected fields.

relations of these properties to depth and temperature are shown on Figure 27. The considerably improved dependence on temperature is particularly noticeable for the plots of oil gravity, the boiling-point fractions of the crude, and the viscosity. The wax content and pour point of the untransformed oils in the higher temperature range are extremely variable, as these presumably depend on the quantity of plant wax in the original source rock, clearly an inconstant parameter. By contrast, the transformed oils characteristically contain less than 5 percent wax and have pour points below 0°F (−18°C) owing to the bacterial destruction of wax.

Late migration deduced from oil properties—The regional distribution of transformed oils in the Niger delta shallower than the 180°F (80°C) isotherm implies that after migration there has been little or no subsidence of such oil-bearing reservoirs, leading, therefore, to further deposition. Had subsidence taken place after earlier migration and bacterial alteration, transformed oils would have been carried deeper, where geotemperatures are in excess of that tolerated by bacteria. This is not the situation in the Niger delta.

Two explanations may be offered for the coincidence of the present interface between transformed and unaltered oils and the 180°F (80°C) isothermal surface. The first explanation is that, for each fault trend, migration occurred after deposition had achieved the present overburden. The alternative solution is that migration took place during

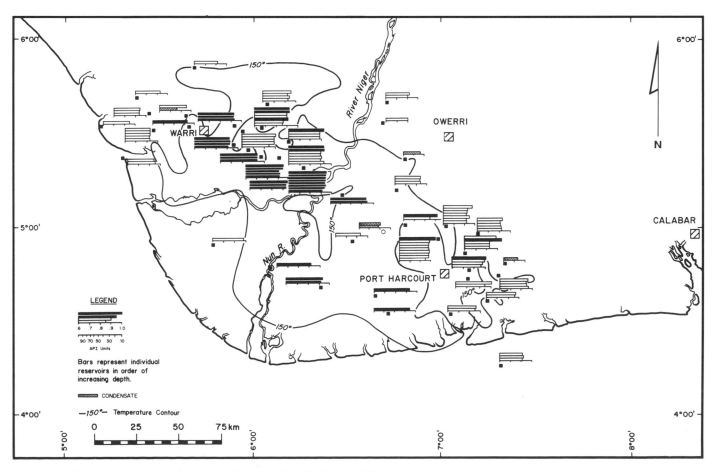

Figure 25. Oil gravity within depth range 8,000 to 9,000 ft (2,400 to 2,700 m).

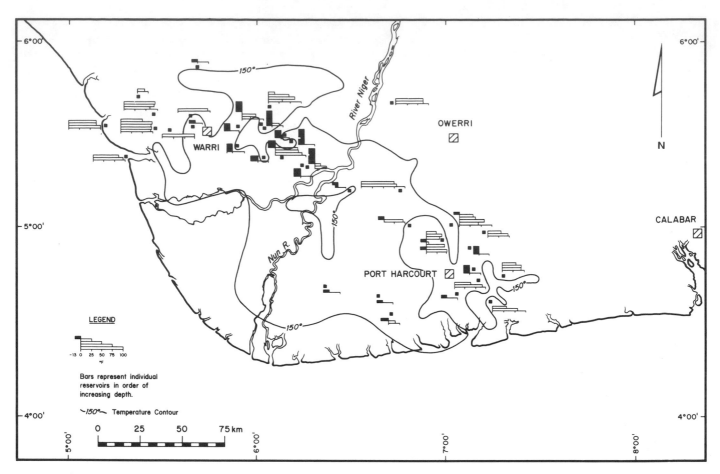

Figure 26. Oil pour point within depth range 8,000 to 9,000 ft (2,400 to 2,700 m).

active subsidence (i.e., in the depocenter phase), but the oil remained at temperatures higher than 180°F (80°C) and thus was protected from bacterial alteration. Subsequently, when deposition had attained the present situation, remigration (e.g., along faults) caused a redistribution of the oil, some finding its way into the shallow low-temperature traps, where it was transformed bacterially.

The first and simpler alternative is preferred because the conclusion reached, advocating late primary migration, is the same as that independently derived from studying the evolution of the oil "kitchen" through time.

Comparison of eastern and western Niger delta—It has been shown that the boundary between the paralic sequence and the continuous marine shales reached the level of oil maturity much later in the east than in the west. Indeed, in the east, even at present the paralic sequence is only locally within the oil-generating zone, implying that the eastern oil was derived, not only late, but also from source rocks mainly in the continuous marine shales. Such source-rocks can be expected to have a more uniform composition than the shales of the heterogeneous paralic and marine to paralic sequences in the maturity zone in the west. A difference in source-rock uniformity, such as theorized, should be reflected in the properties of the resulting oil and also in the proportion of gas to oil generated.

A comparison of the oil properties, considered east and west of the Niger and Nun Rivers, has been made (Figure

28). Below the level of transformation, the gradual diminution of oil gravity into the condensate range relative to temperature (and indirectly to depth) is expressed more regularly in the east than the west. Second, there appear to be greater extremes of pour point (and thus waxiness) of the western crudes than those of the east. A similar pattern is found with respect to the tank-oil viscosity of untransformed crudes. The segregation of hydrocarbons into oil-rich trends and gas belts appears to be more extreme in the west. A greater variety of source rock in the west not only would given credence to the regional distribution of segregated oil and gas in the west, but also would explain the less uniform properties of crude oil from the west.

Late generation and migration from upthrown geopressured shales—R.G. Precious (personal communication, 1973) has postulated that hydrocarbons are generated in the relatively shallow overpressured shales on the upthrown sides of major structure-building and crestal faults. This geopressured section is in contact with the hydropressured section on the downthrown side, and thus sufficient differential pressure is created to enable migration of the hydrocarbons into the fault zone, which then acts as a conduit.

The distribution of hydrocarbons within a megaunit depends on which of the several faults, penetrating a different source-rock province, acted as a conduit during primary migration. The sealing capacity of these and other

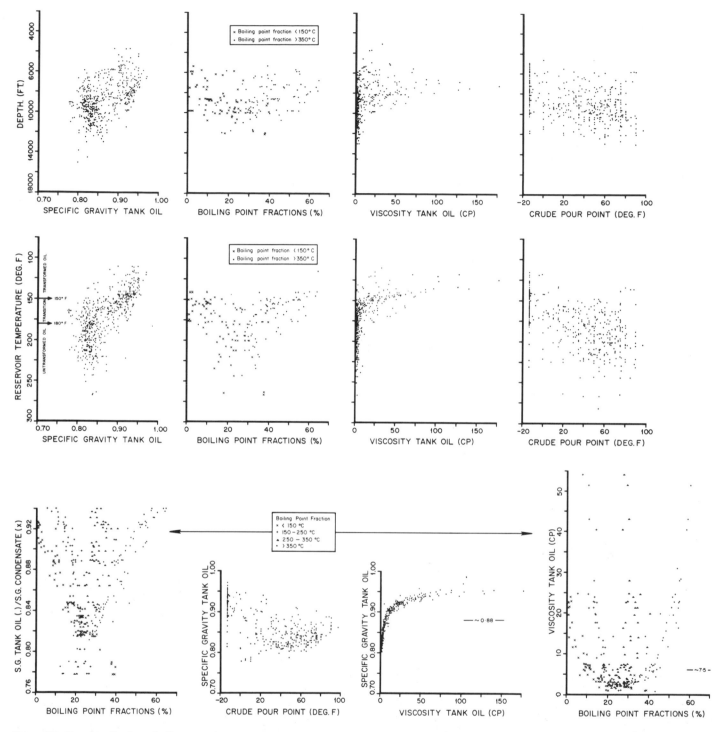

Figure 27. Graphic display of oil properties.

subsidiary faults with respect to oil and gas also could have been a factor in hydrocarbon segregation during secondary remigration.

The general hypothesis explains the barren nature of many structures on the extensive south flanks. Unlike structure-building faults and crestal faults, flank faults do not cause any appreciable difference in the level of the geopressured shales (Figure 9). Their associated structures, furthermore, are too far down flank to be reached by

hydrocarbons migrating along the faults at the apex of the macrostructure.

CONCLUSIONS

1. The Niger delta built out in a series of depocenters, extending over the continental edge onto oceanic basement. Hydrocarbons are concentrated along the updip edge of these depocenters, adjacent to the major growth faults,

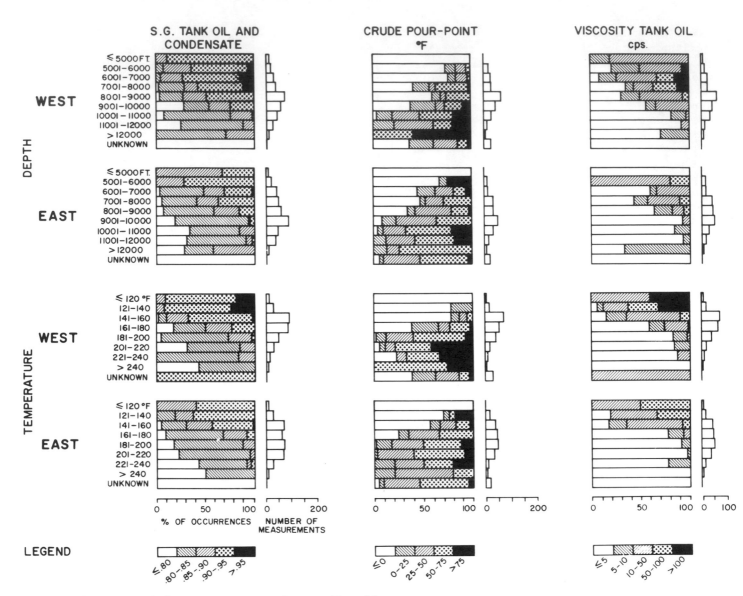

Figure 28. Comparison of oil properties, eastern and western Niger delta.

where viable structural traps predominate. The overall shape of the deltaic wedge, particularly during its initial stages, was controlled by downfaulting in the basement.

2. The delta has been growing from Paleocene time to the present. It consists of a predominantly regressive sequence with massive continental sandstones typically overlying an alternation of paralic sandstones and shales; these in turn grade downward into dominantly marine shale with some turbidite units.

3. The delta is comprised of megaunits which are entities with respect to stratigraphy, structural-building, and hydrocarbon distribution. The development and style of the separate megaunits are related to the balance between the rate of sediment supply and the rate of subsidence.

4. Megaunits are bounded north and south by major breaks in the regional dip. Important counter-regional faults commonly define their southern limits. Megaunits usually are made up of several macrostructural units. Each of the latter is bounded updip by an important structure-building growth fault, against which most of the sedimentary

expansion has taken place. Macrostructures may be simple or complicated by one or several crestal faults, both synthetic and antithetic. Hydrocarbons are trapped typically in the dip-closed crestal areas or against one or more of the faults, especially on their upthrown sides. Southerly shifts of culmination with depth are characteristic.

5. Data available from Shell-BP fields at the end of 1971 showed that, out of 1,400 reservoirs, 70 percent have oil-column heights of 50 ft (15 m) or less and only 5 percent are greater than 150 ft (45 m). Sealing potential is critical for an above-average column height and thus for a sizeable hydrocarbon accumulation.

6. In a given macrostructure, the ratio of volume of gas-bearing rock to volume of oil-bearing rock tends to increase down plunge and in a generally seaward direction. Likewise, the more downdip a macrostructure is within a megastructure the greater is the probability of finding mainly gas.

7. The hydrocarbon source rocks of the Niger delta are

generally poor though volumetrically very important. They are of the humic and mixed types, which yield gas and light oil.

8. The evolution through time of the oil- and gas-generated zones (oil and gas "kitchens") in the delta shows that hydrocarbons probably migrated into the reservoir section after sedimentation had reached the present surface in any one macrotrend, and thus after the structural geometry had been determined. This conclusion is corroborated through other independent evidence.

9. The light waxy indigenous crude of the delta is transformed bacterially to a medium-gravity, low pour-point crude at geotemperatures less than 150 to 180°F (65 to 80°C).

10. It was hypothesized initially that older structures associated with the main growth faults might have preferentially trapped oil as the product of early migration, whereas the younger structures farther south might have received progressively more gas as the source rock subsided from the oil-generating into the gas-generating zone. This hypothesis has been rendered untenable by the conclusion that migration took place after the structural deformation in a given trend. Now it is believed that heterogeneity of source rock, coupled with segregation during migration and remigration, is responsible for the observed hydrocarbon distribution.

ACKNOWLEDGMENTS

The writers are indebted to Shell International Petroleum Company, British Petroleum Company, and Nigerian National Oil Corporation for permission to publish this work.

REFERENCES CITED

Bruce, C.D., 1973, Pressured shale and related sediment deformation; mechanism for development of regional contemporaneous faults: AAPG Bulletin, v. 57, p. 878-886.

Burke, K., 1972, Longshore drift, submarine canyons, and submarine fans in development of Niger delta: AAPG Bulletin, v. 56, p. 1975-1983.

——— , T.F.J. Dessauvagie, and A.J. Whiteman, 1972, Geological history of the Benue Valley and adjacent areas, *in* 1st conference on African geology, Ibadan, 1970, Proceedings: Ibadan, Nigeria, Ibadan University Press, p. 287-305.

Curtis, D.M., 1970, Miocene deltaic sedimentation, Louisiana Gulf Coast, *in* Deltaic sedimentation modern and ancient: Society of Economic Paleontologists and Mineralogists Special Publication 15, p. 293-308.

Emery, K.O., et al, 1975, Continental margin off western Africa—Angola to Sierra Leone: AAPG Bulletin, v. 59, p. 2209-2265.

Frankl, E.J., and E.A. Cordry, 1967, The Niger delta oil province—recent developments onshore and offshore: Mexico City, Proceedings, 7th World Petroleum Congress, v. 1B, p. 195-209.

Hedberg, H.D., 1968, Significance of high-wax oils with respect to genesis of petroleum: AAPG Bulletin, v. 52, p. 736-750.

Hospers, J., 1965, Gravity field and structure of the Niger delta, Nigeria, West Africa: Geological Society of America Bulletin, v. 76, p. 407-422.

——— , 1971, The geology of the Niger delta area, *in* The geology of the East Atlantic continental margin: Great Britain, Institute of Geological Science Report 70/16, p. 121-142.

Mascle, J.R., B.D. Bornhold, and V. Renard, 1973, Diapiric structures off Niger delta: AAPG Bulletin, v. 57, p. 1672-1678.

Merki, P.J., 1972, Structural geology of the Cenozoic Niger delta, *in* 1st conference on African geology, Ibadan, 1970, Proceedings: Ibadan, Nigeria, Ibadan University Press, p. 287-305.

Murat, R.C., 1972, Stratigraphy and paleogeography of the Cretaceous and lower Tertiary in southern Nigeria, *in* 1st conference on African geology, Ibadan, 1970, Proceedings: Ibadan, Nigeria, Ibadan University Press, p. 251-266.

Short, K.C., and A.J. Stäuble, 1967, Outline of geology of Niger delta: AAPG Bulletin, v. 51, p. 761-779.

Stoneley, R., 1966, The Niger delta region in the light of the theory of continental drift: Geological Magazine, v. 105, p. 385-397.

Weber, K.J., 1971, Sedimentological aspects of oil fields in the Niger delta: Geologie en Mijnbouw, v. 50, p. 559-576.

——— , and E.M. Daukoru, in press, Petroleum geology of the Niger delta: Tokyo, Proceedings, 9th World Petroleum Congress, 1975.

Weeks, L.G., 1958, Habitat of oil and some factors that control it, *in* Habitat of oil: AAPG, p. 1-61.

Middle East: Stratigraphic Evolution and Oil Habitat

R.J. Murris
Shell Internationale Petroleum Maatschappij B.V.
The Hague, Netherlands

The post-Hercynian sequence of the Middle East is dominated by carbonate sedimentation on a stable platform flanked on the northeast by the Tethys ocean. Two principal types of depositional systems alternated in time: (1) ramp-type mixed carbonate-clastic units and (2) differentiated carbonate shelves. The first type was deposited during regressive conditions, when clastics were brought into the basin and resulted in "layer-cake" formations. The second type was formed during transgressive periods and is dominated by carbonate cycles separated by lithoclines, time-transgressive submarine lithified surfaces. Differentiation is marked, with starved euxinic basins separated by high-energy margins from carbonate-evaporite platforms.

The tectonic development of the Middle East can be divided into several stages. The first stage, which ended with the Turonian, was characterized by very stable platform conditions. Three types of positive elements were dominant: (1) broad regional paleohighs, (2) horsts and tilted fault blocks trending NNE-SSW, and (3) salt domes. All three influenced deposition through synsedimentary growth. The subsequent stage, from Turonian to Maestrichtian, was one of orogenic activity, with the formation of a foredeep along the Tethys margin and subsequent ophiolite-radiolarite nappe emplacement. From the Late Cretaceous to the Miocene the platform regained its stability, only to lose it again at the close of the Tertiary, when the last Alpine orogenic phase affected the region, creating the Zagros anticlinal traps.

Source rocks were formed in the starved basins during the transgressive periods. Marginal mounds, rudist banks, oolite bars and sheets, and regressive sandstones form the main reservoirs. Supratidal evaporites and regressive shales are the regional seals. The spatial arrangement of these elements and the development of source maturity through time explain the observed distribution of the oil and gas fields.

REGIONAL SETTING AND GENERAL PRINCIPLES

Stratigraphic Frame

The late Carboniferous to Miocene sequence of the Middle East oil province is the world's richest hydrocarbon habitat. The depositional history is dominated by carbonate sedimentation on a very stable, broad platform, bounded on the east by the open Tethys ocean. The carbonates were replaced westward by clastics, which had their source in the uplifted highlands of the Arabo-African continent.

Figure 1 gives a generalized picture of the stratigraphy of the Middle East region and the step-wise evolution of the post-Hercynian sedimentation pattern. From the late Carboniferous to the early Jurassic a very shallow carbonate platform covered large parts of the Middle East region, commonly with evaporitic central depressions and with clastic incursions from the west. From the Middle Jurassic to the Turonian the platform became more differentiated, with intrashelf basins breaking up the shallow carbonate platform. Clastics were still coming from the west, reaching their maximum development during the early to middle Albian.

A major change in the depositional system occurred in the late Turonian to early Senonion, related to the Alpine orogeny affecting the Tethyan realm. A shale-filled open-marine foredeep was formed, with flysch-type sediments coming from the rising orogene in the east. After the late Campanian to early Maestrichtian paroxysm resulting in large-scale overthrusting and ophiolite emplacement, stable conditions returned, leaving a stable carbonate platform in the west and a shaly successor basin in the east. Clastic material was then supplied from the isostatically rising orogenic belt on the eastern flank of the basin.

The depositional basin narrowed considerably during the early to middle Oligocene, probably in connection with a pronounced worldwide drop in sea level (Vail et al, 1977). Late Oligocene to middle Miocene carbonate rocks and

W/SW   **E/NE**

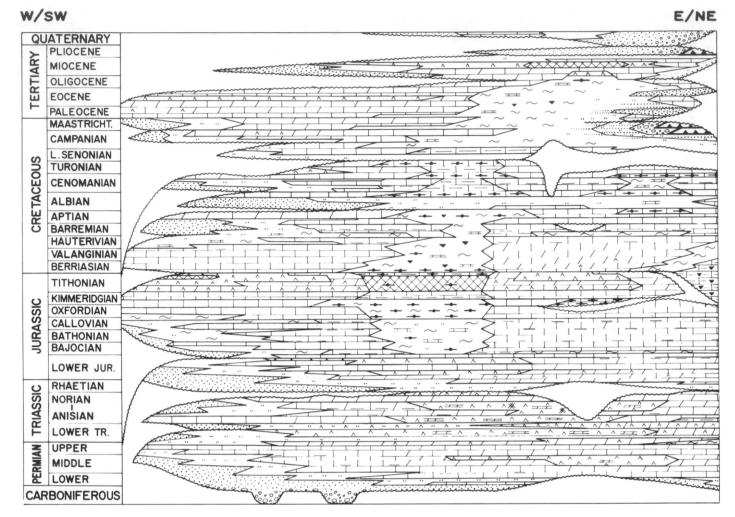

Figure 1. Generalized stratigraphy of Middle East.

evaporites covered the basin, to be replaced by clastic rocks during the younger Miocene and the Pliocene. This clastic supply was connected with the late Alpine phase which created the present Zagros and Palmyra foldbelts, with overthrusting evident along the northern part of the Zagros crush zone.

Basic Patterns of Carbonate Deposition

Two basic types of carbonate depositional systems alternated through time: (1) the so-called carbonate ramp (Ahr, 1973) and (2) the differentiated carbonate shelf. These two types of carbonate deposition are comparable to those distinguished by Bay (1977) in the Lower Cretaceous of the Gulf Coast. This two-fold subdivision into "ramp" versus "differentiated shelf" may well be applicable to other, if not most, shallow-water depositional realms.

Carbonate ramp—The carbonate ramp is characterized by the alternation of more and less clayey units. Each cycle can be correlated over large distances—in the order of hundreds of kilometers—parallel with the depositional strike as well as perpendicular to it. The individual cycles are remarkably constant in thickness and lithology; changes, if any, occur very gradually. The purer carbonate units are typically pelletoidal-bioclastic wacke-packstones

with local ooidal grainstones. The more argillaceous units are rich in pyritized pellets to micropellets and grade laterally into marls and shales. Core investigations show that the cycles, 30 to 100 m thick, can be divided into numerous subcycles, each of which can again be correlated over surprisingly wide areas.

Deposition of the ramp-type formations coincided with periods of increased clastic influx onto the carbonate shelf from the western hinterland. The Middle Jurassic is a good example of this type of deposition with the concomitant excellent correlatability of the individual cycles, as shown in the lower part of Figure 2 (Dhruma Formation).

Differentiated carbonate shelf—The differentiated carbonate shelf conforms more closely to the standard type of carbonate platform described by Wilson (1975, p. 22-27). The formations falling into this category were formed during relatively high stands of sea level, when clastic supply was pushed far back to the west. On the shallower parts of the platform rather pure carbonates were deposited, typically algal-foraminiferal wacke-packstones and ooidal-pelletoidal pack-grainstones. The deeper parts of the inundated platform became sediment starved, with reduced deposition of lime mud and marl, commonly under euxinic conditions.

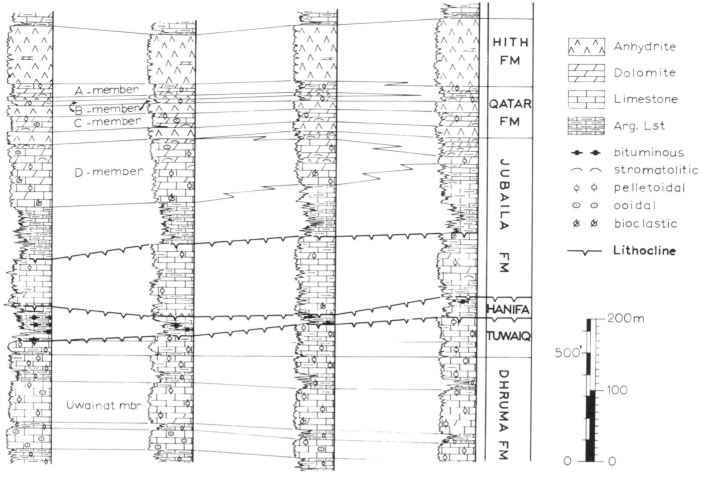

Figure 2. Carbonate ramp (Dhruma Formation) versus differentiated shelf (Tuwaiq-Hith Formations), Qatar area.

Owing to the critical water-depth dependency of carbonate production, the carbonate cycles formed under these conditions are much less constant in thickness and lithology than those of the ramp-type units. Characteristic time-transgressive lithification surfaces (Purser, 1969, 1972) separate the cycles. These were formed by submarine lithification of the slope in front of shelf margins. When the sea level kept rising, these hardgrounds were progressively covered by basinal muds. At the same time, the shallow shelf margin was gradually drowned and subject to lithification and sometimes to slope erosion. Depending on the balance between the rate of carbonate production in the shallow waters and the speed of sea-level rise and slope erosion, the carbonate margin was outbuilding, upbuilding, or inbuilding. When outbuilding, the slope prograded with the formation of numerous minor hardgrounds. Under upbuilding and in particular inbuilding conditions, a single well-defined hardground was formed, which is older in the more basinal part than near the margins. The basinal muds above the lithified surface in the centers of the depressions commonly are demonstrably older than the shallow-water carbonates below the same surface further upslope, which may lead to confusion in regional correlations.

The middle part of Figure 2 shows an example of a differentiated shelf with hardgrounds separating the Tuwaiq, Hanifa, lower Jubaila, and upper Jubaila–Hith cycles (Late Jurassic). Figure 3 is a photomicrograph of the lithocline surface separating the high-energy carbonates at the top of the Tuwaiq cycle (below) from the basinal muds of the Hanifa cycle (above).

Climate has been another important factor controlling the lithofacies development in the Middle East. Since Permian time the area was situated in the tropical to subtropical belts as it migrated from southern latitudes past the equator (Cretaceous) to northern latitudes in conjunction with the northward drift of the Afro-Arabian continent (e.g., reconstruction by Smith and Briden, 1977). Superimposed on this gradual change due to global position were climatic changes of shorter periodicity, reflected in the alternation of more humid and more arid conditions. These were probably related in a very complex way to global mean-temperature variation, continental configuration, presence of mountain ranges, etc.

The interplay of changes in sea level, epeirogenetic crustal movements, and climatic variations controlled to a large extent the depositional patterns on the vast carbonate platforms which characterize the Middle East oil province. The relative role of each of these factors and the way in

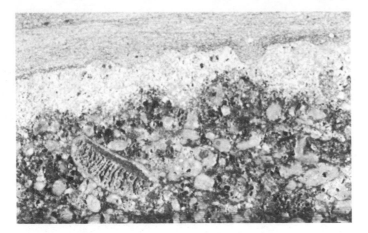

Figure 3. Photomicrograph of lithoclinal surface between bioclastic-pelletoidal grainstone (Tuwaiq) below and lime-mudstone (Hanifa) above, Qatar area. Large object in lower left-hand quarter is about 3 mm long.

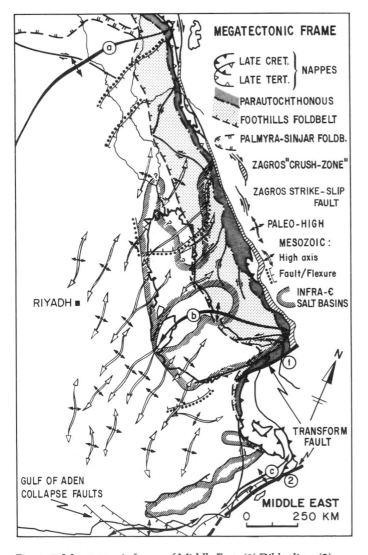

Figure 4. Megatectonic frame of Middle East: (1) Dibba line, (2) Masirah line, (a) Jauf-Gáara-Mosul arch, (b) Qatar-South Fars arch, (c) Huqf-Dhofar arch.

which they interact are not fully understood.

Megatectonic Frame

Figure 4 shows the megatectonic frame of the Middle East as it is today. The area is bounded on the northeast by the Zagros "crush-zone," a major right-lateral strike-slip fault zone of late Tertiary to recent age, which separates the Zagros foldbelt from the tectonically very complex Hamadan-Sirjan zone of interior Iran. In front of this fairly rectilinear fault system are outposts of nappes created during the two orogenic upheavals which affected the region during the late Campanian to early Maestrichtian and late Miocene to Pliocene.

On the south the crush zone terminates against the so-called Dibba line (1 in Figure 4), a NNE-SSW-striking lineament which separates continental basement on its west from the oceanic realm of the Gulf of Oman-Makran on the east. This line acted in all probability as an ocean to continent transform fault during the Mesozoic along the boundary between the Afro-Arabian plate and oceanic Tethys. Southeast of the Dibba line the ophiolite-radiolarite nappe complex of the Oman Mountains was emplaced during the late Campanian to early Maestrichtian (Glennie et al, 1974, p. 393). This complex is bounded on the southeast by another continent to ocean transform fracture zone, the Masirah line (2 in Figure 4), which separates the Arabian continent from the Indian Ocean and which is also of Mesozoic origin.

The stable Arabian platform is dominated by two types of structural elements: (1) broad regional highs such as the Jauf-Ga'ara-Mosul arch in the north (a in Figure 4), the Qatar-South Fars arch in the central Gulf area (b in Figure 4), and the Huqf-Dhofar arch in the south (c in Figure 4); and (2) mostly NNE-SSW-striking anticlinal trends and flexures reflecting deep-seated basement faults. All of these structural elements have influenced sedimentation from the early Mesozoic onward, with periods of increased activity alternating with quiet intervals. They were most active from the late Turonian to the early Campanian, the aforementioned period of major change in basin configuration and the first phase of regional compressive tectonics.

Another major structure former is the Infra-Cambrian salt. Four distinct salt basins can be distinguished, from south to north: (1) South Oman-Dhofar, (2) Fahud, (3) the Southern Gulf, and (4) the Northern Gulf. The limits of the first three basins are fairly well known, but the northern extension of the Northern Gulf basin is highly conjectural. The edge of the salt, as determined from outcrops and/or from geophysical records, may be depositional, that is, by a lateral change into a carbonate-anhydrite platform, or owing to later erosion or solution. The Qatar arch and the Central Oman platform between the Fahud and South Oman salt areas are probably examples of intervening carbonate platforms.

The four salt basins each have their own style of halokinesis probably reflecting differences in original salt thickness and in overburden thickness and lithology. The South Oman-Dhofar basin shows all the classical elements: piercements, domes, pillows, turtlebacks, and rim synclines of different ages. Its southern part is a half-graben bounded on the northwest by the Ghudun-Khasfah fault, a northeast-to-southwest-striking fault zone of early Paleozoic age. The eastern edge of the basin is erosional because of the repeated uplift along the flank of the Huqf-Dhofar high. The small Fahud salt basin contains only a few halokinetic structures, modified in its eastern part of regional tectonics connected with the Oman Mountains.

The large Southern Gulf salt basin is characterized by numerous piercements which are surrounded by Cretaceous to late Tertiary rim synclines. The diapirs are separated by very gentle pillows with flanks generally dipping only a few degrees, many of which contain oil accumulations. No undisputed turtleback structures have been reported. If present, such structures must occur below depths of 4,000 to 5,000 m, where seismic resolution is still insufficient. The Northern Gulf salt basin contains very large, elongate pillows and salt swells, such as Dukhan, Bahrain, and Rumailah and only a few piercements. Some of the largest oil accumulations such as Burgan occur in combined halokinetic-basement horst-block structures.

The Zagros foldbelt, with its famous giant "whale-back"

structures, extends between the undisturbed Arabian platform and the nappes or crush zone. Recent deep drilling and seismic work have shown that the rather simple surface folds become much more complex at depth, where the acute space problem is resolved by repeated reverse faulting and even overthrusting. The Infra-Cambrian salt and the principal shale formations such as the Jurassic Sargelu and the Cretaceous Garau and Khazhdumi provide the necessary detachment zones. The shortening of the sedimentary cover is considerable, increasing eastward. In the High Zagros, designated as parautochthonous in Figure 4, the shortening may well be two to three-fold or more. Unfortunately data are lacking for a reliable palinspastic reconstruction, which therefore has not been attempted when constructing the layer maps (Figures 6-23).

The Zagros fold chains are offset by several strike-slip faults of regional significance. These commonly appear to follow preexisting Mesozoic faults which acted during the compressive phases as boundaries between the differentially folded segments. Near the High Zagros, the faults seem to have been rotated dependent on their angle of incidence

DEPOSITIONAL ENVIRONMENT	LITHOLOGY		SOURCE	RE-SER-VOIR	SEAL
	MAIN	OTHER			
Humid Climate:					
Alluvial - U. Coastal Plain	Sandstone	Silt/Claystone Lignite	(G)	X	
Lower Coastal Plain	Sandstone Siltst/Claystone	Lignite Carbonate	(G)	X	X
Clastic	Shale	Silt/Sandstone Limestone	(X)	(X)	☒
Mixed — Shallow Shelf	Shale/Marl Limestone	Silt/Limestone	(X)	(X)	
Carbonate	Limestone	Shale/Marl Dolomitic Lst		X	
Clastic	Shale/Marl	Siltstone Bituminous Shale	(X)		X
Mixed — Deeper Shelf - Intrashelf Basin	Shale/Marl Arg. Limestone	Bituminous Lst Siltstone	X		(X)
Carbonate	Arg. Limestone Bituminous Lst	Marl/Shale	☒		(X)
Shallow Carbonate Shelf prograding over Basin	Limestone Arg. Limestone	Marl/Shale Bit. Limestone	X	X	
Arid Climate:					
Alluvial - U. Coastal Plain	Sandstone Red-gn Claystone	Siltstone Anhydrite		X	(X)
L. Coastal Plain — Clastic Platform	Red-gn Claystone Siltstone	Anhydrite/Dol. Sandstone	(X)		X
Mixed	Anhydrite/Dolomite Red-gn Claystone	Siltstone Sandstone	(X)		X
Carbonate — Shallow Platform	Dolomite Dolomitic Limest.	Anhydrite Shale/Marl		X	(X)
Evaporite (Sebkha)	Anhydrite Shale/Marl	Dolomite Rocksalt	(X)	(X)	☒
Evaporite Basin	Anhydrite Rocksalt	Shale (bitum.) Arg. Limestone	X		☒
Deep Open Marine:					
Little/No Clastic Influx	Arg. Limestone Siliceous Lst	Marl/Shale	(X)		(X)
With Clastic Influx	Marl/Shale	Arg. Limestone Siltstone			X
Synorogenic Basin:					
Distal	Shale Siltstone	Sandstone Limestone	(X)		X
Proximal	Breccia/Conglom. Claystone	Olistostromes			

☒ = High quality regional source/seal   (G) = Gas source only

Figure 5. Key to layer maps, Figures 6 through 23. X, Well developed, mostly present. (X), doubtful, scattered. Square indicates high quality (source/seal only).

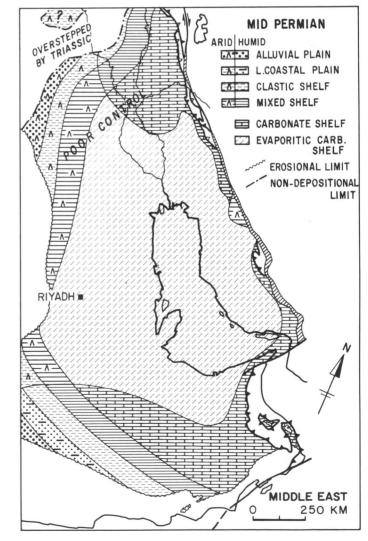

Figure 6. Mid-Permian environments of deposition.

with the axis of maximum compression which was roughly northeast-southwest. At their southwestern end the faults tend to splay off into reverse faults in front of the anticlinal folds, producing a horsetail pattern.

Not all strike-slip faults are confined to the Zagros belt. On the Arabian platform, and in particular on its northern promontory in Syria, northwest Iraq, and southeast Turkey, dextral and sinistral strike-slip faults and fault zones were formed during the Late Cretaceous orogenic period. These fault zones are commonly accompanied by narrow grabens, with a thick Upper Cretaceous sedimentary fill, which were inverted during the late Tertiary compressive phase, giving rise to en echelon foldbelts.

In the southernmost Arabian Peninsula a fault system is related to the collapse of the Gulf of Aden. Though initiated in Jurassic times, this fault system only became fully active in its present form during the early Tertiary. From the late Oligocene onward major rifting occurred in the Gulf of Aden.

Depositional Environment and Oil Habitat

The depositional patterns resulting from the interplay of sea-level changes, epeirogenetic movements plus rejuvenation of relief, climatic variations, and synsedimentary structural growth have a direct bearing on the regional distribution of habitat-controlling parameters such as source rock, reservoir, and seal, and hence are of great interest to the petroleum geologist. Figure 5 serves as a legend to the layer maps of Figures 6 through 23. It lists the depositional environments present in the Middle East, with the lithologies typical of the various environments. Also indicated is the translation of these lithologic units into habitat terms such as source, reservoir, or seal potential.

Periods of low sea-level stand plus increased clastic influx, when ramp carbonates were dominant, produced good reservoirs in the form of coastal and alluvial sands, whereas coastal and platform shales formed effective regional seals. Within the realm of carbonate deposition moderately good reservoirs were laid down during these periods in the form of pelletoidal-bioclastic packstones and, less commonly, ooidal-pelletoidal grainstones. Seal development within the carbonate realm was modest and resulted in marls and argillaceous limestones with limited sealing capacity, especially in the less gentle structures.

During high sea-level stands and consequently suppressed clastic influx, excellent source beds commonly were formed in the euxinic intrashelf basins, whereas the algal boundstones and ooidal-bioclastic pack-grainstones which commonly cap the cycles form good reservoirs. Under evaporitic conditions, excellent and widespread anhydrite-salt seals were formed, both on the shallow carbonate platforms and in the intrashelf basins. In the latter, however, though the seals overlie or are interbedded with euxinic source rocks, there is little reservoir to seal.

The open marine offshelf realm is not a very favorable oil habitat. Reservoirs and source rocks are few or nonexistent, and the calcareous shales and marls are only moderately effective seals, especially when fractured. The synorogenic and postorogenic clastic sediments of the Zagros contain some reservoirs and seals, but they are restricted to the most disturbed part of the foldbelt and are of limited interest.

BASIN EVOLUTION

Pre-Late Carboniferous

The sequence deposited prior to the Hercynian unconformity will be dealt with briefly here, for data are limited. The sequence does not form part of the rich Middle East oil habitat. Commercial oil fields have been found only on the fringes of the basin (e.g., southern Oman-Dhofar). Elsewhere, the pre-Hercynian sequence is buried below thick upper Paleozoic, Mesozoic, and Tertiary sedimentary rocks, offering therefore only speculative deep gas prospects.

The first unmetamorphosed sedimentary rocks covering the Arabian shield are the carbonates, clastics, and evaporites of the Huqf Group. The salt of this depositional megacycle causes halokinetic structures in parts of the basin (Figure 4). The age of the salt is somewhat problematic. Well and outcrop evidence suggest a late Precambrian to early Cambrian age, as the salt is separated from dated Middle Cambrian carbonate rocks by a thick red-bed sequence.

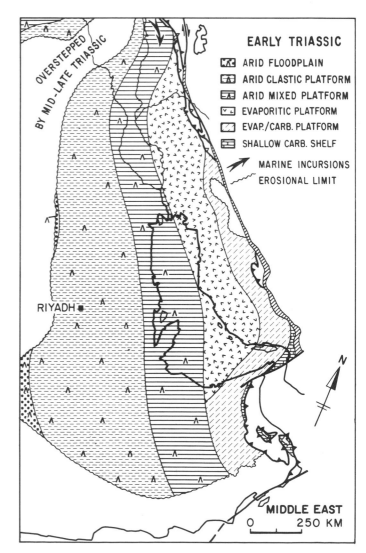

Figure 7. Early Triassic environments of deposition.

Carbonate-evaporite sedimentation of this "Infra-Cambrian" is followed by a predominantly clastic regime during the early Paleozoic. Vast aprons of terrestrial to very shallow-marine sandstones with subordinate shales and silts were laid down over the entire shield area (Helal, 1965). Carbonate rocks are rare and restricted to the middle-late Cambrian, early-middle Devonian, and early-middle Carboniferous. Several regional unconformities related to epeirogenetic movements affect the sequence of which the most important are pre-late Cambrian, pre-Devonian, and early Carboniferous. In general the formations become more sandy toward the south, and the unconformities become more pronounced. As a result, the lower Paleozoic along the southern rim of the basin is composed almost exclusively of sandstone and conglomerate formations, separated by unconformities representing wide time gaps.

In late Carboniferous time the Middle East craton was affected by the main phase of Hercynian movements, which is reflected in upwarp and concomitant erosion, locally cutting as deep as the Cambrian or even the Precambrian.

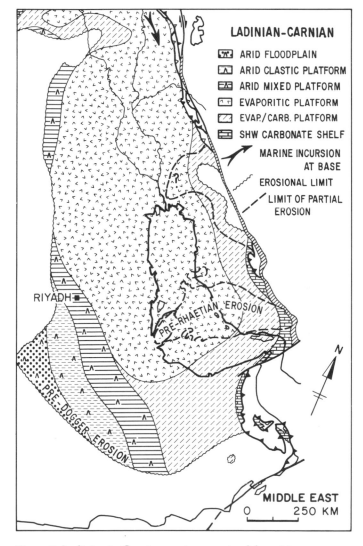

Figure 8. Ladinian to Carnian environments of deposition.

Late Carboniferous to Liassic

The first sedimentary rocks laid down after the Hercynian epeirogeny are clastics of glacial origin, dated as late Carboniferous to earliest Permian by palynology. Hudson and Sudbury (1959) were the first to draw attention to the glacial origin of this formation and Helal (1964) described similar strata in southern Saudi Arabia. The glacial origin of these beds has been corroborated by subsurface data from southern Oman and Dhofar, where they contain oil accumulations in the Marmul area now under development. The glacial nature of these deposits is firmly established in the south only. In Oman the glacial influence can be seen to decrease northward, and beds of similar age in the Iranian Zagros and in eastern Turkey are nonglacial coastal deposits (Szabo and Kheradpir, 1978). The late Carboniferous to early Permian glacial beds of the southern part of the Arabian shield belong to the Dwyka glaciation of the Gondwana continent.

During the Permian the climate became gradually warmer and more arid. By mid-Permian time a carbonate platform was established over most of the region (Figure 6). Evaporites are present along a central belt, whereas clastic material was mainly provided by the western hinterland, with local supplies from the east in the High Zagros. The latter probably represent local sources in the uplifted Hercynian ranges along the margin of Paleotethys.

In the early Triassic (Figure 7) hot arid conditions prevailed over the whole basin. Increased clastic influx from the western hinterland is evident, restricting the carbonate-evaporite platform to a rather narrow northwest-southeast-trending belt. There is evidence in the Oman Mountains that the margin with the open Tethys ocean was somewhere in the eastern part of the present mountain range (Glennie et al, 1974, p. 355-386). The evaporites deposited during this period form, together with younger Triassic evaporites, an effective regional seal to major gas accumulations in the underlying Permian carbonate-rock reservoirs.

Figure 8 shows the situation as it had evolved by Ladinian to Carnian times. During the Anisian a major inundation of the platform had occurred, and clastic influx from the west was again much subdued. Most of the basin was occupied by an evaporitic platform with very little differentiation. In the northern tip of the basin more open-marine incursions suggest a connection with Tethys.

Near the close of the Triassic there was a marked change. The climate became less arid, and there was apparently a relative drop in sea level, caused either by a eustatic lowering of the sea or a rise of the shield. That the craton was tectonically active is shown by the activation of regional highs such as the Qatar arch, as witnessed by some erosion of the Triassic (Figure 8; Szabo and Kheradpir, 1978). The Rhaetian layer map (Figure 9) shows the effects of this late Triassic change with restriction of arid conditions to the northern part of the platform, and nondeposition across the Qatar arch. The fluviatile and coastal sandstones belonging to this period have good reservoir properties, but are so far unproductive, probably owing to lack of access to adequate hydrocarbon charge.

During the subsequent early Jurassic, carbonate deposition became gradually more widespread, and by late

Liassic time (Figure 10) there was again a vast carbonate-evaporite platform which, however, still received substantial clastic influx from the west. Evaporitic conditions were limited to the northwestern half of the carbonate platform.

Middle Jurassic to Turonian

Near the end of the early Jurassic a major change in the regional depositional pattern occurred. The climate became more humid, so that the hitherto ubiquitous evaporites became rare, while at the same time a more pronounced tectonic differentiation of the platform led to the creation of an intrashelf depression, the Lurestan basin, in the north.

Figure 11 shows the depositional pattern during the Bathonian—a wide carbonate platform with an intrashelf basin in its northwestern part. Clastic material was supplied from the west and also from the south, where the Gulf of Aden area was an active positive element. The carbonates deposited on the platform during this period are good examples of the ramp model: wide sheets of pelletoidal-ooidal pack-grainstones cyclically alternating

with argillaceous pelletoidal-bioclastic mud-wackestones (Figure 2, Dhruma Formation). The higher energy, cleaner carbonate rocks form oil and gas reservoirs in the central part of the basin.

Sometime during the late Callovian a major inundation of the Middle East platform was initiated, which continued until late Tithonian-earliest Berriasian time. This event had a profound impact on the paleogeography of the area and is responsible for much of the oil found in the central part of the province.

The environments of deposition by late Oxfordian-early Kimmeridgian times are depicted in Figure 12. Clastics were absent and the Lurestan basin had been extended southward by flooding of the platform. For the first time the continental margin between platform and open ocean is visible. East of this margin, little or no deposition occurred, and a submarine lithification surface was formed, covered by a thin veneer of pelagic crinoid debris. Under the influence of the steadily rising sea level this margin was gradually eroded back so that, westward, increasingly younger pelagic deposits rest on increasingly younger shelf

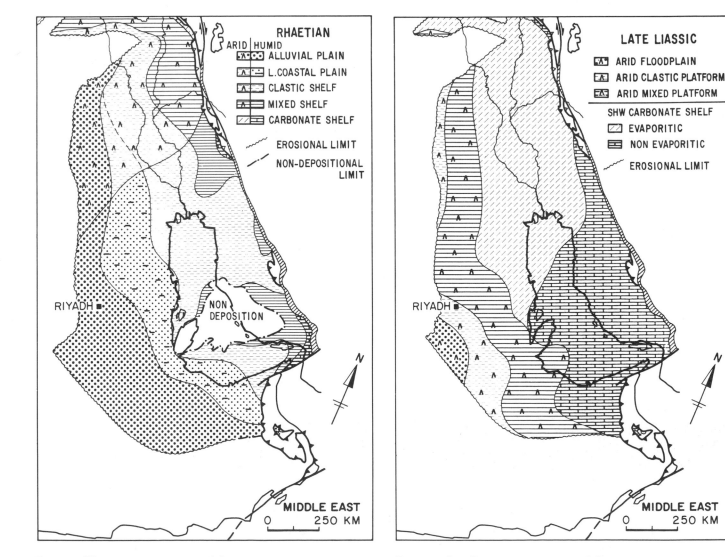

Figure 9. Rhaetian environments of deposition.   **Figure 10.** Late Liassic environments of deposition.

carbonate rocks: a perfect example of a lithocline. As shown by Figure 12, the top of the marginal complex was, near the end of the cycle, subjected to nondeposition and erosion in a high-energy environment, probably in conjunction with a temporary minor drop in relative sea level. Maximum erosion occurred where the Qatar-South Fars arch intersects the marginal complex.

On the flooded platform and in the permanent Lurestan basin with euxinic conditions, laminated bituminous lime muds and marls were deposited, which later became the prolific source of the oil found in the late Jurassic Arab reservoirs. Though the center of the Lurestan basin was by this time several hundreds of meters deep, the starved intrashelf basin on the flooded platform had probably a depth of only a few tens of meters, as reconstructed from the sedimentologic study of cores and from regional isopach and facies data. On the correlation diagram of the Qatar area (Figure 2) this cycle is represented by the Hanifa Formation, here primarily developed in its basinal facies.

During the remainder of the late Jurassic, sea level appears to have risen steadily, as evidenced by the continued recession of the continental margin. On the platform, however, carbonate deposition kept pace with and finally superseded the flooding, reestablishing very shallow depositional conditions over the southern part of the Middle East. In Tithonian time (Figure 13) the climate again became arid, so that extensive evaporites were deposited on the very shallow southern platform in a sabkha environment, in the Lurestan basin as basinal salt and laminated anhydrite and shale (Gotnia Formation). Figure 2 indicates the spatial relation of the Hanifa source, Arab A to D reservoirs, and the sealing anhydrites of the Qatar and Hith formations, the result of the sequence of events sketched previously. It is this source-reservoir-seal triplet which constitutes the world's richest single oil habitat, containing well over 100×10^9 bbl of recoverable reserves and possibly even as much as several times that amount.

During the early Cretaceous there was a gradual return to a more humid climate, and evaporites disappeared from the record. Relative sea level dropped, and ramp-type

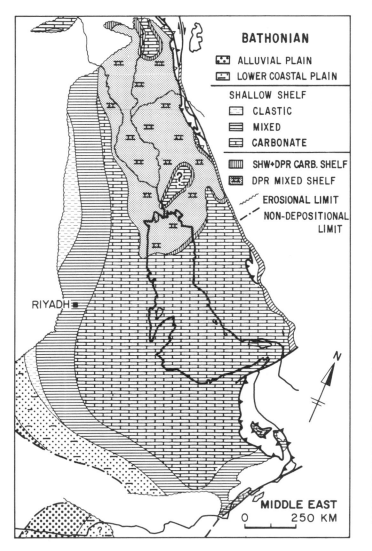

Figure 11. Bathonian environments of deposition.

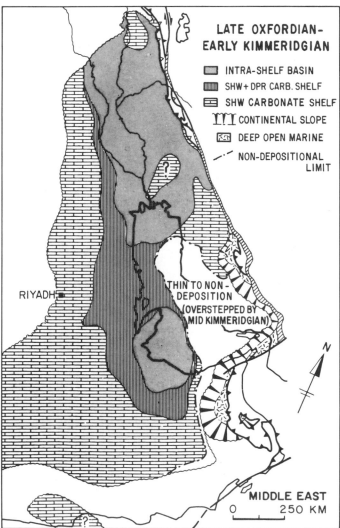

Figure 12. Late Oxfordian to early Kimmeridgian environments of deposition.

carbonate deposition replaced the differentiated shelf. The early-mid-Valanginian layer map (Figure 14) shows the wide carbonate platform which existed at the time, building out into the Tethys ocean in the east, where the prograding carbonate front is preceded by tintinnid-bearing slope and basinal marls. In the south the Gulf of Aden swell is well expressed as a broad arch with nondeposition and erosion. Clastic influx is still limited, and restricted to the far southwestern corner.

Through the Hauterivian and Barremian the trend toward increased clastic influx continued. By mid-late Barremian time (Figure 15) the clastic regime occupied the western half of the basin pushing the carbonate realm, which has a model ramp-type development, in front of it toward the east. The Lurestan basin was much reduced through progradation of the shelf, whereas the margin with the open Tethys was again located beyond today's mountain front. The Gulf of Aden swell was still in evidence, but had apparently been subsiding since the earlier Cretaceous, whereas the Qatar arch was now the locus of marked thinning, nondeposition, and even pre-Aptian erosion.

The extensive sheets of pelletoidal-bioclastic pack-grainstones which form the higher energy parts of the cycles contain oil and gas accumulations in the southern part of the Gulf, where they are sealed by the interbedded marls and argillaceous limestones. Farther north, the coastal plain sandstones of the Zubair Formation contain major oil reserves in several fields. Hydrocarbon charge has been provided by the euxinic late Jurassic to early Cretaceous deposits of the Lurestan basin, whereas interbedded coastal shales form the seals.

The Aptian was a period of renewed basin-wide inundation, and by mid-late Aptian a shallow carbonate shelf had spread westward over the previously deposited clastic rocks (Figure 16). The Lurestan basin again expanded and a new subbasin was formed in the Khuzestan area of Iran, providing the source for at least part of the oil in the giant Iranian fields. In the southern Gulf another intrashelf basin is evident, occupying about the same location and area as the late Oxfordian-early Kimmeridgian basin (compare Figures 12 and 16). On the fringing shelf margins rudist "reefs" grew where conditions were favorable, creating the prolific Shuaiba reservoirs (see also

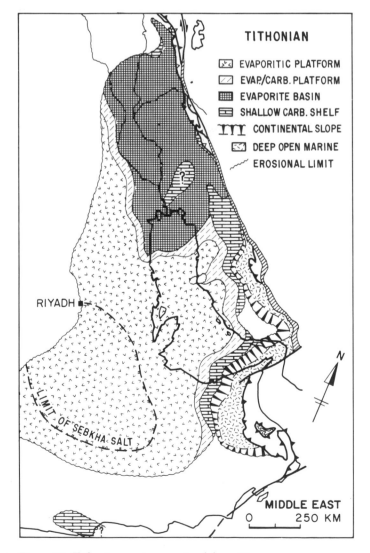

Figure 13. Tithonian environments of deposition.

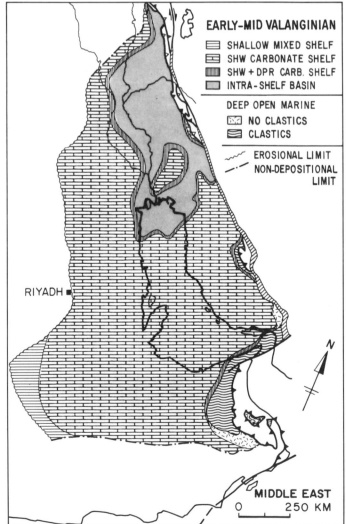

Figure 14. Early to mid-Valaginian environments of deposition.

Wilson, 1975, p. 342-347.).

The Aptian rise in sea level, though very marked in its effect on the paleogeography, was of lesser magnitude and duration than the late Jurassic sea-level rise, with no evidence of an encroaching continental margin such as that on Figures 12 and 13. The climate was also more humid than during the late Jurassic, though evaporites were again deposited on the northern end of the platform.

The temporary flooding during the Aptian was followed by the most pronounced regression since the Rhaetian. As shown on Figure 17, by mid-Aptian time the clastic regime had spread across the whole platform except for a narrow belt in the northeast. The Lurestan basin was now at its smallest, and separated by a shallow mixed-carbonate shelf from the Khuzestan subbasin (K in Figure 17) which was filled by shales deposited in front of the Burgan delta lobe. The Ga'ara-Mosul, Qatar, and Gulf of Aden highs were areas of little or no deposition. The coastal and alluvial sandstones of this interval (the Burgan Formation) are very rich reservoirs in the northern Gulf area. They are sealed by interbedded shales, and have access to charge from the early Cretaceous euxinic deposits of the Lurestan basin.

Figures 18 to 20, late Albian, early Cenomanian, and late Cenomanian, demonstrate how during the remainder of this megacycle which lasted until the late Turonian, the carbonate and clastic realms waxed and waned. Post-Turonian erosion removed part of the section not only over the regional paleohighs and the NNE-SSW-striking axes, but also along the Zagros crush zone, where slope erosion may have occurred along the continental margin fronting the encroaching ocean. The intrashelf basins are typically filled with oligosteginal (calcispheres) marls and radiolarian lime mudstones (in the deeper parts), which are potential source rocks. On the shallow shelf, foraminiferal-algal wacke-packstones and rudist pack-grainstones were deposited and these now form the reservoirs.

Early Senonian to Recent

Near the end of the Turonian a major change occurred with the onset of the Alpine orogeny. The older Mesozoic structural elements were strongly reactivated, with concomitant erosion and nondeposition during the early part of the new megacycle. A foredeep was formed along

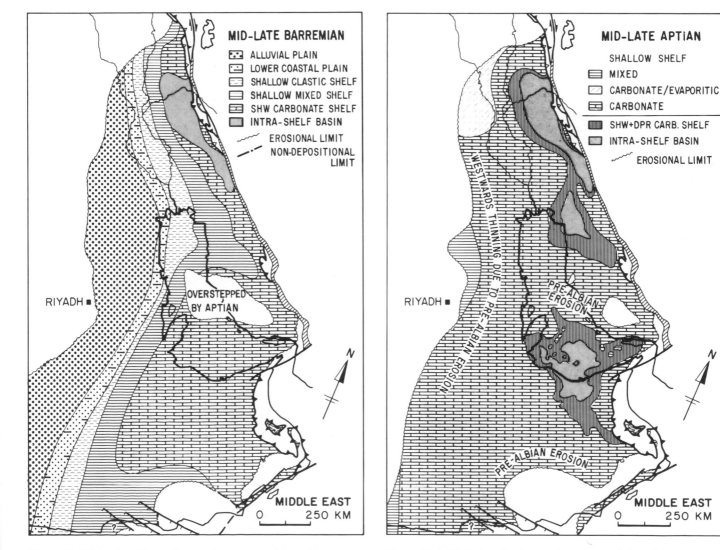

Figure 15. Mid-to late Barremian environments of deposition.

Figure 16. Mid- to late Aptian environments of deposition.

the rising orogene, trending northwest-southeast, the now dominant direction. Figure 21 shows the situation in late Campanian time. Along the orogenic front in the northeast, nappes were emplaced from northeast to southwest. Where they originated from the open Tethys ocean, they are composed of radiolarites and ophiolites—the Oman Mountain complex (Glennie et al, 1974) and the southern Zagros salient in the Neyriz area (Ricou, 1974). In the northern Zagros the late Cretaceous nappes are made up of radiolarites and shelf carbonate rocks of Jurassic age, without genuine ophiolites (Berthier et al, 1974). The lithology, facies relations, and palinspastic reconstructions suggest that these northern nappes represent the old marginal realm of the Arabian plate, underlain by continental crust. There is no evidence for true oceanic conditions such as in the Oman Mountains.

As shown in Figure 21, the nappes are accompanied by a fringe of synorogenic deposits in the form of wildflysch-type boulder clays, conglomerates, and olistostromal masses near the nappe front, and as regular shale-silt-sandstone flysch away from it. As usual, the nappes tended to overrun their own preceding debris, and along the thrust front the complications are manifold and not always resolvable. Seismic data and sedimentologic studies point to an eastern source of the synorogenic deposits, the rising orogene. Between the orogenic front and the shelf in the west was an open-marine basin, in which marls and calcareous shales with globotruncanids and other planktonic forms were deposited. From the shelf in the west shallow-water carbonates prograded over the slope shales into this marine basin.

The shelf had become much more varied. In the south a major source of clastics was present, in the north was a carbonate platform with a central evaporitic pan. In the far north, narrow restricted grabens were formed in conjunction with ENE-WSW-trending wrench-fault systems. The synsedimentary fill is mostly marl and shale with common bituminous limestone, phosphate, and chert beds. At the northeastern end of the plunging Mosul arch in northwest Iraq, near the edge of the area mapped, the presence of Campanian shallow-shelf carbonate rocks with intraformational conglomeratic and brecciated zones was

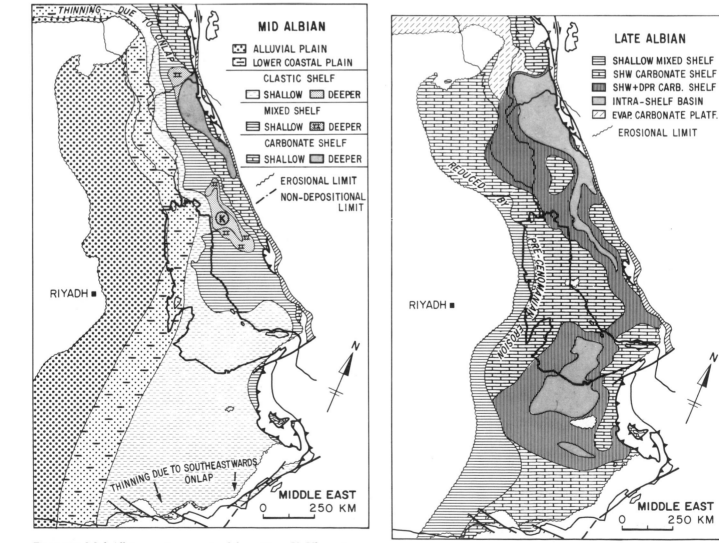

Figure 17. Mid-Albian environments of deposition: K, Khuzestan subbasin.

Figure 18. Late Albian environments of deposition.

reported from the Kurdistan mountain front area (Hadiena and Bekhme limestone) by Dunnington et al (1959). These formations appear to be recemented carbonate slope breccias, later taken up in the thrust front, and as such are indicated on Figure 21 as proximal synorogenic deposits.

By late Maestrichtian time, orogenic upheaval had abated and more quiet depositional conditions returned. The Oman Mountains, loaded with heavy ultrabasic rocks, rose little and were the site of late Maestrichtian shallow-water carbonate deposition. Most of these conditions existed also in the southwestern end of the Zagros thrust belt, but in the north, where ultrabasic rocks are absent and continental crust is present, the orogenic zone started to rise isostatically, producing an apron of clastics which contain reworked radiolarite debris shed by the nappes.

Figure 22 shows the paleogeography in the Paleocene-early Eocene with the open-marine successor basin and the wide shallow evaporitic platform southwest of it separated by a narrow and steep margin. Clastic material came from the rising orogenic belt in the northeast. In the

southern Zagros an isolated carbonate island with a central evaporite pan occupied the middle of the open-marine basin. This shoal was caused by shallowing along the axis of the Qatar-South Fars arch, which still acted as a regional positive element. In the far south, the east-central part of Gulf of Aden swell started to collapse, leading to an incursion of the Indian Ocean.

During the remainder of the Eocene the paleogeography stayed nearly the same, though the climate was less arid and evaporites were therefore virtually absent. The open-marine basin was gradually narrowed through progradation of the western carbonate platform and the growing wedge of clastic and shallow-water carbonate material, generated by the rising lands in the northeast. Across the Arabian platform, beds of late Eocene age are missing, which is partly because of nondeposition on the gradually emerging carbonate shelf. In the early and middle Oligocene this trend was accelerated, aided by an apparent major drop in relative sea level. The drop, which is reflected on the platform by extensive leaching and erosion of the Eocene carbonate rocks, is probably correlative with the global

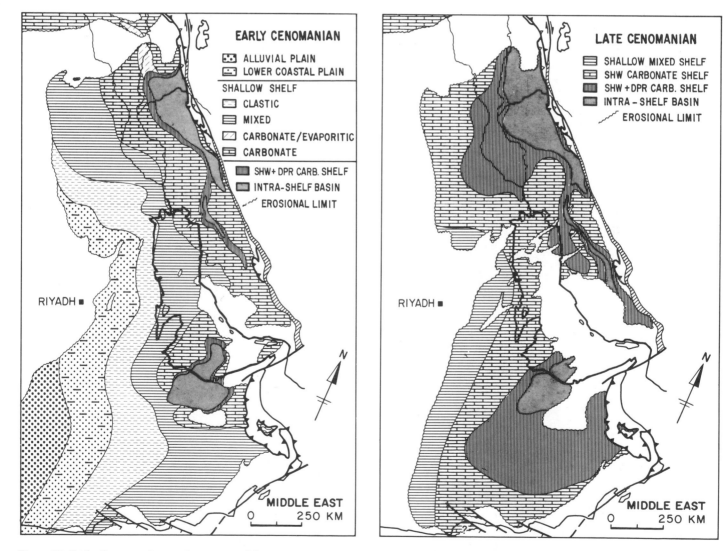

Figure 19. Early Cenomanian environments of deposition.

Figure 20. Late Cenomanian environments of deposition.

mid-Oligocene event (Vail et al, 1977, p. 87), but its precise dating has not been attempted.

Figure 23 shows the situation in Aquitanian time (latest Oligocene to early Miocene). The depositional realm has shrunk as compared to previous maps. Only a rather narrow northwest-southeast-trending evaporitic shelf with central depressions remained, with clastics coming in from the west and east. At the northern tip of the Gulf a major source of sand was present, forming the Ahwaz delta.

The late Oligocene to early Miocene Ahwaz sandstone and Asmari limestone form the principal reservoirs of the giant Iranian fields. Geologic and geochemical data point strongly to a deep source for this oil, that is, from Cretaceous and Jurassic source rocks deposited in the Lurestan and Khuzestan intrashelf euxinic basins (Dunnington, 1967; Thode and Monster, 1970; Murris and deGroot, 1979), Cretaceous carbonate reservoirs thereby commonly served as "way stations" in which oil started to accumulate as early as the late Cretaceous to early Tertiary, in structures which by then still must have been aligned along the old Mesozoic NNE-SSW trends. During the late Tertiary folding phase, which in the Zagros is much

stronger than the Late Cretaceous phase, the present northwest-southeast giant anticlinal traps were formed. The Campanian-Oligocene marls and shales which until then formed an effective seal to the Cretaceous reservoirs were fractured and hydrocarbons escaped upward into the Asmari traps, which are capped by the very efficient sealing Gach Saran evaporites of early to middle Miocene age.

The Miocene evaporites and carbonate rocks were succeeded in the late Miocene and Pliocene by a clastic sequence which gradually coarsens upward and ends with the massive Bakhtiari conglomerates and sandstones, a molasse-type deposit contemporaneous with the later stages of the late Tertiary orogenic event. The Pliocene Zagros folds came into existence and, in Kurdistan and the northern High Zagros, nappe emplacement occurred, which may have started as early as late Miocene (Berthier et al, 1974).

Today, the center of the successor deep again has shifted farther west and coincides with the axis of the Gulf and its northwestern extension, the Mesopotamian depression. The Zagros crush zone which, at least along part of its trace, is a root zone for the late Cretaceous and late Tertiary nappes, now acts as a dextral strike-slip fault (see also Berthier et al,

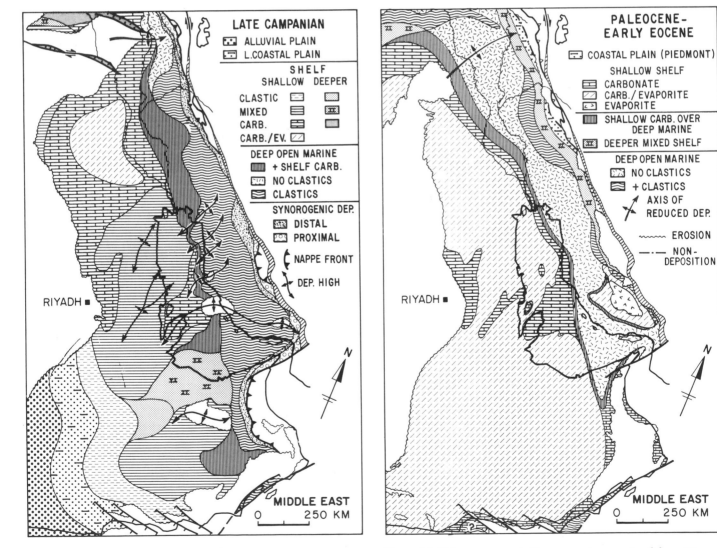

Figure 21. Late Campanian environments of deposition.

Figure 22. Paleocene to early Eocene environments of deposition.

1974, p. 97-101). Earthquake epicenter maps of the region show a concentration of foci 0 to 100 km deep along the Zagros foldbelt (Nowroozi, 1971). In contrast, the zone between the Zagros crush zone and Elburz is seismically remarkably quiet, which indicates that central Iran is today a stabilized block of which the crush zone forms the southwestern frontal boundary. The seismic data suggest that the crustal shortening caused by the active spreading of the Red Sea and the Gulf of Aden is accommodated by an imbrication of the continental crust underlying the Zagros foldbelt. The presence of some very deep foci—more than 100 km—implies that we are not only dealing with a crustal process but that the mantle is involved too.

CENTRAL GULF OIL HABITAT

Bajocian to Albian Sequence

For a clearer understanding of how the depositional environment, in conjunction with structure and maturation through burial, controls the hydrocarbon distribution, a restricted area and section will be described in more detail. Figure 24 depicts the regional development of the Bajocian

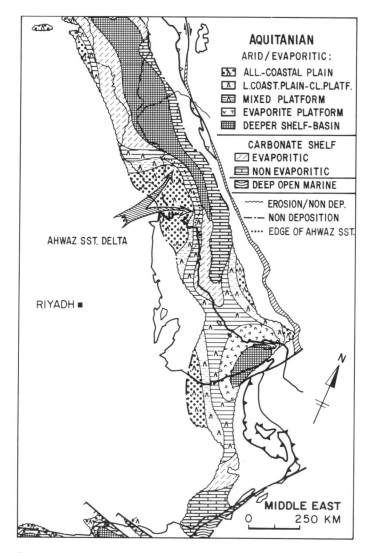

AHWAZ SST. DELTA

RIYADH ■

AQUITANIAN

ARID/EVAPORITIC:
- ALL.-COASTAL PLAIN
- L.COAST.PLAIN-CL.PLATF.
- MIXED PLATFORM
- EVAPORITE PLATFORM
- DEEPER SHELF-BASIN

CARBONATE SHELF
- EVAPORITIC
- NON EVAPORITIC
- DEEP OPEN MARINE

- EROSION/NON DEP.
- NON DEPOSITION
- EDGE OF AHWAZ SST.

N

MIDDLE EAST
0   250 KM

Figure 23. Aquitanian environments of deposition.

to Albian sequence in the central Gulf area, around the Qatar Peninsula. The section is about 700 km long, and oriented roughly east-west, with the approximate location of the Qatar Peninsula indicated above it.

The mid-Jurassic Dhruma Formation is the lowermost unit: a typical carbonate ramp, of which the margin with the open ocean was located well east of the section. This wide flat platform is the substrate for the subsequent differentiated shelf of the late Jurassic, gently backtilted by differential subsidence.

The late Jurassic inundation which started during the late Callovian is very evident in this section. In the east is the inbuilding carbonate margin, separated from the deep open-marine marls and cherty limestones by lithoclinal surfaces. West of the margin is the intrashelf depression, where in late Oxfordian to early Kimmeridgian time euxinic conditions reigned. At the western end of the section, the Tuwaiq (late Callovian-late Oxfordian) and Hanifa (late Oxfordian-early Kimmeridgian) cycles expand upshelf, separated by lithoclinal surfaces. The mid-Kimmeridgian to Tithonian shallow shelf built out into the intrashelf basin, containing tongues of ooidal-pelletoidal grainstones and algal boundstones which represent the very shallow terminal carbonate units. The late Jurassic closed with sabkha evaporites, which at the margin with the open sea are replaced by algal boundstones and oolite sands.

Above the late Jurassic evaporite platform the early Cretaceous carbonate shelf prograded into the open ocean in the east. Development was cyclical with several shallowing-upward sequences containing ooidal-pelletoidal and bioclastic pack-grainstones as high-energy members. By Barremian time, ramp-type sedimentation prevailed, with carbonate rocks very much like those of the Middle Jurassic in lithology and correlatability. Associated sands came from the west.

A minor unconformity, latest Barremian to earliest Aptian in age, is present over the Qatar arch. Above it the foraminiferal limestones of the Kharaib indicate the onset of the basin-wide Aptian transgression. By mid-late Aptian time, the platform was flooded and was succeeded by an intrashelf basin flanked by shallow platform carbonate rocks which contain rudist "reefs" and banks along the margin. The roof of the section is formed by the regressive Albian sales (Nahr Umr Formation), which contain tongues of Burgan sand in the west.

Translation into Oil Habitat

Figure 25 shows how the stratigraphic section of Figure 24 can be translated into habitat terms—source, reservoir, and seal—and how these control the oil and gas distribution. It is a generalized scheme (with local exceptions) for a substantial part of the central and southern Gulf area.

Based on Shell's method for rapid source rock identification, two potential source formations have been identified within this sequence: the U. Oxfordian-L. Kimmeridgian Hanifa and the Aptian Shuaiba, both were developed in their basinal facies. Regressive shales such as the Nahr Umr, the ramp-type marls, and argillaceous limestones apparently do not possess source characteristics. The open marine realm is also poor in preserved organic

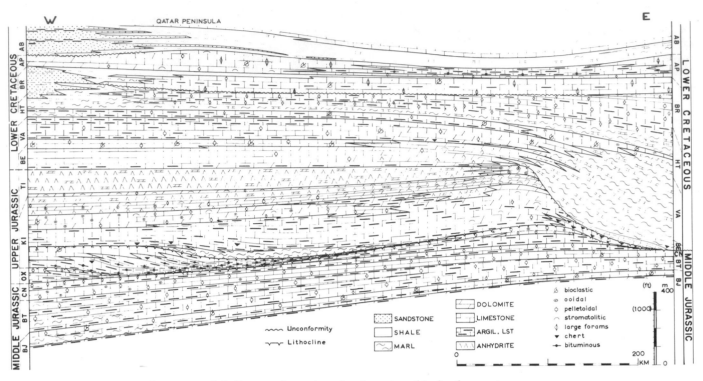

Figure 24. Mid-Jurassic to Albian, central Gulf area: schematic regional stratigraphic development.

matter and is a marginal local source rock at best.

Of the two potential source formations, the Hanifa is the richest within the central Gulf area. It is a thinly laminated lime-mudstone, with a clay content less than 10 percent. Organic carbon content varies between 2 and 6 percent, with some even richer laminae. Oil yield can be as high as 2

volume percent. Figure 26 shows a comparison between gas chromatograms of the saturated hydrocarbons of a typical Arab reservoir oil (above), and the extract of the Hanifa source rock in the same field. The similarity is striking, and also other typing methods, based on diverse geochemical parameters like porphyrins, carbon isotopes,

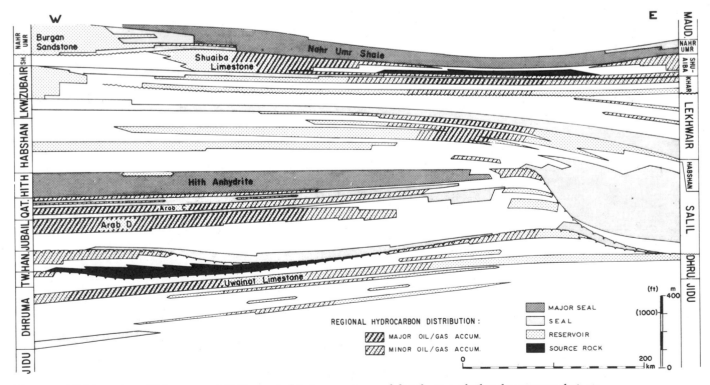

Figure 25. Mid-Jurassic to Albian, central Gulf area: habitat parameters and distribution of oil and gas accumulations.

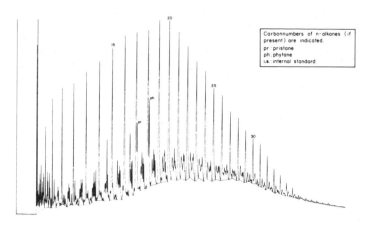

GAS CHROMATOGRAM OF SATURATED HYDROCARBON FRACTION, OIL

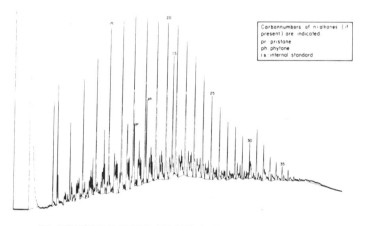

GAS CHROMATOGRAM OF SATURATED HYDROCARBON FRACTION OF (MATURE) HANIFA

Figure 26. Gas chromatogram comparison between an Arab reservoir oil (above) and an extract of the Hanifa source rock in the same field.

phytane-pristane-N alkane ratio's and C_{15}/C_{30} ring number distributions, show a close correlation between reservoired oils and the Hanifa. The high sulphur content of oils and extracts indicates a euxinic carbonate-evaporite environment of deposition (Gransch and Posthuma, 1973), which is fully compatible with the conditions under which the Hanifa was deposited. The importance for oil generation of the basin-wide inundations and the concomitant creation of starved intrashelf basins is obvious.

Reservoirs are present in abundance and are widely distributed through the section. The best reservoirs are formed by the regressive sands and by the high-energy ooidal grainstones terminating the carbonate cycles. Other favorable facies are the rudist "reefs" and the algal boundstones along the shelf margins. Leaching and cementation processes are important factors in determining the final reservoir quality of the carbonate rocks. Some porosity-creating leaching is apparently related to the lowering of sea level and subsequent subaerial exposure (e.g., early Albian) but there is also evidence for late leaching caused by acids in the formation waters.

There are two principal regional sealing formations, the

Hith Anhydrite and the Nahr Umr Shale, which (as shown in Figure 25) control to a large extent vertical migration and hence oil and gas distribution. The Hith is the ultimate cap rock for the prolific Arab reservoirs. Where it is present, the overlying lower Cretaceous reservoirs are devoid of charge, as at the western end of the example section. There are exceptions, however, when the Hith is breached through faulting and Arab oil could escape upward (Idd el Shargi, Bahrain). On the oceanic shelf margin the Hith is absent and hydrocarbons generated in the Hanifa source rock could escape upward to be trapped in Lower Cretaceous reservoirs.

Apart from the two main seals, there are other, less widespread but still effective sealing strata. The Hanifa source rock itself can form a good seal to reservoirs of the Dhruma or Tuwaiq Formations. Upper Jurassic argillaceous lime-mudstones hold modest columns in the Hanifa and Tuwaiq reservoirs in several fields. Marls and argillaceous limestones of the upper Hauterivian to Barremian Lekhwair Formation cap substantial reserves in several fields in the southern Gulf, in the zone where the Hith seal is absent and Hanifa charge is therefore available.

Cross-strata migration of the oil from the source beds to the reservoirs through the intervening section composed of tight limestones, marls, and thin anhydrites is indicated by occasional shows encountered during drilling, but more unambiguously by fluorescence-microscopical investigations performed at Shell's E and P Laboratory in the Netherlands, which show the presence of oil-filled microfractures. In conjunction with the almost identical composition of reservoired oils and source rock extracts, these microscopical observations strongly support the model that oil migrates in the whole phase as thin threads and stringers through fractures, faults, and permeable carrier beds.

Maturity, Kitchens and Regional Oil Distribution
Another important controlling factor is the degree of maturation of the source rocks. There is overwhelming geologic and geochemical evidence that source rocks of the type dominant in the Middle East start to generate oil when they reach a certain thermal maturity, roughly corresponding to a fixed carbon content of 60 percent in coals and to a vitrinite reflectance of 0.60 to 0.65 (Teichmüller, 1971; Hood et al, 1975). In the carbonate environment of the Middle East, reliable vitrinite data are scarce, and the delineation of the mature source rock areas must be based on calculated maturity levels. For this, the Lopatin time-temperature method calibrated to a worldwide set of vitrinite data was used along the lines first published by Waples (1980). Figure 27 gives an example of such a calculation for the Hanifa source rock, which at this location reached a maturity corresponding to a vitrinite reflectance of 0.81, and appears to have reached the generation threshold in the early Tertiary.

The combination of the stratigraphic extent and the maturity limits defines the areas of mature petroleum source or "kitchens," from which the receiving structures and other traps get their proportionate share of the charge depending on their drainage area size (Nederlof, 1980). Figure 28 shows the regional Hanifa kitchen, as well as the extent of the

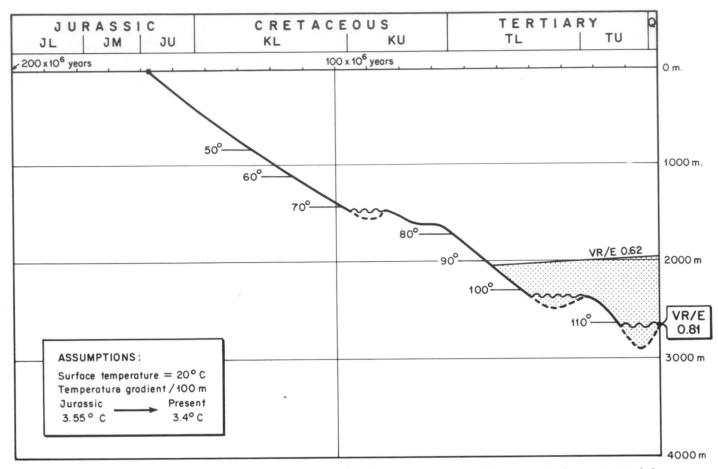

Figure 27. Example of a burial graph of the Hanifa. VR/E = calculated maturity in vitrinite reflectance equivalent units; stippled area = mature zone.

overlying regional Hith Anhydrite seal. The distribution of the Jurassic accumulations is clearly controlled by mature source and seal, whereby it should be noted that the Jurassic targets are hardly drilled in the southern part of the area, where they are deep and gas-prone. As the map shows, the Hanifa source rock is mature over most of the area where it is present, apart from the culmination of the Qatar arch. The absence of sizeable Jurassic-Cretaceous oil discoveries on this broad paleohigh is at least partly due to this lack of maturity within the drainage area of the closure.

Figure 29 shows the kitchen of the Aptian Shuaiba source rock. In the western part of the central Gulf area this potential source formation is poorly-developed and not or barely mature. Major Lower Cretaceous oil fields are concentrated in the eastern part, where the Shuaiba source rock is better-developed and is in the oil generating zone. Part of the reservoired oil—in particular in the earlier Cretaceous Habshan and Lekhwair formations (see Figure 25)—may represent Jurassic oil migrated upwards in this zone where the Hith is thin or absent, as indicated on Figure 29. Detailed geochemical data are lacking to determine, in each case, what the source of the reservoired oil might be. The oil occurring in Lower Cretaceous reservoirs in the northwestern part of the area, outside the Shuaiba kitchen, has in several instances been unequivocally typed to the Jurassic Hanifa source. It can be shown that in these cases

the intervening Hith seal is breached through faulting, often leading to a depletion of the Jurassic reservoirs.

As demonstrated with this example of the Upper Jurassic to Lower Cretaceous of the wider Qatar area, the distribution of oil over the various available reservoirs is closely controlled by the primary development and preservation of source and seal formations, and by source maturity.

CONCLUDING REMARKS

The principles of translating lithologies and depositional environments into hydrocarbon habitat parameters, as given in the example of the Middle Jurassic-Lower Cretaceous of the central Gulf area, can be extended to the other formations and the whole province. Superposition of the layer maps of Figures 6 to 23 permits outlining of other source-reservoir-seal triplets, and their regional extent. With thermal-maturity maps these go a long way in explaining the regional distribution of the oil and gas fields in the Middle East province.

The question remains as to why the Middle East is such an exceptionally rich habitat. Source rocks are not exceptionally thick or rich compared to other basins. Reservoirs are abundant and have generally high porosities and permeabilities, enhanced by fracturing in the Zagros folds, but again they are not a class apart when compared to other basins such as the North Sea, Alberta, or the United States Gulf Coast. Seals, without doubt, are very effective, but not unusually thick or abundant.

The horizontal scale of the basin is extraordinary. Whereas the vertical dimensions are rather average, the horizontal dimensions are greater than in most basins. The pre-erosional Mesozoic depositional platform was 2,000 to 3,000 km wide and at least twice as long. Differentiation was minimal and source rocks, reservoirs, and seals have consequently a very wide extent. The structures are also out of scale horizontally, and closures of 1,000 sq km or more are no exception. Because of large horizontal scale, the structures are very gentle, which has a two-fold effect: loss of seal efficiency through fracturing is minimal and large trap volumes are attached to even modest vertical closures or column heights. Finally, because of the wide extent of the lithologic units, horizontal migration is very efficient and the large structures have a high degree of fill because they could drain very large "kitchens."

REFERENCES CITED

Ahr, W.M., 1973, The carbonate ramp—an alternative to the shelf model: Gulf Coast Assoc. Geol. Socs. Trans., v. 23, p. 222-225.

Bay, T.A., 1977, Some Lower Cretaceous stratigraphic models from Texas and Mexico, in Cretaceous carbonates of Texas and Mexico: Univ. Texas Bur. Econ. Geology, Rept. Inv. 89, p. 12-30.

Berthier F., et al, 1974, Etude stratigraphique, pétrologique et structurale de la région de Khorramabad, Zagros Iranien: Thesis, Univ. Grenoble.

Dunnington, H.V., 1967, Stratigraphical distribution of oil fields in the Iraq-Iran-Arabia basin: Inst. Petroleum Jour., v. 53, p. 129-161.

——, R. Wetzel, and D.M. Morton, 1959, Iraq (Mesozoic and Paleozoic), in L. Dubertret, ed., Lexique stratigraphique international, Asie, fasc. 10a.

Glennie, K.W., et al, 1974, Geology of the Oman Mountains: Ned. Geol. Nijnbouwkd. Genoot., Verh., v. 31, 423 p.

Gransch, J.A., and J. Posthuma, 1973, On the origin of sulphur in crudes: Adv. Org. Geochem., p. 727-739.

Helal, A.H., 1964, On the occurrence and stratigraphic position of Permo-Carboniferous tillites in Saudi Arabia: Geol. Rundschau, v. 54, p. 193-207.

——, 1965, Stratigraphy of outcropping Paleozoic rocks around the northern edge of the Arabian shield (within

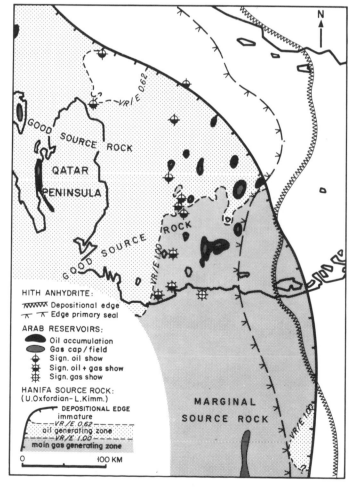

Figure 28. Central Gulf area: Jurassic accumulations in relation to Hanifa kitchen and Hith Anhydrite development.

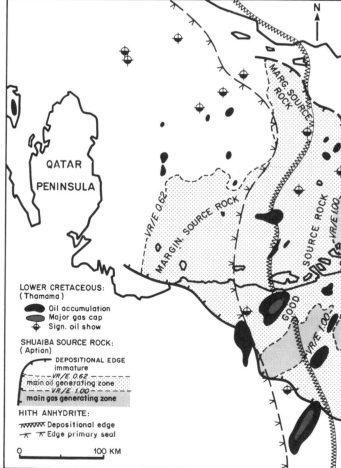

Figure 29. Central Gulf area: Lower Cretaceous accumulations in relation to Shuaiba kitchen and Hith Anhydrite development.

Saudi Arabia): Deutsch. Geol. Gesell. Zeitschr., v. 117, p. 506-543.

Hood, A., C.C.M. Gutjahr, and R.L. Heacock, 1975, Organic metamorphism and the generation of petroleum: AAPG Bull., v. 59, p. 986-996.

Hudson, R.G.S., and M. Sudbury, 1959, Permian Brachiopods from south-east Arabia: Notes et Mém., Moyen-Orient, t. 7, p. 19-55.

Murris, R.J., and K. deGroot, 1979, Oil habitat of carbonate provinces: Cong. Panam. Ing. Petr., Mexico, Proc., Sec. 1, Paper 5.

Nederlof, M.H., 1980, The use of Habitat Models in Exploration Prospect Appraisal: 10th World Petr. Congr., proc. v. 2, p. 13-21.

Nowroozi, A.A., 1971, Seismotectonics of the Persian Plateau, eastern Turkey, Caucasus and Hindu-Kush regions: Seismol. Soc. America Bull., v. 61, p. 317-341.

Purser, B.H., 1969, Syn-sedimentary marine lithification of Middle Jurassic limestones in the Paris basin: Sedimentology, v. 12, p. 205-230.

———, 1972, Subdividion et interprétation des séquences carbonateées: France, Bur. Recherches, Géol. et Miniére Mém. 77, p. 679-698.

Ricou, L.E., 1975, L'étude géologique de la région de Neyriz (Zagros iranien) et l'évolution structurale des Zagrides: Thèse d'état, Paris.

Smith, A.G., and J.C. Briden, 1977, Mesozoic and Cenozoic paleocontinental maps, *in* Cambridge earth science series: Cambridge, England, Cambridge Univ. Press, 63 p.

Szabo, F., and A. Kheradpir, 1978, Permian and Triassic stratigraphy, Zagros basin, south-west Iran: Jour. Petroleum Geology, v. 1, p. 57-82.

Teichmüller, M., 1971, Anwendung kohlenpetrographischer Methoden bei der Erdöl-und Erdgasprospektion: Erdöl u. Kohle, v. 24, p. 69-76.

Thode, H.G., and J. Monster, 1970, Sulfur isotope abundances and genetic relations of oil accumulations in Middle East basin: AAPG Bull. v. 54, p. 627-637.

Vail, P.R., R.M. Mitchum, Jr., and S. Thompson, III, 1977, Global cycles and relative changes of sea level, *in* Seismic stratigraphy—applications to hydrocarbon exploration: AAPG Mem. 26, p. 83-97.

Waples, D.W., 1980, Time and Temperature in Petroleum Formation: Application of Lopatin's Method to Petroleum Exploration: AAPG Bull., v. 64, p. 916-926.

Wilson, J.L., 1975, Carbonate facies in geologic history: New York, Springer-Verlag, 469 p.

Hydrocarbon Habitat of the Cooper/Eromanga Basin, Australia

A.J. Kantsler
T.J.C. Prudence
Shell Development (Australia) Pty Ltd
Perth, Western Australia

A.C. Cook
The University of Wollongong
Wollongong, NSW, Australia

M. Zwigulis
Delhi Petroleum Pty Ltd
Adelaide, Southern Australia

The Cooper Basin is a complex intracratonic basin containing a Permian-Triassic succession which is unconformably overlain by Jurassic-Cretaceous sediments of the Eromanga Basin. Abundant inertinite-rich humic source rocks in the Permian coal measures sequence have sourced some 3 Tcf recoverable gas and 250 million barrels recoverable natural gas liquids and oil found to date in interbedded Permian sandstones. Locally developed vitrinitic and exinite-rich humic source rocks in the Jurassic to Early Cretaceous section have, together with Permian source rocks, contributed to a further 60 million barrels of recoverable oil found in fluvial Jurassic-Cretaceous sandstones.

Maturity trends vary across the basin in response to a complex thermal history, resulting in a present-day geothermal gradient which ranges from 3.0 to 6.0°C/100 m. Permian source rocks are mature to postmature for oil generation and comprise oil/condensate and gas kitchens in separate depositional troughs. Jurassic source rocks generally range from immature to mature but are postmature in the central Nappamerri Trough. The Nappamerri Trough is considered to have been the most prolific Jurassic oil kitchen because of the mature character of the crudes found in Jurassic reservoirs around its flanks.

Outside the central Nappamerri Trough, maturation modeling studies suggest that most hydrocarbon generation followed very rapid subsidence during the Cenomanian. Most syndepositional Permian structures are favorably located in time and space to receive this hydrocarbon charge. Late-formed structures (mid-late Tertiary) are less favorably situated and are rarely filled to spill point.

The high CO_2 contents of the Permian gas (up to 50 percent) may be related to maturation of the humic Permian source rocks and thermal degradation of Permian crudes. However, the high $\delta^{13}C$ of the CO_2 (average, -6.9 per mil) suggests some mixing with CO_2 derived from thermal breakdown of basement carbonates.

INTRODUCTION

The Permian-Triassic Cooper Basin (Figure 1), which is located in the northeast corner of South Australia and extends into the southwest corner of Queensland, contains initially recoverable hydrocarbon reserves estimated at 3 Tcf gas and 250 million barrels natural gas liquids and oil. The basin is one of Australia's most productive onshore hydrocarbon provinces. Recent discoveries in the overlying Jurassic-Cretaceous Eromanga Basin total approximately 60 million barrels initially recoverable oil. Previously, Mesozoic prospects were discounted because of the supposed high risk of flushing associated with the Jurassic aquifer systems.

The source of the hydrocarbons found in the Cooper Basin has long been ascribed (Brooks, Hesp, and Rigby, 1971; Battersby, 1976) to the extensive Permian coal measures with which the reservoirs are interbedded (Figure

2). The origin of hydrocarbons found in the Eromanga Basin however, has been the subject of conjecture (e.g., Poll, 1981) since gas was first discovered at Namur in 1976 and oil at Strzelecki in 1978 (Figure 3). Although several thin, rich, mainly humic source rock horizons occur within the non-marine Jurassic section (Thomas, 1982), a Permian source is equally possible for several of these discoveries.

Despite the pre-eminence of the Cooper/Eromanga Basin among Australia's oil and gas producing provinces, relatively little systematic work of a geochemical nature has been published to facilitate understanding of regional hydrocarbon origin and distribution. This paper examines a large suite of source rock and maturation data, together with regional stratigraphic and tectonic factors, while attempting to develop a regional model for hydrocarbon generation and occurrence.

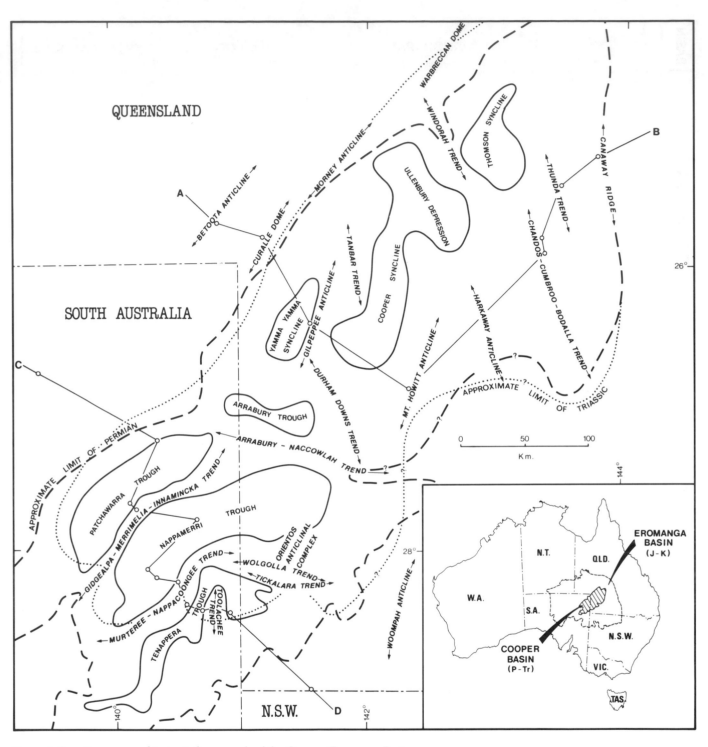

Figure 1. Location map and tectonic framework of the Cooper/Eromanga Basin.

PREVIOUS INVESTIGATIONS

The regional geology of the Cooper/Eromanga Basin is summarized by Battersby (1976), Senior, Mond, and Harrison (1978), Thornton (1979), and Bowering (1982), who provide useful references for many more detailed investigations. Recent advances in knowledge of Eromanga Basin geology are presented in informal proceedings of the Eromanga Basin Symposium (Moore and Mount, 1982). A comprehensive summary of the geology of producing fields and hydrocarbon prospects in the Cooper Basin is given by Battersby (1976) with details of recent discoveries in Mount (1981), Barr and Youngs (1981), Poll (1981), Bowering (1982), Bowering and Harrison (1982), and Moore and Mount (1982).

Introductory work of a geochemical nature began with the pioneering efforts of Brooks (1970) and Brooks, Hesp, and Rigby (1971), and later Powell and McKirdy (1976),

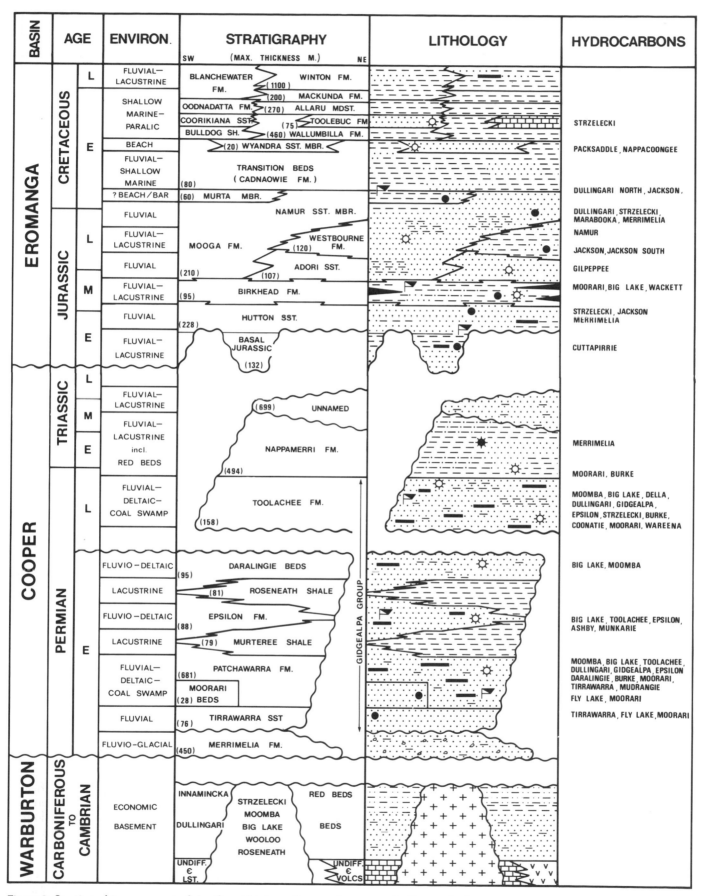

Figure 2. Stratigraphic occurrence of significant hydrocarbons and source rocks (flag symbol).

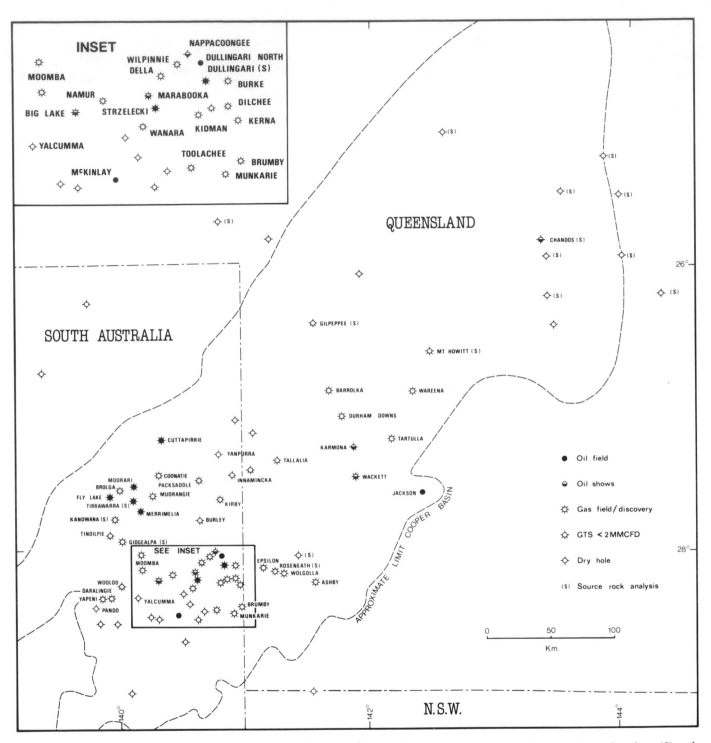

Figure 3. Major hydrocarbon occurrences in the Cooper/Eromanga Basin and location of wells sampled for geochemical analysis (S) and vitrinite reflectance (all wells).

who demonstrated the contribution made by land plant organic matter to Australian oil and gas occurrences and the use of the rank (maturation) concept in hydrocarbon exploration. Similarly broad concepts were outlined by Demaison (1975), Shibaoka and Bennett (1977) and Shibaoka, Saxby, and Taylor (1978). Subsequent maturation surveys (Kantsler, Smith, and Cook, 1978; Kantsler and Cook, 1979) related coal rank variation and

thermal history to the nature and distribution of Permian hydrocarbon occurrences. These were followed by detailed studies of regional variation in gas composition (Schwebel, Devine, and Riley, 1980; Rigby and Smith, 1981) and petroleum geochemistry (McKirdy, 1982).

These papers are supplemented by source rock descriptions (Smyth, 1979; Thomas, 1982; Cook, 1982) and a large body of unpublished analytical data currently on open-file report with various government agencies.

EXPLORATION HISTORY

Exploration began in 1959 and, following a series of six dry holes, commercial quantities of Permian gas were first discovered at Gidgealpa in 1963, followed by Moomba (1966), Daralingie (1967), and Toolachee (1969). Subsequent exploration resulted in the discovery of more than 40 Permian gas fields including three oil/gas fields. Hydrocarbons were first discovered in the Jurassic-Cretaceous sequence in 1976 when Namur-1 tested gas at a rate of 14 MMcf per day from the Namur Sandstone (Figure 2). Nine oil discoveries and three gas discoveries have since been made. To date, some 140 exploration wells and a similar number of development/appraisal wells have been drilled resulting in over 50 discoveries (Figure 3).

REGIONAL SETTING

The Cooper Basin (Figure 1) is a complex, northeast trending Permian-Triassic intracratonic basin covering an area of approximately 127,000 sq km (49,035 sq mi). The basin unconformably overlies earlier Paleozoic basins and ?Carboniferous granites, and is in turn unconformably overlain by Jurassic and Cretaceous sediments of the Eromanga Basin which forms part of the 1.7 million sq km (656,375 sq mi) area of the Great Artesian Basin (Figure 1). The limits of the Cooper Basin are controlled by onlap onto subcropping basement ridges.

Sinuous, narrow, fault-bounded anticlinal zones divide the Cooper Basin into major depositional troughs. The prominent Arrabury-Naccowlah anticlinal trend separates the southern Cooper Basin (an area of thick Permian deposition) from the northern part of the basin (a region of thinner Permian deposition but thick Triassic deposition). In the southern basin, the Permian section in the Nappamerri Trough plunges strongly to the northeast and reaches a maximum thickness of 1,500 m (4,921 ft) (Thornton, 1979) in a depocenter flanking the Arrabury-Naccowlah trend. In the north, Permian thickness reaches a maximum of 500 m (1,640 ft) in the Cooper Syncline but is generally less than 200 m (656 ft). The Permian sequence is overlain by up to 3,200 m (10,500 ft) of Mesozoic and Cainozoic sedimentary rocks.

Major anticlinal structures throughout the southern basin are *en echelon*, steep-sided, fault-controlled basement features which were active during Permian deposition. Incipient rifting (Klemme, 1980; Mount, 1981) appears to have created a series of half-grabens with subsequent right-lateral wrench movements controlling Permian structural development. Many structural trends in the northern part of the basin (e.g., Durham Downs trend) are of different orientation to those in the south but can also be attributed to a right-lateral stress regime. Mesozoic structuration is related primarily to drape and differential compaction, although most structures do show some element of concomitant structural growth. A major phase of left-lateral, wrench-induced, folding took place in the early Tertiary (Wopfner, 1960; Senior and Habermehl, 1980) and created many of the very large anticlines in the northern

basin while enhancing some structures in the southern basin.

STRATIGRAPHY AND BASIN HISTORY

The stratigraphy adopted in this paper is illustrated in Figure 2 and is described in detail by Exon and Senior (1976), Thornton (1979), and Moore and Mount (1982).

Economic basement consists of steeply dipping lower Paleozoic sediments and volcanics, Carboniferous granites, and unprospective glacial sediments of the Early Permian Merrimelia Formation. Up to 1,300 m (4,265 ft) of Early Permian fluvial-lacustrine-coal swamp sediments were deposited in the southern basin with few major Permian river systems reaching the northern basin. This initial phase of sedimentation culminated with two cycles of transgression/regression by an inland sea (Thornton, 1979) which was followed by non-deposition in the south and uplift and erosion in the north. The entire Permian section was removed from some culminations on major anticlinal trends at this time. A coal measures sequence up to 200 m (656 ft) thick (Toolache Formation) was then deposited in a meandering river flood-plain environment. Overall, Permian thickness varies considerably due to onlap, growth faulting, non-deposition, differential compaction, and erosion.

Subsidence continued until the Middle to Late Triassic with continental fluviatile and lacustrine sediments (Nappamerri Formation), including some red beds, being deposited over most of the basin. The Triassic section thickens to the north indicating regional tilt. Late Triassic uplift and erosion resulted in the loss of up to 500 m (1,640 ft) of section in the southeastern part of the basin.

Major epeirogenic downwarping of most of eastern Australia in the Late Triassic/Early Jurassic initiated Eromanga Basin deposition on an irregular erosional surface. Up to 1,000 m (3,281 ft) of Jurassic fluvial-lacustrine-coal swamp sediments were deposited in a broad, gently subsiding, cratonic basin. Marine transgression in the Neocomian-Aptian was accompanied by paralic sedimentation. In the Albian, regression was followed by transgression. Subsidence culminated with the rapid deposition of the Cenomanian Winton Formation, a paralic, lacustrine, and fluvial sequence over 1,000 m (3,281 ft) thick (Figure 2). The Cretaceous section is more than 2,000 m (6,562 ft) thick in several of the major depressions.

Plots of maximum rate of subsidence versus geological time show oscillations related to the Early Permian, Late Permian-Early Triassic, and Jurassic-Cretaceous cycles of deposition. Maximum (compacted thickness) sedimentation rates reach 80 m/million years (262 ft/million years) in the Patchawarra Formation, 30 m/million years (98 ft/million years) in the Nappamerri Formation and up to 200 m/million years (656 ft/million years) in the Winton Formation.

OCCURRENCE OF HYDROCARBONS

Hydrocarbon accumulations have been found in reservoirs which range in age from Early Permian to Early Cretaceous, at depths between 1,100 and 3,300 m (3,609 and

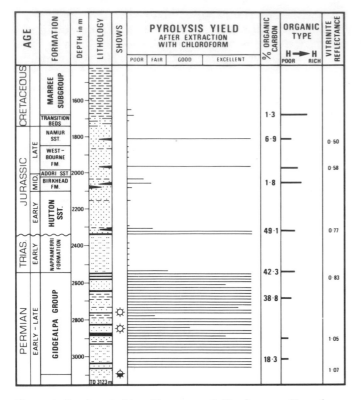

Figure 4. Geochemical log, Kanowana-1, Patchawarra Trough (from Thomas, 1982).

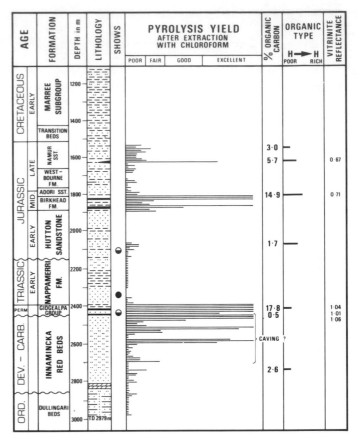

Figure 5. Geochemical log, Chandos-1, far northern Cooper Basin (from Thomas, 1982).

10,827 ft) (Figure 2). Major producing horizons are: Patchawarra Formation (Lower Permian); Toolachee Formation (Upper Permian); Hutton Sandstone (Lower Jurassic); Namur Sandstone (Upper Jurassic).

However, oil and gas are found in reservoir-quality sandstones in virtually all formations below the Neocomian Transition beds, although commercial discoveries of Permian oil are confined to the Tirrawarra Sandstone and Moorari Beds. Most accumulations discovered to date occur in simple anticlinal or fault-bounded anticlinal structures, described by Beddoes (1973), Battersby (1976), and Barr and Youngs (1981). Dip-closed, synthetic fault structures form traps in the Mudrangie, Packsaddle, Wilpinnie and Merrimelia fields. The oil accumulation at Dullingari North occurs in a combination structural/stratigraphic trap which involves drape of a laterally restricted shoreface sand over a basement horst.

The majority of the hydrocarbon discoveries are located in the southern part of the basin and largely reflect higher levels of tenement activity. The distribution of producing fields and discovery wells is restricted to areas flanking the major troughs (particularly the Nappamerri Trough; compare Figures 1 and 3). Toward the southern margin of the basin, where the regional Triassic seal is absent, salinity data suggest that all Permian reservoirs have been invaded by artesian waters of the Eromanga Basin (Young, 1975).

Initial known reserves are presented in Table 1. Commercial Permian oil reserves are exclusively concentrated in the Tirrawarra-Fly Lake-Moorari area of the Patchawarra Trough. The largest Permian gas fields (Moomba, 750 Bcf; Big Lake, 500 Bcf; Della, 500 Bcf) are located in the southern Nappamerri Trough.

Largest Jurassic oil reserves occur in Hutton Sandstone reservoirs at Strzelecki and Jackson in structures flanking the Nappamerri Trough. Largest Jurassic gas reserves are found in a similar structural setting at Namur where they occur within the Namur Sandstone.

SOURCE ROCKS

Fair-to-good, locally rich source rocks occur throughout the Permian section and locally within the Jurassic section (Figures 4 and 5). All are coal measures or land-plant rich sequences and have high pyrolysable carbon and total organic carbon contents (Thomas, 1982). Potential exists for the generation of both oil and gas but specific yields of liquids are likely to be relatively low. This is because of the dominance of vitrinite and inertinite (Type III kerogen of Tissot and Welte, 1978) in the Permian source rocks and the thin and impersistent development of the predominantly vitrinite-rich (mixed Type II/Type III) Jurassic source rocks.

Source rock pyrolysis logs, maturation, maceral, and extract data have been compiled from 20 wells throughout the basin using published material (Beeston, 1975; Smyth, 1979; Senior and Habermehl, 1980; Thomas, 1982), open-file reports at BMR, and commissioned studies. These data, together with the authors' vitrinite reflectance measurements and maceral descriptions from 65 additional wells, provide wide control on the occurrence of oil and gas source rocks (Figure 2).

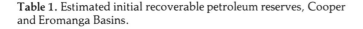

ESTIMATED INITIAL RECOVERABLE PETROLEUM RESERVES

BASIN	OIL (million barrels)	NATURAL GAS LIQUIDS (million barrels)	GAS (SALES) (billion cubic ft)
COOPER	50	200	3000
EROMANGA	approx 60	—	approx 65

Sources:
BMR Petroleum Newsletter
SANTOS Annual Reports

Table 1. Estimated initial recoverable petroleum reserves, Cooper and Eromanga Basins.

Generally, Permian sediments are exinite-poor and inertinite and vitrinite predominate. However, local concentrations of exinite, structureless (biodegraded) organic matter and bacterially-degraded vitrinite (desmocollinite) indicate significant oil potential. The Tirrawarra-Fly Lake-Moorari oil field is undoubtedly sourced from such organic matter.

Schwebel, Devine, and Riley (1980) estimate the amount of coal within the Permian sequence of the southern Cooper Basin to be approximately 1×10^{12} tons. This, together with dispersed organic matter in shales and siltstones, constitutes a vast potential hydrocarbon source. However, the Permian sequence is considerably thinner in the northern Cooper Basin where source potential is correspondingly reduced.

The Triassic Nappamerri Formation is a fine-grained fluviatile sequence with red beds, typically barren of organic matter. It has little regional significance as a source rock.

Jurassic sediments contain a number of thin, rich, source rock intervals, particularly within the Birkhead Formation and also at the base of the Hutton Sandstone and within the Murta Member. As a result of post-Permian floral and climatic changes (Cook, 1981; Thomas, 1982), Jurassic coals and organic matter are dominated by vitrinite (desmocollinite-rich) and exinite, although the basal Jurassic coal-sequence is commonly inertinite-rich. The exinite content typically includes sporinite, resinite, cutinite, and abundant suberinite. Non-coal source rocks also contain structureless organic matter which is believed to be biochemically degraded cellulosic plant tissue (i.e., an equivalent of the vitrinite submaceral desmocollinite) but locally can be attributed to an algal source. Type equivalents of the Jurassic source rocks include the Walloon coal measures of the Surat Basin which have pyrolysis yields of up to 320 liters per ton (Smith, 1980), and the coal measures of the Latrobe group in the Gippsland Basin which appear to have sourced the giant oil fields of Bass Strait (Shibaoka, Saxby, and Taylor, 1978).

The overlying Cretaceous section is commonly organically lean with poor pyrolysis yields, although fair-to-good, marginally-marine source rocks are developed locally within the Murta Member, Transition Beds,

Wallumbilla Formation, and Toolebuc Formation. The better source intervals typically contain minor amounts of bituminite, alginite, and dinoflagellates in addition to bacterially-derived "structureless organic matter," vitrinite, and rare sporinite.

GEOTHERMAL GRADIENTS

The Cooper Basin and the overlying Eromanga Basin have a present-day geothermal gradient ranging from 3.0 to 6.0°C/100 m (Figure 6). Temperature data are derived from bottom-hole temperatures which have been corrected for cooling by circulation of drilling fluids. Geothermal gradients have been calculated assuming a uniform thermal conductivity for the whole sequence and ignore "dog-leg" gradients, which in some instances can be related to artesian flow.

High gradients (greater than 4.5°C/100 m) occur in and around the Nappamerri and Tenappera troughs—particularly in areas underlain by granite basement such as the Strzelecki, Moomba and Big Lake gas fields (Figures 3 and 6). It seems probable that high-level intrusions which are significantly enriched in radioactive elements have enhanced already high levels of regional crustal heat flow through radiogenic decay (Middleton, 1979; Sass and Lachenbruch, 1979). Lowest temperature gradients occur in the Patchawarra Trough, which is presumed to overlie a thick pre-Permian Warburton Basin section of higher thermal conductivity than basement elsewhere.

GENERATION HISTORY

Organic Maturity

The predominance of land-plant organic matter throughout the Cooper/Eromanga Basin allows reliable estimation of organic maturity (rank) by vitrinite reflectance (VR). Data are available from more than 90 wells and provide good control in the southern basin, but a lesser degree of control in the northern basin. For each well, the number of samples examined over the entire section ranges from 5 to 35 (average 17).

Tissot and Welte (1978), Powell and Snowdon (1980), Radke et al (1980), Demaison (1980, personal communication) and Thomas (1982) all conclude that the threshold of significant oil generation from land-plant dominated organic matter occurs between VR = 0.7 and 0.8 percent and reaches a peak around VR = 0.9 percent. Initial naphthenic oil generation from some resinite-rich source rocks may occur at maturation levels as low as VR = 0.4 percent (Snowdon and Powell, 1982) but such source rocks (and crudes) are not typical of the Cooper/Eromanga Basin. Humic organic matter becomes postmature for oil generation between VR = 1.2 and 1.4 percent, at which time the source rocks become mature for gas generation.

The Permian Gidgealpa Group is oil mature to postmature over almost the entire Cooper Basin (Figures 7 and 8). The far south of the basin remains immature to early oil mature due to a combination of shallow burial and lower geothermal gradients. Present-day maturation levels suitable for oil generation occur in the Patchawarra Trough and the Tenappera Trough, in a narrow zone around the

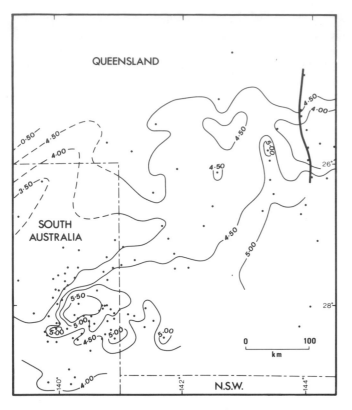

Figure 6. Geothermal gradients (°C/100 m).

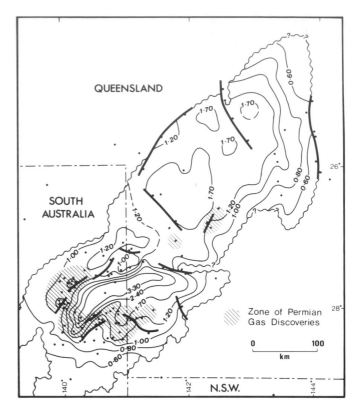

Figure 7. Vitrinite reflectance at top Patchawarra Formation (Lower Permian).

flanks of the Nappamerri Trough, and over much of the northern Cooper Basin. Permian sediments in the central Nappamerri Trough and central Cooper Syncline are currently mature for gas generation and are well located to have charged many of the surrounding gas fields (Figures 9 and 10).

Middle-to-Lower Jurassic source rocks are at peak oil maturity over much of the south-central Cooper Basin and in the major axial depressions (Cooper Syncline, Thomson Syncline, Ullenbury Depression) of the northern Cooper Basin (Figures 11 and 12). They are currently well placed to have sourced many of the Eromanga Basin oils discovered to date around the margins of the Nappamerri Trough. Lower Cretaceous to Upper Jurassic source rocks reach the VR = 0.7 percent threshold in the central Nappamerri Trough (Figures 10 and 13). Locally-developed, humic-to-mixed source intervals of the remainder of the Lower Cretaceous sequence are considered immature to initially mature over most of the basin (VR = 0.50 to 0.75 percent). The vitrinite and exinite-rich coals of the Late Cretaceous Winton Formation are immature.

Figure 14 illustrates the broad geographical variation of rank data from the Cooper and Eromanga Basins. A major feature is the contrast in present-day maturation levels between the "hot," dry gas-prone Nappamerri Trough and the relatively "cool," wet gas-prone Patchawarra Trough. A secondary feature is that vitrinite reflectance of the uppermost coals (Cenomanian) is generally less than 0.4 percent. The latter feature implies only minor Tertiary uplift and erosion and suggests that most source sequences have been at or near maximum burial temperature since the

Cretaceous. Exceptions are young Tertiary inversion structures in the northern basin (e.g., Curalle Dome, Wareena Anticline) where up to 500 m (1,640 ft) of truncation is indicated. *In situ* hydrocarbon generation in these areas has therefore been retarded since early mid-Tertiary uplift.

Data from the southeastern part of the basin fall into the intermediate domain because of the truncation of Permian-Triassic VR profiles at the Jurassic/Triassic unconformity surface (up to 0.25 percent, Figure 15). Such offsets have resulted in thinning of the VR = 0.6 to 1.0 percent window on the southern and southeastern margins of the basin (e.g., Murteree-Nappacoongee Trend—Tenappera Trough area in Figure 10).

Thermal History

Rank variation in the southern Cooper Basin is strongly influenced by high paleo-heat flow in the Nappamerri Trough which has created a maturity "high" persisting into the Lower Cretaceous section (Figures 7 to 13). This is illustrated in the Big Lake-Moomba area of the Nappamerri Trough where early high paleo-heat flow in excess of the modern heat flow regime is required to satisfactorily model maturation (Figure 16).

A Permian to Late Triassic phase of high heat flow is considered likely because of offsets and changes in gradient of plots of VR versus depth at or near the Jurassic/Triassic unconformity surface throughout the basin (Figures 14 and 15). These data also imply significant (up to 500 m or 1,640 ft) Late Triassic erosion. High initial heat flow accompanied

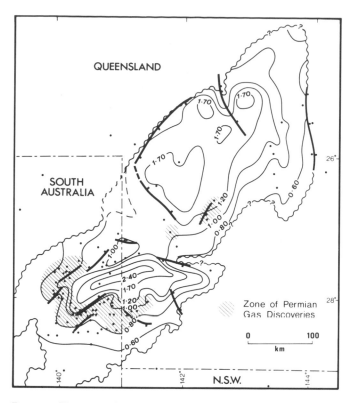

Figure 8. Vitrinite reflectance at Base Mesozoic (Top Permian).

by deep crustal metamorphism (Middleton, 1980) and/or crustal extension (McKenzie, 1978) may be associated with the initiation of basin subsidence. Later, thermal events may have been associated with Tertiary tectonism and/or drift of this part of the Australian continent over a sub-crustal "hot-spot" (Kantsler, Smith, and Cook, 1978).

Outside the Nappamerri Trough, maturation of the Permian section can generally be modeled within the constraints of the present-day thermal regime (Figure 17). However, in some cases, particularly in the Jurassic-Cretaceous section, over-estimation of maturity may occur due to imperfections in the model, model assumptions, a late-rise in temperature, or errors in temperature estimation (Kantsler and Cook, 1979). The latter problems are well-illustrated at Dullingari North where the surface to top Namur geothermal gradient is 5.8°C/100 m compared with a surface to top basement gradient of 4.8°C/100 m. Surface to top basement geothermal gradients established from Permian reservoir temperatures in the Dullingari gas field 5 km (3.1 mi) south are 4.3°C/100 m. Differences in bulk thermal conductivity between the Permian coal measures sequence, the Jurassic aquifer system, and the uppermost Cretaceous mudstone sequence may be responsible for such variation.

A sudden increase (0.1 percent) in vitrinite reflectance observed in the basal Jurassic section of some wells (e.g., Merrimelia, Nappacoongee, Curalle) may be related to convective transfer of heat via major faults. Movement of

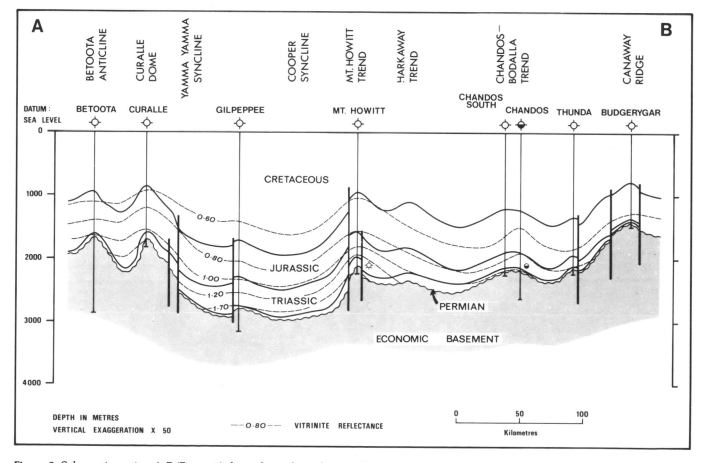

Figure 9. Schematic section A-B (Figure 1) through northern Cooper/Eromanga Basin with profile of isoreflectance surfaces.

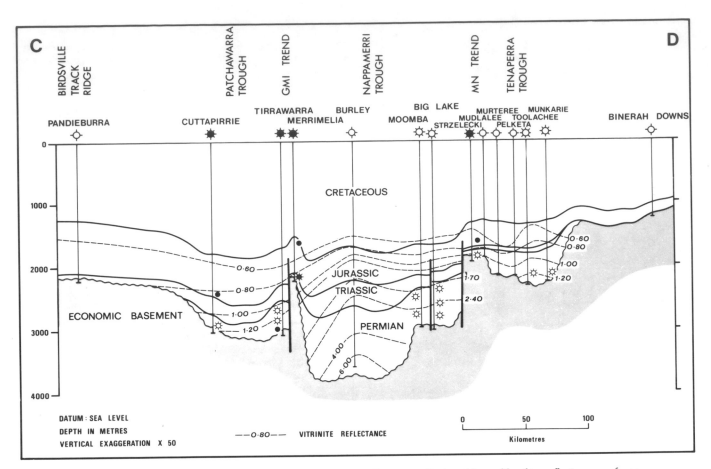

Figure 10. Schematic section C-D (Figure 1) through southern Cooper/Eromanga Basin with profile of isoreflectance surfaces.

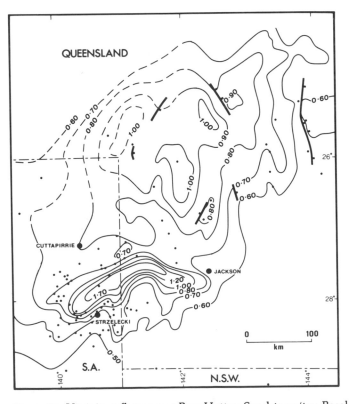

Figure 11. Vitrinite reflectance at Base Hutton Sandstone/top Basal Jurassic (Lower Jurassic).

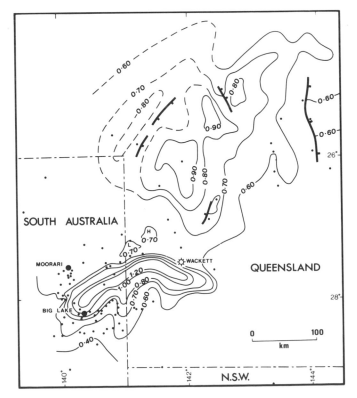

Figure 12. Vitrinite reflectance at top Birkhead Formation (Middle Jurassic).

Locality	Reservoir	API Gravity	Saturates %	Aromatics %	Hetero-compounds %	Prist./Phyt.	Prist./nC_{17}	Phyt./nC_{18}	CPI*	Sulphur %
FLY LAKE	TIRRAWARRA SST.	50	88	10	2	6·1	0·4	0·1	1·10	0·05
TIRRAWARRA	TIRRAWARRA SST.	46	88	10	2	6·0	0·6	0·1	1·11	0·06
DULLINGARI MURTA	MURTA MBR.	54	61	31	8	4·9	0·4	0·1	1·18	0·2
CUTTAPIRRIE	BASAL JURASSIC	50	68	22	10	5·8	0·7	0·1	1·09	0·1
STRZELECKI	HUTTON SST.	43	84	11	5	3·5	0·4	0·1	1·13	0·2
JACKSON	HUTTON SST.	41	90	8	2	4·0	0·3	0·1	1·12	0·04

Table 2. Compositional data of typical Cooper and Eromanga Basin crude oils and condensates.

$$*CPI = \frac{(C_{23}+C_{25}+C_{27}+C_{29})wt\% + (C_{25}+C_{27}+C_{29}+C_{31})wt\%}{2(C_{24}+C_{26}+C_{28}+C_{30})wt\%}$$

both connate and artesian waters in Jurassic sandstones may also have caused temporal variations of the thermal regime.

Timing of Hydrocarbon Generation

Maturation modeling shows that, with the exception of the deeper parts of the Nappamerri Trough, most hydrocarbon generation is likely to accompany or post-date deposition of the Cenomanian Winton Formation (Figures 16 and 17). In the Moomba-Big Lake area of the Nappamerri Trough, Permian source rocks passed through the zone of significant oil generation during Cenomanian subsidence and have subsequently been generating gas (Figure 16). Lower and Middle Jurassic source rocks entered the oil window in the Late Cretaceous and Early Tertiary with peak oil generation taking place in the Late Tertiary. The Upper Jurassic-Lower Cretaceous interval reached initial oil maturity in the mid-Tertiary but has only recently begun generating significant volumes of oil. In the Burley-Kirby area of the central Nappamerri Trough, oil generation from the Lower Permian sequence was complete prior to deposition of the Winton Formation and the final phase of significant gas generation took place in the Early Tertiary. The Upper Permian coal measures sequence continues to generate gas. Lower and Middle Jurassic source rocks passed through the oil generation window in the Late Cretaceous and Tertiary, whereas Upper Jurassic-Lower Cretaceous source rocks are currently nearing peak oil generation.

A different maturation sequence is evident in the Patchawarra Trough. At Cuttapirrie-1, for example, the Permian section entered the significant phase of oil generation following Cretaceous deposition and has remained within the oil window since that time (Figure 17). Lower Jurassic source rocks reached the threshold of significant oil generation in Late Tertiary to Recent time. The generation history of the Patchawarra Trough axis is broadly similar.

Permian source rocks in the central depressions of the northern Cooper Basin passed through the oil window in the Early Tertiary. Gas generation has occurred since that time and is confirmed by the presence of gas accumulations in Tertiary structures such as Wareena. Jurassic source rocks also entered the oil window in the Tertiary but are only now approaching peak oil generation. Jurassic maturation levels decrease toward the margins of the basin, indicating restricted potential for hydrocarbon generation over much of this area.

Most structures in the southern basin predate major hydrocarbon generation and are well-placed in time to have received charge. Younger structures throughout the basin, which either formed or were rejuvenated during the mid-Tertiary, have a charge risk since their development may post-date major phases of generation and migration (e.g., the Curalle Dome, Innamincka Complex). However, some of these young structures contain commercial quantities of gas in Permian reservoirs (Wareena) or good indications of oil in Jurassic reservoirs (Morney). This confirms that late generation and migration has occurred. Few Jurassic reservoirs, including those in early formed structures, are filled to spill point. This suggests that generation is incomplete, but factors such as inadequate source rock volume, dispersal of hydrocarbons during migration, of ineffective caprocks may also have influenced charge volume.

HYDROCARBON TYPE AND DISTRIBUTION

Oil in Permian Reservoirs

The majority of the Permian oils, are light (43° to 56° API), low sulfur, highly paraffinic, condensate-like crudes which have high wax contents and high (3.1 to 10.0) pristane/phytane ratios (see Table 2; Figure 18; Powell and McKirdy, 1976). Differences in n-alkane distribution of the Tirrawarra and Fly Lake crudes (Figure 18) suggest

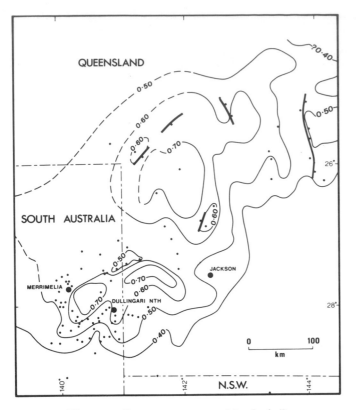

Figure 13. Vitrinite reflectance at top transition beds (Lower Cretaceous).

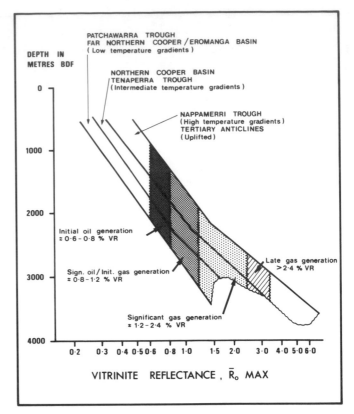

Figure 14. Subdivision of total data vitrinite reflectance versus depth plot by structural province.

differences in either source rock maturation levels at the time of explusion, or subsequent degree of maturation of the crude in the reservoir. All features indicate explusion from a mature source rock rich in land plant organic matter.

Mathews et al (1978) have correlated the Tirrawarra crude with extracts from interseam shales in coal measures of the upper Patchawarra Formation. Smyth (1979) describes this interval as being rich in exinite and vitrinite in neighboring wells and mature for peak oil generation throughout the Patchawarra Trough (Figures 7 and 10). Therefore, it is considered the primary source of the oil and wet gas found within the Permian reservoirs of the trough and may also contribute hydrocarbons to reservoirs on the neighboring Gidgealpa-Merrimelia-Innamincka anticlinal trend, where migration is facilitated by major bounding faults. Minor amounts of Permian oil recovered elsewhere in the basin (Pando, Yanpurra, Toolachee, Karmona) also occur in areas of optimal oil maturity.

Oil in Triassic Reservoirs

Oil occurs in Triassic sandstones at Merrimelia and Chandos. The high-wax Chandos oil (Powell and McKirdy, 1976) is similar to most Permian oils. In the absence of any demonstrable Triassic source, it is believed to be of Permian origin. The recently discovered, waxy Merrimelia oils, however, are water-washed and of lower gravity (30° to 42° API) than the Permian oils of the neighboring Patchawarra Trough. Merrimelia field reservoirs are able to tap Permian source and reservoir rocks (Figure 10) and a Permian origin is considered likely.

Oil in Jurassic Reservoirs

Analyses of the oils found in Jurassic reservoirs (Table 2; Figure 19), together with analyses published by McKirdy (1982), indicate that they are mostly light (39° to 56° API), mature, paraffinic to naphthenic, condensate-like crudes, some with a high wax content. Moderate to low pristane/n-C_{17} ratios, low phytane/n-C_{18} ratios, and intermediate pristane/phytane ratios (Lijmbach, 1975), together with proprietary mass spectrometric typing techniques—indicate a bacterially enriched, mixed (Type II/Type III kerogen) source(s). The lack of gasoline fraction in these waxy oils is attributed to water-washing in an artesian system.

Oils and extracts from Basal Jurassic, and Middle Jurassic Birkhead Formation source rocks are generally dominated by waxy hydrocarbons whereas extracts from Upper Jurassic/Lower Cretaceous source rocks are less waxy and appear to have a higher bacterial component.

The Jurassic oils are all hosted by thermally mature sedimentary rocks (VR = 0.60 to 0.80 percent) and may, in part, be locally sourced. However, their geochemical character indicates generation and migration from more mature source rocks in the trough axes (e.g., the Birkhead Formation oil found at Big Lake; Figure 16). Oils found in the Murta Member are geochemically distinct (Figure 19) from the more waxy crudes found lower in the sequence. Despite their mature appearance they are believed to be generated in the axis of the Nappamerri Trough (Figure 13) from bacterially-enriched, initially mature, Upper Jurassic/Lower Cretaceous source rocks which contain a

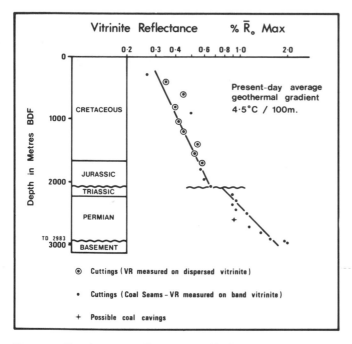

Figure 15. Depth versus reflectance profile for Yalcumma-1 well, southern Nappamerri Trough.

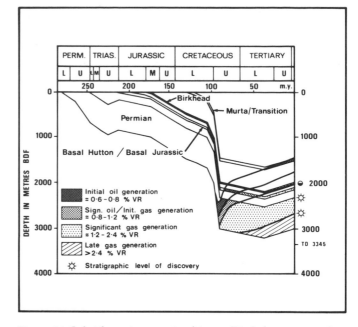

Figure 16. Subsidence/maturation history Big Lake area, southern Nappamerri Trough. Present-day geothermal gradient 5.6°C/100 m. Maturation successfully modeled with time variant dog-leg temperature gradients up to 7.0°C/100 m (see Lopatin, 1971, and Waples, 1980, for details of model technique). VR data measured from well.

minor marine component.

Gas in Permian Reservoirs

The composition and distribution of Permian gases has previously been described by Kantsler and Cook (1979), Schwebel, Devine, and Riley (1980) and Rigby and Smith (1981).

Dry gas is found in the axis of the overmature Nappamerri Trough. Content of wet gas increases away from this overmature zone and reaches a maximum in the oil mature Patchawarra Trough (Figure 20). The increase of wet gas content with depth in the Patchawarra Trough reflects increasing condensate or wet gas generation but this pattern may be enhanced by preferential vertical migration of methane. Methane $\delta^{13}C$ isotopic composition varies from -42.9 to -27.4 per mil and increases with increasing host reservoir maturity. These values are compatible with a source from mature Permian coals and freshwater organic matter.

Carbon dioxide (CO_2) content chiefly ranges from 10 to 20 percent and is concentrated in the trough axes (Figure 21), reaching a maximum of 51 percent at Kanowana in the southern Patchawarra Trough. Schwebel, Devine, and Riley (1980) and McKirdy (1982) believe that the CO_2 originates from the thermal degradation of humic organic matter within the Permian coal measures. These authors demonstrate a correlation between increasing CO_2 content and $\delta^{13}C$ of CO_2, and increasing temperature and vitrinite reflectance. Rigby and Smith (1981) dispute an origin from organic carbon on the basis of the $\delta^{13}C$ isotopic composition of the CO_2, which has a mean value of -6.9 per mil that is far greater than the values of -20 to -30 per mil expected from coal. They attribute the high CO_2 content to either emplacement or metamorphism of basement granites ($\delta^{13}C$ of potentially available CO_2 would be -4.0 to -10.5 per

mil). However, the trend of increasing CO_2 content with depth throughout the basin (irrespective of maturation level, Figure 21), conflicts with the conclusions above. A further inconsistency is the absence of granite basement in the low temperature, but CO_2-rich, Patchawarra Trough. The erratic geographical distribution of subsurface CO_2 may therefore result from multiple sources, high solubility in formation fluids and high reactivity (as discussed by Hunt, 1979).

An alternative origin is the thermal breakdown of widespread Cambrian carbonates of the underlying Warburton Basin through deep seated igneous intrusion, metamorphism, or volcanic activity. Farmer (1965) and Tiratsoo (1972) claim that most large accumulations of CO_2 are sourced in this fashion. Isotopic composition of the CO_2 is not compatible with that of the carbonates on the western flank of the Patchawarra Trough ($\delta^{13}C$ of 0 to $+1$ per mil; Rigby and Smith, 1981) but mixing with isotopically lighter CO_2 derived from humic Permian source rocks could explain this difference. Deep-seated thermal activity may relate to the thermal events postulated from the maturation data.

Gas in Triassic Reservoirs

Gas found in Triassic reservoirs at Merrimelia and Moorari is moderately wet and of high CO_2 content. In the absence of demonstrable Triassic source rock quality in these areas, a Permian source is considered most likely.

Gas in Jurassic Reservoirs

Gas found in Jurassic reservoirs at Namur (Namur Sandstone), Wackett (Birkhead Formation), and Packsaddle

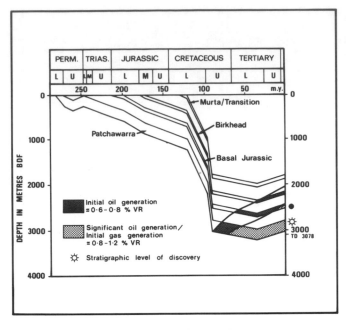

Figure 17. Subsidence/maturation history Cuttapirrie-1, northwestern Patchawarra Trough. Present-day geothermal gradient 3.5°C/100 m. Maturation successfully modeled using time variant gradients up to 3.8°C/100 m. VR data measured from well.

(Murta Member) is dry with low CO_2 content (Rigby and Smith, 1981). The lean character and high levels of maturity indicated by the isotopic composition of ethane and methane appear incompatible with regional maturation trends for the Jurassic and suggest a more mature Permian source. Leakage from the Permian is facilitated by faulting of Permian and Triassic seals on the flanks of each of these structures (for example, Namur). At Packsaddle and Dullingari North, the occurrence of low CO_2, dry-gas in the Upper Jurassic Murta Member reservoir above high CO_2 gas in Triassic and Permian reservoirs suggests migration processes which favor methane rather than CO_2 and wet gas.

ORIGIN, MIGRATION AND ENTRAPMENT OF HYDROCARBONS

Permian gas reservoirs are closely interbedded with source sediments and intraformational seals, resulting in extremely favorable migration and entrapment conditions. Vertical and lateral migration paths are facilitated by communication of the predominantly channel and point bar sandstones and local syndepositional faulting of the regional Murteree Shale and Roseneath Shale seals. Where regional seals remain unbreached, extensive lateral migration toward the structurally high trends flanking the depositional troughs is likely to have occurred. Regional log correlations (Thornton, 1979) show that erosion or non-deposition of regional seals over these trends, together with faulting, allowed vertical migration into the Toolache Formation. Therefore, charge is considered to have occurred within a reasonably continuous migration system. Unsuccessful structural tests either altogether lack Permian strata (Packsaddle), lack good reservoir development (Mount

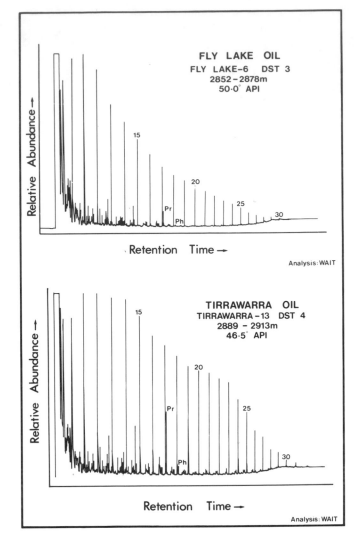

Figure 18. Gas chromatograms, Fly Lake and Tirrawarra oils.

Howitt), suffer fault breaching of intra-Permian and Triassic seals (Namur, Nappacoongee) or are without effective Triassic seal (southern margin of basin).

In the southern basin, the Triassic Mappamerri Formation is a shale-dominated unit, but in the north it is comprised of an upper shale member with seal potential and a lower sandstone section offering reservoir potential. Local reservoir development also occurs within the sealing sequence of the southern basin, as illustrated at Merrimelia (Bowering and Harrison, 1982). This prominent fault-bounded structure (Figure 10) is capable of focusing migration from large downdip or downthrown Permian and Triassic drainage areas. At both Merrimelia and Moorari, Triassic reservoir rocks immediately overlie mature Permian source rocks.

The oil in Basal Jurassic fluvial sandstones at Cuttapirrie is probably sourced by intraformational shales and siltstones. The contrasting maturity of the Cuttapirrie oil (Table 2; Figure 19) and the downdip source sequence (VR = 0.8 to 0.9 percent; Figures 10 and 11) suggests a Permian origin but this is not supported by other geochemical evidence.

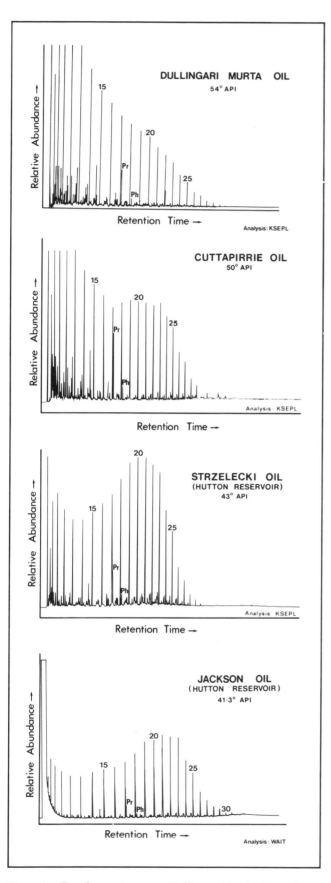

Figure 19. Gas chromatograms, Dullingari North, Cuttapirrie, Strzelecki, and Jackson oils.

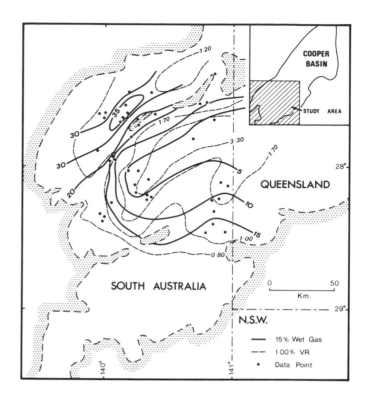

Figure 20. Distribution of wet gas in the Patchawarra Formation (modified from Schwebel et al, 1980).

$$\text{Wet Gas} = \frac{(C_2 + C_3 + C_4)}{(C_1 + C_2 + C_3 + C_4)} \%$$

The overlying, Jurassic Hutton and Namur Sandstones are regional, thick, braided-stream aquifers which until recently were believed to be entirely flushed. However, hydrodynamic studies by Bowering (1982) indicate that the present-day regional hydraulic gradient is largely incapable of displacing existing oil accumulations. Hydrodynamic gradients may have been greater in the past, but the overall light character of the crudes (40° to 50° API) suggests that they are unlikely to be affected by artesian flow. However, artesian flow probably does affect modern migration and charge. The effects are likely to be minimal since most charge models rely on substantial hydrocarbon generation and migration prior to the onset of maximum artesian flow during the Pliocene-Pleistocene.

Jurassic sandstones become shaley or wedge-out to the east and northeast of the basin where the potential for stratigraphic or combination traps is considered high. Several accumulations (Birkhead Formation at Wackett, Murta Member at Dullingari North, Westbourne Formation at Jackson) occur in such restricted sandstone bodies. They are probably sourced by mature sequences in the axis of the Nappamerri Trough.

Largest reserves of Jurassic oil are found in Hutton Sandstone reservoirs in simple structural traps at Strzelecki and Jackson. A downdip Birkhead Formation (Big Lake) source appears likely for Strzelecki, where migration may be assisted by major faults (Figure 10). Bowering (1982) has suggested that oil within the Namur Sandstone at Strzelecki

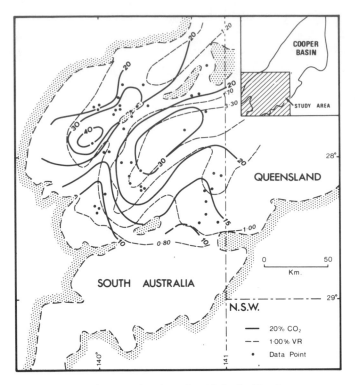

Figure 21. Distribution of carbon dioxide in the Patchawarra Formation (modified from Schwebel et al, 1980).

was displaced from traps further downflank (Namur, Marabooka) by the influx of gas into the lower structures (Gussow effect). The gas-filled Namur structure is probably sourced by gas leaking from the Permian via major faults. At Jackson, the Hutton Sandstone directly overlies the mature Permian sequence immediately updip of the wedge-out of the Triassic seal. A Permian oil source is therefore considered likely. Subsequent migration is hindered by the sealing facies of the Westbourne Formation, which itself contains a small accumulation of oil geochemically similar to that of the underlying Hutton Sandstone.

The Namur Sandstone at the top of the Jurassic sequence is sealed regionally by widespread lacustrine shales of the Transition Beds and/or Murta Member and is therefore capable of trapping hydrocarbons either generated in surrounding formations or leaking/spilling from reservoirs lower in the section. In the latter situation, intervening lithologies are dominated by sandstone and present few barriers to migration. Thin, regressive shoreline sandstones in the upper part of the Murta Member, which provide stratigraphic traps for oil at Dullingari North and Jackson, occur in close stratigraphic proximity to the probable source-rock horizons. Again, migration is not likely to be hindered by intraformational shales.

Valid but dry Jurassic tests are attributed to lack of charge due to erratically developed or marginally mature Jurassic source rocks which have generated only a limited amount of oil and gas.

CONCLUSIONS

1. Fair-to-good, locally rich, source rocks for oil and gas occur throughout the Cooper/Eromanga sequence. Most

of the organic matter is land plant related and gas prone (Type III), but oil source potential is enhanced by local concentrations of exinite or bacterially degraded plant tissue (mixed Type II/Type III). A marine source component of minor significance is locally present in the Lower Cretaceous section.

2. Gas and oil discoveries in the Permian-Triassic Cooper Basin are sourced from thick interbedded Permian coal measures which are mature to postmature for oil generation throughout the basin. The nature and distribution of Permian hydrocarbons is closely related to maturation. Dry gas is found in the postmature Nappamerri Trough whereas wet gas and oil occur in the oil-mature Patchawarra Trough. The potential of the northern Cooper Basin is reduced by a considerably thinner Permian source sequence than that found in the southern basin.

3. Most of the oils discovered in Eromanga Basin reservoirs are of high API gravity and have a mature character. Waxy crudes are sourced from mature basal Jurassic and Middle Jurassic Birkhead Formation source rocks but a Permian source is equally possible for some discoveries. Upper Jurassic-Lower Cretaceous Murta Member oils are geochemically distinct from oils in stratigraphically lower reservoirs and are believed to have been sourced from initially mature source rocks rich in bacterially degraded organic matter.

4. Mapping of rank variation throughout the Cooper and Eromanga basins indicates:
   (1) Evidence of early high heat flow during the Permian-Triassic (manifested by an offset or pronounced change in slope of VR versus depth plots at the unconformable Jurassic/Triassic boundary), and;
   (2) Evidence of a more recent thermal event, particularly in the south of the basin, where isoreflectance contours and present-day isotherms show poor correlation.

5. The high CO_2 content of the Permian gas fields appears to be derived from several sources including Permian coal measures and basement rocks. Deep-seated thermal events related to Miocene tectonism and/or migration of the Cooper Basin over a sub-crustal "hot spot," may have caused thermal breakdown of carbonates in these basement rocks.

ACKNOWLEDGMENTS

Maturation data were obtained as part of a doctorate program (AJK) at the University of Wollongong, supported by Delhi Petroleum Pty Ltd and its co-venturers. Permission to publish, and provision of much basic information and valuable discussion by Delhi and partners, is gratefully acknowledged. Source rock and oil analyses were the work of geochemists at Koninklijke/Shell Exploratie en Productie Laboratorium, Rijswijk, The Netherlands (KSEPL), who also provided detailed interpretation of the data. G.W. Woodhouse (WAIT) performed the analyses of the Jackson, Tirrawarra, and Fly Lake crudes and undertook source rock investigations in the northern basin. The authors thank B.M. Thomas, K.S. Jackson, and D.M. McKirdy for stimulating and enlightening discussion. Thanks also to K.

Spence who provided the modified Lopatin program, and
W.G. Townson, K.S. Jackson, and D.M. McKirdy who
critically reviewed the manuscript. This paper is published
with the support and permission of management of both
Shell Development (Australia) Pty Ltd and Delhi Petroleum
Pty Ltd.

REFERENCES CITED

Barr, T.M., and B.C. Youngs, 1981, Cuttapirrie-1 an oil
discovery in the Early Jurassic of the Eromanga Basin:
Australian Petroleum Exploration Association Journal,
v, 21, n. 1, p. 60-70.

Battersby, D.G., 1976, Cooper Basin gas and oil fields, *in*
R.B. Leslie, H.J. Evans, and C.L. Knight, eds., Economic
geology of Australia and Papua New Guinea—3.
petroleum: Australasian Institute of Mining and
Metallurgy, Monograph 7, p. 321-368.

Beddoes, L.R., Jr., 1973, Oil and gas fields of Australia,
Papua New Guinea and New Zealand: Sydney, Tracer
Petroleum Mining Publications, 391 p.

Beeston, J.W., 1975, Some coal rank determinations in the
Cooper Basin: Queensland Government Mining Journal,
v. 76, p. 266-268.

Bowering, O.J., 1982, Hydrodynamics and hydrocarbon
migration—a model for the Eromanga Basin: Australian
Petroleum Exploration Association Journal, v. 22, n. 1,
p. 227-236.

——, and D.M. Harrison, 1982, The Merrimelia oil and
gas field, *in* P.S. Moore and T.J. Mount, compilers,
Eromanga Basin Symposium, summary Papers:
Adelaide, Geological Society of Australia and Petroleum
Exploration Society of Australia.

Brooks, J.D., 1970, The use of coals as indicators of the
occurrence of oil and gas: Australia Petroleum
Exploration Association Journal, v. 10, n. 1, p. 35-40.

——, W.R. Hesp, and D. Rigby, 1971, The natural
conversion of oil to gas in sediments in the Cooper Basin:
Australian Petroleum Exploration Association Journal,
v. 11, n. 1, p. 121-125.

Cook, A.C., 1981, Temporal variation of type in Australian
coal seams: Bulletin des Centres de Recherches
Exploration—Production Elf-Aquitaine, n. 5, p. 443-459.

——, 1982, Organic facies in the Eromanga Basin, *in* P.S.
Moore and T.J. Mount, compilers, Eromanga Basin
Symposium, summary papers: Adelaide, Geological
Society of Australia and Petroleum Exploration Society
of Asutralia.

Demaison, G.J., 1975, Relationships of coal rank to
palaeotemperatures in sedimentary rocks, *in* B. Alpern,
ed., Pétrographie de la matière organique des sediments,
relations avec la paleotemperature et le potentiel
pétrolier: Paris, Centre National de la Recherche
Scientifique, p. 217-224.

Exon, N.F., and B.R. Senior, 1976, The Cretaceous of the
Eromanga and Surat basins: Bureau of Mineral Resources
Journal of Australian Geology and Geophysics, n. 1, p.
33-50.

Farmer, R.E., 1965, Genesis of subsurface carbon dioxide, *in*
A. Young and J.E. Gulley, eds., Fluids in subsurface
environments: AAPG Memoir 4, p. 378-385.

Hunt, J.M., 1979, Petroleum geochemistry and geology:
San Francisco, W.H. Freeman and Company.

Kantsler, A.J., and A.C. Cook, 1979, Rank variation in the
Cooper and Eromanga basins, central Australia: Report
for Delhi Petroleum Pty. Ltd., unpublished.

——, G.C. Smith, and A.C. Cook, 1978, Lateral and
vertical rank variation; implications for hydrocarbon
exploration: Australian Petroleum Exploration
Association Journal, v. 18, n. 1, p. 143-156.

Klemme, H.D., 1980, Petroleum basins—classification and
characteristics: Journal of Petroleum Geology, n. 3, p.
187-207.

Lijmbach, G.W.M., 1975, On the origin of petroleum, *in*
Proceedings of the 9th World Petroleum Congress, v. 2,
Geology: London, Applied Science Publishers, p.
357-369.

Lopatin, N.V., 1971, Temperature and geologic time as
factors in coalification: Akademi Nauka SSSR Izvestiya
Seriya Geologicheskaya, v. 3, p. 95-106 (in Russian)
(English translation by N.H. Bostick, Illinois State
Geological Survey, 1972).

Mathews, R.T., et al, 1978, Genetic studies on the
Tirrawarra oil field, Cooper Basin, central Australia:
Melbourne University (unpublished).

McKenzie, D., 1978, Some remarks on the development of
sedimentary basins: Earth and Planetary Science Letters,
n. 40, p. 25-32.

McKirdy, D.M., 1982, Aspects of the source rock and
petroleum geochemistry of the Eromanga Basin, *in* P.J.
Moore and T.J. Mount, compilers, Eromanga Basin
Symposium, summary papers: Adelaide, Geological
Society of Australia and Petroleum Exploration Society
of Australia.

Middleton, M.F., 1979, Heat flow in the Moomba, Big
Lake, and Toolachee gas fields of the Cooper Basin and
implications for hydrocarbon maturation: Bulletin of the
Australian Society of Exploration Geophysicists, n. 10,
p. 149-155.

——, 1980, A model of intracratonic basin formation,
entailing deep crustal metamorphism: Geophysical
Journal of the Royal Astronomical Society, n. 62, p.
1-14.

Moore, P.S., and T.J. Mount, compilers, 1982, Eromanga
Basin Symposium, summary papers: Adelaide,
Geological Society of Australia and Petroleum
Exploration Society of Australia.

Mount, T.J., 1981, Dullingari North 1—an oil discovery in
the Murta Member of the Eromanga Basin: Australian
Petroleum Exploration Association Journal, v. 21, n. 1,
p. 71-77.

Poll, J.J.K., 1981, The significance of the southwest
Eromanga Basin oil and gas discoveries (central
Australia): Australian Petroleum Exploration
Association Journal, v. 21, n. 2, p. 33-38.

Powell, T.G., and D.M. McKirdy, 1976, Geochemical
character of crude oils from Australia and Papua New
Guinea, *in* R.B. Leslie, H.J. Evans, and C.L. Knight, eds.,
Economic geology of Australia and Papua New Guinea,
3, Petroleum: Australasian Institute of Mining and
Metallurgy, Monograph 7, p. 18-29.

——, and Snowdon, L.R., 1980, Geochemical controls on
hydrocarbon generation in Canadian sedimentary
basins, *in* A.D. Miall, ed., Facts and principles of world

petroleum occurrence: Calgary, Canadian Society of
Petroleum Geologists Memoir 6, 1003 p.

Radke, M., et al, 1980, Composition of soluble organic
matter in coals; relation to rank and liptinite
fluorescence: Geochimica et Cosmochimica Acta, n. 44,
p. 1787-1800.

Rigby, D., and J.W. Smith, 1981, An isotopic study of gases
and hydrocarbons in the Cooper Basin: Australian
Petroleum Exploration Association Journal, v. 21, n. 1,
p. 222-229.

Sass, J.H., and A.H. Lachenbruch, 1979, Thermal regime of
the Australian continental crust, in M.W. McElhinny,
ed., The earth—its origin, structure and evolution:
London, Academic Press, p. 301-351.

Schwebel, D.A., S.B. Devine, and M. Riley, 1980, Source,
maturity and gas composition relationships in the
southern Cooper Basin: Australian Petroleum
Exploration Association Journal, v. 20, n. 1, p. 191-200.

Senior, B.R., A. Mond, and P.L. Harrison, 1978, Geology
of the Eromanga Basin: Australia Bureau of Mineral
Resources Bulletin, n. 167.

——— , and M.A. Habermehl, 1980, Structure,
hydrodynamics, and hydrocarbon potential, central
Eromanga Basin, Queensland, Australia: Bureau of
Mineral Resources Journal of Australian Geology and
Geophysics, n. 5, p. 47-55.

Shibaoka, M., and A.J.R. Bennett, 1977, Patterns of
diagenesis in some Australian sedimentary basins:
Australian Petroleum Exploration Association Journal,
v. 17, n. 1, p. 58-63.

——— , J.D. Saxby, and G.H. Taylor, 1978, Hydrocarbon
generation in Gippsland Basin, Australia—comparison
with Cooper Basin, Australia: AAPG Bulletin, v. 62, p.
1151-1158.

Smith, I.W., 1980, The flash pyrolysis method for
converting coal to oil: Coal Geology, v. 1, n. 3, p.
133-138.

Smyth, M., 1979, Hydrocarbon generation in the Fly
Lake-Brolga area of the Cooper Basin: Australian
Petroleum Exploration Association Journal, v. 19, n. 1,
p. 108-114.

Snowdon, L.R., and T.G. Powell, 1982, Immature oil and
condensate, modification of hydrocarbon generation
model for terrestrial organic matter: AAPG Bulletin, v.
66, p. 775-788.

Thomas, B.M., 1982, Land plant source rocks for oil and
their significance in Australian basins: Australian
Petroleum Exploration Association Journal, v. 22, n. 1,
p. 164-178.

Thornton, R.C.N., 1979, Regional stratigraphic analysis of
the Gidgealpa Group, southern Cooper Basin, Australia:
Geological Survey of South Australia Bulletin, n. 49.

Tissot, B.P., and D.H. Welte, 1978, Petroleum formation
and occurrence: Berlin-Heidelberg, Springer-Verlag, 538
p.

Tiratsoo, E.N., 1972, Natural gas, 2nd edition:
Beaconsfield, England, Scientific Press.

Waples, D.W., 1980, Time and temperature in petroleum
formation; application of Lopatin's method to petroleum
exploration: AAPG Bulletin, v. 64, p. 916-926.

Wopfner, H., 1960, On some structural development in the
central part of the Great Australian Artesian Basin:
Transactions of the Royal Society of South Australia, v.
83, p. 179-193.

Young, B.C., 1975, The hydrology of the Gidgealpa
formation of the western and central Cooper Basin:
Geological survey of South Australia, Report of
Investigation, n. 43.

Hydrocarbons, Source Rocks, and Maturity Trends in the Northern Perth Basin, Australia

B.M. Thomas
Shell Development (Australia) Pty Ltd
Perth, Western Australia

All known commercial hydrocarbon accumulations in the Perth Basin, Western Australia, occur within the Dandaragan Trough or along its flanks. Land plant-rich source rocks are widely distributed throughout the Permian, Triassic, and Jurassic sections of the basin. Hydrocarbon accumulations are mainly dry gas and gas/condensate, although secondary occurrences of light, waxy oil are also of economic significance. The Lower Jurassic Cattamarra Coal Measures provide both source and reservoir for gas/condensate accumulations in the central Dandaragan Trough (Walyering, Gingin). Gas at Dongara, Mondarra, and Yardarino may have been generated from both the Lower Triassic and Permian, although there is some evidence that the Permian is the principal source. The associated thin oil leg encountered in parts of these fields is attributed to the oil-prone basal Kockatea Shale (Lower Triassic). Regional studies indicate a Neocomian uplift of the western flank of the Dandaragan Trough centered on the Beagle Ridge. The extent and magnitude of truncation is reflected in systematic variations in sandstone porosity trends. Vitrinite reflectance data suggest that the uplift and erosion of the Beagle Ridge was accompanied by higher geothermal gradients than measured today in exploratory boreholes. Modern gradients of up to 5.0°C/100 m were measured on the Beagle Ridge and possibly represent this waning geothermal anomaly. In contrast, low geothermal gradients are found in the Dandaragan Trough (around 2.0°C/100 m), and hydrocarbon generation occurs at great depths where sandstone reservoir properties are often inadequate for commercial production. Extensive vertical migration of hydrocarbons to shallower levels with better reservoir characteristics may be a prerequisite for producible accumulations in areas with low gradients.

TECTONIC SETTING

The Perth Basin (Jones, 1976; Playford, Cockbain, and Low, 1976; Jones and Pearson, 1972) is an elongate sedimentary trough which straddles the southwest Australian coastline for nearly 700 km (435 mi) between Geraldton and Augusta (Figure 1). It contains a thick sedimentary record of mainly continental clastic sediments ranging in age from Silurian to Recent. The basin has a history of rifting which culminated in the complete separation of India from Western Australia in the Early Cretaceous (Neocomian). The structural history and stratigraphy of the Perth Basin from the Triassic onward are dominated by major down-to-the-west displacements on the Darling-Urella fault system which forms the eastern margin of the basin (Figure 1). The greatest movement on this fault system occurred throughout the Late Jurassic and Early Cretaceous (Neocomian), with an earlier major pulse in the Late Triassic-Early Jurassic. The structural elements into which the basin can be divided are essentially horsts and grabens related to this fault system.

This study is restricted to the northern Perth Basin, and particularly to the Dandaragan Trough, a major graben in the northeastern, onshore portion of the Perth Basin. The axis of the trough extends from the Precambrian Northampton Block in the north to the Harvey Ridge in the south. Its eastern boundary is the Darling-Urella fault system, whereas on the West lies the Beagle Ridge, a shallow basement feature which has persisted throughout much of the basin history. The Cadda Shelf forms an intermediate terrace between the Beagle Ridge and the Dandaragan Trough. The Dongara Saddle is a hinge region between the southern nose of the Northampton Block and the Beagle Ridge. It is the site of the major commercial hydrocarbon reserves of the Perth Basin. Presently, the Dandaragan Trough is a half-graben, with the maximum sedimentary section (over 14,000 m, or 45,932 ft, based on reflection seismic data) adjacent to the Darling Fault, thinning westward onto the Beagle Ridge by convergence and truncation (Figure 2).

STRATIGRAPHY AND BASIN DEVELOPMENT
OF THE DANDARAGAN TROUGH
(Figure 3)

In the extreme north of the basin, Precambrian
crystalline basement is overlain by a fluviatile red-bed
sequence, the Tumblagooda Sandstone. This formation is
believed to be of Silurian age and is more widely-known in
the Carnarvon Basin to the north (Thomas and Smith,
1976). Over most of the northern Perth Basin the record
begins with an Early Permian (Sakmarian) glacial phase
represented by the shallow-marine Nangetty Formation,
which consists of tillite with boulder erratics set in an
unsorted matrix of sandstone, siltstone, and shale.
Overlying the Nangetty Formation is the fossiliferous
marine Holmwood Shale. Each of these units onlaps onto
the Northampton Block-Beagle Ridge and is absent on its
crest. The Artinskian sequence is more widespread,
consisting of interbedded coal seams and sandstone (Irwin
River Coal Measures) overlain by marine shales and
limestones of the Carynginia Formation. Late Permian
tectonic activity resulted in tilting and erosion of the
Artinskian and older sequences. The Upper Permian
Wagina Sandstone was then deposited on the irregular
erosion surface.

The coarse Dongara Sandstone, whose distribution is
controlled by the relief of the pre-Triassic land surface,
marks the onset of a regional marine incursion. It is the
main reservoir in the Dongara, Mondarra, and Yardarino
fields. The overlying Lower Triassic Kockatea Shale is a
regressive sequence consisting of a basal black marine shale
which grades upward into interbedded shale, siltstone, and
fine sandstone. The consistent character of the Lower
Triassic throughout the northern Perth Basin suggests that
subsidence was gentle and seldom localized along faults.
Regression continued into the Middle Triassic (Woodada
Formation). The Lower-Middle Triassic section thickens
toward the center of the Dandaragan Trough, implying
broad axial downwarping. The Upper Triassic Lesueur
Sandstone is a thick (2,200 m [7,218 ft] +) continental
sequence of coarse kaolinitic arenites with minor siltstone
interbeds. Such a large volume of coarse detrital material
implies rejuvenation of the source area and probably
indicates the first major rifting in the Perth Basin.

The earliest sediments of the Lower Jurassic Cockleshell
Gully Formation are interbedded multicolored claystones
and coarse-grained sandstones comprising the Eneabba
Member. As fault activity diminished, alluvial flood plain
sedimentation was succeeded by fluvial-coal
swamp-lacustrine environments in which the Cattamarra
Coal Measures Member was deposited. Cattamarra
sandstones are finer-grained and better-sorted, indicating a
topographically more mature source terrain. A basinwide
marine incursion in the Middle Jurassic (Cadda Formation)
flooded the extensive coal swamps of the Cockleshell Gully
Formation in trough areas. Elevation of the Precambrian
Shield during the Late Jurassic produced floods of coarse,
poorly-sorted sand which poured into the rapidly subsiding
graben. The Yarragadee Formation is up to 4,500 m (14,764
ft) thick, mainly continental in aspect, with localized
lacustrine sedimentation in the central Dandaragan Trough.

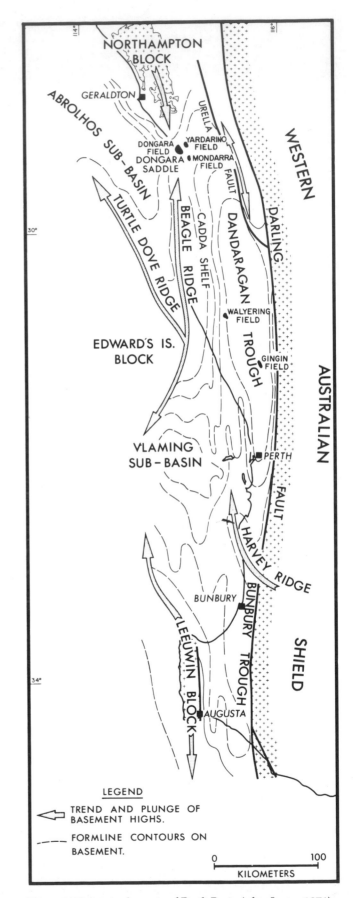

Figure 1. Tectonic elements of Perth Basin (after Jones, 1976).

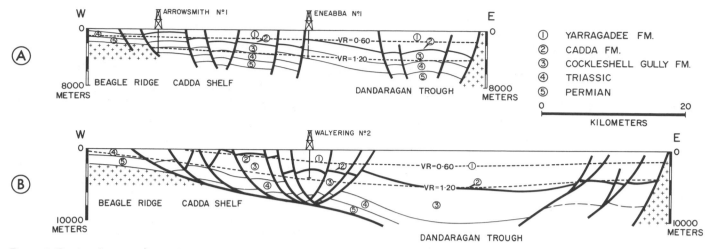

Figure 2. Regional cross sections (see Figure 4 for locations).

During the Neocomian, subsidence was interrupted by a major tectonic event which was centered on the Beagle Ridge. The uplift and truncation of the western flank of the Dandaragan Trough is thought to mark the end of rifting and the onset of drifting as the continent of India moved away from what is now Western Australia. By the end of the Neocomian, the present form of the Dandaragan Trough had largely been established although several minor marine incursions occurred during the Late Cretaceous (Coolyena Group).

HYDROCARBON PRODUCTION

The Dandaragan Trough and the Dongara Saddle are the only areas of the Perth Basin which have yielded commercial hydrocarbon production. Two distinct areas of hydrocarbon occurrence are known.

Dongara Saddle and Adjacent Areas

The Dongara gas field (0.4 Tcf recoverable) was discovered in 1966 and now provides natural gas to the city of Perth 300 km (186 mi) to the south (Figure 4). The field produces relatively lean gas (97 percent methane) from interconnected reservoirs in the Dongara Sandstone (main producer), Wagina Sandstone, and Irwin River Coal Measures. The field is structurally complex (see Jones, 1976; Hoseman, 1971) and some wells produce minor amounts of highly paraffinic crude oil of 35° API gravity. The Dongara reservoir is capped by the Triassic Kockatea Shale. Adjacent to Dongara is the Mondarra gas field, which also produces from the Dongara Sandstone. Minor oil (40.9° API) was recovered from one well and is similar in composition to the Dongara crude. The small Yardarino field is now linked to the Dongara-Perth pipeline and produces gas from the Dongara Sandstone. One well has produced minor amounts of light (44° API) paraffinic crude.

In 1965, oil was recovered from sandstone lenses within the upper Kockatea Shale in Mt Horner-1 (Figure 4) where a production test yielded 50 barrels of oil/day of 38° API crude with a 50 percent water cut. An appraisal well (Mt Horner-2) failed to recover hydrocarbons. However, renewed exploration in the area (1981) led to further drilling

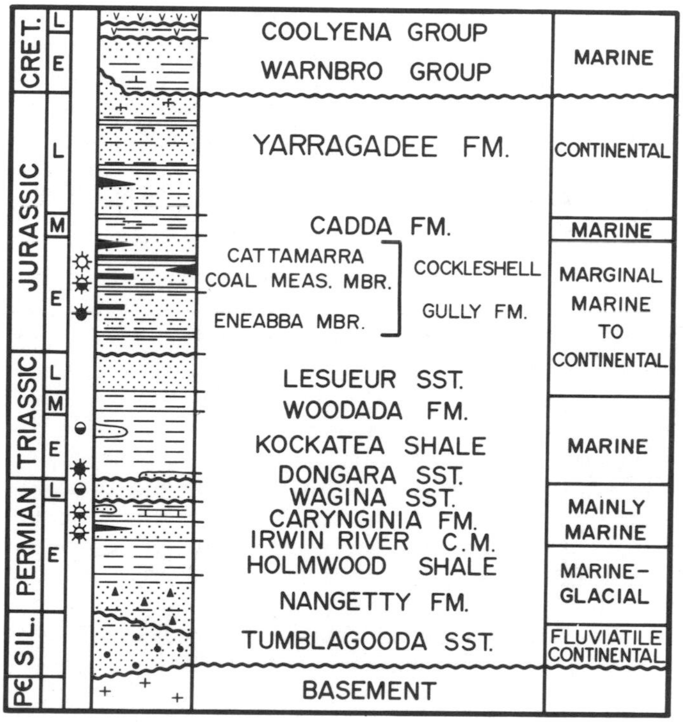

Figure 3. Stratigraphy of the Dandaragan Trough.

and recent wells, Mt Horner-3 and -4, located with the aid of modern seismic data, appear to confirm a small but possibly commercial oil field.

Recent development drilling (1981) in the Dongara field has revealed additional gas reserves within thin lenticular sandstones of the upper Kockatea Shale (Arranoo Member). These reservoirs are stratigraphically equivalent to the oil-productive levels at Mt. Horner.

Gas was discovered in 1980 at Woodada-1 in a fractured carbonate reservoir of the Lower Permian Carynginia Formation. Flow rates of up to 33.4 MMcf/day were reported from the discovery well, after acid stimulation (Norlin, 1981). Further drilling confirmed the discovery and revealed the possible presence of an oil leg. Prolonged gas production testing is planned.

Minor oil and gas have also been tested from the Jurassic Eneabba Member in Erregulla-1, but an appraisal well, drilled in 1980 only 200 m (656 ft) from the discovery well, was dry. A sandstone near the top of the Permian Caryanginia Formation in Arrowsmith-1 has produced gas, but production tests indicated that the reservoir is limited. Oils and condensates from the Dongara Saddle and adjacent areas are similar in composition. They are light (35 to 47° API), highly paraffinic, with a low pristane to phytane ratio (1.0 to 1.3). The wax content of the oils is very high and they are commonly solid at room temperature (Powell and McKirdy, 1975).

Dandaragan Trough

A second area of hydrocarbon accumulation lies in the central Dandaragan Trough. Gas/condensate has been produced from the Gingin and Walyering fields (Figure 4) although both are now depleted. Multiple pay sands are found in both fields within the Jurassic Cattamarra Coal Measures Member. Condensates are paraffinic/naphthenic, varying from 38.5 to 45.6° API gravity with an intermediate pristane/phytane ratio (2.8 to 5.5). The limiting factor in these fields is reservoir quality, which gradually deteriorates with increasing depth. Both fields were produced during 1972 but a rapid decline in production rates and pressures occurred after only a few months.

SOURCE ROCKS

This study aims to identify the major source rocks of the northern Perth Basin, determine their organic character, and link them to the known hydrocarbon accumulations in the basin. Mapping of organic maturity trends has also provided a valuable tool in understanding the structural and geothermal history of the basin. Throughout this discussion, the distinction is made between "potential source rocks"—those with sufficient organic content to produce hydrocarbons at generative temperatures—and "effective source rocks"—those potential source rocks which can be shown to have reached generative temperatures.

Source rock richness was routinely determined by a total pyrolysis technique and by measurement of total organic carbon (TOC). Both core and cuttings material was used, with a sampling interval of 15 m (49 ft) or less, when possible. Organic type was determined by elemental analysis of acid residues (Tissot et al, 1974). To calibrate this chemical method, duplicate preparations were examined microscopically in transmitted light (Staplin, 1969). Results are supplemented with additional optical typing data published by Kantsler and Cook (1979), who used coal petrographical techniques. All fine-grained units in the northern Perth Basin have been sampled and three groups of potential source rocks have been identified: Permian, Triassic, and Jurassic.

Permian (Holmwood Shale, Irwin River Coal Measures, Caryanginian Formation)

The Holmwood Shale consists of gray-green shale and siltstone and was deposited in a marine environment. Geochemical data are available in four wells which indicate

some source potential (average TOC = 1.2 percent) but mainly for gas. Although rich in coal, the Irwin River Coal Measures is predominantly a sandstone unit, with subordinate siltstone, shale, and coal. Organic content is high (up to 14.2 percent TOC) but hydrogen content is low (H/C = 0.65). Microscopic studies indicate a predominance of inertinite and vitrinite (Kantsler and Cook, 1979), and the unit is considered gas-prone. Overlying the Irwin River Coal Measures is the Caryanginia Formation, a marine, predominantly fine-grained unit which has been penetrated by many wells in the northern Perth Basin. Almost invariably organic content is high (up to 11.4 percent TOC) buy pyrolysis yields are low as the organic type is hydrogen-poor (H/C = 0.62 to 0.76) and includes much inertinitic material. In one well, some oil potential is indicated (H/C = 0.95) but in general it is concluded that the Caryanginia Formation is predominantly a source for gas. Of all Permian units in the northern Perth Basin, the Caryanginia Formation seems to offer the best source potential.

Triassic (Kockatea Shale)

The Kockatea Shale consists predominantly of dark gray-green to black calcareous shales and fossiliferous limestones, which become more silty in the middle and upper sections. The thickest interval penetrated is at Woolmulla-1 (Figure 4) on the west flank of the Dandaragan Trough where the formation is 1,061 m (3,481 ft) thick. The Kockatea thins northward by onlap onto the Northampton Block, and in the area of the Dongara gas field it is about 250 m (820 ft) thick. In most wells, the Kockatea Shale is organically rich at its base (over 2 percent TOC), declining upward to less than 0.5 percent TOC. The rich, basal Kockatea contains kerogen with a high hydrogen content (H/C = 1.1 to 1.3), indicating potential for oil generation. Microscopic studies reveal abundant phytoplankton and finely-divided exinite of probable land plant origin (Kantsler and Cook, 1979). In many wells this rich interval may only be about 15 m (49 ft) thick, and overlying leaner sediments are dominated by hydrogen-poor land plant material.

Jurassic (Cattamarra Coal Measures Member, Cadda Formation, Yarragadee Formation)

The Cattamarra Coal Measures Member is a thick (up to 2,000 m [6,562 ft]) fluvial—coal swamp—lacustrine unit which has widespread distribution throughout the Dandaragan Trough. In such a coal-rich sequence, organic content is frequently high (up to 27.2 percent TOC) although rich horizons are sporadic. The abundance of land plant material and a moderate hydrogen content (H/C = 0.73 to 0.84) suggest that the unit is predominantly gas-prone. However, exinite contents of up to 30 percent were reported (Kantsler and Cook, 1979) which may indicate some potential for liquid hydrocarbons.

The Cadda Formation marks a broad marine transgression into the Dandaragan Trough during the Middle Jurassic. The unit consists of marine shales and limestone in the north, becoming progressively sandy and continental to the south. Where studied, the Cadda is organically lean (average 0.6 percent TOC) and hydrogen-poor (H/C = 0.72) and is considered to have only marginal source potential.

Over most of the basin the Yarragadee Formation is characterized by a predominance of coarse-grained, poorly-sorted sandstones. However, up to 4,500 m (14,764 ft) of a fine-grained fluvial-lacustrine facies is developed in the central Dandaragan Trough which has source potential. Coals are rare and concentrated toward the base of the unit, but shales and siltstones are rich in organic matter (average 1.8 percent TOC) with a predominance of land plant material (H/C = 0.67 to 0.95). Although mainly gas-prone, local concentrations of exinite (Kantsler and Cook, 1979) suggest some potential for liquid hydrocarbons.

It is evident that although organically-rich sediments are widely distributed throughout the geological section in the northern Perth Basin, they contain mainly land plant material. This is probably a result of the proximity of the Precambrian Shield to the east (a source of land plant debris) and the predominance of continental sedimentation in the basin. With the exception of the marine Kockatea Shale, all source rocks in the northern Perth Basin contain kerogen of Type III (Tissot et al, 1974) and are mainly gas-prone. The basal Kockatea Shale is a Type II source rock and is oil-prone. Permian source rocks of the northern Perth Basin contain mainly vitrinite and inertinite, and are likely to generate only dry gas. In contrast, Jurassic land plant source rocks are relatively rich in exinite (sporinite, cutinite, resinite, and suberinite) and have potential to generate gas, condensate, and possibly some light oil.

GEOTHERMAL GRADIENTS

Present-day average geothermal gradients in the northern Perth Basin range from 1.9 to 5°C/100 m (Figure 5). Where possible, these data have been derived from extrapolated bottom hole temperatures (using a Horner Plot) although it is found that a good estimate can often be made by adding 10 percent to recorded temperatures from the first log of a suite to be run (usually IES). Although the temperature variation with depth in some wells is not linear, in general the average gradient gives a good indication of depth/temperature relationships assuming a mean surface temperature of 20°C.

There is a strong correlation between the thickness of the sedimentary section and geothermal gradient in the northern Perth Basin. High gradients are associated with shallow basement (Beagle Ridge, Northampton Block) whereas low values are found in the Dandaragan Trough. Intermediate values occur in areas of moderate sedimentary thickness (Dongara Saddle, Cadda Shelf). The variation in geothermal gradients in the basin may be partly related to the relative conductivities of the sediments, depending on predominant lithologies. For example, the section in the Dandaragan Trough is mainly sandstone (low gradient), while shale predominates on the Beagle Ridge (high gradient).

ORGANIC MATURITY

The degree of organic maturity of source rocks has been determined from coal rank data, vitrinite reflectance measurements (VR), and spore discoloration. Solvent

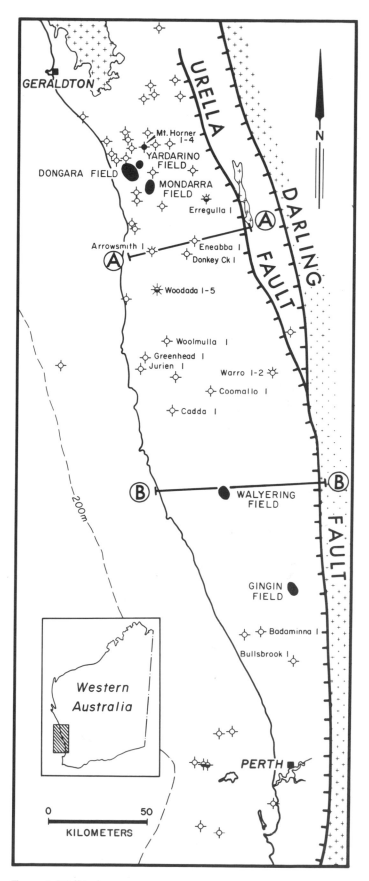

Figure 4. Well index map.

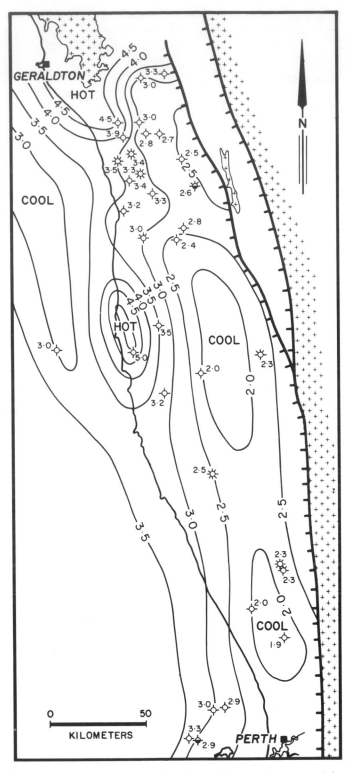

Figure 5. Average geothermal gradients (°C/100 m).

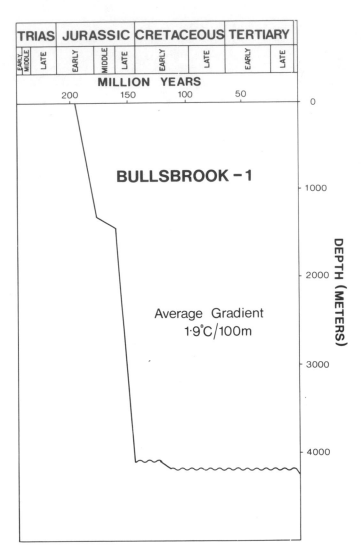

Figure 6. Burial history, Bullsbrook-1.

extracts of source rocks provide additional data on chemical maturity as well as providing a means of oil-source correlation. All of these methods have practical deficiencies, and an integrated approach to maturity estimation has been adopted.

Spore discoloration can be a useful indicator of generative or post-mature conditions, provided brown zone (TAI 3) levels of maturity are reached. At lower levels of maturity (yellow or orange zones), subtle color changes limit the method's precision. Vitrinite reflectance offers a potentially more precise measurement of organic maturity. However, reflectance data from the Perth Basin often has an unusually broad spread, even when core material is used. Kantsler and Cook (1979) note wide variations in vitrinite type and the presence of suberinite and bitumen in some samples, which tends to depress vitrinite reflectance and leads to false readings. A scarcity of dispersed vitrinite in the Kockatea Shale often precludes the direct measurement of maturity by reflectance of this significant source unit.

Detailed vitrinite reflectance profiles have been measured for over 30 wells in the northern Perth Basin. This work is supplemented by data from Kantsler and Cook (1979). Maturation levels and rank gradients vary dramatically over the basin. Despite the wide range of present-day geothermal gradients (Figure 5), the observed variations cannot be attributed to this alone (Thomas, 1979).

Measured vitrinite reflectance trends have been compared with those derived from a mathematical model based on burial history and geothermal gradient (Figure 6).

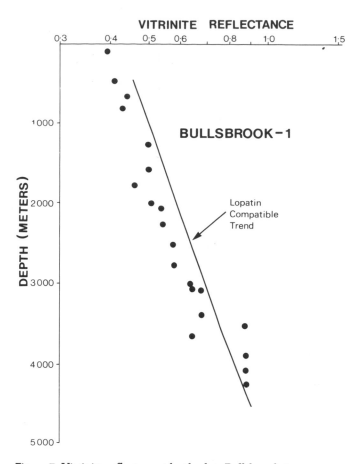

Figure 7. Vitrinite reflectance/depth plot, Bullsbrook-1.

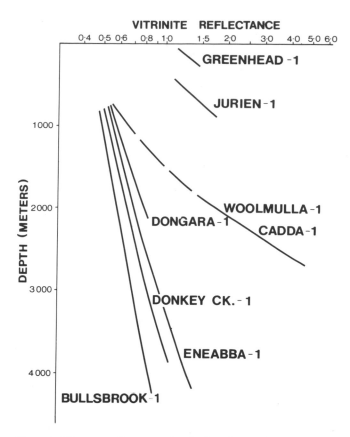

Figure 8. Vitrinite reflectance/depth plots from selected wells.

The method of Lopatin (1971) was used, after careful recalibration (see Waples, 1980). Maturity measurements from deep, synclinal wells in the central Dandaragan Trough (Bullsbrook-1, Gingin-1, Warro-2) are largely compatible with their present burial depth and temperature. The low geothermal gradients measured in the wells appear to be also representative of the geological past, and as a result, sediments at a depth of 4,000 m (13,123 ft) have a vitrinite reflectance of only 1 percent or less (Figure 7). Maturity levels associated with gas generation (VR greater than 1.2) must occur at depths of 5,000 m (16,404 ft) or more in the central Dandaragan Trough.

Maturity levels in the northern Dandaragan Trough (Eneabba-1, Donkey Creek-) and in the Dongara Saddle appear to be compatible, although some Neocomian truncation may have occurred. In contrast, maturity levels and rank gradients in wells on the Cadda Shelf (Woolmulla-1, Cadda-1) and Beagle Ridge (Jurien-1, Greenehead-1) are strongly incompatible, suggesting a complex geothermal history (Figure 8).

REGIONAL TRUNCATION

A number of regional geological factors must be considered in the interpretation of the observed maturity trends. The Beagle Ridge has been structurally prominent since at least the Early Triassic. The Kockatea Shale and most subsequent units thicken eastward into the

Dandaragan Trough, although some stripping may have occurred on the ridge. In addition to the trend of depositional thinning on the Beagle Ridge, there is also evidence of uplift and erosion of the ridge during the Neocomian. This has affected broad areas of the basin to varying degrees depending on distance from the uplift center. Estimation of the magnitude of relative uplift is difficult since well control is limited mainly to the margins of the syncline where some erosion has occurred. Seismic records also indicate that the Yarragadee Formation is composed of huge wedges of sediment which are apparently the result of shifting depocenters during the Late Jurassic. This makes estimation of truncated section hazardous, since it is not clear whether missing section from uplifted areas was of a similar thickness to that preserved elsewhere in synclinal areas. However, it is likely that at least 2,000 m (6,562 ft) of sediment was eroded from the area of the Beagle Ridge during the Neocomian.

POROSITY TRENDS

Sandstone porosity decline in the Dandaragan Trough has been studied in some detail, since this factor often determines the economic viability of gas accumulations in the area. The great variability of sandstone porosity characteristics caused by clay content, sorting, secondary cementation, and other factors is well documented, and tends to obscure regional trends in reservoir parameters. Following the example of Maxwell (1964), an attempt was made to simplify porosity data by restricting it to clean,

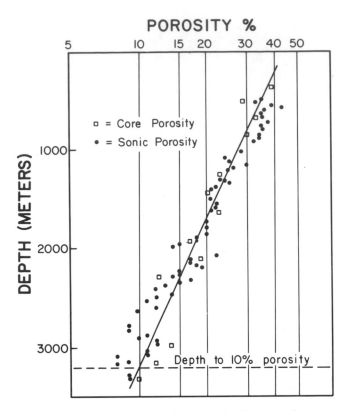

Figure 9. Sandstone porosity/depth plot for Gingin-1. Sonic porosities represent best 6 m (20 ft) of clean sandstone in every 30 m (98 ft) of section. Core porosities are average values.

relatively well-sorted quartzose sandstones.

All data were derived from a single lithological unit with a wide areal distribution and relatively uniform environment of deposition (Yarragadee Formation). Sandstone porosity was determined from sonic logs using the standard Schlumberger chart. Porosity values plotted were the average sonic porosity of the best 6 m (20 ft) of clean (gamma ray log) sandstone in every 30 m (98 ft) of section. This technique provides maximum values (i.e., cleanest, best-sorted sands with no secondary cementation) while avoiding any sonic "shale" effect. Average core porosities, if available, were also plotted. With porosity plotted on a logarithmic scale, the relationship to depth was found to be approximately linear for most wells. The depth to 10 percent porosity (arbitrary gas-production cutoff) was calculated, extrapolating data where necessary beyond the limits of well control. A typical well is Gingin-1 (Figure 9).

The depth to 10 percent porosity within the trough varies systematically and a map of this parameter clearly outlines the synclinal area seen on seismic records (Figure 10). Porosity retention is greatest in the deepest parts of the trough adjacent to the Darling Fault, and porosity characteristics deteriorate progressively westward toward the Beagle Ridge. It seems likely that this is a direct reflection of regional uplift and erosion of overburden and the inferior reservoir characteristics of the western margins of the trough are probably the result of pre-Neocomian deep burial diagenesis. This evidence supports regional observations that sediments of the Beagle Ridge and Cadda Shelf are now at less than maximum depth of burial.

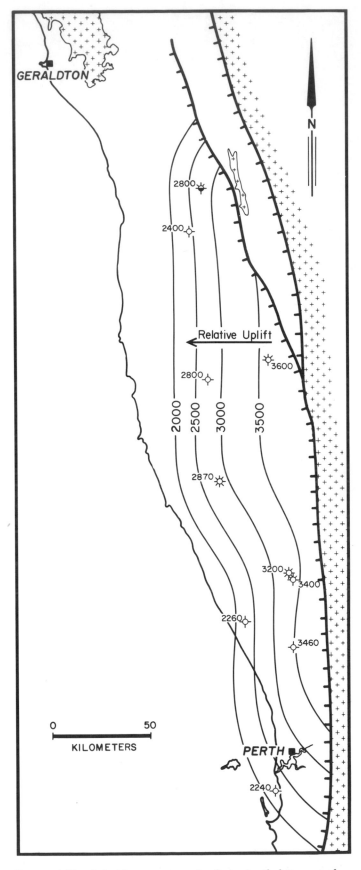

Figure 10. Depth to 10 percent porosity (in meters below ground level) in clean sand in Yarragadee Formation.

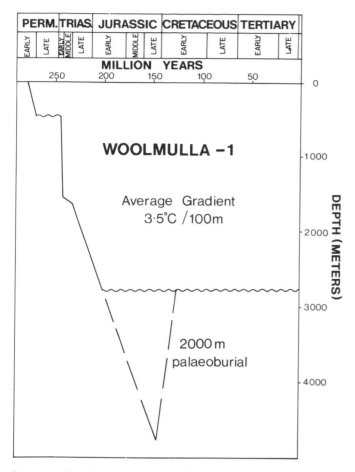

Figure 11. Burial history, Woolmulla-1.

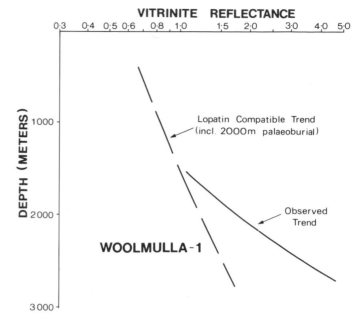

Figure 12. Vitrinite reflectance/depth plot, Woolmulla-1.

THE BEAGLE RIDGE GEOTHERMAL ANOMALY

Geothermal data from wells indicate that there is a strong temperature anomaly associated with the Beagle Ridge. Average gradients as high as 5.0°C/100 m occur at Jurien-1, which is near the crest of the ridge (Figure 5). In more synclinal areas of the basin, gradients of 2.5°C/100 m are common and figures as low as 1.9°C/100 m have been recorded.

Vitrinite reflectance trends measured in two deep wells on the Cadda Shelf have been compared with Lopatin reconstructions based on burial history and temperature. Despite the relatively high modern geothermal gradients in the area, the measured maturity levels are clearly incompatible, even when generous account is taken of paleoburial prior to Neocomian uplift (Figure 11). The steep rank profiles observed (Figure 12) are indicative of even higher paleo-gradients, combined with deeper burial.

The age and cause of this thermal event are subject to speculation. Kantsler and Cook (1979) suggested that higher heat flow occurred between the Permian and the Jurassic, associated with the initiation of rifting in the basin. The author has previously proposed that the thermal event was of Neocomian age, associated with the uplift and truncation of the Beagle Ridge, and that the present day geothermal anomaly is a remnant of that event (Thomas, 1979). The cause of the event appears to have been igneous intrusion,

as cores from Woolmulla-1 consist of metamorphosed sedimentary rocks, still recognizable as Permian Irwin River Coal Measures, which are invaded by quartz veins. The only known igneous event in the history of the Perth Basin is Neocomian in age. It is represented by the Bunbury Basalt (Playford et al, 1976) which was extruded onto the eroded surface of the Neocomian unconformity, although intrusive igneous horizons in the Sue-1 and Whicher Range-2 wells (southern Perth Basin) are probably synchronous.

The Beagle Ridge has suffered greater Neocomian truncation than the Cadda Shelf. In the Greenhead-1 well, maturity levels of VR = 1.4 were encountered in Permian sediments at a depth of 216 m (709 ft). But, since the record is incomplete (through severe truncation) and the well only shallow, it is not possible to separate the effects of higher paleo-gradients from those of deeper burial.

GENERATIVE AREAS OF THE NORTHERN PERTH BASIN

Maturity levels, expressed in terms of vitrinite reflectance (VR), have been mapped on three levels in the northern Perth Basin corresponding to the major source units. Areas of mature source rocks may then be related to the known hydrocarbon accumulations (Figures 13, 14, and 15).

According to Tissot and Welte (1978), the onset of oil generation occurs between VR = 0.5 and 0.7, depending on organic type. Wet gas and condensate become dominant at VR = 1.1 to 1.3. Only dry gas is preserved at over VR = 2.0. In the northern Perth Basin it seems that the effective gas "floor" is primarily determined by loss of porosity and permeability which occurs long before conditions of thermal breakdown are reached.

Dongara Saddle and Adjacent Areas
Permian source rocks in the Perth Basin are land plant-rich, contain hydrogen-poor kerogen and have

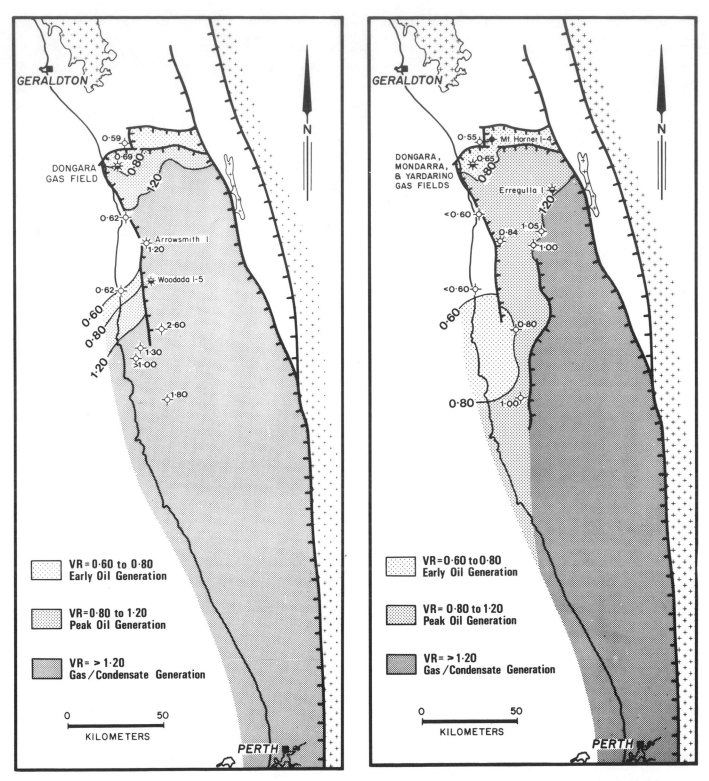

Figure 13. Maturity map, top of Permian.

Figure 14. Maturity map, Triassic (top of Kockatea Shale).

potential mainly for gas generation. These sediments are mature for oil generation (VR = 0.7 to 0.9) within the Dongara gas field and are gas-generative in adjacent areas to the southeast (Figure 13). Although the main reservoir of the Dongara field is Triassic in age, some of the reserves are trapped within the Permian. It is likely that the field is at

least partially sourced from Permian coals and shales of the Carynginia Formation and Irwin River Coal Measures. The generative potential of the Permian is further supported by a test in Arrowsmith-1 where gas was produced from lenticular sands within the Carynginia Formation. The maturity level of the Permian in this well is VR = 1.2 or

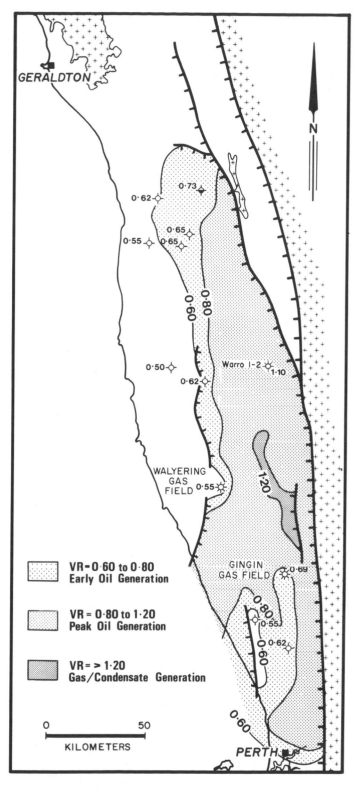

Figure 15. Maturity map, Jurassic (top of Cattamarra Coal Measures Member).

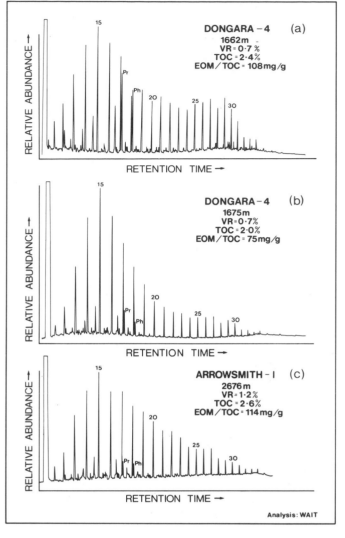

Figure 16. Extract chromatograms (saturated hydrocarbons), basal Kockatea Shale, Dongara-4 and Arrowsmith-1.

greater, which is consistent with in situ gas generation. The principal Triassic source unit is the Kockatea Shale, which forms the top seal of the Dongara, Yardarino, and Mondarra gas fields. The lower Kockatea is enriched in marine phytoplankton and land plant exinite, and is oil-prone.

Solvent extracts of Permian and Triassic source rocks in the Dongara area have been compared with Dongara oils. Although shales and coals of the various Permian units are organically-rich (1.5 to 31 percent TOC) they are characterized by a low extract to total organic carbon ratio (less than 50 mg/g), suggesting that these mature rocks are gas-prone. The basal Kockatea Shale consists of two distinct lithologies, a flakey, laminated, carbonaceous shale, and more massive calcareous shale or limestone. The laminated shales contain up to 3 percent total organic carbon. Extract chromatograms from most samples of Kockatea Shale are waxy, suggesting a high land plant contribution (Figure 16a), although some horizons are rich in aquatic organisms as indicated by an abundance of normal alkanes less than C_{20} (Figure 16b). Both types of source rock are characterized by high levels of extractable organic matter (EOM/TOC = over 100 mg/g). Dongara crude (Figure 17) is highly waxy and paraffinic, and a low pristane/phytane ratio (1.3) indicates that its source rocks were deposited in an aquatic, reducing environment (Lijmbach, 1975). It resembles the

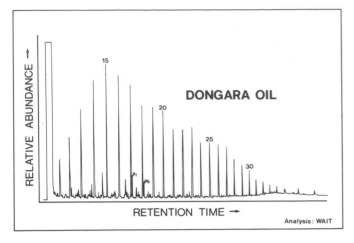

Figure 17. Gas chromatogram of crude oil (saturated hydrocarbons), Dongara-8.

waxy Kockatea extracts, but probably contains contributions from both source rock types and intermediate lithologies. A highly-mature (VR = 1.2) Kockatea extract from Arrowsmith-1 (Figure 16c) is closely comparable with Dongara oil. It is concluded that the probable source of the Dongara oil is the Kockatea Shale, and particularly the laminated, carbonaceous shales developed at its base. Similarly, oil within sandstones of the upper Kockatea Shale at Mt. Horner is probably of Kockatea origin.

Vitrinite reflectance data suggest that the base of the Kockatea Shale is early-mature for oil generation in the Dongara area (VR = 0.7) and reaches peak maturity in adjacent areas (Figure 13). The Kockatea Shale is not mature for major gas generation within the Dongara, Yardarino, and Mondarra fields, but it is probably gas-generative to the south and east where updip migration could readily charge the productive structures.

Powell and McKirdy (1975) suggest that the oil at Dongara, Yardarino, and Mondarra is not in chemical equilibrium with the overlying gas, which is dry. This may indicate that the bulk of the gas was generated from the vitrinite and inertinite-rich source rocks of the Permian sequence, independently of the Kockatea-sourced oil leg.

The Woodada gas discovery, in a fractured Carynginia carbonate reservoir, lies in an area where the Permian is mature for gas generation and is probably locally sourced. Oil shows reported from Woodada-3 may originate from mature Kockatea Shale which is downthrown against the Carynginia Formation along a normal fault trend to the east (Norlin, 1981; Figure 5).

Oil is also reported in fine-grained sandstone lenses within the Kockatea Shale on the Beagle Ridge, suggesting that generation occurred there. The absence of suitable reservoirs on the top of the ridge and structural complexity are possible explanations for the lack of success to date. An unusual occurrence of oil and gas at Erregulla-1 within the Lower Jurassic Eneabba Member—itself a poor source rock—may be sourced from the Triassic. The Kockatea Shale is mature in the Eregulla area and appears to have fed the Eneabba reservoir sands across a major fault to the west of the structure.

Central Dandaragan Trough

The major effective hydrocarbon source in the Dandaragan Trough is the Lower Jurassic Cattamara Coal Measures Member. Limited commercial gas production has been obtained at Walyering and Gingin, although producibility problems were encountered due to low permeabilities and discontinuous reservoirs. The Walyering and Gingin accumulations differ from those in the Dongara area as they contain wet gas and condensate. A small amount of oil was also recovered at Gingin. The hydrocarbons are believed to be generated within the Cattamarra Coal Measures and the presence of condensate and light oil is attributed to the relatively high proportion of exinite in the source sequence. This is in accord with trends observed in other Australian basins and worldwide (Thomas, 1982). Land plant source rocks from the Jurassic, Cretaceous, and Tertiary are often exinite-rich, with potential for oil generation, while those of older periods contain mainly vitrinite and inertinite and may yield only dry gas. Source sequences petrologically similar to the Cattamarra Coal Measures are found in the Jurassic of the Eromanga Basin and the Upper Cretaceous-Lower Tertiary of the Gippsland Basin, both of which are oil-productive.

The Lower Jurassic Cattamarra Coal Measures are thermally mature (VR > 0.6) over an area of 15,000 sq km (579 sq mi) in the central Dandaragan Trough (Figure 15).Due to the low geothermal gradients (see Figure 5) generation occurs at relatively great depths where the permeability of associated sandstone reservoirs is often inadequate for commercial production. Extensive vertical migration of hydrocarbons to shallower levels with better reservoir characteristics may be a prerequisite for producible accumulations in areas with low geothermal gradients. The extent to which vertical migration occurs is unpredictable, although from a geological standpoint, lateral continuity of intraformational seals is not great in the continental sediments of the Dandaragan Trough. It is estimated that perhaps 1,000 m (3,281 ft) of vertical migration from source to reservoir has occurred at Walyering and Gingin.

CONCLUSIONS

Potential hydrocarbon source rocks are widely distributed throughout the Permian, Triassic, and Jurassic sections of the Dandaragan Trough. Most are dominated by land plant material and the basin is largely gas-prone, although a secondary occurrence of light, paraffinic oil is of economic significance.

The Lower Jurassic Cattamarra Coal Measures provide both source and reservoir for gas/condensate accumulations in the central Dandaragan Trough. Due to the low geothermal gradients in this area, hydrocarbon generation occurs only at great depths where sandstone reservoir properties may be inadequate for commercial production.

Gas at Dongara, Mondarra, and Yardarino may have been generated from both the Lower Triassic and Permian, although there is some evidence that the Permian is the principal source. The associated thin oil leg encountered in parts of these fields is attributed to the oil-prone basal Kockatea Shale (Lower Triassic).

The Neocomian was a period of extensive unloading, particularly in the area of the Beagle Ridge and Cadda Shelf. The extent of uplift and erosion appears to be reflected in systematic variations in sandstone porosity trends in the basin.

Reflectance of vitrinite data suggest that the area of the Beagle Ridge and Cadda Shelf has experienced higher geothermal gradients than are measured today in exploratory boreholes. The thermal event may be of Neocomian age, caused by an igneous intrusion associated with uplift of the basin which occurred prior to continental breakup.

ACKNOWLEDGMENTS

This is a revision of an earlier paper published in the AAPG Bulletin in 1979, based on the author's studies while with West Australian Petroleum Pty. Ltd. (WAPET). The data were reviewed and supplemented with more recent published information and analytical results. The contributions of many early WAPET workers in the Perth Basin are gratefully acknowledged. The author thanks Shell Development (Australia) Pty. Ltd. for releasing recent extract data from the Dongara area and for additional drafting. WAPET provided samples and geochemical analyses were performed by the Western Australian Institute of Technology (WAIT). Thanks also to K. Spence for his modified Lopatin program, to A.J. Kantsler for vitrinite reflectance data, and to W.G. Townson, J.C. Parry, and A.J. Kantsler for review of the manuscript. This paper is published with the permission of the management of West Australian Petroleum Pty. Ltd. and Shell Development (Australia) Pty. Ltd.

REFERENCES CITED

Hosemann, P., 1971, The stratigraphy of the Basal Triassic Sandstone, North Perth Basin, Western Australia: Australian Petroleum Exploration Association Journal, v. 11, n. 1, p. 59-63.

Jones, D.K., 1976, Perth Basin, *in* R.B. Leslie, J.H. Evans, and C.L. Knight, eds., Economic geology of Australia and Papua New Guinea, part 3—Petroleum: Australasian Institute of Mining and Metallurgy, Monograph Series n. 7, p. 108-126.

——— , and G.R. Pearson, 1972, The tectonic elements of the Perth Basin: Australian Petroleum Exploration Association Journal, v. 12, n. 1, p. 17-22.

Kantsler, A.J., and A.C. Cook, 1979, Maturation patterns in the Perth Basin: Australian Petroleum Exploration Association, v. 19, n. 1, p. 94-107.

Lijmbach, G.W.M., 1975, On the origin of petroleum, *in* Proceedings, 9th World Petroleum Congress, v. 2: London, Applied Science Publishers, p. 357-369.

Lopatin, N.V., 1971, Temperature and geologic time as factors in coalification: (in Russian), Akademiya Nauk. SSSR Izvestiya Seriya Geologicheskaya, n. 3, p. 95-106.

Maxwell, J.C., 1964, Influence of depth, temperature, and geologic age on porosity of quartzose sandstones: AAPG Bulletin, v. 48, p. 697-709.

Norlin, W.K., 1981, Woodada-1—a discovery in the Perth Basin: Australian Petroleum Exploration Association Journal, v. 12, n. 1, p. 55-59.

Playford, P.E., A.E. Cockbain, and G.H. Low, 1976, Geology of the Perth Basin, Western Australia: Western Australia Geological Survey Bulletin, n. 124.

Powell, T.G., and D.M. McKirdy, 1975, Geologic factors controlling crude oil composition in Australia and Papua New Guinea: AAPG Bulletin, v. 59, n. 7, p. 1176-1197.

Staplin, F.L., 1969, Sedimentary organic matter, organic metamorphism, and oil and gas occurrence: Bulletin of Canadian Petroleum Geology, v. 17, p. 47-66.

Thomas, B.M., 1982, Land plant source rocks for oil and their significance in Australian sedimentary basins: Australian Petroleum Exploration Association Journal.

——— , 1979, Geochemical analysis of hydrocarbon occurrences in the Northern Perth Basin, Australia: AAPG Bulletin, v. 63, p. 1092-1107.

——— , and D.N. Smith, 1976, The Carnarvon Basin, *in* R.B. Leslie, H.J. Evans, and C.L. Knight, eds., Economic geology of Australia and Papua New Guinea, part 3—Petroleum: Australasian Institute of Mining and Metallurgy, Monograph Series, n. 7, p. 126-155.

Tissot, B., et al, 1974, Influence of the nature and diagenesis of organic matter in formation of petroleum: AAPG Bulletin, v. 58, p. 499-506.

——— , D.H. Welte, 1978, Petroleum formation and occurrence: Berlin, Springer Verlag, 538 p.

Waples, D.W., 1980, Time and temperature in petroleum formation; application of Lopatin's method to petroleum exploration: AAPG Bulletin, v. 64, p. 916-926.

Hydrocarbon Generation in the Taranaki Basin, New Zealand

W.F.H. Pilaar
Shell Internationale Petroleum Maatschappij B.V.
The Hague, Netherlands

L.L. Wakefield
Thailand Shell Petroleum Development Ltd.
Bangkok, Thailand

The Taranaki Basin, situated on the west coast of the North Island of New Zealand, contains the only commercial gas and condensate fields found in New Zealand to date, Kapuni and Maui. These were discovered in 1959 and 1969 respectively.

The Upper Cretaceous and Tertiary infill of the Taranaki Basin represents a depositional mega-cycle affected by two tectonic phases: initial rifting and foundering during the Late Cretaceous and Early Tertiary, followed by wrench faulting, graben formation, and basin inversion during the Miocene and Pliocene. The Taranaki Basin consists of a Western Platform and an eastern Graben Complex. The former was relatively stable throughout most of the Tertiary and was affected by normal block faulting only in Late Cretaceous to early Eocene times. The latter comprises a tectonically complex graben system, which was particularly active during the Miocene. This Taranaki Graben Complex closes to the south and is bounded to the east by the major Taranaki fault zone, whereas to the west, the Graben Complex generally shallows across a series of steep, normal and reverse en echelon arranged faults.

Late Cretaceous half-grabens accumulated thick sediment wedges ranging from coarse clastics to fine-grained coal measures. A marine incursion during the Paleocene was interrupted by an Eocene regressive phase, during which coal measures were deposited in southern and eastern parts of the basin. Quartzose sandstones in these coal measures form the reservoirs in the Kapuni and Maui fields. Regional subsidence resumed during the Late Eocene and Oligocene and resulted in the deposition of mainly marls and pelagic carbonates in the west and neritic limestones, sandstones, and/or mudstones in the east and south. During the Miocene development of the Taranaki Graben Complex, thick flysch-type deposits within the graben spilled over onto the Western Platform. From then on, through deposition of a thick sequence of "giant foresets," the continental shelf prograded across the Western Platform to its present limit.

Upper Cretaceous and Lower Tertiary coal measures form the main objectives in the Taranaki Basin; they provide source, reservoir, and cap rocks. The "Lopatin" method calibrated against measured vitrinite reflectance values was used for calculating source-rock maturation to outline potential hydrocarbon kitchen areas and to estimate the degree of maturity. The Taranaki Graben Complex is interpreted as a fully mature gas province, whereas the Western Platform shows decreased maturity to the west and northwest. Source rocks are confined to the coals and intervening carbonaceous shales. The coals are mainly hydrogen-poor, but intercalated lacustrine shales with resin and bitumen-rich layers may provide a more kerogenous source rock. Determination of the effective burial time of these source rocks is important in understanding the generation of hydrocarbons in the Taranaki Basin.

INTRODUCTION

The Taranaki Basin is situated on the western margin of the North Island. It extends north from the northwestern tip of the South Island to about the latitude of Auckland and includes the Taranaki Peninsula (Figure 1).

The exploration history of the Taranaki Basin dates back to 1859, when the Moturoa oil field was discovered through oil seepages occurring around a volcanic plug near the present day harbor of New Plymouth. Much later, a number of shallow holes were drilled in the vicinity of these seepages, and these have produced, and are still producing a few barrels of oil a day since early in the present century. During World War II some shallow holes were drilled on land in Taranaki, but without success.

Extensive exploration for hydrocarbons in the Taranaki Basin started in the mid-1950s, initially on land, when the Shell, BP and Todd Group (SBPT) was granted the total land acreage. This was extended to the offshore continental shelf area in 1965. During this period of exploration activity, which lasted to about the end of 1977, five deep land wells and thirteen offshore tests were drilled. They

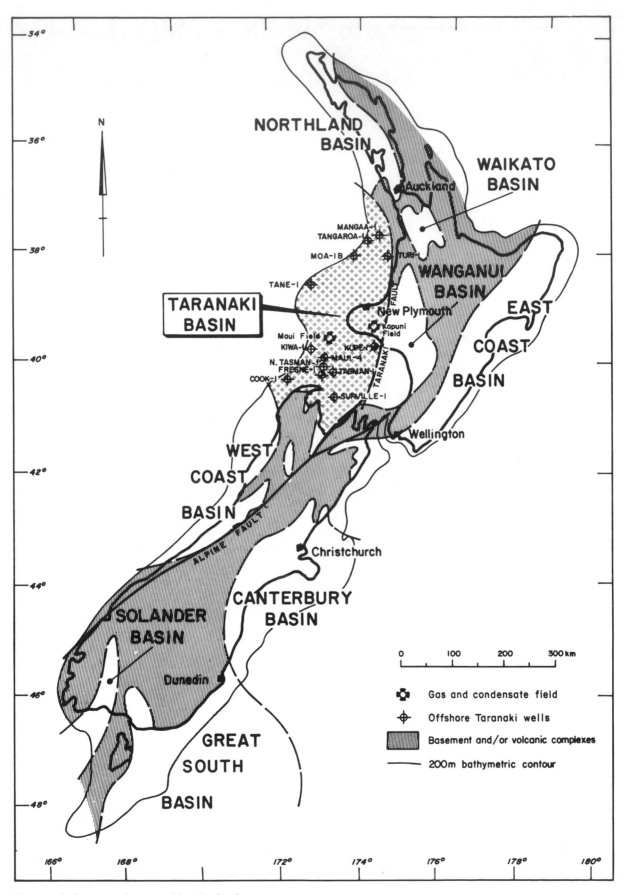

Figure 1. Sedimentary basins in New Zealand.

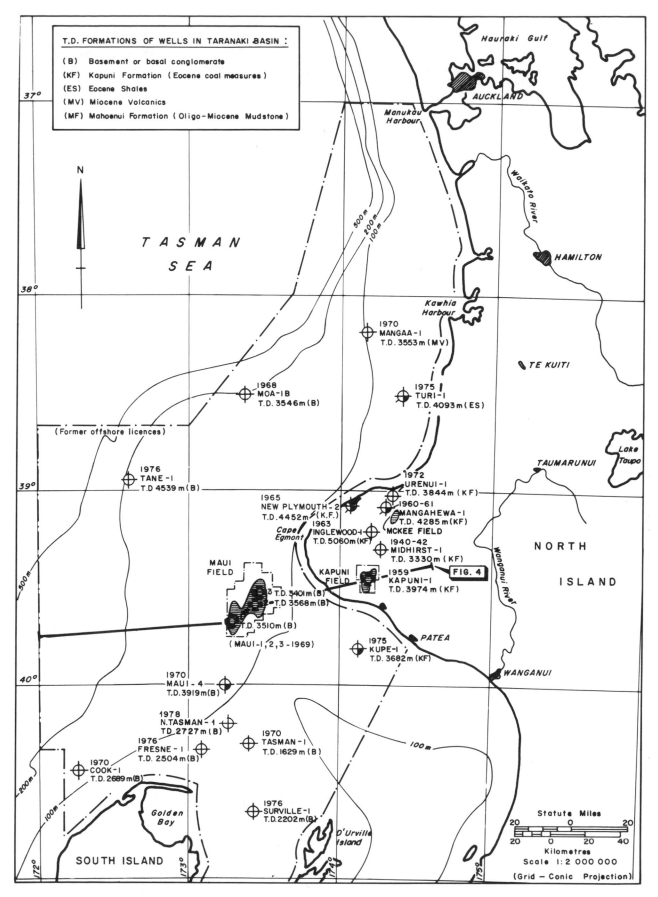

Figure 2. Location of Taranaki wells referred to in the text.

resulted in the discovery of two gas and condensate fields: the onshore Kapuni field in 1959 and offshore Maui field in 1969 (Figure 2). Ultimate recoveries of the Maui field, which went on stream in 1979, are estimated at about 200 billion cu m (7 Tcf) of gas with some 30 million cu m (200 million barrels) of condensate. Recoveries of the Kapuni field, which has been producing since 1970, are about 30 million cu m (1 Tcf) of gas with nearly 10 million cu m (50 million barrels) of condensate.

Exploration activities ceased temporarily on land upon the expiration of the then existing licences before the State Oil Corporation took out an exploration licence in 1977 covering the whole of the Taranaki Peninsula. Since then, Petrocorp drilled another four deep land wells, two of which discovered a possibly commercial oil accumulation at McKee (Figure 2). To date, exploration activities both onshore and offshore still continue within the Taranaki Basin.

REGIONAL SETTING

In New Zealand, prospective sedimentary sequences are restricted to Cretaceous and younger sediments. Older sediments are metamorphosed and form the economic basement.

Late Cretaceous to early Tertiary coal measures form the primary objectives for exploration in the Taranaki Basin. They are regionally well-developed and provide source, reservoir and cap rocks.

The initial rifting and foundering of the New Zealand part of the Tasman Sea probably began in early Upper Cretaceous and continued to late Eocene times. This was followed by a major Tertiary transgression which affected the whole of New Zealand and culminated in the Oligocene. During the early Tertiary, local grabens and fault-angle depressions or half grabens formed under an extensional stress regime and accumulated thick sedimentary sequences ranging from coarse clastic to fine-grained coal measure deposits. The intervening elongated high blocks commonly served as sources of sediment supply. These differential block movements resulted in abrupt and pronounced thickness and facies variations of the infill of these initial and irregularly distributed half grabens. This rift-fill period was succeeded by tectonic quiescence during the Oligocene and in turn followed by a second interval of major tectonic activity during Miocene time, which continued into the Present (Pilaar and Wakefield, 1978). The sedimentary sequence deposited during this time represents the regressive phase of a depositional Upper Cretaceous-Tertiary mega-cycle, shallowing up from deep water and associated turbidite conditions to the present shelf situation.

The Taranaki Basin is part of a complex rift system, located north of the major transcurrent Alpine fault zone (Figure 1). The basin's tectonic history correlates well with periods of activity that occurred along this fault system.

A major Graben Complex formed within the basin during the Neogene probably under a dextral wrench regime. It extended from the Westcoast Basin in the south to the Northland Basin in the north (Figure 1).

STRUCTURE

The Taranaki Basin comprises two structural provinces: the *Western Platform* and the eastern *Taranaki Graben Complex* (Figures 3 and 4). The former remained a relatively stable block throughout most of the Tertiary, and was affected only by Upper Cretaceous to Eocene normal block faulting. Similar synsedimentary fault activity probably occurred in the eastern areas also, but its traces have nearly been completely obliterated by the later north-south trending Taranaki Graben Complex. The north and south Taranaki grabens are separated by a northeast-southwest trending oblique fault, which corresponds to the northern coast of the Taranaki Peninsula with its Pleistocene to Recent volcanics (Figure 3).

The Taranaki Graben contains a sedimentary/igneous infill, locally more than 6,000 m (19,685 ft) thick, in which the volcanics are associated with the axial parts of the grabens and their boundary faults. Especially in the north Taranaki graben, extensive Miocene-Pliocene volcanics obscure the configuration of its deeper part in the north. In the south Taranaki graben, volcanism appears to have been confined largely to the Lower Tertiary or to the pre-Tertiary basement. These volcanics probably were associated with the rift phase of this graben.

To the east, the Taranaki Graben Complex is bounded by the Taranaki fault zone, which separates the graben from the narrow, elongated Patea, Tongaporutu, and Herangi basement highs (Figure 3). The Taranaki fault zone has probably been active throughout the Tertiary, with the earliest movements concentrated in the south Taranaki graben. Major movements along this fault, however, occurred during the Miocene and Pliocene, when a thick sedimentary wedge accumulated at its downthrown side (more than 6,000 m, or 19,685 ft, in the deeper parts; Figure 4). To the west the Graben Complex generally shallows across a series of normal faults and finally abuts against the Cape Egmont fault zone. The latter zone consists basically of a series of steep, normal-to-reverse en echelon faults, which over short distances often show drastic changes in the amount of throw in both a vertical and a horizontal sense (Pilaar and Wakefield, 1978).

North-south and northeast-southwest tectonic trends control the structural configuration of the Taranaki Basin. At their intersections, the north-south trend is usually offset by the younger, oblique trend. The fault features and fault angle relationships can be explained as a first order wrench system (for example, Wilcox, Harding, and Seely, 1973; Harding, 1974) with associated synthetic strike-slip or normal shear faults (Figure 5). Paleotectonic reconstructions presented in Figure 6 illustrate the structural development of the south Taranaki graben.

STRATIGRAPHY

Upper Cretaceous

The oldest prospective sediments, the Late Cretaceous Pakawau formation, represent an onlapping sequence on an erosional surface of igneous and metamorphic rocks (Figure 7). Their distribution was initially known from onshore areas of northwestern South Island and from the wells Maui-4, Fresne-1, Cook-1, and Tane-1. From seismic evidence, Late Cretaceous sediments appear to be widely distributed over the Western Platform. In the southern areas they mainly fill fault-angle depressions and reach thicknesses of 800 to 2,000 m (2,625 to 6,562 ft).

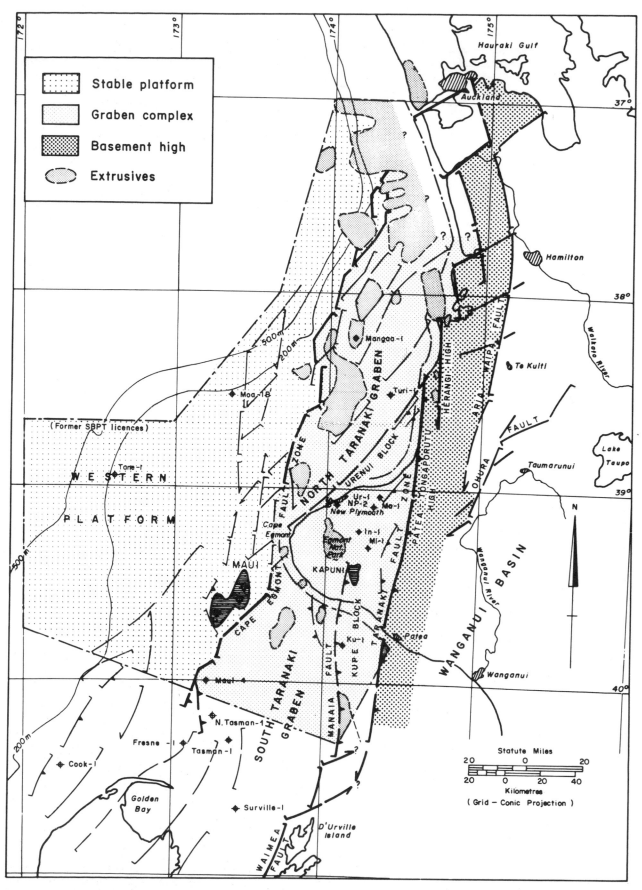

Figure 3. Taranaki Basin—major structural elements.

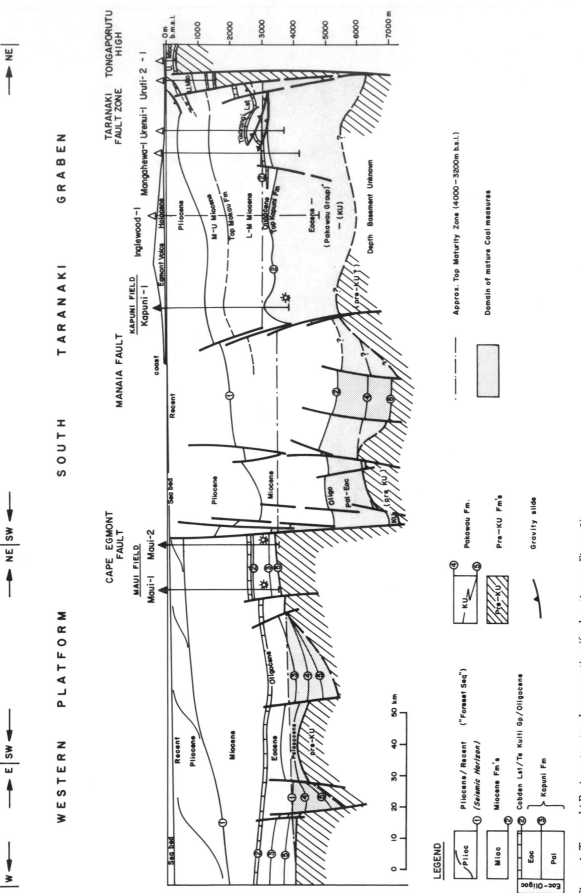

Figure 4. Taranaki Basin—structural cross section (for location see Figure 2).

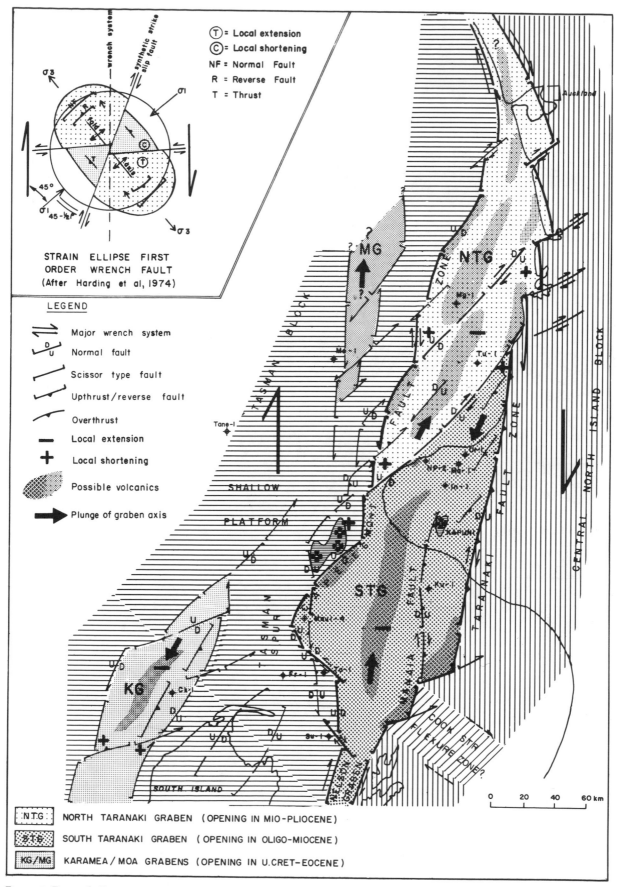

Figure 5. Taranaki Basin—structural concept.

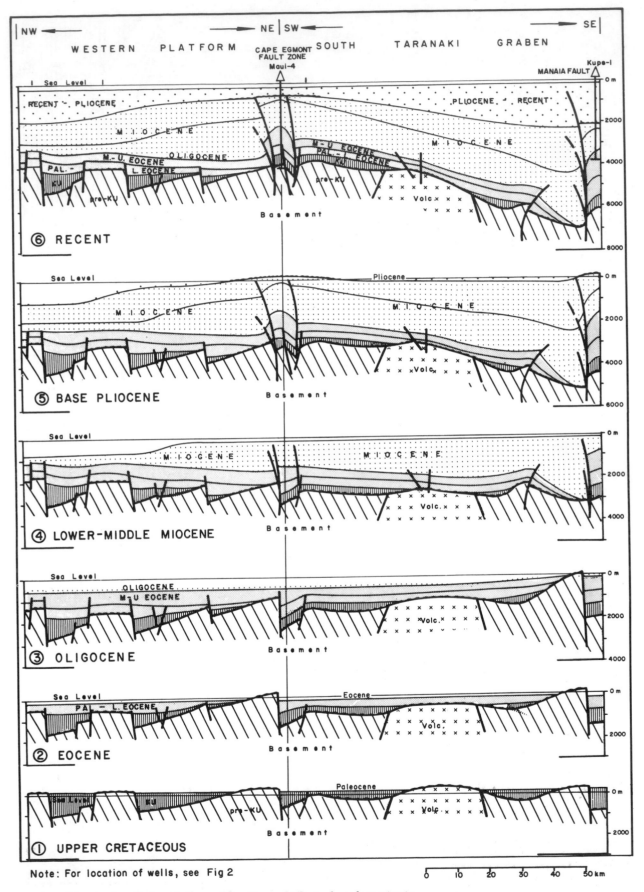

Figure 6. Structural evolution of the southern Taranaki Basin, based on seismic.

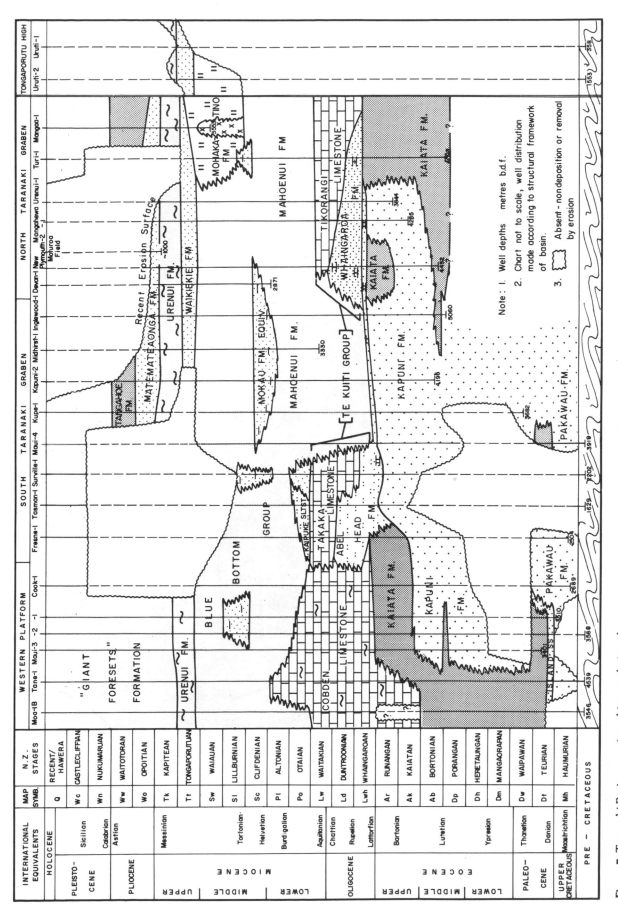

Figure 7. Taranaki Basin—stratigraphic correlation chart.

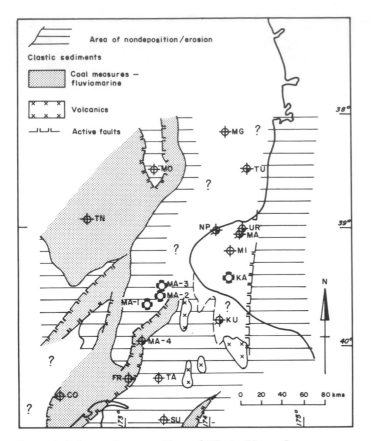

Figure 8. Paleo-environment Taranaki Basin: Upper Cretaceous.

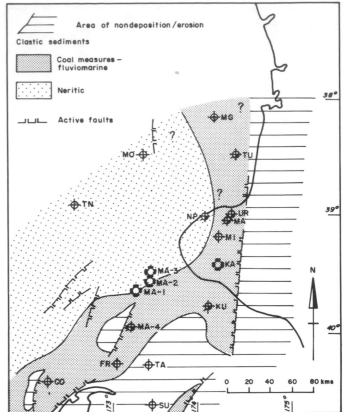

Figure 9. Paleo-environment Taranaki Basin: Upper Cretaceous.

The Pakawau formation comprises coal measures throughout the basin (Figures 8 and 9). Lithologies range from basal polymict conglomerates and coarse sandstones to alternating kaolinitic quartzose sandstones, carbonaceous mudstones, and humic coals.

Paleocene-Eocene

During the Paleocene, deposition of the Pakawau coal measures (only locally interrupted by erosion) continued in the southern parts of the basin. To the north, in the area of the Maui field and further northwest, marginal marine or holomarine environments prevailed. Well-sorted, glauconitic sandstones (Island Sandstone) overlain by massive brown-gray mudstones and fine sandy siltstones (Kaiata formation) of Paleocene to Eocene age were deposited (Figures 9 and 10).

The late Paleocene-early Eocene was apparently a phase of non-deposition or erosion in the southern half of the basin. By middle-late Eocene times, however, renewed deposition of coal measures (Kapuni formation) is evident, particularly in the eastern part of the basin where sedimentation was tectonically controlled and fault angle depressions locally contain more than 1,000 m (3,281 ft) of fill. In the southeastern part of the Taranaki Peninsula, the Kapuni formation contains a lower unit that is dominated by sandstones and overlain by shale and coal-rich sequences. To the northwest, in the well New Plymouth-2, marine calcareous and muddy, fine-to-medium sandstones predominate, showing an interdigitation of the Kapuni and

Kaiata formations. North of the Taranaki Peninsula, the fault-controlled Eocene sequence is thick, but made up entirely of mudstones of the Kaiata formation as seen in the well Turi-1.

During the middle and late Eocene, the area of the Maui field formed part of a northeas-tsouthwest trending fluviomarine facies. Here, the lower stratigraphic units contain thin coals, carbonaceous shales, and quartzose sandstones. The sandstones become dominant toward the top of the Kapuni formation and finally are overlain by gray-brown calcareous mudstones of the Kaiata formation.

Oligocene-Miocene

The Oligocene sequence is characterized by calcareous lithologies of the Te Kuiti Group. The Western Platform is covered by the Cobden limestone representing a deep water marine environment (Figure 11). In the southwestern and northeastern parts of the Taranaki graben, highly calcareous sandstones, mudstones, and bioclastic limestones of the Tikorangi limestone formation were deposited under shallow-marine conditions. In the southeastern parts of the basin, however, there is no limestone development and, instead, highly calcareous mudstones gradually merge into blue-gray silty mudstones and fine-grained flysch deposits of the Mahoenui formation.

In the Miocene, a thickening of the clastic sequences in the eastern parts of the basin heralds the development of the Taranaki Graben Complex. Neritic sandstones, siltstones, mudstones and localized coal measures (Mokau formation)

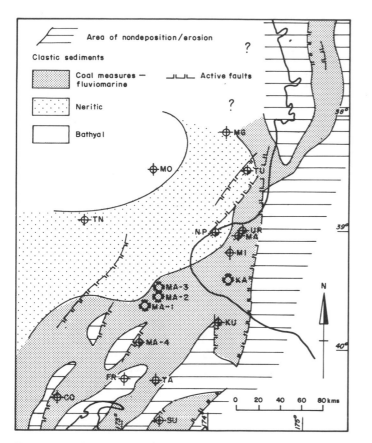

Figure 10. Paleo-environment Taranaki Basin: Middle Eocene.

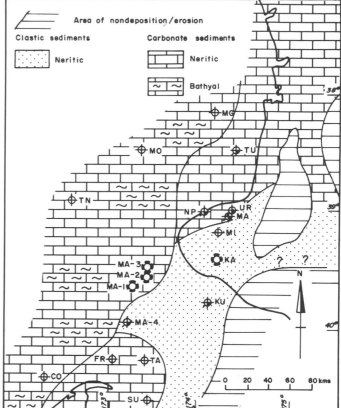

Figure 11. Paleo-environment Taranaki Basin: Upper Oligocene.

were deposited in the Early Miocene on the Patea, Tongaporutu and Herangi Highs, while deeper-water graben-fill sequences, mainly containing monotonous mudstones (Mahoenui formation), were deposited to the west (Figure 12). By contrast, the Western Platform and southern areas were not yet affected by tectonic activity and continued to receive fine-grained sediments rich in planktonic faunas (Cobden limestone and Blue Bottom Group).

With the acceleration in tectonic activity, which reached its climax in late Miocene-Pliocene times, the eastern and southern onshore areas became sources of abundant lithoclastic detritus, which accumulated in the Taranaki graben and formed an actively prograding sedimentary wedge, the precursor of the modern continental shelf. Throughout the Taranaki Basin, middle to late Miocene sequences are dominated by massive mudstones, poorly to moderately sorted sandstones and muddy siltstones typical of turbidite deposits.

A massive silty to marly mudstone sequence, the Urenui formation, of late Miocene age marks the last transgressive phase of the Taranaki Basin. In the east, it is overlain by massive shelly sandstones and conglomerates of the Matemateaonga formation. In the southern areas it is eroded.

Major volcanic activity occurred in the north Taranaki graben, where andesitic plugs intruded into the graben-fill and produced tuffaceous sequences (Mohakatino formation) in middle and late Miocene times.

Pliocene-Recent

A widespread Pliocene erosion surface marks the climax of tectonic activity and basin inversion in the eastern and southern areas of the Taranaki Basin, while over the Western Platform, the continental shelf prograded outward to the northwest toward its present limit, forming the Pliocene-Recent "giant foresets" sequence (Figure 13). The north Taranaki graben continued to receive a thick infill of neritic to bathyal mudstones and minor turbidites, although the graben was considerably reduced in width. In the south Taranaki graben, neritic fossiliferous sandstones, siltstones, and mudstones with thin lignite layers and local conglomerates were deposited.

HYDROCARBON GENERATION

Whenever exploring for hydrocarbons, type, richness, and maturity of source rocks are important factors, particularly in complex basins with distinct tectonic and/or sedimentary provinces, such as the Taranaki Basin. Here, the Graben Complex has reached the stage of a proven gas province whereas the Western Platform has failed so far in demonstrating its hydrocarbon potential. Is there any specific reason for this?

The Late Cretaceous and Eocene coal measure sequences are the only proven source rocks in the Taranaki Basin and their presence can be confidently established over large parts of the basin from seismic and supporting well data. Despite their common abundance of leaf-cuticles, pollen, spores,

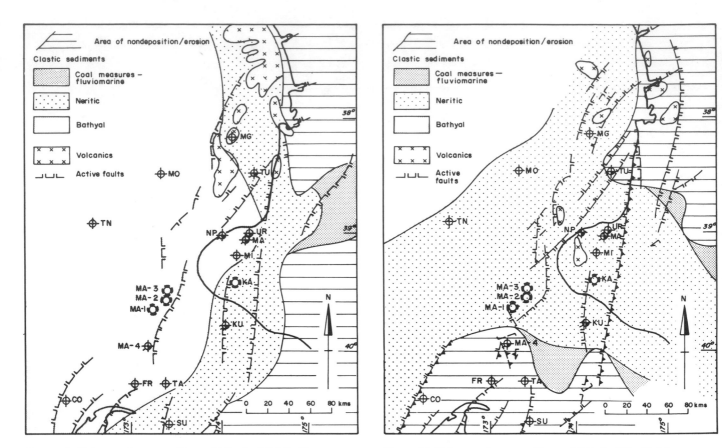

Figure 12. Paleo-environment Taranaki Basin: Middle-Upper Miocene.

Figure 13. Paleo-environment Taranaki Basin: Plio-Pleistocene.

Table 1. Maceral composition of coal measures in Tane-1, New Zealand (Analysis by Shell Exploration and Production Research).

DEPTH (M. BDF) Cuttings Cores (C) Sidewall Spl. (SWS)	LITHOLOGY	Total Org. Carb. (TOC) (weight %)	Pyrolysis Yield (weight %)	Structureless Organic Matter[1]	Vitrinites	Liptinites			Exsudatinite	Inertinites	% Vitr. Refl. (Vr/M)	Average Vr/M
						Sporinite	Cutinite	Resinite				
3637-3640	Coal	9.3		P	*A*	P	P	P	P	P		
3640-3643	Carb. Sh.	12.0	4.4	P	TR	S	S	S	P	P		
3675 (SWS)	Coal											0.57
3687 (C)	Carb. Sh.	44.0	10.9	P	P	S	P	P	S	P	0.61	
3689 (C)	Coal				*A*	P	P	P		S	0.67	
3706-3709	Coal	16.2	5.0	P	P		TR	S	P	P		
3721-3724	Carb. Sh. +Coal			P	*A*	P	P	S	S	P		0.60
3769-3772	Carb. Sh.			S	P	S	S	S	S	S		
3883-3886	Coal + Carb. Sh.			S	S	S		S	S	P		
3912 (SWS)	Coal											
3977 (SWS)	Carb. Sh.											0.62
4243-4246	Carb. Sh.	4.8	0.4	P	P	P	TR	TR	P	P	0.77	0.72
4411-4414	Coal	30.0	3.8								0.80	0.80
4417-4420	Carb. Sh.	21.5	3.2	S	*A*	P	S	S	P	P		0.84

Notes: *A*: Abundant   P: Present   S: Spare   TR: Trace

[1] Residual bacterial mass; no algal material observed.

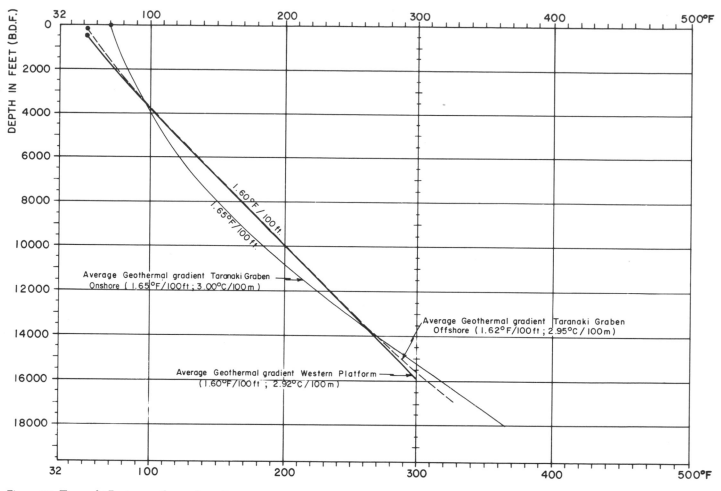

Figure 14. Taranaki Basin, geothermal gradients.

and resins relative to woody material, the coals are classified as hydrogen-poor or gas-generative. Maceral analyses run on ditch cuttings and core material from the well Tane-1 (Table 1) suggest that kerogen-rich, but lean source rocks may be present in the resinous laminae of the coals and intervening carbonaceous shales.

The influence of depth of burial, that is to say the subsurface temperature, on coalification or hydrocarbon maturation processes has long been recognized and is now well understood (Suggate, 1973; Stach, 1975; Hood, Gutjahr, and Heacock, 1975; and others), but the effect of geologic time has not been always sufficiently considered. Although Hood, Gutjahr, and Heacock (1975) defined an "effective" heating time, this was not directly related to the geological age of the "effective" overburden (that is, the time when maturity was reached at a certain level) since the thickness of the overburden remained the only factor to be considered and not its age. In relatively young sedimentary basins, however, the age of the "effective" overburden may be the critical factor in determining the maturation history. This has been borne out by the Taranaki Basin study.

In a first attempt, the approximate maturity ranges existing within the Taranaki Basin were extrapolated as a function of depth, using present-day geothermal gradients calculated from BHT's of wells drilled (Figure 14). For this

purpose, approximate formation temperatures were calculated from BHT's using a time ratio correction method, which takes into account the "cooling time" effect of the drilling mud. This, and other correction methods (Roux et al, 1980) based on the same principle, are generally accepted to provide the best means for estimating formation temperatures.

In plotting the vitrinite reflectance values (that is, VR/M or R_o Max) measured in the various wells against depth, two "best fit" maturation curves can be drawn; one for wells drilled on the Western Platform and another for those drilled in the Taranaki Graben Complex (Figure 15). From this it follows that the assumed top maturity level and, consequently, the zones of generation and expulsion of hydrocarbons lie at a greater depth on the Western Platform than in the Graben Complex.

This difference in depth cannot be readily explained by the small differences observed in the geothermal gradients calculated for the two provinces (Figures 14 and 15). It seems more likely to be the result of the age difference of the "effective" overburden in the graben and on the platform, being Miocene in the former and Plio-Pleistocene in the latter province. The major half-grabens within the platform area, however, may form an exception in that they contain a thick, pre-Pliocene sediment fill and, therefore, may

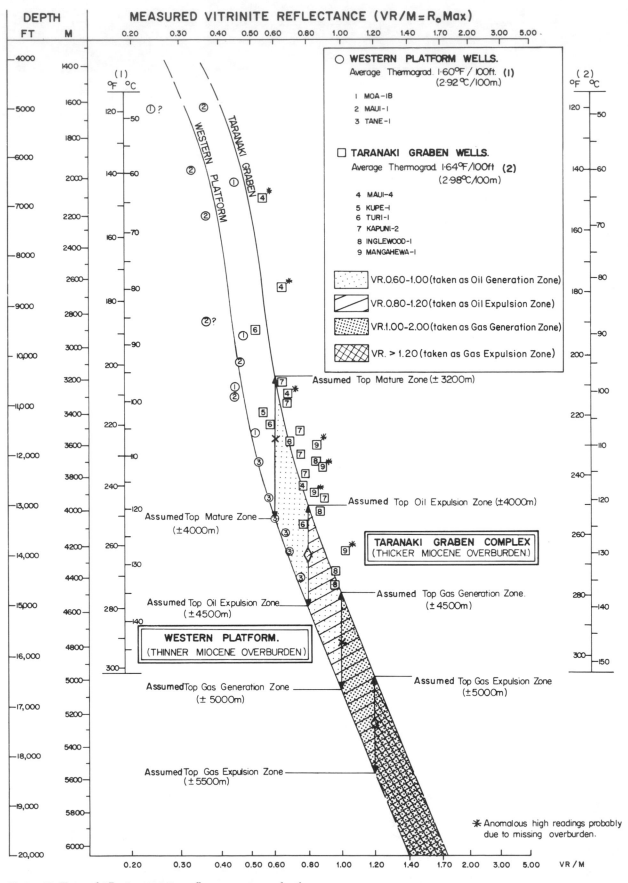

Figure 15. Taranaki Basin, vitrinite reflectance versus depth.

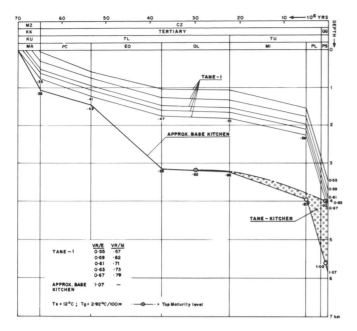

Figure 16. Burial graph Tane-1, Western Platform (west).

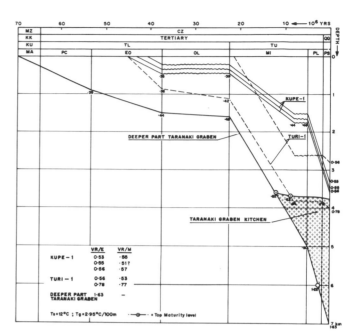

Figure 18. Burial graph Taranaki Graben Complex.

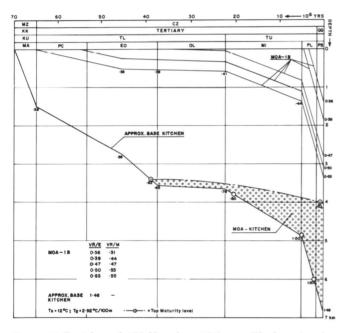

Figure 17. Burial graph "Half-graben," Western Platform (east).

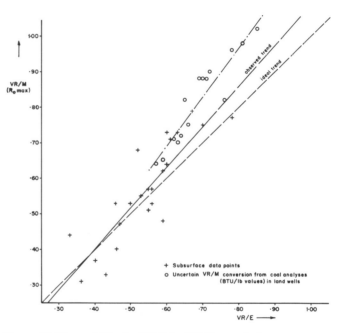

Figure 19. Taranaki Basin, VR/M—VR/E correlation.

follow more closely the graben trend.

To check this hypothesis, burial graphs giving the relationship between the depth of burial and geologic time were constructed for each vitrinite reflectance value (VR/M) measured in Graben Complex wells and Western Platform wells. These values can be compared to the calculated vitrinite reflectance (VR/E) based on the Lopatin method (Lopatin, 1971; Waples, 1980). The results for some of the key wells are given in Figures 16, 17, and 18, together with the burial graphs constructed for the floor or deeper parts of the related hydrocarbon kitchens. These graphs infer the influence that geologic time has on the maturation process.

The degree of correlation that exists between the measured (VR/M) and the calculated (VR/E) vitrinite values is shown in Figure 19. The vitrinite values obtained through conversion from coal analyses (expressed in BTU/lb calorific values), made by the New Zealand Coal Research Association on material from the land wells, significantly deviate from the observed and ideal VR trends. Otherwise, the observed trend follows the ideal trend rather closely, although the VR/E values appear to be consistently too high in the lower range (that is, VR values below 0.40) and slightly too low in the higher values (VR above 0.60; Figure 19). This may be due to measurement and/or instrument

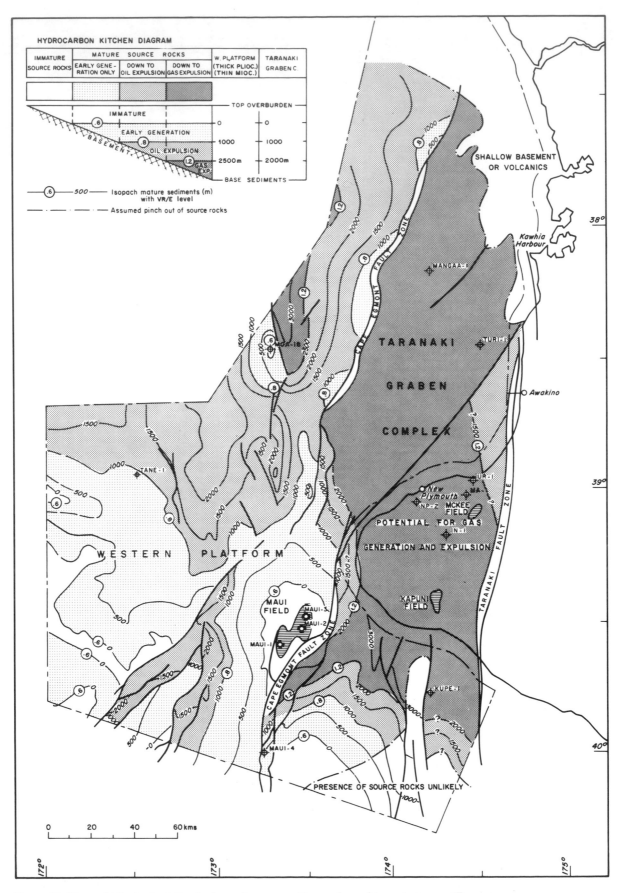

Figure 20. Taranaki Basin, tentative hydrocarbon generation and expulsion provinces at Present.

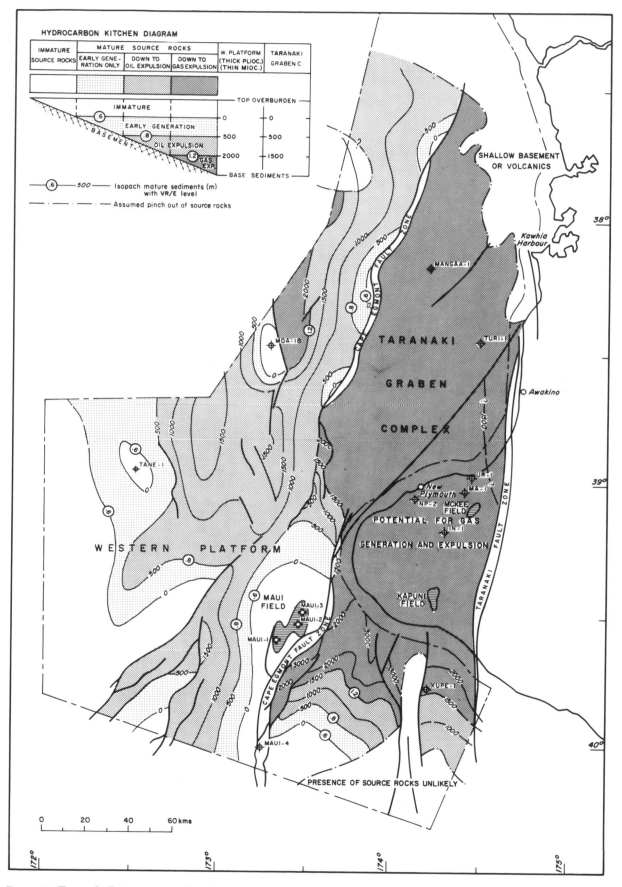

Figure 21. Taranaki Basin, tentative kitchen map at the beginning of the Pleistocene.

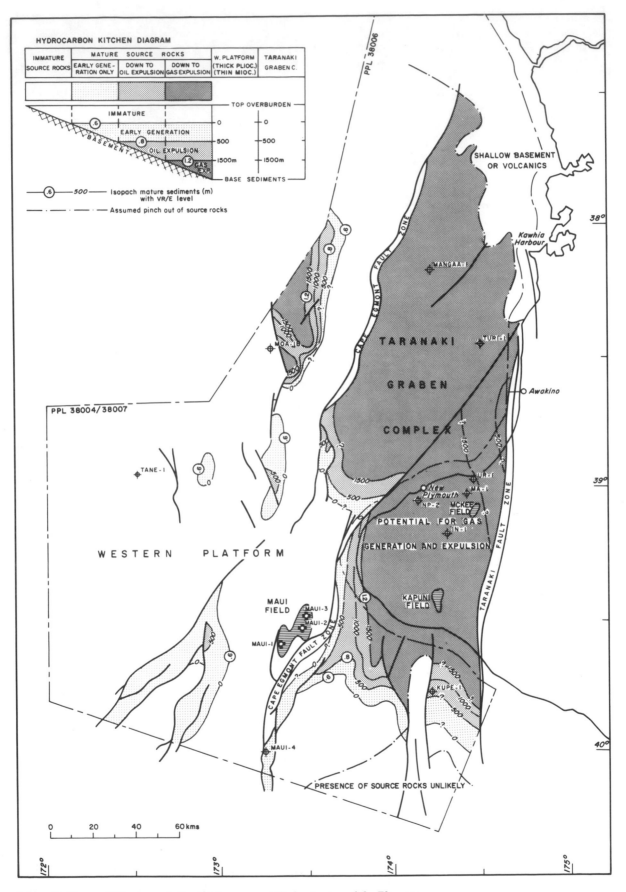

Figure 22. Taranaki Basin, tentative kitchen map at the beginning of the Pliocene.

errors, or to differences between the calculated (present day) and the paleo-geothermal gradients. No corrections were made for the unknown amount of overburden missing at present. Nevertheless, the prediction method provides a measure of confidence, especially in the important middle range of vitrinite values.

The effect of geologic time on the maturation process in the Taranaki Basin is illustrated in Figures 20 to 22, in which the mature kitchen areas are outlined as they are thought to be today (Figure 20), to have been at the beginning of the Pleistocene (Figure 21) and of the Pliocene (Figure 22). It appears that generation and expulsion of hydrocarbons may have started as early as the upper Miocene in the Graben Complex and in the half-grabens in the eastern part of the platform area, but have hardly commenced in the western part of the latter province. Here, the "effective" overburden is either too young to have activated expulsion and migration of the hydrocarbons that may have been generated in the deeper, more mature parts of the kitchen areas, or, the concentration of the kerogen within the source rocks was too lean for any generated hydrocarbon product to have been expelled.

If correct, the constraints imposed by the geologic time factor on the maturation process in the Taranaki Basin may well explain why the Tane well, located in the westernmost part of the platform domain, remained dry. The same constraints, however, were less critical in charging the Kapuni and Maui fields from deeper kitchens in the Graben Complex (Figures 20, 21, and 22). In particular, the Kapuni field appears well placed relative to its underlying kitchen area, whereas Maui relies on deeper kitchens located in the Taranaki Graben to the north and the east of the field for its charge (Figure 20).

OIL AND GAS POTENTIAL

Potential reservoirs exist within the sandstones of the Upper Cretaceous to lower Tertiary coal measures (Pakawau and Kapuni formations) or are associated with the immediately overlying basal transgressive marine sands (Island sandstone). In addition, Miocene sands, when present, may provide secondary objectives such as located over shallow volcanic plugs in the central part of the north Taranaki graben or in turbidites in the eastern Graben Complex.

Sealing lithologies are amply provided by the marine Paleocene to Miocene shales overlying both the coal measures sequences and "basal" transgressive sands, as well as the shales intercalated within the coal measures themselves.

As already mentioned, Upper Cretaceous to lower Tertiary coal measure series, providing the only known source rocks for gas and possibly for oil, are present in large parts of the basin. The lower maturity levels observed in the western part of the basin when compared to those measured at similar depths in the eastern part, probably reflect differences in their respective burial histories. When going from east to west across the basin, the overburden becomes gradually younger in age, which equally shortens the time span during which the burial or heating time becomes "effective" in this direction or even fails to reach this.

Consequently, the eastern Graben Complex can be

regarded as a gas and condensate province, whereas the Western Platform shows a progressive decrease in maturity in a westerly to northwesterly direction, with a concomitant decrease in prospectivity in that direction.

In conclusion, this study emphasizes that temperature and burial time are critical factors when considering the maturation process of source rocks, particularly in basins with young burial histories or with an "effective" overburden that becomes progressively younger in age across the basin, such as is the case in outbuilding deltas and/or continental shelf margins.

ACKNOWLEDGEMENT

This paper is based mainly on pre-1979 data and work carried out at that time by the staff of Shell, BP and Todd Oil Services Ltd., New Plymouth, New Zealand and Shell Internationale Petroleum Maatschappij B.V., The Hague, Netherlands.

The authors are indebted to Shell Petroleum Mining Company Ltd., BP (Oil Exploration) Company of New Zealand Ltd., and Todd Petroleum Mining Company Ltd. for granting permission to publish this paper. They thank their numerous colleagues who, through their own work and helpful discussion, contributed significantly to the paper. The cooperation of the Shell Research Exploration and Production Laboratory (KSEPL) in the Netherlands, who performed the geochemical analyses, is gratefully acknowledged. The original study was carried out within the exploration department of Shell BP and Todd Oil Services Ltd., New Plymouth; the final draft was prepared at SIPM, The Hague.

REFERENCES

Harding, T.P., 1974, Petroleum traps associated with wrench faulting: AAPG Bulletin, v. 57, p. 74-96.

Hood, A., C.C.M. Gutjahr, and R.L. Heacock, 1975, Organic metamorphism and the generation of petroleum: AAPG Bulletin, v. 59, p. 986-996.

Lopatin, N.V., 1971, Temperature and geological time as factors in coalification: Akademiya Nauk SSSR, Izviesta, Seriya Geologicheskaya, n. 3, (English translation by N.H. Bosteck, 1972, Illinois State Geological Survey, p. 96-106).

Pilaar, W.F.H., and L.L. Wakefield, 1978, Structural and stratigraphic evolution of the Taranaki Basin, offshore North Island, New Zealand: Australian Petroleum Exploration Association Journal, p. 93-101.

Roux, B., et al, 1980, An improved approach to estimating true reservoir temperature from transient temperature data: Society of Petroleum Engineers of AIME, n. 8888.

Stach, E., et al, 1975, Stach's textbook on coal petrology, 2nd revised edition: Berlin-Stuttgart, Gebr. Borntraeger.

Suggate, R.P., 1973, Coal ranks in relation to depth and temperature in Australia and New Zealand oil and gas wells: New Zealand Journal of Geology and Geophysics, v. 17, n. 1, p. 149-167.

Waples, D.W., 1980, Time and temperature in petroleum formation: application of Lopatin's method to petroleum exploration: AAPG Bulletin, v. 64, p. 916-926.

Wilcox, R.E., T.P. Harding, and D.R. Seely, 1973, Basic wrench faulting: AAPG Bulletin, v. 57, p. 74-96.

Index

A reference is indexed according to its important of "key," words.

Three columns are to the left of a keyword entry. The first column, a letter entry, represents the AAPG book series from which the reference originated. In this case, M stands for Memoir Series. Every five years, AAPG will merge all its

indexes together, and the letter M will differentiate this reference from those of AAPG Studies in Geology Series (S) or the AAPG Bulletin (B).

The following number is the series number. In this case, 35 represents a reference from AAPG Memoir 35.